Handbook of Rigging

About the Authors

Joseph A. MacDonald—of San Diego, CA—has more than 45 years of editing experience in the engineering and construction sector, including 18 years with *Engineering News-Record* and *Construction Methods & Equipment*; as well as 10 years engineering field experience on various construction projects throughout New York and New England.

During his years with McGraw-Hill, he was editor-in-chief of the *Encyclopedia of U.S. Building & Construction Technology* (a McGraw-Hill/Chilton joint venture publication for the former Soviet Union); managing editor of *CM&E;* senior editor of *EN-R;* editor of the *Directory of Construction Information Resources* (EN-R); and coeditor of the *Handbook of Rigging* (fourth ed). Since 1980, he has served as chief editor for a number of international publications, including *Private Power Executive, Information Display,* and *Computers for Design & Construction.*

In 1974, he received the industry's top *Technical Article Award* for his chapter *"Will Solid Wastes Bury Us?"*—published originally in *Engineering News-Record's* 100th anniversary publication *Probing the Future*, and then republished (in Japanese) by Kajima Institute Publishing Co. Ltd. (1975).

He earned his degree in civil engineering from Manhattan College, Riverdale, NY; graduated from the U.S. Naval Construction Battalion (Seabee) School, Port Hueneme, CA; and served as a Mobile Construction Battalion project engineer for the construction of a 250-personnel naval facility.

W. E. Rossnagel was a consulting and fire-protection engineer and was a safety engineer with the Consolidated Edison Company of New York.

Lindley R. Higgins was a professional engineer, writer, and consultant in the construction industry for more than 25 years. He was editor of McGraw-Hill's *Handbook of Construction Equipment and Maintenance* and its *Maintenance Engineering Handbook.*

Handbook of Rigging

Lifting, Hoisting, and Scaffolding for Construction and Industrial Operations

Joseph A. MacDonald

W. E. Rossnagel

Lindley R. Higgins

Fifth Edition

McGraw Hill

New York Chicago San Francisco
Lisbon London Madrid Mexico City
Milan New Delhi San Juan
Seoul Singapore Sydney Toronto

The McGraw-Hill Companies

Cataloging-in-Publication Data is on file with the Library of Congress.

Handbook of Rigging, Fifth Edition

The Fourth Edition was published as *Handbook of Rigging: For Construction and Industrial Operations* (McGraw-Hill, 1988).

3 4 5 6 7 8 9 10 DOC/DOC 1 4 3 2 1 0

ISBN 978-1-265-82325-2
MHID 1-26-582325-1

Sponsoring Editor
Larry S. Hager

Acquisitions Coordinator
Alexis Richard

Editorial Supervisor
David E. Fogarty

Project Manager
Somya Rustagi

Copy Editor
Suzanne Lassandro

Proofreader
Bhavna Gupta

Indexer
Broccoli Information Management

Production Supervisor
Pamela A. Pelton

Composition
International Typesetting and Composition

Art Director, Cover
Jeff Weeks

Contents

Section 3 Worksite Preparation

Section 6 Rigging Tools

Section 7 Lifting and Hoisting Machinery

Preface

This fifth edition of the industry acknowledged standard rigging reference, first published in 1950—*Handbook of Rigging—for Construction & Industrial Operations (Lifting, Hoisting, and Scaffolding)*, has been comprehensibly revised, expanded, and updated to include the current best practices for safe rigging procedures on the construction site, as well as in industrial and utility facility environments.

Emphasis in each chapter is on the legal obligation of rigging contractors and project management to comply with safety and health requirements, as set forth in *OSHA 1926* federal regulations; *ANSI/ASTM* safe operation standards for lifting and hoisting equipment—for materials and personnel; as well as specific industry standards and guidelines applicable to various rigging procedures, equipment, and devices.

The handbook is intended to provide a ready reference and guide for:

- Engineers and contractors engaged in rigging design and operations.
- Expert riggers and erectors of buildings and other structures.
- Maintenance mechanics in industrial plants and electric power generation facilities having less frequent rigging jobs to perform.
- Operators of all types of hoists, derricks, and cranes.
- Workers engaged in the erection of scaffolds or signages, or in climbing tall structures (towers and chimneys) for repair or maintenance work.

Accident statistics show that construction is a high-hazard industry to work in. As most of the accidents that occur are the result of human error, it's important for everyone involved in rigging—lifting, hoisting, and scaffolding procedures—to accept responsibility for the part they play in the operations.

Almost all the crane accidents reported each year, however, could have been avoided if the workers controlling the cranes' operation had paid more attention to what they were doing, checked the condition of slings and lifting equipment before use, ensured that loads were not lifted over any public place or work area, and not allowed anyone to ride on a suspended load.

A key aspect of this handbook is to encourage managers and workers to accept their responsibility to minimize the risk of accidents to persons and property through proper training that can significantly improve accident prevention on the construction site.

The handbook attempts to identify the elements of safe rigging practice and expand on the minimum safety regulations promulgated by *OSHA* and the various standards organizations. It should be used as a guide, in conjunction with the applicable safety regulations by everyone concerned with or responsible for the safety of employees at work on a project, as well as for the general public and surrounding property and structures.

Information in specific chapters on rigging equipment, systems, and devices can provide the basis for developing instructional material used in training personnel engaged in rigging operations. It also can be included in standard instructions issued to employees for the safe use of rigging equipment and handling of construction materials.

Safe practices recommendations in the handbook, of necessity, are framed in general terms to accommodate the many variations in rigging practices and the different ways in which rigging is used. Because these recommendations are only advisory in nature, they must be supplemented by strict observance of specific relevant regulations—as well as individual equipment manufacturers' recommendations and requirements.

The author is grateful to the various professional, technical, and trade organizations, government agencies, equipment manufacturers, and materials producers for providing technical information, illustrations, and tables cited throughout the text.

Joseph A. MacDonald:
Construction Writers Association
Society of American Military Engineers

Introduction

The art of rigging may be traced to prehistoric times. Levers were used then, as now, to pry stones, roll logs, and move objects that were too heavy to be moved by hand. The inclined plane, or natural ramp, was in use even then to help move heavy objects up to higher elevations.

The first major rigging job, of which there is not only a record, but also indisputable evidence, was the construction of the three pyramids at Gizeh, near Cairo, Egypt—about 2700 BC. Estimates are that preparation work must have taken about 10 years, and construction about 20 years.

As it stands today, the large pyramid—built to contain the remains of Pharaoh Cheops (Khufu)—is 746 ft (227 m) square at the base and 451 ft (141 m) high. Originally, the structure was encased in fine grain limestone. But at some unknown time during the past 4600 years, or more, this sheath was removed.

The large pyramid contains about 2.3 million stones—weighing from 2 to 30 tons each; a total of about 5.75 million tons—nearly 20 times the weight of the masonry in the 102-story Empire State Building in New York City.

Inclined Planes (Ramps)

Records indicate that a sand ramp, requiring nearly one million tons of sand, transported from the desert, was built-up on one side as the pyramid rose in height. Another million tons of sand were then required to backfill the interior of the pyramid. And, when the job was completed, the ramp had to be removed.

The construction crews had no mechanical equipment. Instead, they used levers, rollers, crude ropes, sledges, plumb lines, and string sightings to get the massive job done. The huge stones were hauled up the ramp on rollers, for an average lift of 100 ft, by the brute strength of 100,000 slaves in teams of 50 workers each, driven by the slave master's whips.

The pyramid remains today as indisputable proof of the ingenuity and brutality of the Egyptian constructors.

Winch and Pulley Hoist

Archaeological records show that as far back as 515 BC, the ancient Greeks were using a crane mechanism for lifting heavy loads using small construction teams—unlike the more labor intensive ramp, requiring large bodies of unskilled labor that had been the norm in the autocratic society of Egypt and Assyria.

Although the art of rigging may be traced to prehistoric times, it was the Greeks who first invented the winch and pulley hoist that lead to a widespread replacement of ramps as the main means of vertical motion.

Greek temples of the classical age like the Parthenon featured stone blocks weighing less than 15 to 20 tons. And, the practice of erecting large monolithic columns was practically abandoned in favor of using several column drums.

Archaeologists point out that the distinctive holes appearing on the stone blocks of Greek temples—either above the center of gravity of the block, or in pairs equidistant from a point over the center of gravity—are positive evidence of the existence of a mechanical lifting machine. The first recorded evidence for the existence of the compound pulley system appears in the *Mechanical Problems*, attributed to Aristotle (382–322 BC).

The Romans adopted the Greek crane and developed it further—enabling the construction of buildings having enormous dimensions. Several extensive recorded accounts, as well as particularly detailed reliefs of Roman cranes, exist from the late first century AD.

Modern Lifting and Hoisting Procedures

The art of rigging has developed to the degree that today manufacturers build 200- to 400-ton traveling cranes for power plants; hammerhead cranes of even greater capacity for shipyards; and mobile cranes and derricks capable of handling trusses and girders weighing up to 200 tons for buildings or bridges.

This handbook, however, does not deal with rigging operations of such magnitude. Rather, it covers conventional rigging operations on construction sites for erection and demolition of buildings and structures; in industrial factories, and electric power plant; in the transporting and handling of heavy machinery; as well as for mining and port facilities operations.

Scaffolding Systems

Included, also, is a section on the erection of temporary scaffolding used for painting, construction, repair, or demolition of buildings and structures (towers and chimneys).

In the period up to 1916, the majority of scaffolding was constructed from timber. The shortages and costs of imported timber in Great Britain at the time of World War I spurred the development of new systems.

The first major development took place in 1918 when British engineers patented the universal coupler for use on steel tubes, providing many advantages over timber not only in strength and stiffness, but also in extensive reuse—as well as the obvious relative speed by which it could be assembled compared to timber scaffolding.

During that period, many different scaffolding systems were being developed around the world, consisting of welded frames that could be slotted or clipped together to form access or support towers, resulting in reduced construction time and complexity, and requiring less skill in assembly.

Continual developments have resulted in stronger systems that incorporate either horizontal restraints (using lasers, ties, or braces) at more levels; or stronger tubes or connections—making scaffolding systems even easier to use.

The handbook's appendices include a bibliography, as well as government agency and industry association information resources.

SECTION 1

Codes, Standards, and Regulations

CHAPTER 1

Codes and Standards

Numerous codes, standards, specifications, and regulations govern all rigging operations. They are promulgated and enforced to protect construction site workers as well as the general public against unsound practices, misuse of equipment, and improper handling of materials—to prevent physical harm, property damage, and possible financial disaster.

Most standards were developed for permanent construction, although also applicable for temporary construction. Some, however, have been written specifically to address temporary structures used in construction such as those for scaffolds and scaffolding.

Standards written for temporary structures in construction are voluntary and generally only guidelines, developed by a consensus procedure through the major standards-writing organizations: American National Standards Institute (ANSI), ASTM International (ASTM); and the National Fire Protection Association (NFPA).

Other standards are written by independent organizations for their specific industries, such as American Institute of Steel Construction (AISC), Aluminum Association (AA), and American Institute of Timber Construction (AITC).

Regulations and requirements promulgated by the federal Occupational Safety & Health Administration (OSHA) as well as state and local jurisdictional agencies, however, are compulsory technical and legal documents. In some cases, provisions of the voluntary standards are adopted, either in part or as a whole, by regulatory agencies thereby becoming mandatory standards. Many ANSI standards, for example, have been incorporated into OSHA Regulations.

1.1 Building Codes

Formerly, three model building codes were used in different regions of the country. In the late 1990s, however, the three organizations formed the International Code Conference (ICC) to jointly develop a single set of comprehensive and coordinated national Model Construction Codes. Since 2000, the *International Building Code* has been the single building code throughout the United States.

Authority for building code enforcement varies from state to state, depending on whether there is a statewide building code with an ultimate central authority, or if codes are the purview of local authorities. Although each state holds the ultimate power to enforce building codes, traditionally these regulations have been adopted entirely, or modified, and enforced at the municipal and the county levels to meet local conditions.

These local jurisdictions may write their own codes or may legally adopt the state promulgated building code, or the *International Building Code*.

Codes Guide

Since 1967, the National Conference of States on Building Codes and Standards (NCSBCS) has coordinated state and local building codes. The organization consists of chief building regulatory officials of the states, local code officials from across the nation, federal officials, and other organizations and professionals interested in the need to strengthen building codes and public safety.

NCSBCS provides an interchange of information and technical services to encourage safe, durable, accessible, and efficient design and construction of buildings throughout the nation. It publishes, among a number of related documents:

- Guide to current building codes and standards, covering local, state, and federal regulations
- Standard safety guidelines (adopted by many states) to help provide uniformity in building construction and to guard against fire hazards and other unsafe conditions in buildings

While national design codes can also be applied to temporary structures in construction, specific commentary on design criteria and construction methods is minimal.

1.2 U.S. Consensus Standards and Specifications

Several subject-specific and nationally recognized design standards have been developed and written to allow their adoption by reference in a general building code, including the following:

ANSI—American National Standards Institute

American National Standard for Construction and Demolition Operations (*A10*)

American Safety Code for Building Construction

American Standard Building Code Requirements for Minimum Design Loads in Buildings and Other Structures (*A58.1-1945*)

Minimum Design Loads for Buildings and Other Structures (*A 58.1-1982*)

Scaffolding (*A10.8*)—covers a broad range of scaffold types, including general requirements and provisions for platforms, tube and coupler scaffolds, and fabricated tubular frame scaffolds

Specifications and Guidelines for Aluminum Structures—Aluminum Association Design Manual

Steel Erection (*A10.13*)—general guidelines on temporary flooring, bolting and "fit-up," building and bridge erection, dismantling, mill work, and working over water; relatively prescriptive; cross references a variety of related standards

Structural Framing, Studs, Joists and Planks (*PS 20*)—American Lumber Standards

ANSI/ASCE—American National Standards Institute/American Society of Civil Engineers

Standard for design loads on structures during construction

Guidelines for design of buildings and the associated temporary structures construction

Performance criteria, design loads, load combinations, and safety factors to be used in the analysis and design of structures during their transient stages of construction, as well as temporary structures used in construction operations

ASTM—ASTM International

Standard ASTM Specifications for Structural Steel Shapes (*A992/992M-06a*)

Nonferrous Metal Products: Aluminum and Magnesium Alloys—Book of Standards (*ASTM Volume 02.02–05*)

CSI—Construction Specifications Institute

Standards and formats to improve construction documents and project delivery

NIST—National Institute of Standards and Technology

Develops and promotes the use of science-based tools—reference standards, measurements, data, models, and protocols—designed to enhance:

- Construction industry through innovations in building materials and construction technology
- Safety, security, and sustainability of buildings and physical infrastructure

The institute provides science-based measurement and predictive tools and protocols for advanced construction technologies—innovative connections, high-performance constructed systems, and process technologies for automated construction.

1.3 Industry Consensus Standards

More than 400 industry organizations produce or promulgate (or both) standards regulating design and construction of buildings, among them:

American Institute of Steel Construction (AISC)

***Steel Construction Manual*, (13th Ed.)**—Update of the *AISC Steel Construction Manual*, combining the previously separate *Allowable Stress Design* and *Load and Resistance Factor Design*. Manual includes:

- *2005 AISC—Specification for Design Fabrication and Erection of Structural Steel Buildings*

  Applies to both permanent construction and temporary structures

  Reflects development of independent Load and Resistance Factor (LRFD) specification and exclusively covers allowable stress design

  Covers fabrication and erection of structural steel, including specific sections related to temporary structures in construction

  Covers temporary Support of Structural Steel Frames; and presents some guidelines on temporary structures

  Identifies characteristics and related requirements for self-supporting and non-self-supporting steel frames—steel frames that require interaction with other elements not classified as structural steel to provide the required stability or resistance to wind or other lateral loads

  Temporary supports such as guys, braces, falsework, cribbing or other elements required for erection that are determined, furnished, and installed by the erector
- *2005 AISC Code of Standard Practice for Steel Buildings and Bridges*
- *2004 RCSC Specification for Structural Joints Using ASTM A325 or A490 Bolts*

This Manual replaces the *ASD Manual* (*9th Ed.*) and the *LRFD Manual* (*3rd Ed.*). Much of the *HSS Connections Manual* has also been incorporated and updated in the Manual.

American Welding Society (AWS)

Structural Welding Code (*AWS D1.1/D1.1 M – 2004*)—Covers the welding requirements for any type of welded structure made from the commonly used carbon and low-alloy constructional steels; includes 12 mandatory and 14 nonmandatory annexes. Sections 1 through 8 provide rules for the regulation of welding in steel construction; this code:

- Contains latest workmanship standards, inspection procedures, and acceptance criteria
- Updates certain ASTM steel specifications and adds new prequalified steels,
- Reflects advances in technology
- Details newest allowable stress range formulae
- Presents latest revision of prequalified base metal list

American Society of Civil Engineers (ASCE)

Design Load on Structures during Construction (*ASCE 57-02*)—Standard provides minimum design load requirements during construction for buildings and other structures; addresses partially completed structures and temporary structures used during construction.

Loads specified are suitable for use either with strength design (such as USD and LRFD) or allowable stress design (ASD) criteria and are equally applicable to all conventional (and temporary) construction methods.

Topics include loads and load combinations, dead and live loads, construction loads, lateral earth pressures, environmental loads, and performance and safety criteria.

1.4 Recommended Practices and Guidelines

Sources of information related to the design, erection, and construction of temporary structures include:

ANSI—*National Standard for Construction and Demolition Operations: Scaffolding Safety Requirements; Steel Erection—Safety Requirements*

AISC—*Code of Standard Practice and Manual of Steel Construction—Allowable Stress Design*

AITC—*Timber Construction Manual*

ASCE—*Minimum Design Loads for Building and Other Structures*

ASSE—*Safety Requirements for Rigging Qualifications and Responsibilities* (American Society of Safety Engineers)

ASTM—*Standards in Building Codes*

AWS—*Structural Welding Code-Steel*

CSA—*Falsework for Construction Purposes* (Canadian Standards Association)

NFPA—*National Design Specifications for Wood Construction*

SSFI—*Recommended Procedures for Compression Testing of Welded Frame Scaffolds and Shoring Equipment* (Scaffolding Shoring and Forming Institute)

TPI—*Recommendations of Handling and Erecting Wood Trusses –Bracing Wood Trusses: Commentary and Recommendations* (Wood Truss Council of America)

UL—*Internal-Combustion-Engine-Power Industrial Trucks–Electric-Battery-Powered Industrial Trucks* (Underwriters Laboratories)

Regulations and Standards

OSHA Regulations

OSHA 29 CFR

1926 *Safety and Health Regulations for Construction*

1910 *Occupational Safety and Health Standards*

U.S. Consensus Standards and Guidelines

ANSI Construction Collection

Personnel and material hoists

Requirements for scaffolding

Temporary and portable space heating devices and equipment

Safety nets

Steel erection

ASTM—International Book of Standards

Section 04 – *Construction*

Wood

Building Construction (I) E 72-E1670

Building Construction (II) E1671

Latest, Property Management Systems, Technology

CHAPTER 2
Government Regulations

A number of federal regulations apply to working conditions during rigging operations such as worker and public safety and health, waste disposal, and water runoff.

2.1 Federal Regulations

The regulations most profoundly affecting temporary structures in construction are those of the Occupational Safety & Health Administration (OSHA), Department of Labor.

Most states administer their own occupational safety and health programs, generally adopting the federal OSHA regulation of similar requirements. When three sets of regulations may apply—federal, state, and local supplementing each other—contractors should follow the strictest requirements.

OSHA

OSHA Regulations—Mandatory requirements to protect employees from the hazards of construction (and rigging) operations are published in the following:

29 CFR Part 1926—*Safety and Health Regulations for Construction*

- E - *Personal Protective and Life Saving Equipment*
- F - *Fire Protection and Prevention*
- H - *Materials Handling, Storage, Use and Disposal*
- I - *Tools (Hand and Power)*
- J - *Welding and Cutting*
- L - *Scaffolds*
- M - *Fall Protection*
- N - *Cranes, Derricks, Hoists, Elevators, and Conveyors*
- O - *Motor Vehicles and Mechanized Equipment (Forklift)*
- W - *Rollover Protective Structures and Overhead Protection*
- X - *Stairways and Ladders*

29 CFR 1910 – *Occupational Safety and Health Standards*
- D - *Walking–Working Surfaces*
- E - *Means of Egress*
- F - *Powered Platforms, Manlifts, and Vehicle-Mounted Work Platforms*
- G - *Occupational Health and Environmental Control*
- H - *Hazardous Materials*
- I - *Personal Protective Equipment*
- J - *General Environmental Controls*
- K - *Medical and First Aid*
- L - *Fire Protection*
- M - *Compressed Gas and Compressed Air Equipment*
- N - *Materials Handling and Storage*
  - *App A to 1910.178 – Stability of Powered Industrial Trucks (Non-mandatory)*
  - App to Paragraph (l) of this section
- O - *Machinery and Machine Guarding*
- P - *Hand & Portable Powered Tools and Other Hand-Held Equipment*
- Q - *Welding, Cutting, and Brazing*
- S - *Electrical Design Safety Standards for Electrical Systems*
  - *Safety Related Maintenance Requirements*
  - *Safety Requirements for Special Equipment*
  - *App A – Reference Documents*

A number of OSHA regulations are patterned after the standards promulgated by various standards organizations. For example:

1926 Subpart L—Scaffolds is patterned after *ANSI A10.8—Scaffolding* and organized into general and type-specific provisions, with general provisions applying to scaffolding systems of all types:

- *Light duty*—designed for a 25 lb/ft^2 (1.2 kN/m^2) maximum working load to support workers and tools only. Storage of equipment or material on the platform is not allowed.
- *Medium duty*—designed for a 50 lb/ft^2 (2.4 kN/M^2) maximum working load for workers and materials, often intended for bricklayers' and plasterers' work.
- *Heavy duty*—designed for a 75 lb/ft^2 (3.6 kN/m^2) maximum working load for workers and material storage, often intended for stone masonry work.

Department of Energy (DOE)

Worker Protection Management for Federal and Construction Employees—Guidance, standards, and orders pertaining to the management of safety and health on DOE construction projects:

10 CFR 851—Worker Safety and Health Program

DOE Order 440.1A—Worker Management for DOE Federal and Contractor

Environmental Protection Agency (EPA)

National Pollutant Discharge Elimination System Regulations—NPDES requires that the construction manager implant:

Stormwater pollution prevention plan
Inspection for construction site operator
Erosion and sediment control
Stabilization of soils after construction

Although EPA delegates authority to implement certain regulatory programs to some states, a state may have requirements that are more stringent than the federal requirements. Be sure to check with state and local agencies before starting a construction rigging project.

Stormwater Runoff

Under the *Clean Water Act* (CWA), a construction project that disturbs one or more acres of land requires obtaining a CWA permit to cover discharge of stormwater runoff from the construction site. Stormwater general permits are issued through either EPA's National Pollutant Discharge Elimination System program, or the state NPDES permitting authority. Obtaining permit coverage requires considering:

- *Notice of Intent*—submitting a notice of intent or permit application as required by the permitting authority, requiring the contractor's certification to not harm federally listed endangered species.
- *Pollution Prevention Plan*—developing and implementing a stormwater pollution prevention plan that describes the physical characteristics of the site, listing potential sources of pollutants, and identifying erosion prevention, sediment control, and stormwater management practices that will be implemented at the site.
- *Notice of Termination*—submitting a Notice of Termination, if required by the permitting authority, when construction site work is completed, or when someone else assumes control of the site.

Solid and Hazardous Waste

Examples of materials at construction sites that may be classified as hazardous wastes include spent cleaner (such as organic solvents), paints (including lead-based paint), used oil, paint thinners, wastes that contain ignitable and corrosive materials, and wastes that contain certain toxic pollutants. A list of hazardous wastes and their allowed concentrations is provided in the regulations that implement the *Resource Conservation and Recovery Act* (RCRA).

- ***Managing, treating, and disposing of hazardous wastes***—listing of quantities of hazardous wastes that determine whether a generator is large or small. Generators of large amounts of hazardous wastes are subject to more regulatory requirements.
    - Storing hazardous wastes at the construction site: if storage time is exceeded, a RCRA permit is required
    - Transport, treatment, and disposal of hazardous wastes: transporters of hazardous wastes must be registered with either EPA or a state as a hazardous waste transporter; the generator is responsible for assuring that the hazardous waste is delivered to a RCRA-permitted treatment or disposal facility and must sign the hazardous wastes manifest used to track the hazardous wastes transport to a permitted treatment or disposal facility
- ***Other wastes subject to EPA***—these, as well as state and local regulations, include lead-based paint wastes; fluorescent lamps that contain mercury; construction/demolition wastes (such as wood, roof material, insulation plaster, or sheet rock), and storage tanks (either above ground or underground) subject to the requirements of either
    - Comprehensive Environmental Response, Compensation, and Liability Act (CERCLA)
    - Oil Pollution Act (OPA)

These regulations dictate how these materials are to be stored, labeled, and disposed of, and plan for spill prevention.

Spill Reporting

Emergency planning and reporting requirements are covered in *The Emergency Planning and Community Right-to-Know Act* (EPCRA), and the *Oil Pollution Act* (OPA), with exact requirements dependent on the type of chemical handled. Each EPCRA hazardous chemical has an associated "reportable quantity."

2.2 State Requirements and Local Ordinances

At the state level, health and safety requirements are important considerations in rigging work, not only for site workers, but also the general public (see Table 2.1).

Internet Law Library
US State and Territorial Laws— provides inks to local ordinances and regulations (*www.priweb.com/internetlawlib/1.htm*)

American Legal Publishers
Websire includes city and county codes for 26 states (*www.amlegal.com/library/*)

County Codes and Ordinances
National Association of Counties web site provides access to county codes and ordinances, arranged by subjects, state, and county (*www.naco.org/Content/NavigationMenu/About, Counties/Codes, and Ordinances/.htm*)

Cyburbia: The Urban Planning Portal
Publishes planning resource directory; contains links to local zoning codes and ordinances (*www.cyburbia.org*)

First Source Building Codes Library
Provides detailed information on local building codes and amendments, plus contact information for local building authorities and public utilities; covers major cities nationwide (*www.reedfirstsource.com/codes/index.asp*)

General Code Publishers
Provides access to more than 170 city, township, village, and borough codes from 19 states in the Northeast, Southeast, Midwest, and West (*www.generalcode.com*)

LexisNexis
Provides free access to municipal and county codes or ordinances from more than 33 states (*www.municipalcodes.lexisnexis.com*)

Municipal Code Corporation
Provides the full text of county, municipal, and local codes from 47 states (*www.municode.com*)

Municipal Codes Online
Seattle Public Library web site provides links to codes and ordinances for municipalities and counties in nearly all 50 states (*www.spi.org/default.asp*)

Municipal Codes Web Library
Provides free access to full text of local ordinances in more than 30 states (*www.ordlink.com*)

Ordinance.com
Subscribers can search zoning ordinances databases by state, county, and town from 10 regions to view and print zoning ordinances, applications and instructions, submission checklists, fees, and applications for variance (*www.ordinance.com*)

Smart Growth Online
Resources include municipal land development ordinances, codes, statutes, fact sheets, guidebooks, and other documents (*www.smartgrowth.org*)

Table 2.1 Resources for Local Ordinances and Regulations

2.3 Licenses, Permits, and Certifications

Each municipal government has the authority to issue its own business licenses, construction permits, and trades certifications; however, states also issue certain licenses and permits.

Because there is no uniformity throughout the country regarding these licenses, permits, and certifications, the contractor *must* consult with the appropriate local and state officials to determine the specific requirements for the proposed project.

Permits and licenses relating to construction include:

- *Licenses*—Architects, interior designers, and landscape architects
- *Licensing*—Electricians; elevators and escalators
- *Licenses*—Plumbers
- *Registration*—Professional land surveyors
- *Registration*—Easement or right-of-way
- *Permits*—Stormwater permits for construction

Licenses

Many states and large municipalities also require that contractors be licensed; and various construction trades *must* be licensed, including:

- Site safety—Site safety manager certification
- Electricians—Electrical contractors, Master and Journeymen electricians, special electricians, and other persons who perform electrical work
- Elevators and related equipment
- Hoisting machine operators—Three classes (A, B, C); three subclasses (C-1, C-2, C-3)
- Riggers—Master riggers, special riggers, climber and tower crane riggers
- Plumbers—Master and Journeyman plumbers
- Welders—Four main classes (1-4); three classes (1R, 2H, 3R)

Building Permits

State and municipal agencies require permits for:

- Construction and renovation work covering foundations and earthwork, demolition, and debris removal/disposal
- Construction equipment operation
- Stormwater runoff

Zoning Requirements

Specific regulations include uses as of right, uses permitted by a special permit, supplementary use regulations, performance standards, signage, bulk restrictions, and off-street parking/loading.

2.4 Compliance Resources

Vital information on compliance with OSHA construction regulations is available on a totally searchable CD-ROM containing *2007 OSHA*

- Construction Industry Safety Standards (29 CFR Part 1926)
- Administrative Standards
- General Industry Safety Standards (29 CFR Parts 1900-1910)
- OSHA posters and forms (required) and more

***CICA* (Construction Industry Compliance Assistance Center)**—Provides information on environmental requirements in federal and state laws, including commercial/industrial building, demolition, stream or other water body; publishes a planning guide for construction explaining in detail the environmental obligations that construction companies can face across the various federal laws (stormwater, wetlands, air, waste, etc.). (*www.cicacenter.org*)

NECAC* (National Environmental Compliance Assistance Clearinghouse)**—Web site provides quick access to compliance assistance tools, contacts, and planned activities from ***EPA and other compliance assistance providers. (*www.cfpub.epa.gov/clearinghouse*)

***EPA* (Environmental Protection Agency)**—Compendium of Compliance Assistance Tools for the Construction Sector. Identifies numerous tools that cover a wide range of environmental issues pertinent to the construction sector—from state-specific permitting guidance to detailed technical guidance that is applicable for all states. (*www.p2pays.org/ref/41/4077*)

***OSHA* (Occupational Safety & Health Administration)**—29 CFR 1926 – *Occupational Safety and Health Regulations for Construction* 29 CFR 1910 – *Occupational Safety and Health Regulations* (*www.osha.gov*)

***USGPO* (US Government Printing Office)**—29 CFR 1926.20 – *General Safety and Health Provisions* (*www.gpo.gov*)

SECTION 2

Engineering Principles

CHAPTER 3

Basic Machines

The first and most important step in any rigging operation is to determine the forces that will affect the job, and then to select and arrange the equipment that will move the loads safely—both horizontally and vertically.

Too often, the load-bearing parts of a rigging system are stressed to a point dangerously close to the breaking point, without the rigger realizing it. This element of chance can be reduced to a minimum by a simple knowledge of how to determine the loads that must be moved and the capacity of the equipment being used.

The forces involved in rigging will vary with the method of connection, the direction of support in reference to the load, and the effects of motion. Thus, the rigger must know something about mechanical laws, the determination of stresses, the effects of motion, the weight of loads, the centers of gravity, and factors of safety.

Also essential is a basic knowledge about the strength of materials.

3.1 Mechanical Laws

Machinery offers a mechanical advantage in moving loads. The simple machines, from which every machine is constructed, include:

- ***Inclined plane***—An inclined plane requires less force to move a load, but the distance traveled is greater. The greater the distance over which the force must be applied, however, the smaller the force required to do the work with the load.

    Ramp—inclined plane; friction is a problem in using a ramp to move heavy objects; if an incline is very steep, steps cut into the incline make it easier to use

    Wedge—two back-to-back inclined planes
- ***Lever***—A bar that turns on a point called a fulcrum.

- ***Wheel and axle***—A lever that rotates through 360°; the circle turned by the wheel is much greater than the circle turned by the axle.
  - *Gear*—wheel with teeth around the outer rim; used to regulate speed and direction of motion in complex machines and increase or decrease the force applied
  - *Crank*—handle of a machine connected at right angle to an axle; used to transmit motion
  - *Pulley and shaft*—wheels with grooves around the rim; fixed pulley doesn't change position; moveable pulley, both load and pulley move.
- ***Screw***—Derived from the inclined plane; wound about a nail; the distance between the screw threads depends on the slope of the inclined plane.
- ***Block and fall***—Apparatus for raising or lowering heavy weights, consisting of a pulley blocks and the rope of a tackle to which the power is applied in hoisting.

Inclined Plane

Riggers frequently make use of the inclined plane when hauling a load on rollers up a ramp, or on skids onto a truck. To estimate roughly the pull required to haul a load of 15,000 lb on rollers up an incline of 4 ft in 20 ft, draw a diagram *ABC* representing the incline. Then draw, to any suitable scale, a vertical line *DE* representing the weight of the load (Fig. 3.1). Assume a scale of 1 in. = 10,000 lb so that a line 1½ in. long will represent the weight of the load. From *D* draw a line *DF* at a right angle to the slope of the incline *AB*.

Using the same scale of 1 in. = 10,000 lb, measure the distance *EF*, which will be the theoretical pull required. This scales to about 0.3 in., or 3000 lb, which is the pull if frictionless rollers were used. To this value the resistance of friction must be added to determine the actual pulling force.

The standard equation for determining the resistive force of friction when trying to slide two solid objects together state that the force

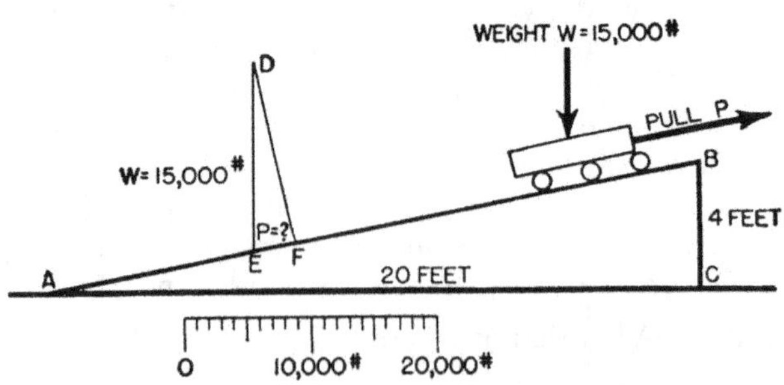

FIGURE 3.1 Force required to move an object up an inclined plane or ramp.

of friction equals the coefficient of friction times the normal force pushing the two objects together:

$$F_r = u \times N$$

where F_r = resistive force of friction (pounds or newtons)
u = coefficient of friction for the two surfaces (number between and infinity)
N = normal or perpendicular force pushing the two objects together (pounds or newtons)

The coefficient of friction is a number related to the two specific surfaces that are in contact with each other that is heavily dependent on the roughness of each surface and how the materials slide against each other.

This equation applies to both static friction—before an object starts to slide—and kinetic friction—when the object is actually moving or sliding. Static friction and kinetic friction have different coefficient of friction values (Table 3.1).

On an incline, the normal force will be reduced by the cosine of the incline angle:

$$N = W \times \cos(a)$$

where N = normal force on the incline
W = weight of the load
a = incline angle
$\cos(a)$ = cosine of angle

Surface #1	Surface # 2	Static friction	Kinetic friction
Steel	Steel (dry)	0.6	0.4
Steel	Steel (greasy)	0.1	0.05
Oak	Oak (parallel grain)	0.62	0.48
Oak	Oak (cross grain)	0.54	0.32
Wood	Wood (clean)	0.25–0.50	0.2
Wood	Wood (wet)	0.2	–
Wood	Metals (clean)	0.2–0.6	–
Wood	Metals (wet)	0.2	–

Source: USDA Forest Service, Forest Products Laboratory.

TABLE 3.1 Coefficient of Sliding Friction for Various Materials, with the Static Coefficient being Larger than the Kinetic Coefficient

Leverage

Another simple application of the mechanical advantage is through leverage, in which one point (the fulcrum) of a rigid bar is fixed, another point is connected with the force (the load) to be acted on, and a third point is connected with the force (the power) to be applied. The crowbar is a typical lever. As commonly used, there are two arrangements of the fulcrum (Fig. 3.2).

- Located at one end of the lever with an upward pull on the other end to lift the load; gravity acts as the downward force (Fig. 3.2a).
- Located under the end of the lever, with a downward push on the other end to lift the load; ground or floor exerts an upward force to resist the downward pressure (Fig. 3.2b).

Assume a weight W of 1000; acting at a distance of 3 in. from the fulcrum F and a person lifting up on the lever at a distance 30 in. from the fulcrum. The force P times the force distance (30 in.) always equals the weight W (1000 lb) times the weight distance (3 in.). Thus,

$$\mathrm{P} \times 30 \text{ in.} = 1000 \text{ lb} \times 3 \text{ in.} \qquad \mathrm{P} = \frac{1000 \text{ lb} \times 3 \text{ in.}}{30 \text{ in.}} = 100 \text{ lb}$$

If the force distance is 10 times the weight distance, the force is one-tenth the weight, and the lever is said to provide a mechanical advantage of 10. This holds true for other ratios of distances also.

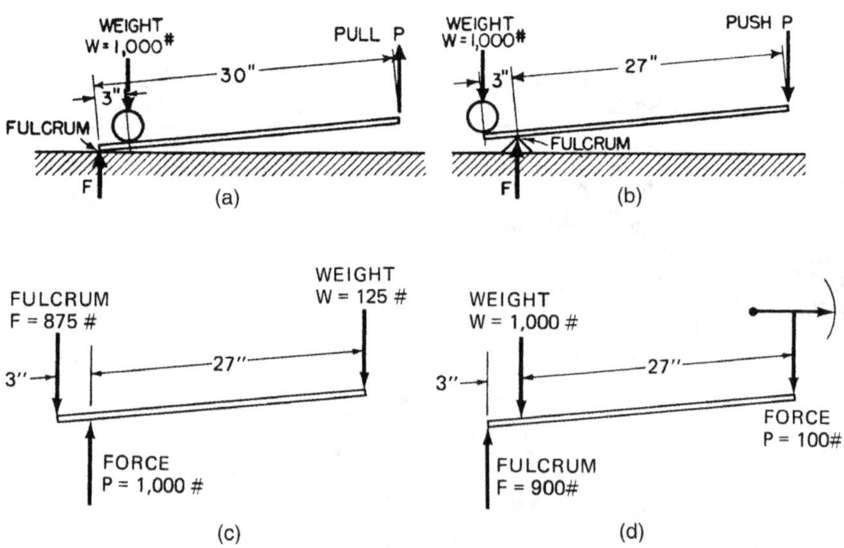

Figure 3.2 Force required to lift or to move a load by means of levers.

If the crowbar is used, however, as shown in Fig. 3.2b where the fulcrum is between the force and the load, the force then is a push *P* in the downward direction. Using the same 30 in., the force *P* times the force distance (27 in.) equals the weight *W* (1000 lb) times the weight distance (3 in.). Thus:

$$P \times 27 \text{ in.} = 1000 \text{ lb} \times 3 \text{ in.} \qquad P = \frac{1000 \text{ lb} \times 3 \text{ in.}}{27 \text{ in.}} \qquad P = \frac{1000 \text{ lb}}{9} = 111 \text{ lb}$$

By observation, then, the crowbar is most efficient, or requires less force when used (Fig. 3.2a), although sometimes it may be necessary to limit the pressure on the fulcrum (Fig. 3.2b). In this latter application, the pressure is always the sum of the weight *W* and the force *P*.

A third form of leverage with fulcrum and load at the ends of a rigid member and the applied force at an intermediate point (Fig. 3.2c) produces a mechanical disadvantage. This arrangement, however, is important in a fishing rod, where a relatively lightweight at the end of the rod must be lifted quickly by a relatively heavy pull of small travel above the butt end. In construction work a heavy load of a very small travel, on a scale used for weighing loads on trucks, is balanced by a small force with considerable travel on the indicator end (Fig 3.2d).

Some form of the simple lever is the basis for almost every other tool; even the simple twisting of two lines with a pipe to pull a load is merely the application of a lever with a long driving arm; and a short resisting arm.

Pulley System

One application of the basic lever is in the use of a pulley, where the block-and-tackle arrangement acts as the lever, with the fulcrum at the side of the block and the rope fixed to the ceiling.

The mechanical advantage obtained with the block-and-tackle arrangement is explained in detail in Chap. 18.

Wheel and Axle

An important adaptation of the lever principle is the wheel and axle (or shaft). Although the movement of the lever is limited, the motion of the wheel may continue indefinitely. For the wheel-and-axle arrangement with belts and pulleys, the belt pulls or tensions, are inversely proportional to the pulley diameters, that is, when one pulley is four times as large as the other, the belt pull on the larger pulley is one-fourth the belt pull on the smaller pulley (Fig. 3.3).

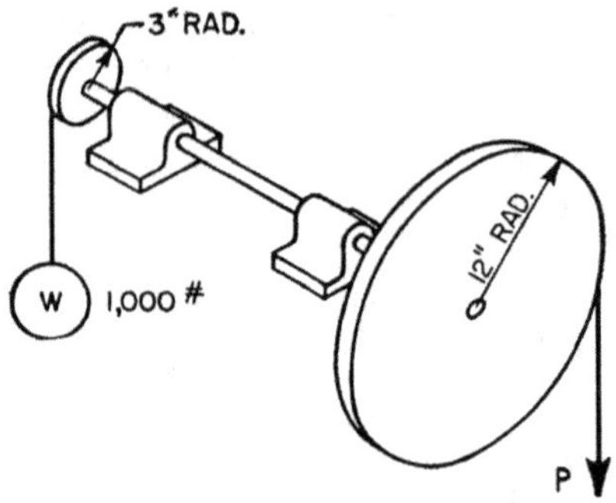

FIGURE 3.3 Force required to lift a load by means of wheels and a shaft.

In this example, a force P is exerted on the radius R (12 in.) of the large pulley, while the weight W (1000 lb) is supported on the radius r (3 in.) of the small pulley. Thus,

$$P \times 12 \text{ in.} = 1000 \text{ lb} \times 3 \text{ in.} \qquad P = \frac{1000 \text{ lb} \times 3 \text{ in.}}{12 \text{ in.}} = \frac{1000 \text{ lb}}{4} = 250 \text{ lb}$$

For gears on the same shaft, the mechanical advantage of the driven pulley relative to the driver is directly proportional, and the speed is inversely proportional to the lengths of the respective radii (Fig. 3.4).

If there is a train of gears, such as on a hand winch consisting of several gear reductions, the theoretical pull that can be exerted on the hoist rope may be readily estimated. Thus,

$$W = \frac{F \times C \times G_1 \times G_2 \times G_3 \times \cdots}{D \times P_1 \times P_2 \times P_3 \times \cdots}$$

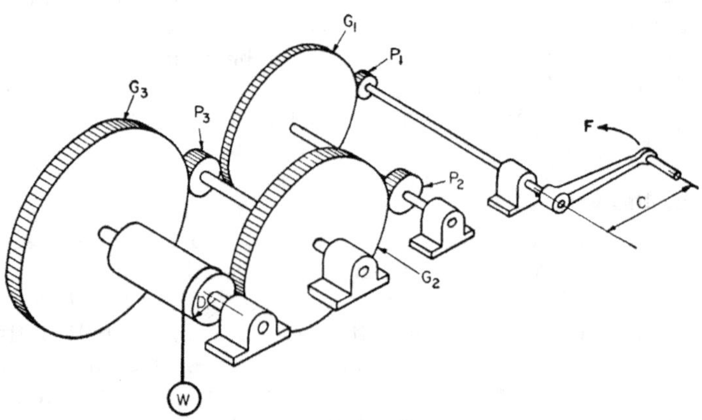

FIGURE 3.4 Force required to lift a load by means of a train of gears.

where W = pull on hoist rope (pounds)
F = force on hand crank (about 17 lb at 2½ ft/sec)
C = radius of hand crank (inches)
$G_1, G_2, G_3, \ldots$ = number of teeth in gear $(G_1, G_2, G_3, \ldots)$
D = radius of drum, to center of rope (inches)
$P_1, P_2, P_3, \ldots$ = number of teeth in pinion $(P_1 \times P_2 \times P_3, \ldots)$

(Deduction *must* be made for friction.)

3.2 Determining Stresses

To figure the stress developed by a load, it is important to know that the forces acting in one direction must be equal and opposite to the forces acting in the other direction—that is, all forces must be in equilibrium.

Swinging Scaffold

When an ordinary outrigger is used on a building to support a swinging scaffold, the reactions at the supporting and hold-down points are easily calculated, once the load is known. The load force multiplied by the length of its lever arm must equal the hold-down force multiplied by the length of its lever arm. Thus the fulcrum reaction is equal and opposite the two forces (Fig. 3.5).

Derrick

Additional factors complicate the determination of these reactions, since the supporting members act at numerous angles and in various combination, a simple graphic method provides adequate results for all members (Fig. 3.6).

This diagram, called a triangle of forces, shows how all the forces at one point must be in equilibrium (Fig. 3.6a).

NOTE: *Conventionally, vector arrows heading toward a point under consideration indicate that the member is under compression; arrows heading away from the point indicate tension.*

FIGURE 3.5 Load distribution for an outrigger when used to support a scaffold.

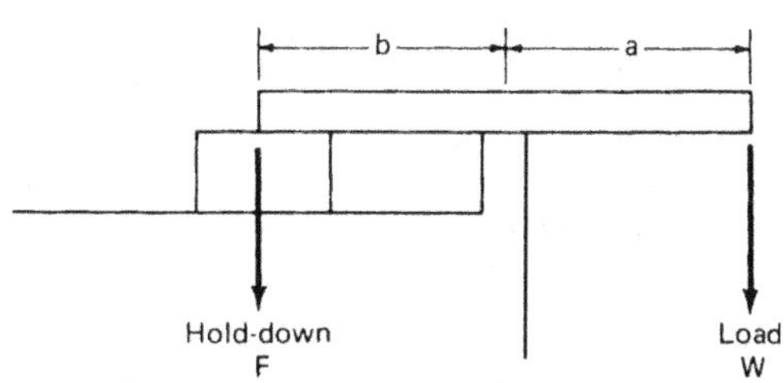

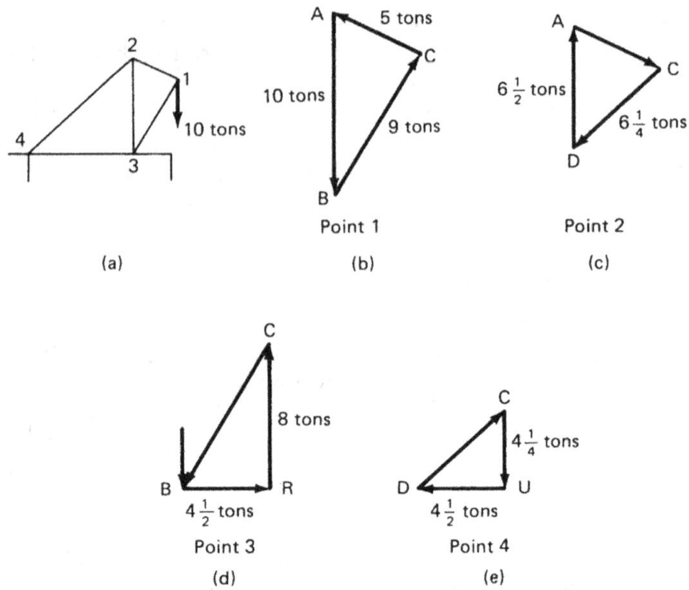

FIGURE 3.6 Force distribution vectors (Scale: 1 in. = 10 tons).

Starting at point *1* (Fig. 3.6b), the derrick's load-line pull must be equally balanced by the boom resistance and the topping-lift pull. Thus, three forces act at this point. Draw line *AB*—at a convenient scale—parallel to the load line. Next, draw lines *BC* and *AC* parallel to the derrick boom and to the topping lift, respectively. These lines will intersect at a point *C*. Measuring the lengths of these two lines will provide directly the force induced in the respective members.

Assume the derrick load is 10 tons. By drawing line *AB* 5 in. long, the diagram will have a scale of 2 tons = 1 in. Then, if line *BC* is 4.5 in. long, the stress in the derrick boom is 9 tons; if line *AC* is 2.5 in. long, then the stress in the topping lift is 5 tons. Any one force in this triangle is the result that balances the other two forces.

To analyze the forces at point 2 (Fig. 3.6c), draw a new force triangle using the same stress and length of the line for the topping lift as at point *1*, but reverse the direction of the arrow. Next, draw line *CD* to represent the guy, and line *DA* to represent the mast. Again, the completed diagram constitutes a triangle of forces with the respective lengths a direct measure of the stresses in the various parts.

Next, to determine the stress at the foot of the mast, note that the boom stress *BC* is made up of a vertical and a horizontal reaction (Fig. 3.6d). The vertical reaction *CR* goes down into the foot block along with the vertical reaction found previously in the mast reaction *AD*. Because *CR* equals 8 tons and *AD* equals 6.5 tons, the total vertical load on the foot block, point *3*, is 14.5 tons. *BR* is the horizontal force of 4.5 tons that must be resisted to keep this point from slipping.

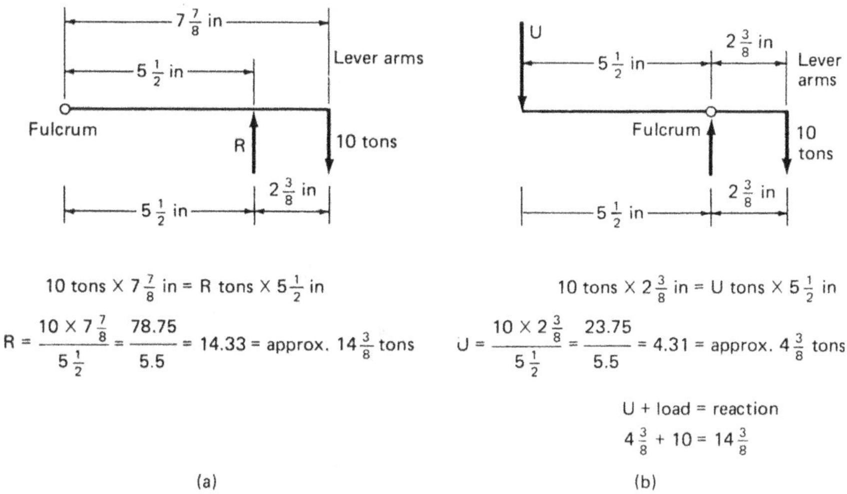

FIGURE 3.7 Checking derrick reactions upon supporting points. (a) Calculating lever loads; (b) calculating vertical positioning.

Similarly, the guy reaction can be found by drawing *DC* equal and opposite to the force noted on the drawing for point *2* (Fig. 3.6e). The horizontal and vertical reactions are necessary to form a closed triangle of forces.

To check the reactions of the derrick on the supporting points, eliminate all of the members that are used to make the mechanical device workable; then simply consider the problem as a rigid lever (Fig. 3.7).

Project the load line down to meet the horizontal line of the lever, because the effective load lever arm and the load are perpendicular to it. Next, consider point *4* as the fulcrum of a lever and point *3* as a resisting load. Then, mathematically calculate the resisting load as 14.33 tons (Fig. 3.8a). This checks close enough with the 14.5-ton reaction previously determined.

By the same method, calculate the vertical reaction at *U* as 4.31 tons, which is close enough to the 4.5-ton force determined to keep point *3* from slipping (Fig. 3.8b).

NOTE: *Differences in the values obtained by the check method result from graphic discrepancies in the triangle of forces method and are small enough to be acceptable.*

Sling

When a sling suspension is being analyzed, using a triangle of forces clearly shows the effect of decreasing the angles of lift. Assume a 10-ton load, sling arrangement (Fig. 3.8).

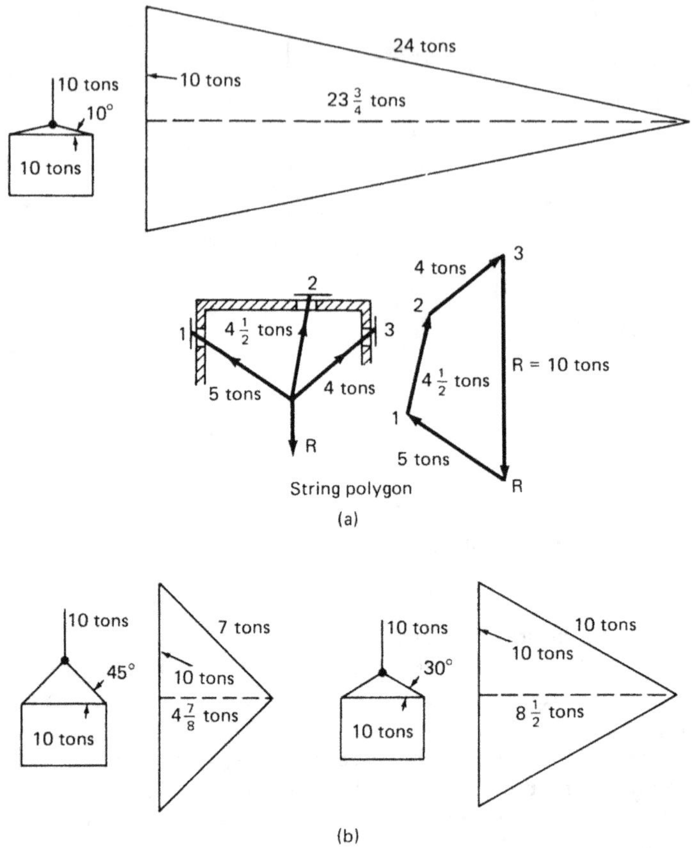

FIGURE 3.8 Analyzing sling suspension and force distribution.

When the sling's lift angle is

- 10°, each arm is stressed to 24 tons, with the resistance reaction reaching 23.2 tons.
- 45°, the load puts a 7-ton stress in each sling arm that must be resisted by a 4.875-ton dividing reaction.
- 30°, each arm is stressed 10 tons, and the dividing reaction of 8.5 tons.

Thus, it is apparent that when a chocker hitch is jammed down for hoisting, the resulting very small lift angle produces sling arm stresses considerably greater than the load. Also, the stress is very high at the point where the hitch bends over the upper corner of the load, and can cause quick damage at this point. Such hitches require the use of a supplementary compression member (such as spreader bar) to provide the cross resistance necessary for the weaker load member.

When three or more known forces act at one point, a string polygon can be used to determine the resultant force. Simply draw a set of lines (to scale) parallel to the known forces and place the lines end to end. The beginning and end points are connected with a line that represents the result of all the forces. The resultant line is opposite in direction to the known forces and of a magnitude equal to the scaled length.

Three or more unknown forces acting on a point constitute an indeterminate problem, and a triangle of forces cannot be drawn for the solution. Such a problem requires a design engineer to use calculus for the solution.

3.3 Effect of Motion

A knowledge of the principles of basic machines described thus far would be quite adequate if the rigger were to deal only with stationary forces. Most rigging operations, however, involve loads in motion. The rigger, therefore, should also be aware of the effect that motion has on loads and the stress placed on a rope, chain, beam, scaffold, ladder, or other load-bearing member because of a suddenly applied load, such as a jerk or impact.

For example, a beam may safely support a concentrated load of 1000 lb at the center. If a wheel, however, carrying a 1000-lb load is rolled at high speed over this beam, it will produce a stress in the beam twice as great as that produced by the static (stationary) load. Similarly, if the 1000-lb load (assumed to be incompressible) is dropped onto the beam from a height of 2 ft, the energy developed at the instant of impact would be 1000 lb × 2 ft, or 2000 ft-lb. This energy must be absorbed by the beam.

If the beam could be deflected by 2 ft, the impact would be

$$\frac{2000 \text{ ft-lb}}{2 \text{ ft}} = 1000 \text{ lb}$$

to which must be added the static load of 1000 lb that would be applied if the load were at rest on the beam.

If the beam deflects only 1 in., the impact would be

$$\frac{2000 \text{ ft-lb}}{1/12 \text{ ft}} = 24,000 \text{ lb}$$

plus 1000 lb, or a total of 25,000 lb.

If the beam deflects only ¼ in., the impact would be

$$\frac{2000 \text{ ft-lb}}{1/48 \text{ ft}} = 96,000 \text{ lb}$$

plus 1000 lb, or a total of 97,000 lb, under which loading the beam would fail.

The load will compress slightly on impact or the sling will stretch, or both, and therefore the stopping distance of the load's center of gravity will be somewhat greater.

When the movement of the load is horizontal, or nearly so (such as when a heavy slab on a pair of rollers gets out of control on a slight downgrade, and the load strikes a fixed object such as building column), the distance the load drops (probably only a few inches or so), is known, and the deflection of the column is assumed.

If, on the other hand, the speed is known, then it is possible to calculate the theoretical vertical drop by using the formula:

$$H = \frac{V^2}{2g} \quad \text{or} \quad H = \frac{V^2}{64.32}$$

where H = theoretical drop (ft)
V = velocity (ft/sec)
g = acceleration due to gravity (ft/sec)

For example, if a 3600-lb automobile, traveling at 60 mph (88 ft/sec) runs head-on into a massive concrete bridge abutment, backed by an earth embankment, the movement of the abutment will be negligible. But, when the car comes to rest, if the rear axle is 3 ft nearer to the wall than when the front bumper first touched it, then the stopping distance of the car's center of gravity is 3 ft (the equivalent drop to produce this speed).

$$H = \frac{V^2}{64.32} \qquad H = \frac{88 \times 88}{64.32} = 120 \text{ ft}$$

At the instant of impact, the energy developed would be

$$E = 3600 \text{ lb} \times 120 \text{ ft} = 432{,}000 \text{ ft-lb}$$

Thus, the impact force would be

$$F = \frac{432{,}000 \text{ ft-lb}}{3\text{-ft } (\textit{stopping distance})} = 144{,}000 \text{ lb } (\textit{nearly } 72 \textit{ tons})$$

The inertia of a rugged object in motion that takes place with a decrease or increase in speed tends to increase the stresses the rigging must bear. As hoist load line starts to move, the load must accelerate from zero to the normal speed of ascent. This requires an additional force over and above the weight of the load, which will vary with the rate of change of the velocity.

A load lifted very slowly by a hand winch induces little additional stress in the load line, whereas a load accelerated quickly by mechanical power may put twice as much stress on the rope as the weight to be lifted.

Friction, too, can impose severe stresses in rigging. Where sheave maintenance is neglected, additional friction in each sheave may be as much as 10 percent of the rope tension for every sheave used. Thus, if a hoist line is reefed up, friction will add considerably to load and acceleration requirements.

In addition, the rope may have been weakened by continuous use, involving small sheaves with worn grooves, reverse bending, or generally harsh construction applications.

Remember: *It is important to apply a load or take up slack on a rope slowly.*

3.4 Determining Weight of Loads

The most important step in any rigging operation is to determine the load's weight. If this information cannot be obtained from design plans, catalog data, or other dependable sources, it may be necessary to calculate such weights.

The weights and properties for all sizes of standard and special structural wood and metal shapes and members—and methods of finding weights of built-up or special members—are contained in these handbooks.

- *Manual of Steel Construction*—America Institute of Steel Construction (*www.aisc.org*)
- *Aluminum Design Manual (2005)*—Aluminum Association (*www.aluminum.org*)
- *Timber Construction Manual (4th Ed.)*—American Institute of Timber Construction (*www.aitc.org*)
- *Manual for Engineered Wood Construction*—American Wood Council (*www.awa.org*)

Because these books may not be available on all job sites, the approximate weight of one square foot by 1-in. thick steel is 40.6 lb; iron is 37.5 lb; aluminum is 13.3; and lumber (pine) is 2.5 lb to 3.6 lb.

It is customary to figure weights based on 1/8-in. increments of material:

One square foot of 1/8-in. thick steel weighs 5.1 lb

One square foot of iron weighs 4.7 lb

One square foot of aluminum weighs 1.8 lb

Any additional thickness of the material is some multiple of those weights.

Structural beam sizes are usually supplied on erection plans together with their weight per foot of length and the length of the member. Thus, it is easy to compute the weight of any member to be lifted. The weights of angles, plates, or built-up members, however, will have to be calculated.

Structural Shape Weights

To compute structural shape weights, separate the parts, or flatten them into rectangles, which in turn become parts or multiples of a square foot of material 1 in. thick. For example:

- Steel Plate—½ -in. thick, 2-ft wide, 20-ft long, weighs 2 × 20 × 4 × 5.1 = 816 lb
- Iron plate—½-in. thick, 2-ft wide, 20-ft long, weighs 2 × 20 × 4 × 4.7 = 752 lb
- Aluminum plate—½-in. thick, 2-ft wide, 20-ft long, weighs 2 × 20 × 4 × 1.8 = 288 lb

If the steel plate were a cover plate on the top and bottom flange of a 24-ft long, wide-flange beam W 30 × 132, the composite weight of the member (not including fastenings) would be (816 × 2) + 132 × 24 = 4800 lb.

To compute angle weights, approximately close enough for safe use consider the angle as flattened out to form a plate.

- Steel angle—L 2 × 3.5 × 5/8, would flatten out to approximately a 7 × 5/8-in plate, thus its weight would be 7/12 × 5 × 5.1 = 14.7 lb. Since the actual weight of this angle is 13.6 lb (*Manual of Steel Construction* tables), the calculated weight is close enough for figuring rigging lifting requirements.
- Tanks or other cylindrical shapes—consider these shapes as rectangular plates to calculate the weight of load to be lifted. Multiply the circumference (distance around the cylinder) × 3.14 × the length (or height) of the cylinder, and add to this value the closed end areas of the cylinder 2 × (1/2 diameter × 1/2 diameter × 3.14).

Thus, a steel boiler 5 ft in diameter has a circumference of 5 × 3.14 = 15.7. If the boiler is 8-ft long and made of a 3/8-in. plate, it would flatten out to a rectangle 15.7 × 8 = 125.6 ft^2, which would weight 3 × 5.1 × 125.6 = 1922 lb.

Because the end area of the boiler is 3.14 × 5/2 × 5/2 = 19.62 ft^2, if the ends were capped with a ½-in. plate the end pieces would weigh 2 × 19.62 × 4 × 5.1 = 800.5 lb. Thus, the boiler's total weight would be 1922 + 800.5 = 2722.5 lb.

3.5 Calculating Approximate Weights

When the weight of a load cannot be obtained from readily available sources, it may be necessary to roughly calculate the weight of such objects.

Plane shape—If the object is of a simple shape, consider it cut up into a number of rectangular geometric shapes, the cubic contents (and the weight) of which then can be calculated with a fair degree of accuracy. The sum of these values will give the total weight (Fig. 3.9).

For example: To find the area of a "flat plate" of irregular shape (Fig. 3.10), draw chalk lines from any corners desired so as to subdivide the shape into a number of triangles. Take the measurement of

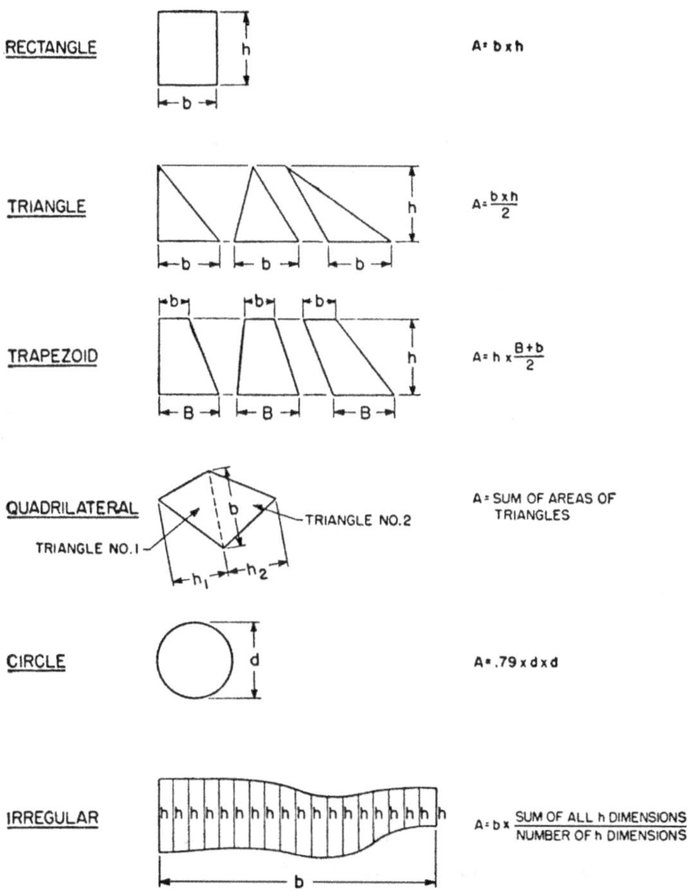

Figure 3.9 Calculating areas of various plane figures.

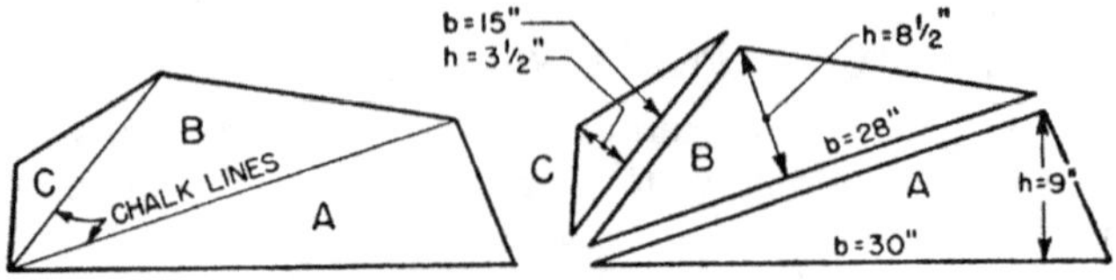

FIGURE 3.10 Calculating area of a complex plane figure.

each triangular shape and apply the formula $A = ½ (b \times h)$. The sum of the triangles equals the area of the plane.

Cut the irregular shape into three triangles A, B, and C.

> Triangle A has a base of 30 in. and a height of 9 in., area is $30 \times 9/2 = 135$ in.^{2}
>
> Triangle B has a base of 28 in. and a height of 8 ½ in., area is $28 \times 8.5/2 = 119$ in.^{2}
>
> Triangle C has a base of 15 in. and a height of 3 ½ in., area is $15 \times 3.5/2 = 26.25$ in.^{2}

The total area of the plate, therefore, is $135 + 119 + 26.25 = 280.25$ in.^{2} If the plate is 3/4-in. thick, the volume of the plate is $20.25 \times 3/4 = 210$ in.^{3}

From a table of material weights, steel weighs 0.28 lb/in.^{2}, thus the plate weights $210 \times 0.28 = 59$ lb.

Complex solid objects—calculating an irregularly shaped solid object, such as a concrete member, begins with dividing the object into smaller simpler geometric shapes, and then determining the volumes of the individual parts (Fig. 3.11).

For example: To calculate the weight of an irregularly shaped concrete block (Fig. 3.12), divide the shape into a rectangular block *A* and a frustum (part) of a solid pyramid *B*. The volume of part *A* is $4 \times 5 \times 3.5$ ft = 70 ft^3. The volume of part *B* is the volume of the entire pyramid (6 ft high) minus the volume of the upper pyramid (3 ft high).

The volume of the large pyramid is the area of the base, times one-third the height $4 \times 5 \times 6/3 = 40$ ft^3. From this, subtract the volume of the small pyramid $2 \times 2 ½ \times 3/3 = 5$ ft^3. Thus, the volume of the frustum is $40 - 5 = 35$ ft^3. The total volume of the concrete structure is $70 + 35 = 105$ ft^3. Since concrete weighs about 144 lb/ft^3, the block weighs $105 \times 144 = 15{,}120$ lb, or about 7 ½ tons.

Because the weight of most materials to be lifted is based on their weight per cubic foot, it is first necessary to determine how many cubic feet of material (its volume) must be hoisted before the weight can be estimated (see Table 7.4a, Weights of Various Materials).

The weights of some common materials should enable a rigger to compute the approximate weight of a given load. When in doubt, *do not hesitate* to seek advice from an engineer or foreman on the job.

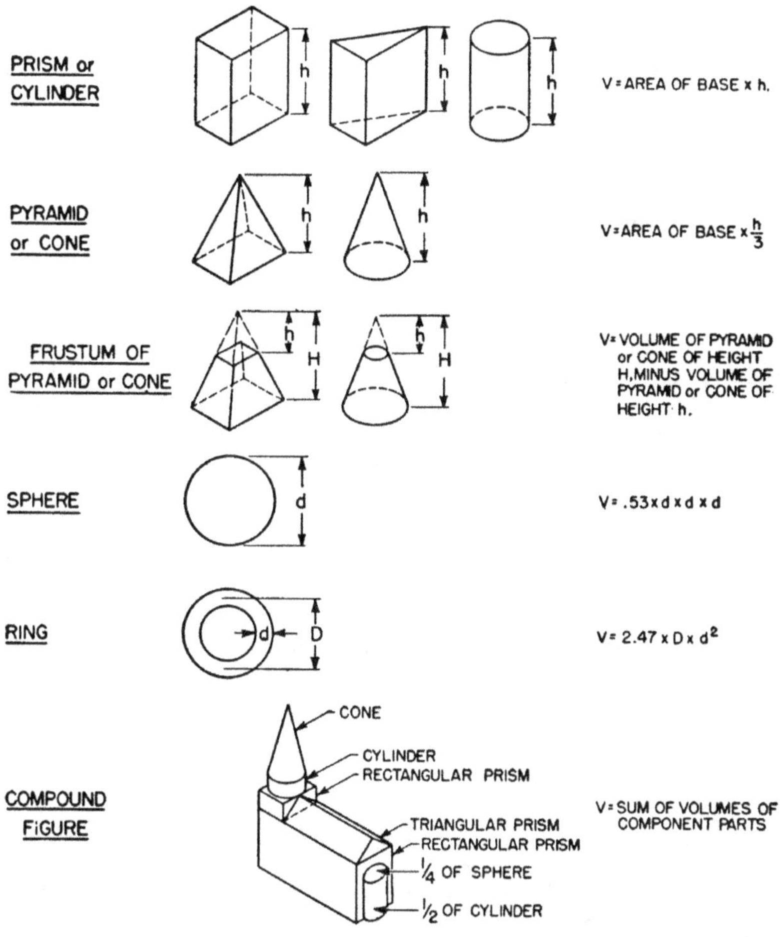

FIGURE 3.11 Areas and volumes of various solid shapes.

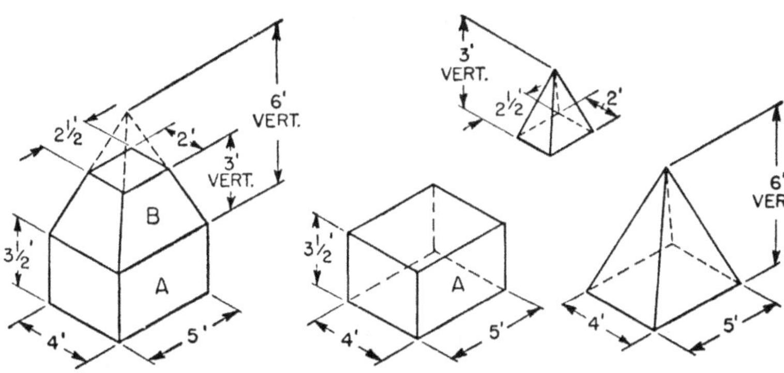

FIGURE 3.12 Calculating volume of a complex solid figure.

REMEMBER: *The time taken to calculate the approximate weight of any object is time well spent; and may prevent a serious accident through overloading of lifting gear.*

3.6 Determining Centers of Gravity

Frequently, the rigger must determine the *center of gravity* of an irregularly shaped load before safely slinging it for lifting. *Center of gravity* is the location where the center of the object's entire weight is theoretically concentrated and where the object will balance when lifted. When the object is suspended freely from a hook, this point will always be directed below the hook. Thus, a load that is slung above and through the *center of gravity* will be in equilibrium and will not tend to slide out of the hitch or become unstable in any manner.

Of course, the *center of gravity* of a solid figure has to be located in three planes or directions: lengthwise, crosswise, and vertically. Finding the exact location of the *center of gravity* of a solid figure requires mathematical calculations. For the average rigging job, however, the *center of gravity* can be estimated closely enough.

For example: Cut out of a piece of cardboard, at any convenient scale, a shape to represent familiar and some irregularly shaped plan figures, representing the length or cross-section of an object to determine the exact location of their *centers of gravity.*

Plane Figures

Most of the *centers of gravity* will lie within an object, with few located outside (Fig. 3.13).

Irregular Objects

Punch pinholes near two adjacent corners of the cardboard shape and then suspend it freely from a pin stuck into a wall, or other vertical surface. Suspend from the pin a small plumb bob or other weight

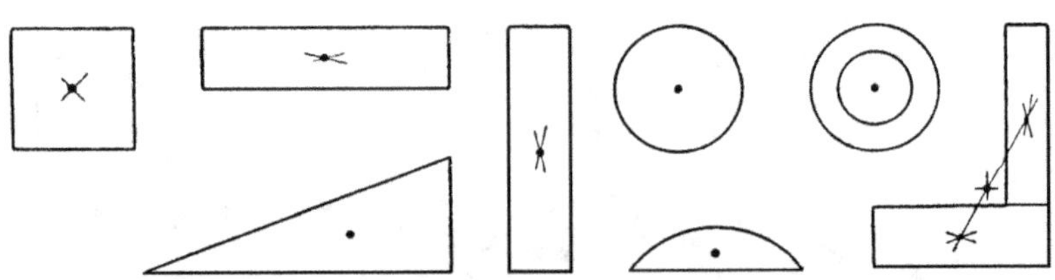

FIGURE 3.13 Approximate location of center of gravity of plane shapes.

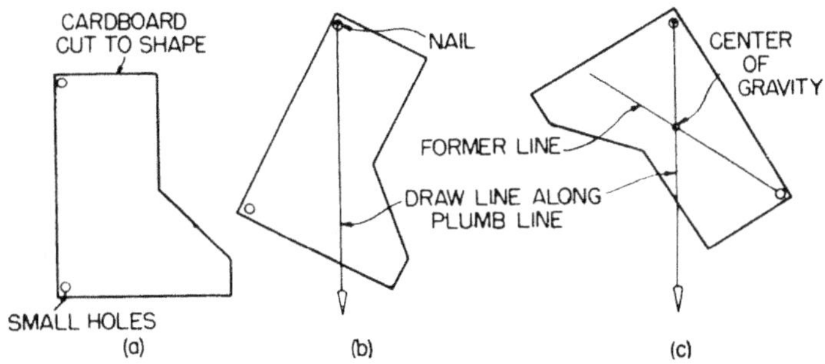

FIGURE 3.14 Locating center of gravity of an irregular plane figure.

attached to a string, and draw a line on the cardboard along the string; repeat, using the other hole. The object's *center of gravity* will lie at the intersection of the shape to be lifted (Fig. 3.14).

Another way to determine the *center of gravity* of an irregularly shaped object is to break up the shape into simple masses, then determine the resultant balancing load and its location at a point where the weights multiplied by their respective lever arms will balance (Fig. 3.15).

In odd-shaped bodies, where the *center of gravity* cannot be calculated readily, the rigger must assume a *center of gravity* then correct it by trial and error, moving the hoist, load, and sling suspension a little at a time until a satisfactory result is obtained (Fig. 3.16).

If any load tilts more than 5 percent after it is lifted clear off the ground, it should immediately be landed and rerigged. A hitch should be arranged to support the load above the center of gravity at all times during the rotation for correct positioning, so that the load will never tend to turn without being under control (Fig. 3.17).

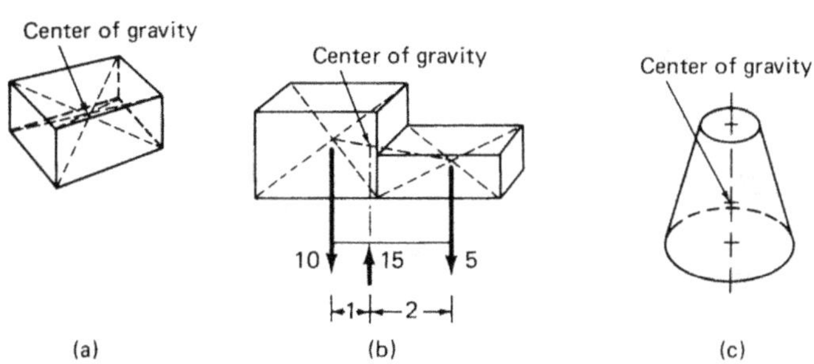

FIGURE 3.15 Locating center of gravity of various irregular solids.

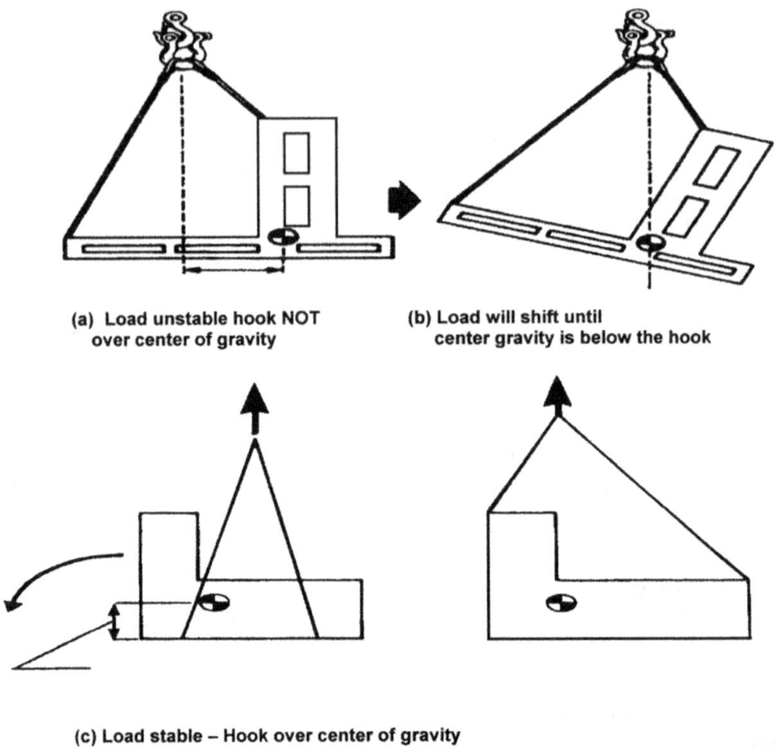

FIGURE 3.16 Location of hook. (a) Load unstable; (b) Load will shift; (c) Load stable with hook over center of gravity.

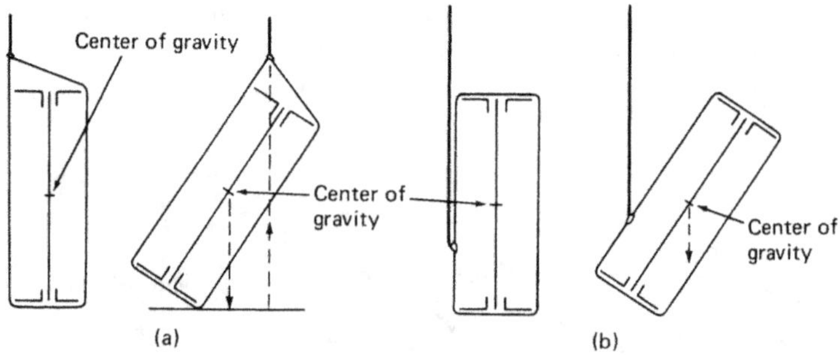

FIGURE 3.17 Making a hitch arrangement for turning a load onto its side. (a) Correct method with the load under control at all times; (b) Incorrect method, load will fall out of control as soon as the center of gravity passes outside of the lower corners.

Remember: *The center of gravity must always be located below the effective point of lift. The need for determining the location of the center of gravity of a load to be hoisted is explained in detail in Chap. 12.*

3.7 Applying Factors of Safety

Although *center of gravity* factors must be considered before selecting the type and size of rope to be used for moving any particular load, it is virtually impossible for the rigger to evaluate all the variables that can affect rigging and lifting equipment. Therefore, to compensate for any unforeseen influences, a *factor of safety* is usually applied to the material being used.

This factor is defined as the usual breaking strength of a material divided by the allowable load weight. In the case of plow-steel cable, the factor is 10; for manila rope, it is 5. Factors of safety, however, can vary with a manufacturer's experience of what is safe for the use and material under consideration.

For example: If a particular 1/4-in.-diameter, plow-steel cable has a breaking strength of 50,000 lb, a rigger might use 10,000 lb as the maximum load weight that such a cable should lift, whereas, for 3/4-in.-diameter manila rope having a breaking strength of 5000 lb, the rigger might restrict the loading to 500 lb.

In hoisting operations, where many cumulative effects are often encountered, the actual stress in a rope may come close to the actual breaking strength, even if a factor of safety is applied. Before determining the safe working load of a particular rope, check the rope manufacturer's ratings. If the manufacturer provides rope capacity in terms of breaking strength, be sure to divide that value by the factor of safety to get the safe working load.

Never consider the factor of safety as reserve strength. It is *not* available for additional capacity. Moreover, it must never be reduced, under any circumstance.

CHAPTER 4

Wood Technology

In scaffolding, wood has been all but superseded by tubular steel structures. Yet, wood is still commonly used by riggers in many applications, such as for structural members in rigging and shoring timbers, as scaffold planks, and for ladders.

It is essential, then, that the rigger is able to ensure that the correct species and quality of wood are used, and also to check depreciation and loss of strength of wood members (Table 4.1).

Most riggers can readily identify spruce, yellow pine, Douglas fir, and a few other kinds of wood, while the average carpenter has a much broader knowledge of the various species. But, identification is more or less by general appearance, rather than by any technical knowledge.

When it comes to determining the actual condition of a piece of wood that has been in use for some time and exposed to the elements, only a competent inspector examining a specimen with extreme care can even roughly provide such information.

4.1 Physical Properties

Physical properties are the quantitative characteristics of wood and its behavior to external influences other than applied forces. Some physical properties of wood are influenced by species as well as by variables such as moisture content; other properties tend to be more independent of species (Table 4.2).

The weight, shrinkage, and strength of wood depend on the moisture content, which is expressed as a fraction or percentage of the weight of over-dry wood. Tree moisture content may range from about 30 to over 200 of the weight of the wood substance, with the sapwood portion usually having the highest moisture content. While the heartwood moisture content is usually much less than that of sapwood, in some species it can be greater.

Although the moisture content values given in Table 4.2 are considered typical for various species, there is considerable variation within and between trees. This variability often exists in wood as water or water vapor in cell cavities, and as water bound chemically within cell

Size	Class	Grading property	Primary standard	Applicable
Boards	Visual	Clear wood	ASTM D245	All species
Dimension	Visual	Clear wood	ASTM D245	Redwood, cedars, minor western softwoods, hardwoods
		Full size	ASTM D1990	Other softwoods
	Mechanical	Full size	MSR policy rating by specific gravity policy	All species
Timbers	Visual	Clear wood	ASTM 245	All species

Source: American Wood Council, Washington, DC (*www.aws.org*)

TABLE 4.1 Standards and Specifications for Determining Design Values for Solid-Sawn Structural Lumber under PS 20 Procedure

walls. Thus green wood is often defined as wood in which the cell walls are completely saturated with water vapor in cell cavities.

The moisture content at which the cell walls are completely saturated without water being present in the cell cavities is called the *fiber saturation point*. This averages about 30 for most wood. It is often considered that the moisture content below which the physical and mechanical properties of wood begin to change is a function of both the relative humidity and temperature of the ambient air.

Wood in service usually is exposed to both long-term (seasonal) and short-term (daily) changes in the relative humidity and temperature of the ambient air. Thus, wood is always undergoing at least slight changes in moisture content. The long-term changes are usually gradual, while the short-term fluctuations tend to influence only the wood surface.

The practical objective of all wood seasoning, handling, and storing methods should be to minimize moisture content changes in wood during use. Protective coatings, such as varnish, lacquer or paint, help retard these changes.

Shrinkage

Wood tends to shrink mainly tangentially, that is, in the direction of the annual growth rings; about one-half as much radially, across the rings; and only slightly longitudinally, across the grain. The combined effects of radial and tangential shrinkage, however, can distort

Moisture content, percent		
Species	Heartwood	Sapwood
Hardwoods		
Elm		
American	95	92
Cedar	66	61
Rock	44	57
Maple		
Silver	58	97
Sugar	65	72
Oak		
California black	76	75
Northern red	80	69
Southern red	83	75
White	64	78
Softwoods		
Cedar		
Alaska	32	166
Eastern red	33	249
Western red	58	249
Douglas fir, coast-type	37	115
Fir		
Grand	91	136
Noble	34	115
Pacific silver	55	164
White	98	160
Hemlock		
Eastern	97	119
Western	85	170
Pine		
Longleaf	31	106
Ponderosa	40	148
Red	32	134
Shortleaf	32	122
Sugar	98	219
Western white	62	148
Redwood	86	210
Spruce		
Eastern	34	128
Sitka	41	142

Table 4.2 Moisture Content of Wood

the shape of wood pieces because of the difference in shrinkage and the curvature of the annual rings.

Wood shrinkage is affected by a number of variables. In general, though, wood having a greater density will shrink more. The size and shape of a piece of wood may also affect shrinkage; in some species, the temperature and rate of drying may influence shrinkage.

Transverse and volumetric shrinkage can vary by as much as 15, while longitudinal shrinkage is generally quite small. Average values for green to oven-dry shrinkage are between 0.1 and 0.2 for most species of wood. Certain types of wood, however, exhibit excessive longitudinal shrinkage and should be avoided in use where longitudinal stability is important, including:

- *Reaction wood*—whether compression wood in softwoods, or tension wood in hardwoods, tends to shrink excessively along the grain, and this longitudinal shrinkage can cause serious warping, such as bow, crook, or twist; cross breaks may develop in the zones of high shrinkage.
- *Center wood*—eliminates from the center of some species of wood exhibiting lengthwise excessive shrinkage.
- *Cross-grain wood*—subject to increased shrinkage along the longitudinal axis of the piece.

Density and Specific Gravity

Two primary sources of variation affect the weight of wood members: the wood's density and its variable moisture content

Density—The density of wood, exclusive of water, varies greatly, both within and between species. For most it falls between about 20 lb and 45 lb of mass per cubic foot and can be as high as 65 lb of mass per cubic foot.

Although calculated density values should always be considered as approximations, because of the variations in wood structure, moisture content, and the ratio of heartwood to sapwood found in the different species, these values are sufficiently accurate to permit estimation of structural loads and calculation of lifting weights.

Specific gravity—To compare species or products and to estimate weights, specific gravity is used as a standard reference basis, rather than density. The specific gravity of wood is usually based on the oven-dry weight and the volume at some specified moisture content. Specific gravity of oven-dry wood does not change at moisture contents above approximately 30, the wood fiber saturation point.

If the specific gravity of a piece of wood is known, based on its over-dry weight, its volume at a specified moisture content between 9 and 30 can be approximated.

Weathering

Freshly cut wood exposed to weathering, without first being treated for protection, undergoes certain physical changes, such as change in color, warping, loss of surface fibers, surface roughening, and checking.

The most severe chemical degradation of untreated wood occurs from exposure to ultraviolet light—the sun's rays.

In most woods, cycles of wetting and drying produce checks or cracks that are easily visible. Moderate- to low-density woods develop fewer checks than do high-density woods; and vertical grain woods check less than flat-grain boards.

4.2 Common Wood Defects

No matter what the species of timber, some pieces will contain defects—naturally occurring or man-made.

Growth-related

- *Spiral grain*—Helical orientation of grain giving a twisted appearance; when severe, will affect lumber dimensional stability; widespread in softwoods tending to be left spiral; in hardwoods—tending to be right spiraled; affected by the environment
- *Knots*—Fiber density and orientation increase likelihood that knots will star-check or crack when drying; if annual ring continuity is maintained the knot will be solid; if the branch dies, then trunk diameter growth will encase the bark and a loose knot results
- *Reaction wood*—Greatly affects dimensional stability as wood dries; *compression wood* (conifers)—produced on the underside of an eccentric branch in order to "push" the branch back against the pull of gravity; *tension wood* (hardwoods)—produced on the upper side to "pull" the branch back up; noted as "woolly" surface in green lumber
- *Pitch defects*—Various conditions found in softwood having resin canals; *pitch streaks*—excessive amounts of resin in localized areas of wood; *pitch pockets*—cavities in growth rings filled with liquid or crystallized sap

Seasoning-related

- *Checks*—Typical problem with green wood (log) drying

Wood Deterioration

- *Fungus*—living organism that breaks down living wood cells to extract nourishment; ultimately, the wood is destroyed as a useful product; and won't burn; sapwood is most susceptible and generally becomes unusable; heartwood is more durable
- *Brown Rot (conifers)*—due to speed of fungus attack and more severe damage

- *White Rot (hardwoods)*—spalt, a by-product of the rotting process by a vast array of stain, mold, and decay fungi causing many different and beautiful patterns in rotting wood (well known result in maple evidenced by black lines representing forward progress of fungus)

Natural defects can be due to many reasons, such as environmental factors, growth patterns, and soil composition. Man-made defects can occur at many points—from the falling of the tree, transport, storage, sawing, drying, etc.

Boards having knots usually do not pose a serious problem, since work usually can be done around them and splits can be cut off; however, boards that are heavily twisted, bowed, cupped or crooked are not usable (Fig. 4.1).

Knots

The most common defect in wood is the knot—that portion of a branch that become incorporated in the hole of the tree. Its influence on the mechanical properties of wood is primarily due to the interruption of continuity and changes in direction of wood fibers around the knot.

The influence of a knot on the performance of lumber depends on the size, location, shape, soundness, accompanying local slope of grain, and type of stress to which the know it subjected. Knots have a much greater effect on strength and axial tension than in axial compression. The effect on bending is somewhat less than that in axial tension.

A branch or limb extending at about a right angle from a tree forms a round knot in the face of a piece of lumber. When it extends at an angle, the knot in the face of the lumber will be an oval; when it extends more nearly parallel to the surface, a spike will be formed. When estimating the strength of a piece of wood that contains a knot, remember that the axis of the knot extends to the pith of the tree and forms a cone with the knot's apex at the pith.

Knots are further classified as intergrown and encased .The intergrown kind is formed where a limb is alive and growth is continuous at the junction of limb and trunk. The encased knot, however, is formed after the branch dies, with new growth of the trunk enclosing the dead limb, thus the fibers of the trunk no longer connect continuously with the fibers of the knot.

Pitch Pockets

A pitch pocket is a well-defined opening that contains free resin, extends parallel to the annular rings, and is almost flat on the pith side and curved in the bark side. These pockets are characteristic of pines, spruces, and Douglas firs. The effect of pitch pockets on strength depends on their number, size, and location in the piece of wood.

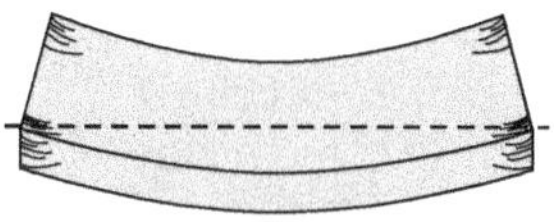

Bow – Warp along face of a board, frequently end-to-end

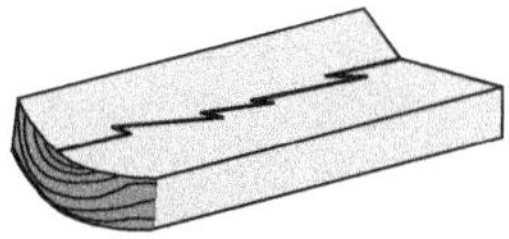

Cup – Hollow across face of board's width

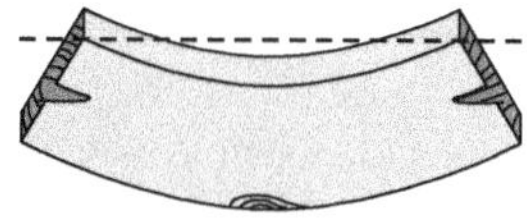

Crook – Warping along edge line (crown) from end-to-end

Twist – Multiple bends in board where ends twist in opposite directions directions

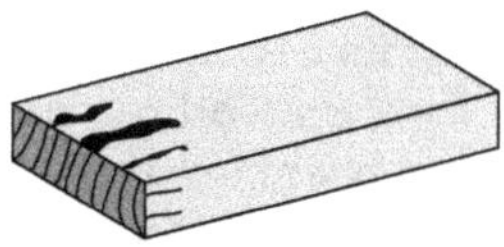

Checking – Crack in the wood structure of a piece, usually running lengthwise, restricted to the end of a board, and not penetrating the opposite side of a piece of sawn timber

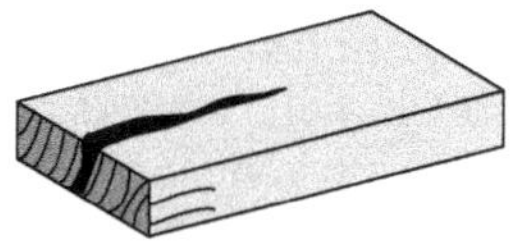

Split – Longitudinal separation of the fibers that extends to the opposite face of a piece of sawn timber

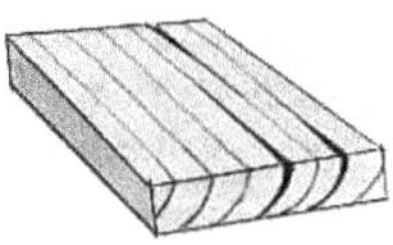

Shake – Separation of grain between growth rings, often extending along board's face and sometimes below its surface

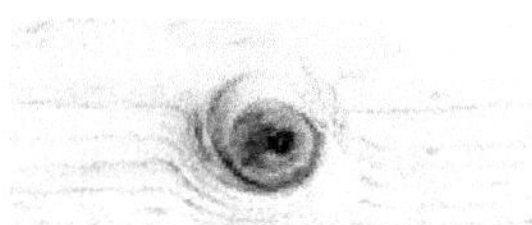

Tight knot – Knot fixed by growth or position in the wood structure so that it firmly retains its place in the surrounding wood..

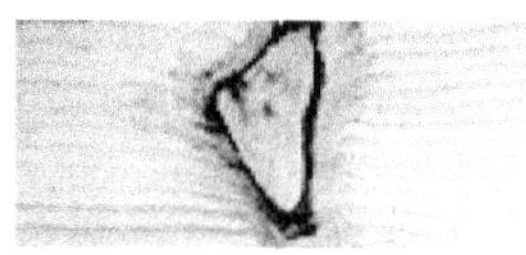

Loose knot – Knot that cannot be relied upon to remain in the pierce of wood; caused by a dead branch not fully integrated into the tree before it was cut down

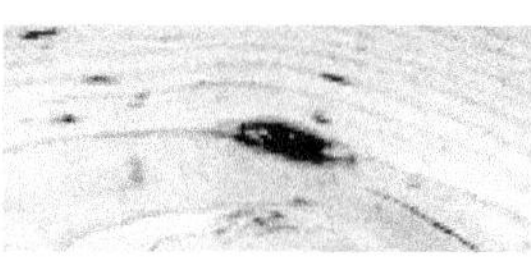

Pitch – Accumulation of resinous material (gum or sap) on the surface or in pockets below the surface of wood

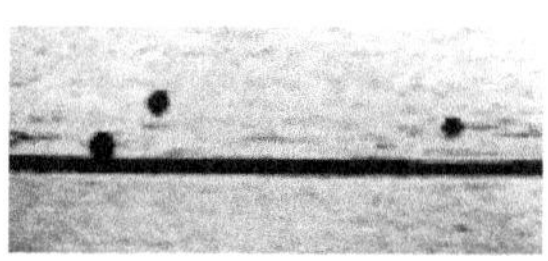

Wormholes – Small holes in wood caused by insects and beetles

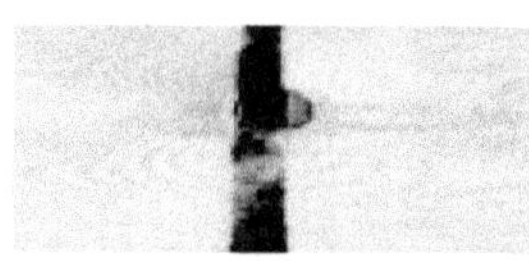

Machine burn – Darkening of wood due to overheating by machine knives or rolls when pieces are stopped in a machine

FIGURE 4.1 Common wood defects.

The pitch pocket defect in large timber is not too serious provided the dip grain that accompanies it is not excessive. However, a large number of pitch pockets indicate a lack of bond of the wood and the probable presence of a shake.

Checks

The next most common wood defect is the check, which develops because of uneven shrinkage when the wood is dried. This results in the pulling apart or separating of cell rows in a radial split, usually starting at the bark or outside the timber. A check of almost imperceptible width is as detrimental to the wood member as a wider one.

Checks are most harmful if located near the ends of a beam's vertical faces, at about the middle of the beam's height. At this point, the horizontal shearing stress is at its maximum, and the beam tends to split in half. Checks at this location tend to reduce resistance to this horizontal shear, with resultant beam failure under small loads.

Failure from a check is particularly true of beams having a depth greater than one-twelfth of their span (Fig. 4.2). In most woods, cycles of wetting and drying produce checks or cracks that are easily visible. Moderate- to low-density woods develop fewer checks than do high-density woods, and vertical-grain woods check less than flat-grain boards.

Shakes

These are lengthwise separations of wood that occur between and parallel to the annular rings. Like checks, shakes are most objectionable when occurring at the points of maximum horizontal shear. Occasionally, in an old timber, splits may be observed longitudinally on one face and diagonally on the adjacent face. Inserting a knife blade into the split and then running it lengthwise should show if it has a tendency to rotate around the pith. If it does, the split is a check; if not, it probably is a shake—assuming that the rings cannot be observed at the end of the timber.

Warping

When wood dries, it shrinks more tangentially than radially, and an almost insignificant amount longitudinally. Unless influenced by

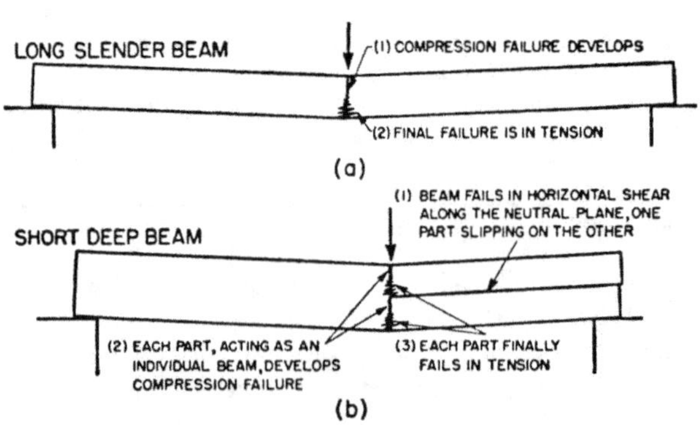

FIGURE 4.2 Typical failures. (a) Long, slender beam; (b) short, deep beam.

outside factors, dimensional changes caused by shrinkage follow a definite pattern (Fig. 4.3).

- Edge-grain planks shrink more on their thickness.
- Flat-grain planks shrink more on their width.
- Boxed heart timbers cause all faces to become slightly convex. If large checks develop, the faces will become somewhat concave.

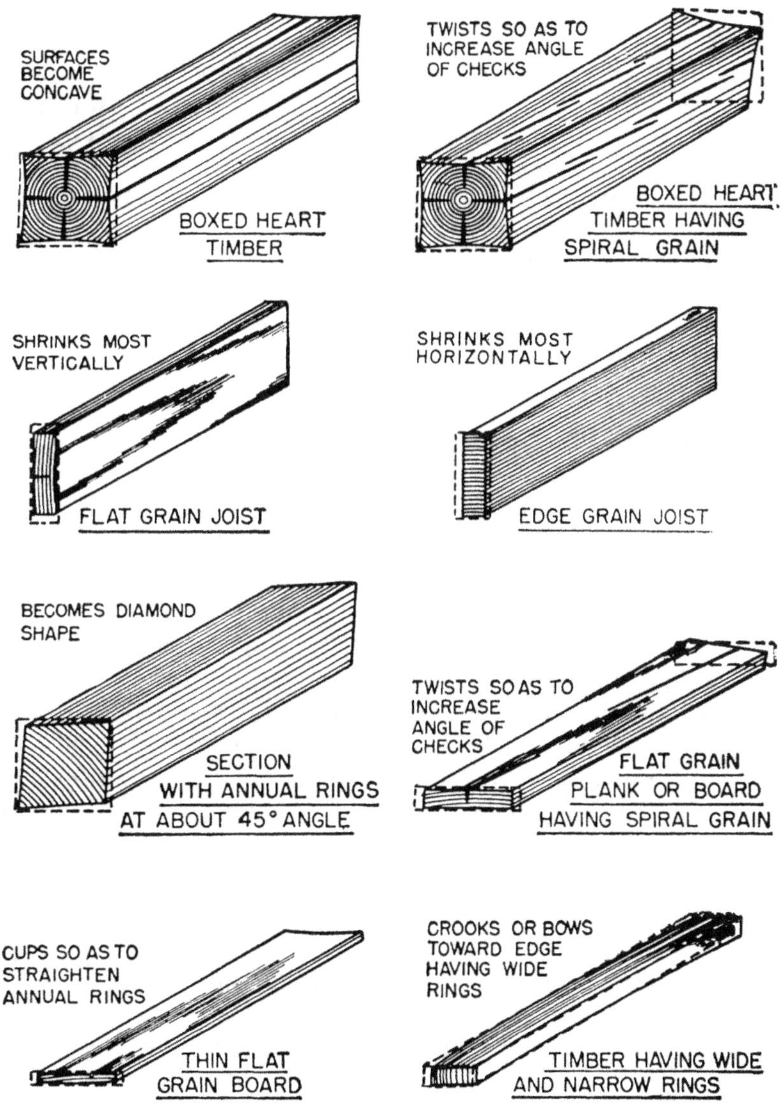

FIGURE 4.3 Predetermined manners of wood warping.

As a result of weathering, boards tend to warp (particularly cup) and pull out of their fastenings. To resist this action, it is best to use boards having widths no greater than eight times their thickness.

Reaction Wood

Abnormal woody tissue, frequently associated with leaning trunks and crooked limbs of both conifers and hardwoods, is termed *reaction wood*. In softwoods this tissue is found on the lower side of an inclined member and it is called *compression wood*. In hardwoods, it is located on the upper side of the inclined member and is known as *tension wood*.

Both types of reaction wood undergo excessive longitudinal shrinkage when subjected to moisture loss below the fiber saturation point—up to ten times normal in compression wood and five or more times normal in tension wood. When either of these woods is present in the same board with normal wood, unequal longitudinal shrinkage causes internal stresses that result in warping.

Pronounced compression wood can be detected by ordinary visual examination. It is characterized by eccentric growth of annual rings, which are of a dense nature and are predominantly in summerwood.

4.3 External Causes of Damage

Although wood generally is highly resistant to many chemicals, mild acids, and solutions of acidic salt, various chemical solutions will have different effects on the mechanical properties of wood. Two general types of action result from contact with chemical solutions:

- *Swelling of wood*—Liquids such as water, alcohols, and some other organic liquids swell the wood and thereby lower its mechanical properties. These liquids do not degrade the wood structure chemically. This action is completely reversible.
- *Chemical alteration*—Action of acids or acidic salts results in the hydrolysis of celluloses and hemicelluloses; oxidizing agents result in oxidized wood substances; and alkalies or alkaline salt solutions result in delignification and dissolved hemicelluloses. Acidic conditions tend to make the wood brittle.

These types of actions are irreversible and involve permanent changes in the wood's structure because of alteration of one or more of its chemical constituents. In general, heartwood of such species as Douglas fir, Southern pint, and white oak is quite resistant to attack by dilute mineral and organic acids. Oxidizing acids such as nitric acid, however, have a greater degradative action than nonoxidizing acids; and alkaline solutions are more destructive than acidic solutions.

Hardwoods are more susceptible to attacks by both acids and alkalies than softwoods.

Wood products are sometimes treated with decay preservatives or fire-retarding chemicals, usually in water solutions, and then are kiln-dried. While wood properties are affected to some extent by the combined effects of such water-soluble chemicals, treatment methods, and kiln drying, overall the mechanical properties of such woods remain essentially unchanged.

Temperature

Effect on wood properties may be either immediate (reversible) or permanent (irreversible). In general one immediate effect is that mechanical properties tend to decrease as the temperature is increased. There is an interaction with moisture content because dry wood is less sensitive to temperature change than is green wood.

Increases in temperature, however, are usually accompanied by a reduction in moisture content. Permanent loss in mechanical properties can occur if wood is subjected to high temperatures over long periods. The magnitude of this effect depends on temperature, duration of exposure, wood moisture content, and wood property. (In 1999, the U.S. Department of Agriculture Forest Service published a report on the permanent effects of increased temperature on wood.)

Decay or Rot

Nearly all wood is susceptible, in varying degrees, to common decay fungus. This is not an inorganic oxidizing process, such as the rusting of steel or the crumbling of stone, but rather a disease of the wood—wood-destroying fungi that seriously reduce the strength of wood members.

Decay fungi remove wood substance, thus lowering the gross specific gravity of the material. Strength losses can occur rapidly, even during the early stages of decay before noticeable loss in mass has occurred. For wood to be susceptible to decay, four conditions must be present.

- Wood must provide food for the fungus—unless chemically treated, most woods are able to do this to some extent.
- There must be a sufficient amount of moisture present—when wood is kept constantly dry, or when it contains less than 20 percent moisture, it will not decay; and when wood is submerged continuously in water, it shows very little decaying.
- There must be air present—moist stagnant air being the most effective.
- There must be warmth—wood will not decay in extremely cold climates.

Everyone is familiar with the fruiting bodies produced by decay fungi on fallen trees and tree stumps, including toadstools, shelves, and crusts. On the underside of these fruiting bodies, millions of tiny spores or germs are produced which, when matured, become free and are blown by the wind.

When deposited on susceptible wood, they will germinate and form minute fibrous strands. As the strands grow, they puncture the wood's cell walls and not only feed on the contents of the cells, but also actually devour the cell walls. Even apparently sound wood adjacent to obviously decayed parts may contain hard-to-detect early (incipient) decay that decidedly weakens wood, especially in shock resistance.

The term *dry rot* is frequently used, but this is a misnomer, for no wood that is dry can rot. It usually refers to decay where there is no visible evidence of contact with moisture, such as in portable ladders. Moisture, however, had to be present at some time, though not necessarily continuously. For example, wood that has been painted in the green, or unseasoned, state may decay under the coating of paint.

In its incipient stage, decay may not be too advanced to take remedial action, but the symptoms must first be recognized. Among the surface indications of incipient decay:

- Small bleached or otherwise discolored areas on the surface wood
- Zigzag zone lines not far from the ends of structural members
- Oozing of extractive liquids from the joints between wood structural members
- Dark zones in the wood separated by zones of lighter-color tissue caused by fungi
- Persistently moist appearance of freshly cut sections

Further decay of such wood can be arrested by keeping the wood dry. Under no circumstances should wood showing evidence of incipient decay be used where members might be subject to shock or impact loads.

In its advanced stage, decay is more readily recognizable—but the only remedy then is to remove and replace the rotted members. Advanced decay in the interior of wood structural members can be detected by:

- Inspecting structural members for sagging.
- Jabbing a knife blade into the wood and noting the resistance to penetration; rotted wood is very soft.
- Drilling small holes into the timber and observing the resistance to penetration to drilling, and the color of the chips removed; dark wood, powder, or paste is an indication of internal decay.

- Striking the timber with the round end of a ball-pen hammer and noting if the sound is dead and hollow.
- Striking the wood member against a concrete floor to determine loss of resonance.

In addition to structural weakness, decayed wood is much more combustible than sound wood, and therefore presents a greater fire hazard.

It is poor practice to leave the end grain of any wood exposed, since it readily absorbs moisture from the air. The ends of lumber should be heavily painted or tarred. Ordinary paints, varnishes, and similar protective coatings, however, cannot be relied on to preserve wood against decay. These contain no substances that are poisonous to fungi, and they may themselves support the growth of fungus in the presence of dampness. Although protective coatings retard wood's absorption of moisture, they also retard the drying out of wood that has taken up moisture through an uncoated surface or joint.

NOTE: *End grain of oak in contact with galvanized metal will cause local rotting of the wood, while similar contact with steel beams or plates will cause rapid local corrosion of the metal.*

Termites

The termite is perhaps the most troublesome insect to contend with in structural timbers. Superficially, it resembles an ant in size, general appearance, and habit of living in colonies. Although there are numerous species of termites in this country, they can be grouped into two general categories: ground inhabiting or subterranean; and wood-inhabiting or nonsubterranean.

Most of the termite damage in this country results from the subterranean variety, which is more prevalent in the southern than in the northern states, where colder weather does not favor development. These termites develop their colonies and live in the ground, building their tunnels through earth and around obstructions to get at the wood they need for food. They also must have a constant source of moisture.

The most frequently observed termite is the winged adult; however, it is the grayish-white worker termite that is the most destructive. Winged adult termites are brownish or blackish in color, with elongated body and long white wings extending beyond it at the rear (Fig. 4.4).

During certain seasons, usually in the spring and fall, the winged sexual adults migrate in large numbers and at such time may be observed for a short period of several hours. They then lose their wings, enter crevices between timbers, and breed new colonies.

Worker termites are blind, shun the light, and are seldom seen—except when a structure or building is demolished or altered, and

FIGURE 4.4 Principal forms of the subterranean termite. (a) Soldier; (b) tiny, destructive, worker termite; (c) swarmer.

the timber in which they are living is suddenly cut into or the soil is excavated.

Termites frequently eat away the entire inside of a timber, leaving nothing but a thin shell of wood (and possibly paint)—with no exterior evidence of attack until the member ultimately fails structurally. Signs of infestation include the earthlike shelter tubes by-passing masonry to provide communications between ground and wood; migrating termites; the wings that the termites have shed just before reentering the wood; pellets of fine digested, excreted wood similar to sawdust on the floor below joints in the wood; holes in the surface of the wood about the size of BB shot; and the sagging or collapse of structural members. Termite damage often is confused with decay, which is quite different and is caused by fungi.

Combating termites is nearly a hopeless task, unless all the infested wood is removed and burned. When contact with moisture from the earth, a leaking pipe, or a roof leak is cut off, the termites, which depend on such a supply, will die. Among the poisons used in combating termites are orthodichlorobenzene, Paris green, sodium fluorosilicate, and carbon tetrachloride.

4.4 Mechanical Properties

Mechanical properties of wood members are the characteristics of a material in response to externally applied forces. The properties most commonly measured to determine wood strengths include modulus of rupture in bending; maximum stress in compression parallel to the grain; compression strength perpendicular to the grain; and shear strength parallel to the grain. Additional measurements often include work-to-maximum-load in bending, impact bending strength, tensile strength perpendicular to the grain, and hardness

Wood Strength and Stiffness

The ability of wood to resist loads depends on a number factors, including the type, direction, and duration of loading; ambient conditions of moisture content and temperature; and the presence or absence or defects such a knots and checks.

The elastic properties of wood are those produced at low stress levels and are completely recoverable after the loads are removed. When wood is loaded to higher stress levels beyond the elastic range, plastic deformation of failure occurs. Six strength properties commonly measured for design purposes include:

- Bending—cross-section fiber stress
- Bending MOE—edgewise loading orientation
- Compression—parallel to the grain
- Compression—perpendicular to the grain
- Tension—parallel to the grain
- Shear—parallel to the grain

In addition, measurements are sometimes required for tensile strength perpendicular to the grain, and side hardness.

For clear, defect-free wood, the bending test probably provides the most widely available strength property. The *modulus of rupture* (MOR) reflects the maximum load-carrying capacity in bending and is proportional to the maximum moment borne by the specimen. Modulus of rupture is an accepted criterion of strength, although it is not a true stress because the formula by which it is calculated is valid only within the elastic range. The MOR of clear wood is intermediate between tensile and compression strength parallel to the grain.

Compression strength parallel to the grain is much lower than tensile strength. During a bending test, initial yielding occurs on the compression side, followed by visible compression failures and enlargement of the compression zone. The neutral surface shifts toward the tensile side of the specimen as the tensile stress continues to increase. The maximum moment in the member is reached when failure in tension occurs.

Notes:

- Modulus of rupture in bending—*reflects the maximum load-carrying capacity of the member and is proportional to the maximum moment borne by the wood member*
- Work-to-maximum load in bending—*measure of the energy absorbed by the member as it is slowly loaded*

- Impact bending height of drop—*related to the energy absorption due to a rapid or falling load*
- Hardness—*load required to embed a 0.444-in. diameter ball to one-half its diameter in a direction perpendicular to the grain*

The mechanical properties of wood derived from extensive sampling and analytical procedures often are represented as the average mechanical property of species and are used to derive allowable properties for design.

Although only clear straight-grained wood is used to determine basic mechanical properties, wood products vary considerably in their physical properties because of a tree's growth characteristics (Table. 4.3).

Variables such as specific gravity, cross grain, knots, and localized slopes of grain, plus natural defects, such as pitch pockets, must be considered when assessing the mechanical properties or estimating the actual performance of wood products.

Shakes

These are lengthwise separations of wood that occur between and parallel to the annular rings. Like checks, shakes are most objectionable when occurring at the points of maximum horizontal shear.

	Strength of straight-grained* wood (%)	
Slope of grain	**Beams****	**Posts/Columns*****
1:40	100	100
1:20	85–100	100
1:18	80–85	up to 100
1:16	76–80	up to 100
1:15	74–76	87–100
1:14	69–74	82–87
1:12	61–69	74–82
1:10	53–61	66–74
1:8	50–53	56–66
1:6	50–53	50–56

* Cross grain causes a noticeable loss of strength in beams and columns.
** Applies only to middle half of length.
*** Applies to entire length.

Table 4.3 Strength of Straight-Grained Wood as a Function of Grain Slope

Occasionally, in an old timber, splits may be observed longitudinally on one face and diagonally on the adjacent face.

Inserting a knife blade into the split and then running it lengthwise should show if it has a tendency to rotate around the pith. If it does, the split is a check; if not, it probably is a shake—assuming that the rings cannot be observed at the end of the timber.

Pitch Pockets

A pitch pocket is a well-defined opening that contains free resin, extends parallel to the annular rings, and is almost flat on the pith side and curved in the bark side. These pockets are characteristic of pines, spruces, and Douglas firs. The effect of pitch pockets on strength depends on their number, size, and location in the piece of wood.

The pitch pocket defect in large timber, however, is not too serious—provided the dip grain that accompanies it is not excessive. A large number of pitch pockets, though, indicate a lack of bond of the wood and the probable presence of a shake.

Grain Angle

Grain angle produces a profound effect on a wood member's Modulus of Elasticity and strength because of the strong anisotropic nature of wood. Tensile is sometimes 20 to 50 times higher along the grain than in the radial or tangential directions in the softwoods used for structural dimension lumber.

Cross grain is one of the most treacherous defects in a piece of wood and is a general classification that includes various grain patterns such as diagonal, spiral, interlocked, dip, wavy, and curly. Cross grain can have a major effect on the mechanical properties of lumber. Elastic properties in directions other than those along its three mutually perpendicular axes (orthotropic)

- *Longitudinal*—parallel to the grain (parallel to the long axis of the wood fibers
- *Tangential*—perpendicular to the grain but tangent to the annual growth rings
- *Radial*—normal to the growth rings

Diagonal grain results from sawing a straight piece of wood from a crooked log, or from not sawing parallel to the bark in a straight but tapered log (the wood fibers do not run parallel to the edges of the piece). Such a grain may occur on either the edge-grain or the flat-grain face of the piece. More generally, it occurs on the former in which it is usually readily discernible by the angular direction of the annual ring markings. It occurs on the flat-grain face, it will be considered in the same class as spiral grain.

Spiral grain is not as easy to detect. Wood fibers in a log generally run in a longitudinal direction. In some logs, however, the fibers run at a slight angle, thus taking a spiral path around the log. Spiral grain may frequently be observed on telephone poles and flagpoles, where it is definitely indicated by checks. On the flat-grain face of a piece of wood, the markings of the annul rings do not indicate the direction of the grain. If there are a number of herringbone-Vs on this face, they indicate diagonal grain, but the angle must be measured on the edge grain or radial face. If the points of these Vs are not in a line parallel to the edge of the piece, then spiral grain exists.

To determine the direction of the true grain in a piece of wood, first observe the markings or visible grain on the radial face. If these do not run parallel to the edge of the piece, the grain is diagonal, the angle being measured by the length required to produce a deviation of 1 in. Next, observe the tangential or flat-grain.

Also, in hardwoods the direction of the rays, or of the large pores, indicates the grain. In certain softwoods (those that may contain pitch pockets), the resin ducts are positive indicators of the grain. These ducts appear as fine brownish lines similar to tiny pin scratches.

If none of these indicators are observed, apply the point of a fountain pen to the wood and note the direction in which the ink runs in the wood cells. Another test is to jab the point of a sharp knife blade into the tangential face at an angle of about 15° to the surface, with the knife set crossways to the length of the piece. Slowly rotate the knife blade with the cutting edge toward the piece of wood. This will lift up a splinter, with the direction in which the fibers tend to pull out indicating the direction of the grain.

Torn or chipped grain on the radial face also indicates spiral grain, which should always be measured on the flat-grain surface farthest from the pith. If the annual rings on a cross section run diagonally across the piece, or if there are not true flat-grain or edge-grain faces, determining the angle of cross grain becomes more involved (Fig. 4.5).

Diagonal—To measure the angle of diagonal grain on such a piece, locate on the end (the cross section of the piece) that corner which is farthest from the pith. From this corner draw a line radially, or at right angle to the rings. On this line, if the size of the piece will permit, measure 1 in. from the corner. Then, follow the annual ring markings until they intersect that edge farthest from the pith. The ratio *B:A* indicates the diagonal grain

Spiral—To measure the angle of spiral grain in a piece of wood, whose faces are neither radial nor tangential, mark a point some distance from the end on one of the edges that is neither nearest nor farthest from the pith; and follow along the fibers of a check, or parallel to a check, to the end of the piece.

Then draw a line from the end of this line radially toward the pith. The minimum distance of this radial line from the starting edge D,

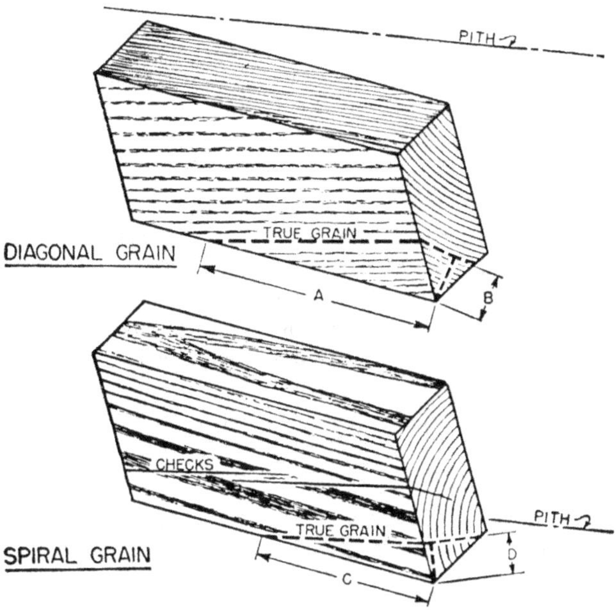

FIGURE 4.5 Method of measuring the angle of diagonal and spiral grain, when the rings are oblique to the edge of a piece of wood.

used in conjunction with the distance of the turning point from the end of the member *C*, will give the angle of the spiral grain.

Some wood may contain both diagonal and spiral grain. To obtain the true angle of the grain in such a piece, first obtain the angles of the spiral grain and the diagonal grain, both being expressed decimally.

Dip grain or burls—small pieces of wood are generally weakened by the presence of dip grain or burls, which are usually the result of defects that may or may not exist in the piece. The cross grain in the dip usually runs at a sever angle and greatly reduces the strength, in particular when this defect is on the tension side of a beam. To determine the strength of a beam of small cross section, such as a ladder side rail containing dip grain at the tension face, assume that the dip cannot take tensile stress, and therefore the depth of the section should be deducted from the depth of the beam.

Brashness

This is an abnormal condition that causes wood in bending to break suddenly and completely across the grain when deflected only a relatively small amount. Brashness is usually associated with slow grown hardwoods, very fast or very slow softwoods, and wood with pre-existing compression failures. Wood that is exceptionally lightweight for its species is usually brash, as is wood that has been exposed to a high temperature for a long time—such as wood ladders used in boiler rooms.

A piece of tough wood and a piece of brash wood may have identical strengths in static bending, but under impact the brash piece will fail at a much lower stress. Under ordinary conditions, the strength of wood in tension is from two to four times its compressive strength, but this ratio is greatly reduced in brash wood.

Creep

Time-dependent deflection, over that resulting from classic deformation, occurs when wood is subjected to a constant load over a period of time. Changes in climatic conditions during a duration of load or creep test may produce a lower load capacity and more creep than that observed under constant conditions of temperature and moisture content. Fortunately, such changes are moderate for most wood structural members in typical service environments.

Compression Failures

Next to cross-grain and brashness, a tendency for compression failure is perhaps the most hazardous defect in structural wood. Although this defect is not too common, the danger is that it is sometimes extremely difficult, if not actually impossible, to detect such a defect with the unaided eye.

NOTE: *Because of the low strength associated with compression failure, many safety codes require certain wood structural members, such as ladder rails and scaffold planks, to be entirely free of such conditions.*

Reaction Wood

Abnormal woody tissue, frequently associated with leaning trunks and crooked limbs of both conifers and hardwoods, is termed *reaction wood,* and includes compression and tension woods. Both types of reaction wood undergo excessive longitudinal shrinkage when subjected to moisture loss below the fiber saturation point—up to 10 times normal in compression wood and five or more times normal in tension wood.

When either of these woods is present in the same board with normal wood, unequal longitudinal shrinkage causes internal stresses that result in warping.

Compression wood—tissue in conifers found on the lower side of an inclined member. Pronounced compression wood can be detected by ordinary visual examination. It is characterized by eccentric growth of annual rings, which are of a dense nature and are predominantly in summerwood.

In finished lumber, compression wood is usually somewhat darker because of the greater proportion of summerwood. Compression wood on the flat-grain and the edge-grain faces is dull and lifeless in appearance. It is the one exception to the rule that the strength of any piece of wood, regardless of its species, can be judged by its weight.

By its very nature, compression wood should be *excluded* from all use where strength under shock or impact is essential. The great longitudinal shrinkage of compression wood (2 ½ to 20 times normal wood), if located near the center of the piece, may cause this pat of the wood to fail in tension, thereby producing cross cracks in the compression wood. If located near the edge of a small piece, the compression wood, in shrinking, will cause crooking or bowing.

Also, because of abnormal shrinkage, spike knots are frequently twisted from their positions and tend to protrude above the face of the wood member. Poles containing compression wood are *hazardous* for linemen to climb, for if cross checks develop at the surface, the shell of the outer wood may peel off, with the result that the points of their climbers will lose their hold on the pole.

Tension wood—tissue in hardwoods located on the upper side of the inclined member. In some species, tension wood may show up on a smooth surface as areas of contrasting colors. Although tension wood is somewhat more difficult to detect, eccentric growth visible on the transverse section frequently indicates its presence. Also, the tough tension wood fibers resist being cut cleanly and result in a wooly condition on the surface of sawn boards.

Reaction wood, compared to normal wood of comparable specific gravity, is definitely weaker.

CHAPTER 5

Wood Structural Members

Wood structural members—beams, columns, joists, and planks—are usually rectangular in cross section and have constant depth and width throughout their span. They should be designed large enough so that loading will not stress the member beyond allowable values for tension, compression, shear, and bending. Under certain conditions, such as to maintain specified clearances, beams, joists, and planks should also be large enough to limit deflection under loading.

A wood member's strength differs on each axis, with the member being strongest when the load is applied parallel to its grain. When properly designed, a wood member may also be safely loaded perpendicular to the grain (Table 5.1).

Notation for Beam Formulas

A = cross-sectional area (in.^{2})
E = modulus of elasticity (lb/in.^{2})
F_b = design value for extreme fiber in bending (lb/in.^{2})
f_b = unit stress (lb/in.^{2})
F_c = design value in compression parallel to grain (lb/in.^{2})
f_c = unit stress (lb/in.^{2})
F_t = design value in tension parallel to grain (lb/in.^{2})
F_v = design value in horizontal shear (lb/in.^{2})
f_v = unit stress (lb/in.^{2})
I = moment of inertia (in.^{4})
L = length of member (ft)
M = bending of resisting moment (in.-lb; or ft-lb)
P = total concentrated load (lb)
Q = static moment of area about neutral axis (in.^{2})
r = radius of gyration (in.^{2})
S = section modulus (in.^{2})
V = vertical design shear (lb)
W = total weight (lb)
w = weight per foot (lb)

Stress parallel to grain	Stress perpendicular to grain
Tension	
 Tension parallel to grain	
Wood fibers tend to stretch along the grain.	Wood fibers tend to separate from the grain.
A wood member's highest strength property is its natural resistance to tension parallel to the grain.	A wood member's least strength is when the fibers are under tension perpendicular to the grain. As a result, design values are seldom given.
Knots or holes reduce a member's cross section, thereby substantially reducing fiber resistance to the applied force.	Not applicable.
A force applied at an angle to the grain reduces the fiber resistance.	Not applicable.
Compression	
 Compression parallel to grain	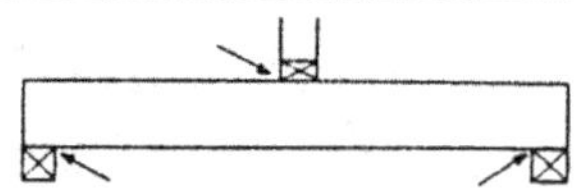 Compression perpendicular to grain
Wood fibers tend to compress lengthwise.	Wood member surfaces tend to compress, resulting in a displacement of the member.
Resistance to compression parallel to the grain is affected by the presence of knots or holes, as well as by the angle of loading.	Not applicable.
Shear	
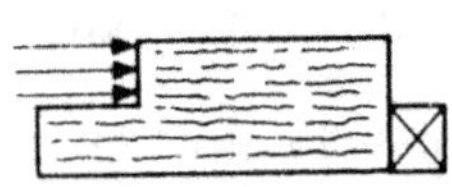 Shear parallel to grain	
Where loading induces a compression stress on one side of a wood member and tension on the other side, shear stress tends to occur parallel to the grain, the largest occurring along the neutral axis on the plane at which the induced stress changes from compression to tension.	This factor need not be considered, since limits on design stresses in shear parallel to the grain, and in compression or bearing perpendicular to the grain, control the design of a wood member.
Shear resistance is reduced by excessive checks and splits in the wood.	Not applicable.

Table 5.1 Stress Effects for Load Conditions Parallel and Perpendicular to Grain (a) Tension (b) Compression (c) Shear (d) Bending

Stress parallel to grain	Stress perpendicular to grain
Bending	
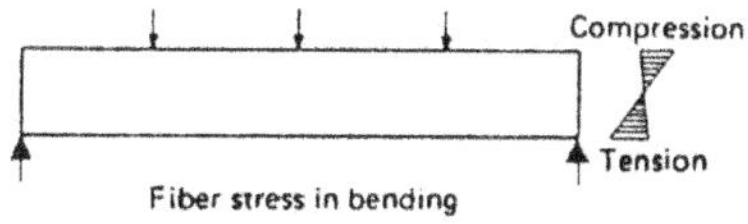	
Perpendicular loading induces compression stresses in the extreme fibers on the loaded side of a wood member, and tension stresses on the opposite side.	Not applicable.
These stresses diminish in intensity from the outside face fibers to zero at the center of the member's neutral axis.	Not applicable.
Resistance to bending in the extreme fibers is reduced by knots or holes, as well as by deviations in the grain slope.	Not applicable.

TABLE 5.1 (*Continued*)

5.1 Structural Properties

Certain mathematical expressions of the properties or elements of wood sections are used in computing the values of structural members of various shapes for the various conditions under which they are subjected to stress.

- *Cross section.* Section taken through the member, perpendicular to the member's longitudinal axis (Fig. 5.1).
- *Cross sectional area.* Area of a section taken through the member, perpendicular to the member's longitudinal axis; calculated by multiplying the section's base by its depth A in square inches.

$$A = bd \text{ (in.}^2\text{)}$$

- *Neutral axis.* Line in the cross section of a member on which there is neither tension nor compression.

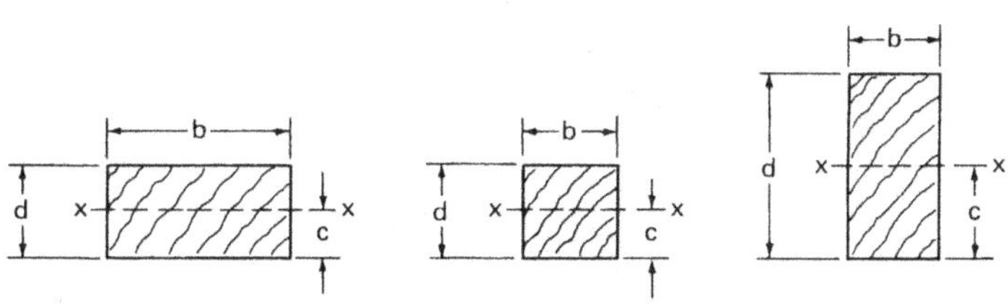

FIGURE 5.1 Typical cross sections of timber beams.

- *Modulus of elasticity.* Ratio of a wood member's tensile strength to tensile strain, in its deformation region.

$$E = \frac{stress}{strain} \text{ (lb/in.}^2\text{)}$$

- *Moment of inertia.* Sum of each member's basic cross-sectional area, multiplied by the square of its distance from the neutral axis of that particular area.

$$I = \frac{bd^3}{12} \text{ (in.}^4\text{)}$$

- *Radius of gyration.* Square root of a section's moment of inertia divided by its area.

$$r = \sqrt{\frac{I}{A}} \text{ (in.}^2\text{)}$$

- *Section modulus.* Moment of inertia of a section, divided by the distance from the neutral axis of the cross section.

$$S = \frac{I}{c} = \frac{bd^2}{6} \text{ (in.}^3\text{)}$$

where A = cross-sectional area (in.^{2})
b = width of member (in.)
c = distance from neutral axis (in.)
d = depth of member (in.)
E = modulus of elasticity (lb/in.^{2})
I = moment of inertia (in.^{4})
L = distance from face-to-face of supports (ft)
r = radius of gyration (in.^{2})
S = section modulus (in.^{3})
W = total uniform load (lb)
w_D = dead load (lb)
w_L = live load (lb)

- *Effective span.* Distance from face to face of supports, plus one-half the required length of bearing at each end, L in feet.
- *Design load.* Total uniform load, equal to the dead load (weight of structural member) plus live load (workers and materials supported by the member), W in pounds.

$$W = w_D + w_L \text{ (lb)}$$

- *Linear loading.* Uniform load w over the length of the member, wL in pound-feet.

Species	Specific gravity	Weight (lb/ft^3)
Coast Sitka pine	0.39	26.6
Douglas fir—Larch	0.51	34.3
Douglas fir—South	0.48	32.7
Eastern spruce	0.43	28.9
Eastern white pine	0.38	25.5
Hem fir	0.42	28.1
Northern pine	0.46	31.2
Oak - red and white	0.67	47.3
Ponderosa pine (North)	0.49	33.0
Southern cypress	0.48	33.6
Southern pine		0.55
Spruce - pine - fir	0.42	26.9
Western cedars	0.35	24.2

Source: American Wood Council

TABLE 5.2 Properties of Selected Structural Lumber Used in Scaffolding and Ladders

5.2 Structural Members

Beams

When determining stresses and deflections of wood structural members, actual—not nominal—sizes must be used.

The proper design of timber beams involves three steps:

1. Compute the member's section modulus ($S = M/f$) and select an appropriate sized beam from *Wood Structural Design 1962–* American Forest and Paper Association.
2. Test the beam selected for horizontal shear; increase dimensions if necessary.
3. Test the beam selected for deflection.

Most designers simply perform the first computation (section modulus) and then select the proper beam size that they know from experience will meet the requirements for steps 2 and 3. In determining the strength and stiffness of a wood member under transverse loading, it is necessary to consider:

1. Bending moment induced by a load
2. Horizontal shear at beam supports
3. Bearing on supporting members
4. Deflection or deformation caused by load

Any of the following four factors may control the design of a structural wood member.

Bending

For structural safety, a beam's bending moment, induced by the combination of live and dead loads on the beam, may not be greater than the resisting moment of the wood member.

For example: Assume a beam having a span L uniformly loaded with w pounds per foot (Fig. 5.2):

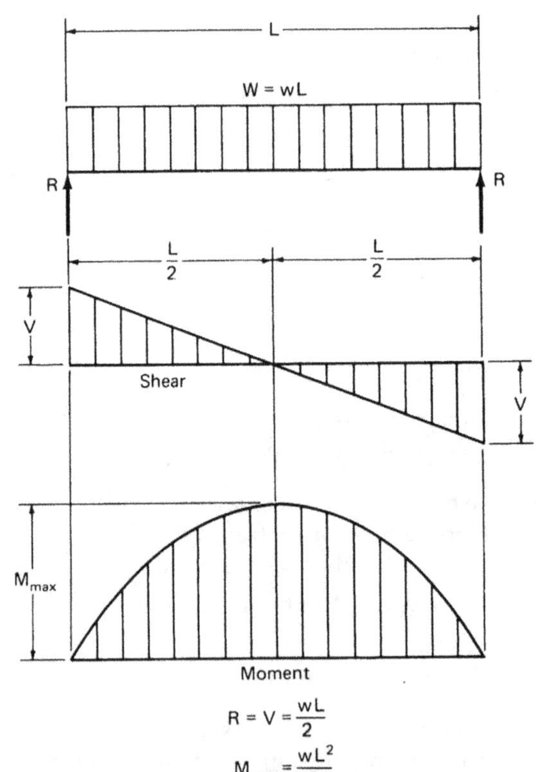

FIGURE 5.2 Forces and stresses developed in a uniformly distributed load on a simple beam.

1. Compute the maximum bending moment M, which occurs at the beam's midpoint $L/2$, using the formula:

$$M = \frac{wL}{8} \text{ (lb-ft)}$$

Convert to lb-in.

$$M = \frac{wL^2 \times 12}{8} = \frac{3wL^2}{2} \text{ (lb-in.)}$$

2. Divide the resisting moment M by the allowable fiber stress in bending Fb for the particular type of wood and grade of lumber, to determine the section modulus S of the structural member.

$$S = \frac{M}{F_b}$$

Since the bending moment may not be greater than the resisting moment, the two formulas can be equated:

$$F_b S = \frac{3wL^2}{2}$$

3. Determine one of the following:

a. *Size of beam required*—when span and load are known:

$$S = \frac{3wL^2}{2F_b}$$

b. *Allowable span*—when size and load per linear foot are known:

$$L = \sqrt{\frac{2F_b S}{3w}}$$

c. *Allowable load per linear foot*—when size of beam and span are known:

$$w = \frac{2F_b S}{3L^2}$$

Alternatively, once the section modulus of a wood structural member is determined, the rigger can select a properly sized beam directly from precomputed tables; however, make sure to select a wood section that has a section modulus equal to or greater than the computed value. (The section modulus of a structural wood member is I/c, which for a rectangular beam is $bd^2/6$.)

Although many beam sizes will meet these requirements, the most practical sizes are those beams whose breadth ranges from one-half to one-third of the beam's depth.

REMEMBER: *Beams that are relatively deep compared to their width tend to bend sideways under loading unless properly braced.*

Rules-of-thumb for beams having the following ratios of depth to width, or thickness:

2:1	No lateral support is needed
3:1 or 4:1	Ends should be held in position
5:1	One edge should be held in line for entire length of beam
6:1	Lateral supports are required at intervals of 8 ft

Although lumber is customarily specified in terms of nominal sizes, always use the actual sizes (net dimensions) for any design computations.

Horizontal Shear

Horizontal or longitudinal shearing forces in a member must be considered in designing a beam subject to vertical shear (Fig. 5.3). When a beam is loaded vertically, the upper part of the beam tends to slide along its lower part. To prevent this sliding action, the shear resistance of wood member must equal or exceed the horizontal shear that the vertical loading induces.

Maximum horizontal shear occurs in a rectangular beam at the neutral axis of the section and depends on the magnitude of vertical shear in the member:

$$f_v = \frac{VQ}{Ib}$$

$$f_v = \frac{3V}{2bd}$$

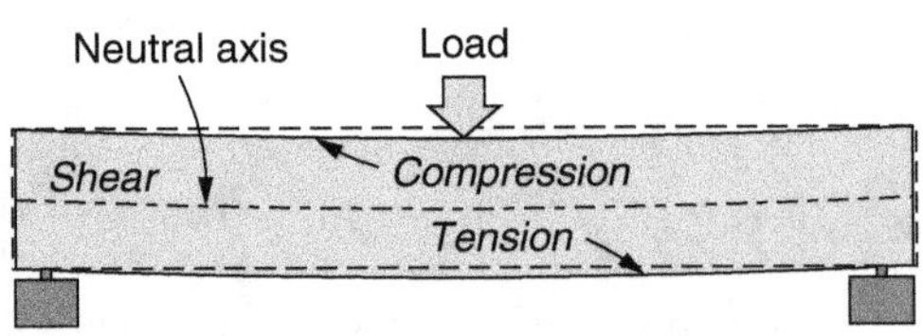

Breaking strength of a beam:
- Directly proportional to the width, but
- Proportional to depth squares, and
- Inversely proportional to the span

Stiffness of a beam:
- Directly proportional to width, but
- Proportional to depth cubes, and
- Inversely proportional to span cubed

(Thus, whether the depth or width of a solid beam is doubled, the mass is the same, but the strength/stiffness/weight ratios for a beam are very much better with increased depth rather than increased thickness)

FIGURE 5.3 Horizontal or longitudinal shear in a simple beam.

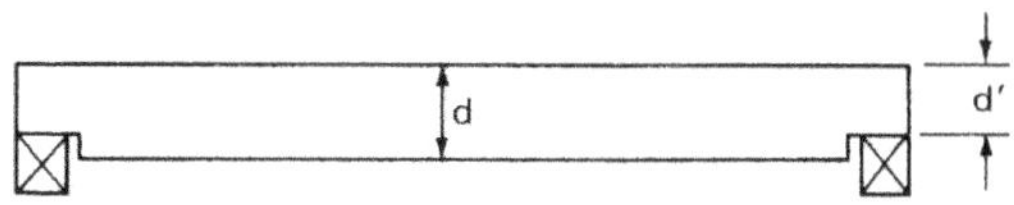

FIGURE 5.4 Notched beam with tapered notches 1/6 the beam's depth.

Where f_v = Unit Stress
V = Maximum Shear Force
Q = First moment of Area (around neutral axis)
I = Moment of Inertia
B = Width of Beam

The unit stress, however, may not be greater than the design value in horizontal shear *Fv* for the particular type and grade of wood used.

Because notching a beam will affect the shear strength of the member, notching should be avoided whenever possible, especially on a beam's tension side. *Never* notch the tension side of a beam that is 4-in. deep or deeper except on its ends. Notches at the ends of a beam, however, should *not* be greater than one-fourth the depth of the wood member (Fig. 5.4).

To reduce stress concentration, where notching a beam is necessary, use a gradually tapered notch instead of square cornered one. Notches in wood beams should not be deeper than one-sixth the depth of the wood member. And *never* put a notch in the middle one-third of a span.

Always check the desired bending load of a beam having square cornered notches at its ends against the load obtained by the formula:

$$V = \left(\frac{d'}{d}\right)\left(\frac{2F_v bd'}{3}\right)$$

Bearing on Supports

When a wooden beam is loaded, its fibers tend to compress at the beam support points. Make sure there is sufficient end-bearing area for the beam to transfer the load without damage to the fibers. Determine the bearing area by dividing the beam's end reaction by the design value of compression perpendicular to the grain F_c from the particular type and grade of wood being used. If the bearing area, from the end of a wood beam, is less than 6 in. long, it is safe to use higher stresses in compression perpendicular to the grain. For such bearing areas, if located 3 in. or more from the ends of a beam, increase the compression values as indicated in Table 5.3.

Deflection

A beam's deflection Δ is the amount of deformation that results from a load applied to the beam. As long as the induced bending stress does not exceed the design value, deflection does not seriously affect

Length of bearing (in.)	Factor
½	1.75
1	1.38
1½	1.25
2	1.19
3	1.13
4	1.10
6 or more	1.00

TABLE 5.3 Factors by Which to Increase Compression Values Based on Bearing Length

the safety of the beam, and deflection can be ignored, with the design of the beam based on strength alone. Deflection, however, becomes a critical factor for the rigger when a specified clearance is required under the beam.

Joists

These wood members are small beams spaced closely together to support floor (or roof) loads. They are designed in the same manner as wood beams. Maximum spans for various sizes of floor loads and stresses are based on the bending strength using the live load indicated on each heading, plus a dead load of 10 lb/ft^2 (span tables for Joists and Rafters–American Wood Council, Washington, DC).

Planks

The span of a scaffold plank is the distance it runs between supports. The longer the span, the more deflection (bend) it will have, and therefore, the less its load-bearing capacity will be.

For scaffolding, wood planks 2 in. thick, or thicker, are laid on the flat spanning from beam to beam (similar to the way a floor would be laid). The beams are spaced 5 ft to 12 ft on center, depending on the load to be supported and the kind of wood used (Tables 5.4 and 5.5).

To compute the correct thicknesses of a plank, assume a width of 12 in. and apply beam design methods. Scaffold planks should be unsurfaced, and actual dry dimensions should *not* be less than specified. For example:

2×9 $1^7/_8 \times 8^3/_4$

2×10 $1^7/_8 \times 9$

2×12 $1^7/_8 \times 11^3/_4$

Maximum intended load	Maximum permissible span using full thickness lumber	Maximum permissible span using nominal thickness lumber
25 lb/ft^2	10 ft	8 ft
50 lb/ft^2	8 ft	6 ft
75 lb/ft^2	6 ft	—

Source: OSHA, Washington, DC.

Table 5.4 Allowable Spans (for a given load) for 2 × 10 (nominal) or 2 × 9 (rough) Solid Sawn Wood Planks

The longer the span, the more deflection (bend) it will have, and therefore, the less its load-bearing capacity will be (Table 5.6). Because nominal thickness lumber is not cut to its exact dimensions, nominal 2 × 10 is really more like 1½ × 9¼. As a result, it does not have the same load-bearing capacity of full thickness lumber.

For fabricated planks and platforms, maximum spans are to be recommended by the manufacturer. To assure that scaffold planking remains within its safe load-bearing capacity, it may not be allowed to deflect more than 1/60th of its span between supports [OSHA 1926.451(f)(16)].

Rated load capacity	Intended load
Light duty	25 lb per ft^2 applied uniformly over the entire span area
Medium duty	50 lb per ft^2 applied uniformly over the entire span area
Heavy duty	75 lb per ft^2 applied uniformly over the entire span area
One person	250 lb placed at the center of the span (total 250 lb)
Two person	250 lb placed 18 in. to the left and right of the center of the span (total 500 lb)
Three person	250 lb placed at the center of the span and 250 lb placed 18 in. to the left and right of the center of the span (total 750 lb)

Source: OSHA, Washington, DC.

Table 5.5 Allowable Spans (for a given load) for Fabricated Planks and Platforms, Maximum Spans to be Recommended by the Manufacturer Based on this Table

Span of plank between supports	Calculation	Maximum permissible deflection
10 ft	120 in./60 in.	2 in.
7 ft	84 in./60 in.	$1\frac{3}{8}$ in.
5 ft	60 in./60 in.	1 in.

Source: OSHA, Washington, DC.

TABLE 5.6 Maximum Permissible Planking Deflection

Planks should have strength ratios of at least 80 percent of the strength of a theoretically flawless plank.

- Planks, when fully accepted, should be immediately branded on their ends—permanently marked suited for scaffolding.
- Inspected planks should be carefully handled and not dropped in bulk from the delivery truck.
- Use of planks other than those specifically accepted for scaffolding should be forbidden.

REMEMBER: *Deflection can indicate when a platform is overloaded. To assure that scaffold planking remains within its safe load-bearing capacity, it may not be allowed to deflect more than 1/60th of its span between supports [OSHA 1926.451(f)(16)]. A competent person should not allow employees to occupy a platform that exceeds the 1/60th ratio.*

Columns

Most timber columns, or posts, have square or rectangular cross sections and are single lengths of timber. Although columns built up from pieces of timber and joined by nails, bolts, or other mechanical fasteners are often used, they *do not* have the same strength as a one-piece member of comparable material and dimensions. The strength of a built-up column must be reduced by precomputed percentages to provide a section equal to the strength of a one-piece column of the same dimension and quality.

The slenderness ratio of a column is a measure of its stiffness and has an important bearing on the load a column will support. The strength of built-up columns must be reduced by the percentages in the table to provide a section equal to the strength of a one-piece column of the same dimension and quality (Table 5.7).

The slenderness ratio is computed by dividing the column's laterally unsupported length by the appropriate cross-sectional dimension,

Slenderness ratio (*L/D*) of equivalent solid column	Ratio of built-up column strength to equivalent solid column strength
6	82
10	72
14	71
18	65
22	74
26	82

TABLE 5.7 Adjustment Factors for Built-Up Columns with Cover Plate

both in inches (Fig. 5.5). To determine the laterally unsupported length of a column, measure the column's distance parallel to its longitudinal axis—between the supports that restrain the column against any lateral movement.

A short column, one having a slenderness ratio less than 11, will support a load equal to the area of the cross section multiplied by the full allowable compressive strength of the particular type and grade of wood. If the length is increased, however, then the cross-sectional area must also be increased.

Numerous timber column formulas are used in finding the permissible unit stress. A formula recommended by the National Forest Products Association is

$$F = C\left(1 - \frac{I}{80d}\right)$$

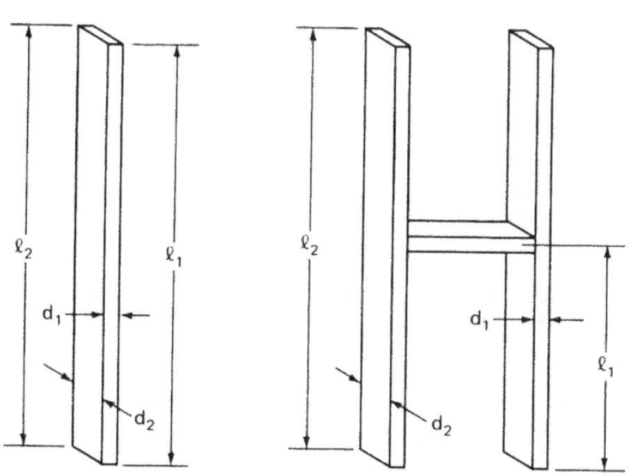

FIGURE 5.5 Slenderness ratio l/d for simple solid columns.

where F = maximum permissible unit stress for column cross section (lb/in.^{2})
C = maximum allowable unit stress parallel to grain for short blocks (lb/in.^{2})
l = unsupported length of column (in.)
d = least width or diameter of cross section (in.)

Timber columns are designed by trial and error, first assuming a cross section and then testing to see whether or not it meets the specified requirements. For example:

1. Assume a cross-sectional area of approximate proper dimensions.
2. Solve for F in the timber column formula, substituting for l the length of the column, in inches; and for d the least width or diameter of the column, in inches.
3. The computed F value is in units of pounds per square inch, and is multiplied by the number of square inches in the assumed column's cross-sectional area to obtain the allowable safe load for the column.
4. Compare the allowable load found in step 3 with the actual load to be supported.
5. If the actual load is greater, the assumed cross section is too small.
6. Assume a larger cross section, and test again.

While lumber columns may be designed by use of a formula, the rigger will find it more convenient to use precomputed tables to make the proper selection of column sizes (National Forest Products Association).

5.3 Rules-of-Thumb

When designing wood structural members:

1. The safe load on a beam varies directly with the beam's width; doubling the width of a beam doubles the safe load it can carry.
2. The safe load on a beam varies as the square of the beam's depth; doubling the depth of a beam increases by four times the safe load it can carry.
3. Placing two planks on top of each other provides twice the strength of one plank.
4. Securely nailing or doweling plank ends together to prevent slipping of one plank on the other provides four times the

strength of one plank, or twice the strength of the two planks placed on top of each other (unsecured).

5. The safe load on a beam varies indirectly with the beam's span; doubling the span cuts the safe load by one-half.
6. A beam of a given size, material, and span will carry a uniformly distributed load twice as great as a concentrated load applied at the center of the span (provided the unbraced span is relatively short).
7. The deflection of a beam of a given size, material, and span varies directly as the load placed on the beam; doubling the load doubles the deflection.
8. The deflection of a beam of a given size, material, and loading varies as the cube of the beam's span; doubling the span increases the deflection $2 \times 2 \times 2 = 8$.

References

Annual Book of Standards, Vol. 04.10 Wood, ASTM International (*astm.org*).

Guide to Good Practice for Handling, Installing & Bracing of Metal Plate Connected Wood Trusses (BCSI 1-03), Wood Truss Council of America (*woodtruss.com*).

Lumber Storage (TG-5), Western Wood Products Association (*wwpa.org*).

Manual for Engineered Wood Construction, American Wood Council (*amc.org*).

National Design Specifications for Wood Construction, (ANSI/AF&PA NDS-2005) American Forrest & Paper Society (*afandpa.org*).

Recommended Design Specification for Temporary Bracing of Metal Plate Connected Wood Trusses, DSB-89, Truss Plate Institute (*tpinst.org*).

Southern Pine Use Guide, Southern Pine Council (*southernpine.com*).

Storage and Erection, AITC 111, American Institute of Timber Construction (*atc.glulam.org*).

Timber Construction Manual (4th Edition), *American Institute of Timber Construction* (*atc.glulam.org*).

Wood Handbook: Wood as an Engineering Material, FPL-113 - USDA Forest Service, Forest Products Laboratory (*fpl.fs.fed.us*).

Industry Standards Resources and Guidelines

American Institute of Timber Construction, Centennial, CO (*www.aitc-glulam.org*)

APA – The Engineered Wood Association, Tacoma, WA (*www.apawood.org*)

American Forest and Paper Association, Washington, DC (*www.afandpa.org*)

CHAPTER 6

Metal Structural Members

Uniform cross-sections include tees, angles, and beams manufactured through extrusion, continuous casting, roll forming, and other manufacturing processes. Typically, profiles and structural shapes are formed securely to satisfy various strength requirements and precise cross-sectional dimensions of the shape or profiles.

Profiles and structural shapes are available in a variety of shapes and sizes, including solid, hollow, and semihollow structural members. Ultimately, the characteristics of a structural shape are determined by the cross-sectional view. Selecting profiles and structural shapes requires an analysis of how cross sections can be adapted to construction.

Selecting an adequately sized steel or aluminum section for a rigging or scaffolding operation is a relatively simple task using the well-established manuals that offer vital data for selecting structural members that have sufficient strength to ensure safe, adequate loading.

- *Steel Construction Manual.* Furnishes complete tabular data and methodology for choosing and dimensioning every variety of standard structural steel shape, covering even pipe and tubing, bars, and plating (American Institute of Steel Construction)
- *Aluminum Design Manual.* Furnishes comparable facts and figures for structural aluminum (Aluminum Association)

The use in rigging of steel or aluminum beams today is almost universal for construction or maintenance of facilities and structures. The highly practical and useful wood beam is, of course, still well-suited to many applications. The steel and aluminum manuals, however, are essential to any installation where either of these metals is to be part of the rigging activity (Table 6.1).

Property	Steel	Aluminum
Extrudability	Not practical	Very good
Weldability	Good (no reduction in strength)	Fair (reduction in strength)
Corrosion resistance	Fair	Good
Tensile yield strength	35–50 ksi	35 ksi
Stiffness	29,000 ksi	10,000 ksi
Elongation	20%	8–10%
Density	0.283 psi	0.098 psi
Strength-to-weight ratio	1.0–1.41	2.0

TABLE 6.1 Comparison of Common Structural Shapes and Grades

It is to the rigger's advantage to understand the basic terminology and design fundamentals used in selecting required metal beams or columns.

6.1 Structural Metal Shape Properties

Structural steel and aluminum is manufactured in a wide variety of cross-sectional shapes and sizes:

- *W-shape.* Most widely used structural member for beams, columns, and other load-bearing applications. Parallel inner and outer flange surfaces of a constant thickness provide greater cross-sectional area in the flanges, resulting in greater strength than is provided by the S-shape member. The distance between W beams is measured from the center of the sections' webs.
- *S-shape.* Has relatively narrow flanges, which taper in thickness, approximately 17° toward the flange's base. The distance between S beams is measured from the center of the sections' webs.
- *C-shape.* Channel sections resemble half an S-beam, with flanges on only one side of the web. Inner flange surfaces slope approximately 17°, making the C-shape member especially useful in locations where a single flat surface on one side is required. When used alone, the C-shape is not very efficient as a beam or column. Efficient built-up

members, however, may be constructed of channels assembled together with other structural shapes and connected by rivets or welds. The distance between channels is measured from the back of the web or the heel of the flanges.

Structural steel member W-, S-, and C-shapes are designated by their nominal depth, in inches, along the web and the weight in pounds, per foot of length. A W14 × 30, for example, indicates a W-shape that is 14 in. deep along its web and weighs 30 lb per linear foot.

- *HP-shape (bearing pile).* Almost identical to the W-shape; the only difference is that the web and flange thicknesses are equal.
- *Angle.* L-shaped section with leg lengths either equal or unequal, identified by dimensions and thickness of its legs, such as L6 × 4 × ½. The dimension of each leg is measured along the outside of the angle, and for unequal-leg angles, the dimension of the wider leg is always give first. The third dimension applies to the thickness of the legs, which always have equal thicknesses. Angles are used primarily to support, brace, or connect other structural steel members. They may be used as single members, or they may be used in combinations of two or four to form main structural members.
- *Steel plate.* Structural member that has a width greater than 8 in. and a thickness of ¼ in. or more. Generally used as connections between other structural steel members; may also be used as component parts of built-up structural members, such as columns.

6.2 Selecting Structural Metal Beams

When selecting a metal structural beam, keep in mind these rules:

- A beam of a given size and span will carry a uniformly distributed load twice as great as a load concentrated at the center of the span. Therefore, if the table indicates that a certain beam on a given span will safely carry a concentrated load of 7500 lb, if the load is spread out over the entire length of the beam, 15,000 lb may safely be applied. Likewise, on a cantilever beam the allowable distributed load is twice the allowable concentrated load.
- The safe load on a simple beam varies indirectly as the span. Therefore, if the span of a beam of given size is doubled, the safe load is reduced to one-half. On a cantilevered beam, if the lower arm is reduced to one-half, the load may be doubled.

- The deflection, or bending, of a simple or cantilevered beam under load varies directly as the load. Therefore, doubling the load will ordinarily double the deflection.
- The deflection of a simple beam of a given size and load varies as the cube of the span. For example, if the span is multiplied by 2, the deflection is multiplied by 2^3 ($2 \times 2 \times 2 = 8$); if the span is multiplied by 2½, the deflection is multiplied by $2½^3$ ($2½ \times 2½ \times 2½ = 15⅝$).
- The load applied to a beam must include not only the useful load to be lifted, but also the weight of the slings, the hoist tackle or chain hoist, the pull on the hauling line or hand chain, and, if the beam is very heavy, the weight of the beam itself. For scaffolding, it would be necessary to consider not only the weight of the workers, but also that of scaffold planking and any material stored on the planks as well as any equipment being used.
- The span of a simple beam, supported at or near its ends, is the distance between the supports measured in inches. The length of span depends to a degree on the nature of the material on which the beam rests.
- If the supports are very firm, such as heavy steel or concrete (Fig. 6.1a), the effective span is measured from edge to edge of the supports.
- If the beam rests on timbers that compress slightly under the bearing load (Fig. 6.1b), or if the steel beams on which it rests can rotate slightly under the beam (Fig. 6.1c), the span is measured from the center of one bearing to the center of the other bearing.

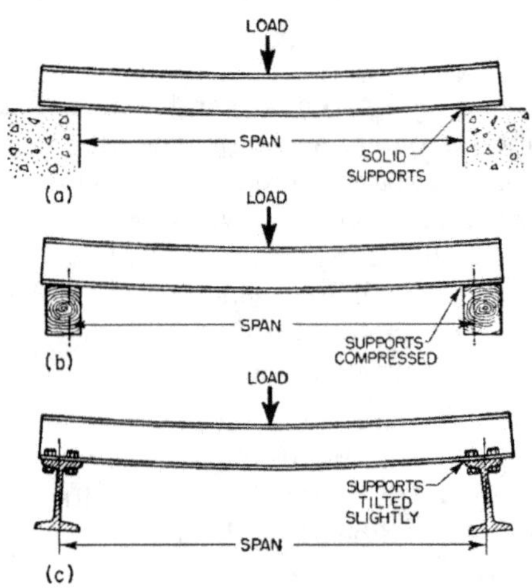

FIGURE 6.1 Measuring a beam's length of span between various supports.

6.3 Design Calculations

Although the manuals on steel and aluminum design will provide most of the required data for selecting a structural member, it is necessary to follow certain procedures.

To calculate	Need to know
Safe load	Span of beam, size, material
Beam size	Span of beam, load, material
Maximum beam span	Size of beam, load, material

The same unit stress of 36,000 psi is used for all steel beams and 14,000 psi for all aluminum beams. Tables in the respective manuals list the product of unit stress times section modulus under the heading of maximum bending moment, thus simplifying the calculations.

Bending Moment

A key factor in proper beam selection is the bending moment of a structural member (Fig. 6.2). To determine the bending moment when the size of the beam is to be calculated or, if the beam size is known, to find the maximum span or load, use these bending formulas:

where M = bending moment (lb-in.)
P = concentrated load (lb)
W = total uniformly distributed load (lb)
L = length of span (in.)

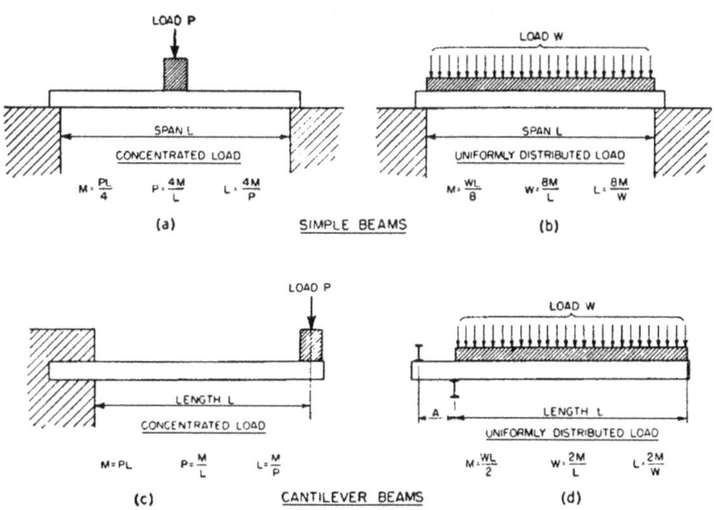

FIGURE 6.2 Calculating the bending moment of a beam under various types of loading.

Unless the beam is very narrow relative to the span, all calculations are quite simple. On beams with a high slenderness ratio, additional investigations must be made. The slenderness ratio is the length of the span between lateral bracing of the compression flange, divided by the width of the beam flanges.

In a W-beam or S-beam used as a simple beam, the upper flange resists compressive stresses and the beam has the same tendency to buckle as does a loaded column. In selecting a beam, make sure that the member has a cross-sectional area large enough to resist adequately all bending, shear, and deflection.

It is essential to take the slenderness ratio into consideration; if the beam is very slender, there is a tendency for it to deflect sideways, then twist and roll over. Because the beam is much weaker on its side, it will fail and allow the load to drop. Bracing the compression flange against deflection sideways so that the compression flange of the longest unbraced part of the span is 15 in. or less will prevent this from happening, and the full calculated load can be safely carried.

Bending Design

To design a metal beam for bending use the formula:

$$\frac{M}{f} = S$$

First, determine the type of metal member being used and its unit stress, plus the laterally unsupported length of span. Then complete the maximum bending moment (as previously explained) of the beam. Next, divide this moment by the allowable extreme fiber unit stress. (Building codes permit 22,000 psi for *A36 steel*; 14,000 psi for *6061-T6 alloy aluminum*.)

This gives the required section modulus S of the beam. Referring to the tables of properties of steel shapes, simply select a beam with a section modulus equal to or greater than that which is required. Generally, the section with the lightest weight per linear foot is the most economical.

Warning: *In using the flexure formula,* do not *mix units of measure: if f is in pounds per square inch, then M must be in pound-inches,* not *pound-feet.*

This method of calculating the beam size is applicable to any type of loading and any material. In most applications, however, beams have uniformly distributed loads. Therefore, when designing for bending, it is more convenient to use the tables covering this type of loading to select the proper size of beam directly—all that is needed is the load and span.

Shear

Since most beams are large enough to resist bending are also large enough to resist shear, this step is usually omitted.

Deflection

To avoid floor and ceiling cracking in building construction, beam deflection should not exceed 1/360 of the span. For rigging and scaffolding applications, however, the only serious problem with beam deflection might be one of having to provide sufficient clearance for equipment. Usually, because beams used for this type of operation have short lengths, this computation can be disregarded. A long-span beam designed to carry a relatively light load, however, should be investigated to make sure that deflection does not exceed 1/360 span.

Because simple beams with uniformly distributed loads occur so frequently in practice, tables of maximum loads for specific spans are of great convenience to the designer. The loads in these tables are given in 1000 lb (kips), and the extreme fiber stress used in computing the load is 24,000 psi. (Refer to the appropriate manuals to determine maximum load deflection values.)

Equivalent Distributed Loads

Although common practice generally dictates the use of tables covering uniform loading, such tables can be abridged for beams subjected to direct loading (Table 6.2).

Since equivalent distributed loads found by this method do not include the weight of the beam, beam sections determined this way should also be checked for shear and deflection when used for structural applications.

Bearing Plate

A steel bearing plate is usually required to distribute the beam load over the bearing area of a masonry wall, or concrete pier support to prevent the beam from crushing the masonry or concrete support. The plate uniformly distributes the beam load on the support

Equivalent uniform load	Deflection coefficient
2.00*P*	0.80
2.67*P*	1.02
4.00*P*	0.95
4.90*P*	1.01

Source: Manual of Steel Construction, American Institute of Steel Construction.

Table 6.2 Types of Loading—Equal Loads, Equal Spaces

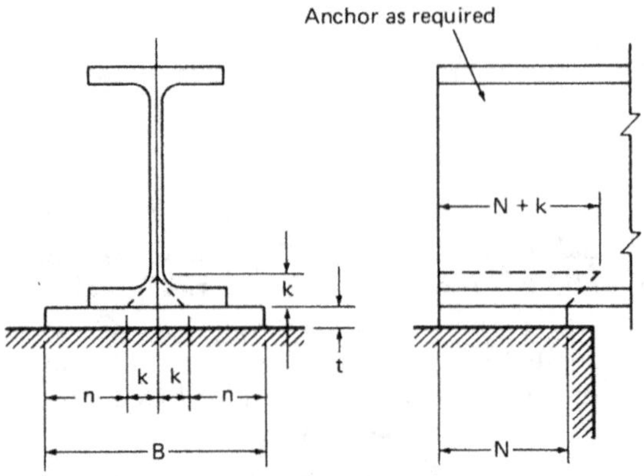

FIGURE 6.3 Bearing-plate design data.

structure and also helps seat the beam at its proper elevation (Fig. 6.3). The size of the plate required is found by dividing the beam load on the support area by the allowable bearing unit stress.

$$A = \frac{P}{F_p}$$

where A = area pf bearing plate B × C (in.^{2})
P = load from beam (lb)
F_p = allowable bearing stress of support material (lb/in.^{2})

The wall thickness will determine the plate dimension parallel to the beam length. The plate dimension parallel to the length of the wall is B. Both dimensions are usually given in even inches (see Table 6.3).

Material	Bearing capacity (psi)
Stone concrete, depending on quality	600–800
Common brick, lime mortar	100
Common brick, lime-cement mortar	200
Common brick, cement mortar	250
Hard brick, cement mortar	300
Rubble, cement mortar	150
Rubble, lime-cement mortar	100
Hollow T.C blocks, cement mortar	80
Hollow cinder blocks, cement mortar	80

TABLE 6.3 Allowable Bearing Capacity of Masonry Walls (psi)

To determine the thickness of the plate, take the maximum bending moment at a distance n from the edge of the plate. Then use the formula:

$$t = \sqrt{\frac{3pn^2}{F_b}}$$

where t = thickness of bearing plate (in.)
p = actual bearing pressure of plate on masonry (psi)
F_b = allowable bending stress in bearing plate (psi)
(AISC gives $F_b = .75F_y$ where F_y is the yield-point stress of the steel plate)
$n = (B - k)/2$ (in.)
k = distance from bottom of flange to web toe of fillet (in.)
(Values of k may be found in the AISC manual under dimensioning for detailing)

Since *A36* steel is commonly used for bearing plates, the formula reduces to:

$$t = \sqrt{\frac{3pn^2}{27{,}000}} = \sqrt{\frac{pn^2}{9000}}$$

6.4 Steel Structural Sections

Steels used for building construction in the United States use standard alloys identified and specified by ASTM International. These steels have an alloy identification that begins with *A*, followed by two, three, or four numbers.

AISC 360 – *Specification for Structural Steel Buildings*
AISC 303 – *Code of Standard Practice for Steel Buildings and Bridges*
AISC 325 – *Steel Construction Manual*

(The four-number AISI steel grades, commonly used for mechanical engineering, machines, and vehicles, are a completely different specification series.)

The most commonly used structural steel used today is the *ASTM A36* type, which has a high yield point (36,000 psi) (Table 6.4).

Because of this strength, smaller lighter sections can be used for many applications. All structural members made of *A36* steel may be riveted, bolted, or welded for erection. Structural steel members include:

Carbon steels
A36 – structural shapes and plate
A53 – structural pipe and tubing
A500 – structural pipe and tubing
A501 – structural pipe and tubing
A529 – structural shapes and plates
High strength low alloy steels
A441 – structural shapes and plates
A572 – structural shapes and plates
A618 – structural pipe and tubing
A992 – W shapes beams only

Source: ASTM International

TABLE 6.4 Standard Commonly Used Structural Steels

Beams

W	Wide flange
WT	Structural tee
S	American standard flange (I-shape)
ST	American standard tee
L	Angle
C	American standard channel
HP	Bearing pile
PL	Plate
M	Miscellaneous shapes

Open-Web Joists

Theses lightweight steel trusses are used primarily to support floors and roof panels between main supporting beams, girders, trusses, or walls. They may also be used to support wood decking for scaffolding. Manufactured joists used in construction are standardized in depths, spans, and load-carrying capacities. The primary series include *K, KCS, LH, DLH,* and *JG* (joist girder), ranging in sizes from 8 in. to 30 in. deep (in 2-in. increments); and in lengths 8 ft to 60 ft.

Columns

The load a steel column can carry safely depends not only on the number of square inches in its cross section, but also on the shape of the column section. *(This is not true for a wood column, which is always a solid shape.)* The cross section of a steel column, however, is seldom symmetrical with respect to both major axes; therefore, when axially loaded, such a column tends to bend in a plane perpendicular to the axis of the cross section about which the moment of inertia is the least.

In an I-beam the moment of inertia about the axis parallel to the web is considerably smaller than that about the axis parallel to the flanges. Therefore, despite the amount of material in such a cross section, an I-beam is not an economical column section.

Failure of short columns occurs by crushing; that of long slender columns, by stresses resulting from bending. In short members, the average unit stress is $f = P/A$. But for small cross-section steel columns, long enough to have a tendency to bend, the stresses are not equally distributed over the cross section and therefore the average unit stress must be less than 20,000 psi. This average stress is dependent on the slenderness ratio of the column, its end conditions, and its cross-sectional area.

While the slenderness ratio for timber columns is l/d (the unbraced column length divided by the dimension of the least side, both in inches), for steel columns it is l/r (the unbraced column length divided by the least radius of gyration, both in inches). The least radius of gyration describes a steel column's measure of effectiveness in resisting bending. Given the least moment of inertia and the area of the cross section, the radius of gyration r may be readily computed. Tables of properties of steel sections, however, provide their value, thus eliminating the need to compute it.

Pipe Columns

Steel pipe and tubing, with their symmetrical shapes, have equal moments of inertia about each axis, and therefore are frequently used as columns, especially for scaffolding. Properties and allowable loads for various sizes of standard steel pipe columns—designated as HSS sections—are given in the *AISC* handbook.

In addition, although *AISC 2000 Specification for Design of Hollow Structural Sections* covers load and resistance factors for any square, rectangular, or round shape produced in accordance with a pipe or tubing specification. The information also applies to steel pipe.

Column Base Plates

All steel columns must be supported at their bases on a rolled steel slab or billet to distribute the column load over an adequate area and prevent crushing of the supporting concrete or masonry foundation. The base of the column and the steel slab must be in absolute contact, attached by welding or by angle sections secured to the masonry footing by anchor bolts (Fig. 6.4).

The area of the plate or billet is found by dividing the column load by the allowable unit compressive stress for the concrete or masonry footing. Determine the thickness of the plate or billet by assuming the plate to be an inverted cantilever with the maximum bending moment at the edge of the column.

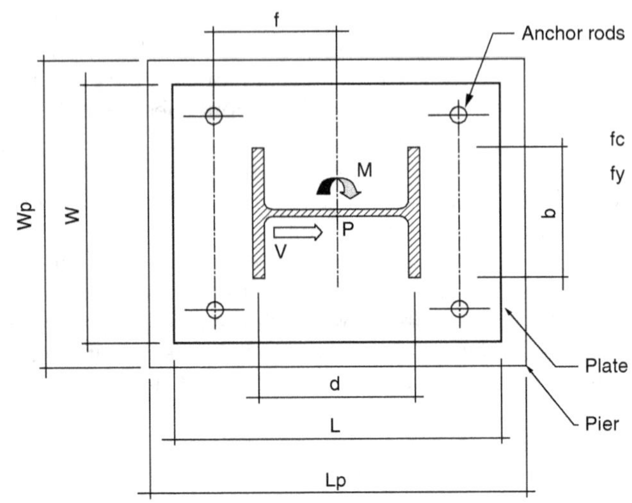

Figure 6.4 Typical end-of-column base plate to distribute concentrated load of the column.

The design of a column base plate involves two main considerations:

- Spreading the load so as to maintain the bearing pressures under the allowable values.
- Connecting or anchoring the base plate and the column to the concrete or masonry foundation.

(For frame analysis it may be important to consider the degree of fixity of the entire anchorage, so the design of the base plate and the anchor bolts should consider the interaction of axial load and bending moment.)

6.5 Aluminum Structural Sections

One of the most widely used structural aluminum alloys today is *ASTM Aluminum-Alloy 6061-T6*, a general-purpose structural aluminum alloy, having moderate strength with a yield point of 39,900 psi, good dimensional stability, and excellent weldability. *Alloy 6061-T6* structural aluminum is popular for medium to high strength requirements; has good toughness characteristics; and is easily welded and joined by various commercial methods. *Excercise caution*, however, when using *6061-T6* sections because direct contact with dissimilar metals can cause galvanic corrosion.

The *Aluminum Design Manual* covers types, characteristics, and use of aluminum structural sections—I-beams, channels, angles, tees, and zees—including:

- *Section properties*. Shapes similar to the American standard shapes, but weighing 11 percent less

- *Y-axis.* Dramatic increases, ranging from 40 percent to 150 percent
- *Flanges.* Thicker than the web; section modulus (resistance to bending) in the Y-axis is approximately 11 percent greater for the same weight per foot
- *Wider flanges.* Make extruded aluminum beams more stable
- *Straight flanges.* Make joining aluminum structural members easier
- *Radius of gyration.* Shows similar improved properties

Use only a section's nominal dimension when calculating structural aluminum members. Safety factors (or resistance factors, if you use load and resistance factor design) account for the fact that aluminum beams might be smaller than the nominal dimension.

The allowable tensile stress for a structural aluminum member is the tensile strength divided by a safety factor of 1.95. The tensile strength is the lesser of the tensile strength of the filler alloy and 90 percent of the weaker of the welded tensile ultimate strengths (Table 6.5).

	English	**Metric**
Physical properties		
Density	0,0975 psi	2.7 g/cc
Mechanical properties		
Tensile strength (Ultimate)	45,000 psi	310 MPa
Tensile strength (Yield)	39,900 psi	275 MPa
Elongation at break	12%	12%
Modulus of elasticity	10,000 ksi	69.GPa
(Average of tension and compression. In Aluminum Alloys, the compressive modules is typically 2% greater than the tensile modulus)		
Notched tension strength	47,000 psi	324 MPa
Ultimate bearing strength	88,000 psi	607 MPa
Bearing yield strength	56,000 psi	366 MPa
Fatigue strength	13,800 psi	96 MPa
Fracture toughness	4,207 psi	29 MPa

Pa = unit of measure pascal is equal to one newton per square meter; MPa = mega (million) pascals; GPa = giga (billion) pascals.
Source: Aluminum Design Manual, Aluminum Association.

Table 6.5 Physical and Mechanical Properties of *Aluminum 6061-T6: 6061-T651*

Web

To calculate the web crippling strength use the inside radius.

- For extruded structural shapes (where the corner between the web and flange is filler) use R1 = 0.
- For cold-formed steel beams, use R1 (the bend radius at the web-flange juncture).

No general deflection limit is given for aluminum beams, since the *Specification for Aluminum Structures* is used in many different applications, each with its own performance requirements.

References

American Institute of Steel Construction (1999), Load and Resistance Design Specification for Structural Steel Buildings, AISC, Chicago, IL.

American Institute of Steel Construction (2000), Load and Resistance Factor Design Specification for Steel Hollow Structural Sections, AISC, Chicago, IL.

American Institute of Steel Construction (2000), Load and Resistance Factor Design Specification for Single-Angle Members, AISC,Chicago, IL.

American Society of Civil Engineers (2006), Minimum Design Loads for Buildings and Other Structures, ASCE, Reston,VA.

ASTM International (2000), Pipe, Steel, Black, and Hot-Dipped, Zinc-Coated Welded & Seamless, ASTM/ANSI, West Conshohocken, PA.

ASTM International (2000), Standard Specification for Low and Intermediate Tensile Strength Carbon Steel Plates, West Conshohocken, PA.

ASTM International, Standard Specification for Cold-Formed Welded & Seamless High-Strength, Low Alloy Structural Tubing with Improved Atmospheric Corrosion Resistance, West Conshohocken, PA.

Regulations and Standards

OSHA Regulations

1926 Subpart R—*Steel Erection*

1926.750 *Scope*
1926.751 *Definitions*
1926.753 *Hoisting and Rigging*
1926.754 *Structural Steel Assembly*
1926.755 *Column Anchorage*
1926.756 *Beams and Columns*
1926.757 *Open Web Steel Joists*
1926.759 *Falling Object Protection*
1926.760 *Fall Protection*

1926 Subpart R–Appendix G
1926.502 (b)-(e) *Fall Protection Systems Criteria and Practices*

U.S. Consensus Standards

ASTM B308 *Standard Specification for Aluminum Alloy 6061-T6 Standard Structural Profiles*

ATM B429 *Standard Specification for Aluminum and Aluminum Alloy Extruded Structural Pipe and Tube*

AWS D1.2 *Structural Welding Code, Aluminum*

Industry Consensus Standards and Guidelines

Aluminum Association, Arlington, CA *(www.aluminum.org)*

American Institute of Steel Construction, Chicago, IL *(www.aisc.org)*

American Iron and Steel Institute, Washington, DC *(www.steel.org)*

American Welding Society, Miami, Fl *(www.aws.org)*

Association for Iron and Steel Technology, Warrendale, PA *(www.aist.org)*

Steel Joists Institute, Myrtle Beach, SC *(www.steeljoist.org)*

Steel Tube Institute of North America, Coral Gables, FL *(www.steeltubeinstitute.org)*

CHAPTER 7

Useful Formulas, Tables, and Conversion Factors

Areas and Volumes

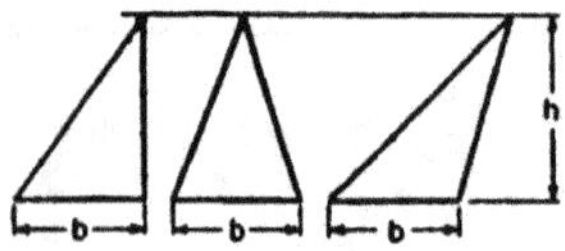

Triangle: Area $= \dfrac{bh}{2}$

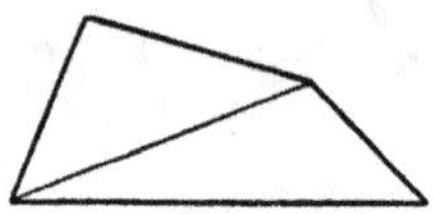

Trapezium: Area = Sum of areas of component triangles

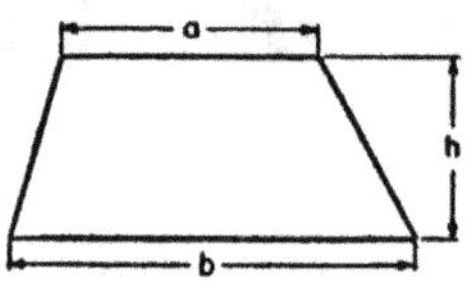

Trapezoid: Area $= \dfrac{(a + b)h}{2}$

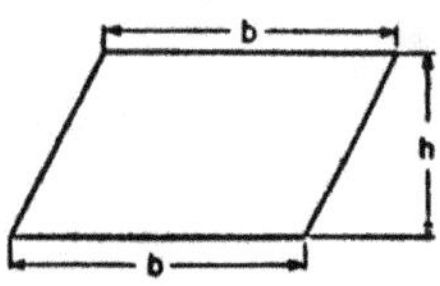

Parallelogram: Area $= bh$

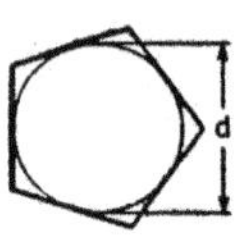

Regular polygon: Area $= \dfrac{d \times \text{perimeter}}{4}$

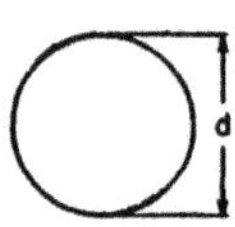

Circle: Area $= 0.7854d^2$

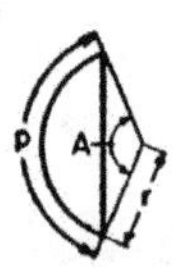

Segment: Area $= \dfrac{r^2}{2}(0.0175A - \sin A)$

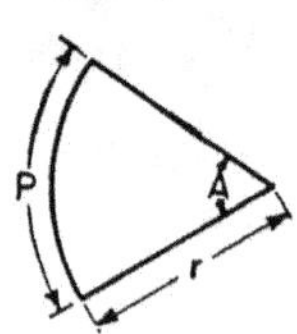

Sector: Area $= 0.00873r^2A = \dfrac{Pr}{2}$

Irregular shape: divide length into parallel strips of equal width.

Area $= b\,\dfrac{h_1 + h_2 + h_3 + \cdots + h_n}{n}$ (approx.)

Table 7.1 Areas of Plane Figures

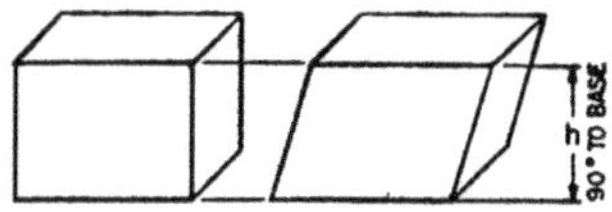

Regular prism: Volume = area of base × h

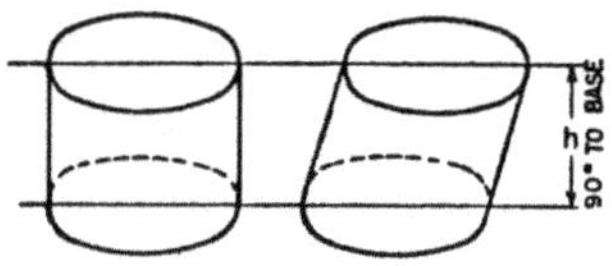

Cylinder: Volume = area of base × h

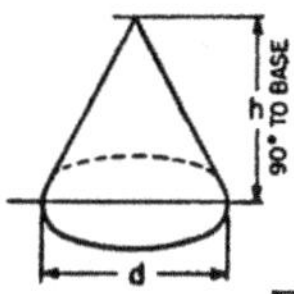

Cone: Volume $= \dfrac{\text{area of base} \times h}{3}$

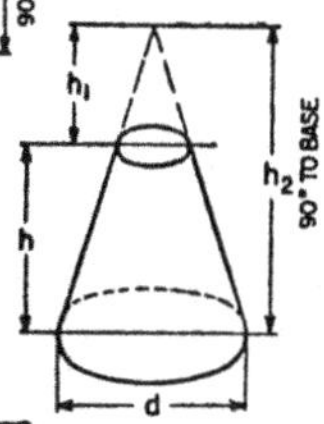

Frustum of cone: Volume = volume of cone of height h_2 − volume of cone of height h_1

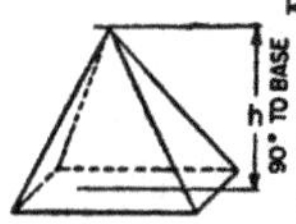

Pyramid: Volume $= \dfrac{\text{area of base} \times h}{3}$

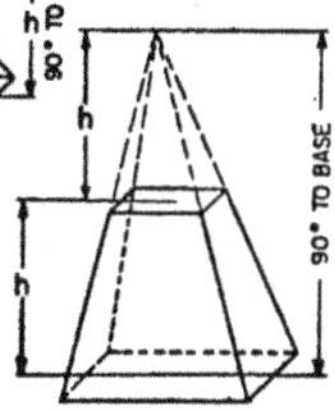

Frustum of pyramid: Volume = volume of pyramid of height h_2 − volume of con of height h_1

Sphere: Volume = $0.524d^3$

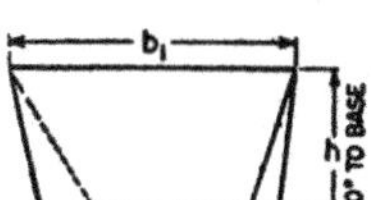

Wedge: (Rectangular base, b_1 parallel to b)

$$\text{Volume} = \frac{ha(2b + b_1)}{6}$$

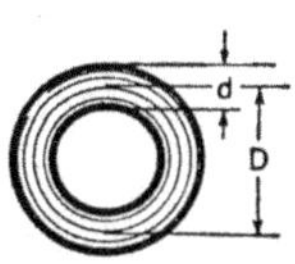

Ring: Volume = $2.46Dd^2$

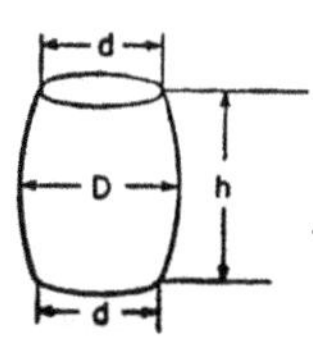

Barrel: Volume = $0.262h(2D^2 + d^2)$

Table 7.2 Volumes of Solid Figures

Diameter	Circum-ference	Area	Diameter	Circum-ference	Area	Diameter	Circum-ference	Area	Diameter	Circum-ference	Area
			7/8	2.749	0.6013	4	12.57	12.57	9	28.27	63.62
1/64	0.04909	.00019	57/64	2.798	0.6230	1/16	12.76	12.96	1/8	28.67	65.40
1/32	0.09817	.00077	29/32	2.847	0.6450	1/8	12.96	13.36	1/4	29.06	67.20
3/64	0.1473	.00173	59/64	2.896	0.6675	3/16	13.16	13.77	3/8	29.45	69.03
1/16	0.1963	.00307	15/16	2.945	0.6903	1/4	13.35	14.19	1/2	29.85	70.88
5/64	0.2454	.00479	61/64	2.994	0.7135	5/16	13.55	14.61	5/8	30.24	72.76
3/32	0.2945	.00690	31/32	3.043	0.7371	3/8	13.74	15.03	3/4	30.63	74.66
7/64	0.3436	.00940	63/64	3.093	0.7610	7/16	13.94	15.47	7/8	31.02	76.59
1/8	0.3927	.01227	1	3.142	0.7854	1/2	14.14	15.90	10	31.42	78.54
9/64	0.4418	.01553	1/16	3.338	0.8866	9/16	14.33	16.35	1/8	31.81	80.52
5/32	0.4909	.01917	1/8	3.534	0.9940	5/8	14.53	16.80	1/4	32.20	82.52
11/64	0.5400	.02320	3/16	3.731	1.108	11/16	14.73	17.26	3/8	32.59	84.54
3/16	0.5890	.02761	1/4	3.927	1.227	3/4	14.92	17.72	1/2	32.99	86.59
13/64	0.6381	.03241	5/16	4.123	1.353	13/16	15.12	18.19	5/8	33.38	88.66
7/32	0.6872	.03758	3/8	4.320	1.485	7/8	15.32	18.67	3/4	33.77	90.76
15/64	0.7363	.04314	7/16	4.516	1.623	15/16	15.51	19.15	7/8	34.16	92.89
1/4	0.7854	.04909	1/2	4.712	1.767	5	15.71	19.63	11	34.56	95.03
17/64	0.8345	.05542	9/16	4.909	1.917	1/16	15.90	20.13	1/8	34.95	97.21
9/32	0.8836	.06213	5/8	5.105	2.074	1/8	16.10	20.63	1/4	35.34	99.40
19/64	0.9327	.06922	11/16	5.301	2.237	3/16	16.30	21.14	3/8	35.74	101.6
5/16	0.9817	.07670	3/4	5.498	2.405	1/4	16.49	21.65	1/2	36.13	103.9
21/64	1.031	.08456	13/16	5.694	2.580	5/16	16.69	22.17	5/8	36.52	106.1
11/32	1.080	.09281	7/8	5.890	2.761	3/8	16.89	22.69	3/4	36.91	108.4
23/64	1.129	.1014	15/16	6.087	2.948	7/16	17.08	23.22	7/8	37.31	110.8
3/8	1.178	.1104	2	6.283	3.142	1/2	17.28	23.76	12	37.70	113.1
25/64	1.227	.1198	1/16	6.480	3.341	9/16	17.48	24.30	1/8	38.09	115.5
13/32	1.276	.1296	1/8	6.676	3.547	5/8	17.67	24.85	1/4	38.48	117.9
27/64	1.325	.1398	3/16	6.872	3.758	11/16	17.87	25.41	3/8	38.88	120.3
7/16	1.374	.1503	1/4	7.069	3.976	3/4	18.06	25.97	1/2	39.27	122.7
29/64	1.424	.1613	5/16	7.265	4.200	13/16	18.26	26.53	5/8	39.66	125.2
15/32	1.473	.1726	3/8	7.461	4.430	7/8	18.46	27.11	3/4	40.06	127.7
31/64	1.522	.1843	7/16	7.658	4.666	15/16	18.65	27.69	7/8	40.45	130.2
1/2	1.571	.1963	1/2	7.854	4.909	6	18.85	28.27	13	40.84	132.7
33/64	1.620	.2088	9/16	8.050	5.157	1/8	19.24	29.46	1/8	41.23	135.3
17/32	1.669	.2217	5/8	8.247	5.412	1/4	19.63	30.68	1/4	41.63	137.9
35/64	1.718	.2349	11/16	8.443	5.673	3/8	20.03	31.92	3/8	42.02	140.5
9/16	1.767	.2485	3/4	8.639	5.940	1/2	20.42	33.18	1/2	42.41	143.1
37/64	1.816	.2625	13/16	8.836	6.213	5/8	20.81	34.47	5/8	42.80	145.8
19/32	1.865	.2769	7/8	9.032	6.492	3/4	21.21	35.78	3/4	43.20	148.5
39/64	1.914	.2916	15/16	9.228	6.777	7/8	21.60	37.12	7/8	43.59	151.2
5/8	1.963	.3068	3	9.425	7.069	7	21.99	38.48	14	43.98	153.9
41/64	2.013	.3223	1/16	9.621	7.366	1/8	22.38	39.87	1/8	44.37	156.7
21/32	2.062	.3382	1/8	9.817	7.670	1/4	22.78	41.28	1/4	44.77	159.5
43/64	2.111	.3545	3/16	10.01	7.980	3/8	23.17	42.72	3/8	45.16	162.3
11/16	2.160	.3712	1/4	10.21	8.296	1/2	23.56	44.18	1/2	45.55	165.1
45/64	2.209	.3883	5/16	10.41	8.618	5/8	23.95	45.66	5/8	45.95	168.0
23/32	2.258	.4057	3/8	10.60	8.946	3/4	24.35	47.17	3/4	46.34	170.9
47/64	2.307	.4236	7/16	10.80	9.281	7/8	24.74	48.71	7/8	46.73	173.8
3/4	2.356	.4418	1/2	11.00	9.621	8	25.13	50.27	15	47.12	176.7
49/64	2.405	.4604	9/16	11.19	9.968	1/8	25.53	51.85	1/8	47.52	179.7
25/32	2.454	.4794	5/8	11.39	10.32	1/4	25.92	53.46	1/4	47.91	182.7
51/64	2.503	.4987	11/16	11.58	10.68	3/8	26.31	55.09	3/8	48.30	185.7
13/16	2.553	.5185	3/4	11.78	11.04	1/2	26.70	56.75	1/2	48.69	188.7
53/64	2.602	.5386	13/16	11.98	11.42	5/8	27.10	58.43	5/8	49.09	191.7
27/32	2.651	.5591	7/8	12.17	11.79	3/4	27.49	60.13	3/4	49.48	194.8
55/64	2.700	.5800	15/16	12.37	12.18	7/8	27.88	61.86	7/8	49.87	197.9

* From Marks, *Mechanical Engineers' Handbook*, 4th ed., McGraw-Hill, 1941.

Table 7.3 Circumferences and Areas of Circles*

Diameter	Circum-ference	Area	Diameter	Circum-ference	Area	Diameter	Circum-ference	Area	Diameter	Circum-ference	Area
16	50.27	201.1	19 ½	61.26	298.6	23	72.26	415.5	29	91.11	660.5
⅛	50.66	204.2	⅝	61.65	302.5	⅛	72.65	420.0	¼	91.89	672.0
¼	51.05	207.4	¾	62.05	306.4	¼	73.04	424.6	½	92.68	683.5
⅜	51.44	210.6	⅞	62.44	310.2	⅜	73.43	429.1	¾	93.46	695.1
½	51.84	213.8	20	62.83	314.2	½	73.83	433.7	30	94.25	706.9
⅝	52.23	217.1	⅛	63.22	318.1	⅝	74.22	438.4	¼	95.03	718.7
¾	52.62	220.4	¼	63.62	322.1	¾	74.61	443.0	½	95.82	730.6
⅞	53.01	223.7	⅜	64.01	326.1	⅞	75.01	447.7	¾	96.60	742.6
17	53.41	227.0	½	64.40	330.1	24	75.40	452.4	31	97.39	754.8
⅛	53.80	230.3	⅝	64.80	334.1	¼	76.18	461.9	¼	98.17	767.0
¼	54.19	233.7	¾	65.19	338.2	½	76.97	471.4	½	98.96	779.3
⅜	54.59	237.1	⅞	65.58	342.2	¾	77.75	481.1	¾	99.75	791.7
½	54.98	240.5	21	65.97	346.4	25	78.54	490.9	32	100.5	804.2
⅝	55.37	244.0	⅛	66.37	350.5	¼	79.33	500.7	¼	101.3	816.9
¾	55.76	247.4	¼	66.76	354.7	½	80.11	510.7	½	102.1	829.6
⅞	56.16	250.9	⅜	67.15	358.8	¾	80.90	520.8	¾	102.9	842.4
18	56.55	254.5	½	67.54	363.1	26	81.68	530.9	33	103.7	855.3
⅛	56.94	258.0	⅝	67.94	367.3	¼	82.47	541.2	¼	104.5	868.3
¼	57.33	261.6	¾	68.33	371.5	½	83.25	551.5	½	105.2	881.4
⅜	57.73	265.2	⅞	68.72	375.8	¾	84.04	562.0	¾	106.0	894.6
½	58.12	268.8	22	69.12	380.1	27	84.82	572.6	34	106.8	907.9
⅝	58.51	272.4	⅛	69.51	384.5	¼	85.61	583.2	¼	107.6	921.3
¾	58.90	276.1	¼	69.90	388.8	½	86.39	594.0	½	108.4	934.8
⅞	59.30	279.8	⅜	70.29	393.2	¾	87.18	604.8	¾	109.2	948.4
19	59.69	283.5	½	70.69	397.6	28	87.96	615.8	35	110.0	962.1
⅛	60.08	287.3	⅝	71.08	402.0	¼	88.75	626.8	¼	110.7	975.9
¼	60.48	291.0	¾	71.47	406.5	½	89.54	637.9	½	111.5	989.8
⅜	60.87	294.8	⅞	71.86	411.0	¾	90.32	649.2	¾	112.3	1003.8

TABLE 7.3 Circumferences and Areas of Circles* (*Continued*)

Feet	Inches											
	0	1	2	3	4	5	6	7	8	9	10	11
0	.0000	.0055	.0218	.0491	.0873	.1364	.1963	.2673	.3491	.4418	.5454	.6600
1	.7854	.9218	1.069	1.227	1.396	1.576	1.767	1.969	2.182	2.405	2.640	2.885
2	3.142	3.409	3.687	3.976	4.276	4.587	4.909	5.241	5.585	5.940	6.305	6.681
3	7.069	7.467	7.876	8.296	8.727	9.168	9.621	10.08	10.56	11.04	11.54	12.05
4	12.57	13.10	13.64	14.19	14.75	15.32	15.90	16.50	17.10	17.72	18.35	18.99
5	19.63	20.29	20.97	21.65	22.34	23.04	23.76	24.48	25.22	25.97	26.73	27.49
6	28.27	29.07	29.87	30.68	31.50	32.34	33.18	34.04	34.91	35.78	36.67	37.57
7	38.48	39.41	40.34	41.28	42.24	43.20	44.18	45.17	46.16	47.17	48.19	49.22
8	50.27	51.32	52.38	53.46	54.54	55.64	56.75	57.86	58.99	60.13	61.28	62.44
9	63.62	64.80	66.00	67.20	68.42	69.64	70.88	72.13	73.39	74.66	75.94	77.24
10	78.54	79.85	81.18	82.52	83.86	85.22	86.59	87.97	89.36	90.76	92.18	93.60
11	95.03	96.48	97.93	99.40	100.9	102.4	103.9	105.4	106.9	108.4	110.0	111.5
12	113.1	114.7	116.3	117.9	119.5	121.1	122.7	124.4	126.0	127.7	129.4	131.0
13	132.7	134.4	136.2	137.9	139.6	141.4	143.1	144.9	146.7	148.5	150.3	152.1
14	153.9	155.8	157.6	159.5	161.4	163.2	165.1	167.0	168.9	170.9	172.8	174.8

TABLE 7.4(a) Circle Diameters (ft and in.) and areas (sq ft)

Inches	1	2	3	4	5	6	7	8	9	10	11
Feet	0.0833	0.1667	0.2500	0.3333	0.4167	0.5000	0.5833	0 6667	0.7500	0.8333	0.9167
Inches	⅛	¼	⅜	½	⅝	¾	⅞				
Feet	0.0104	0.0208	0.0313	0.0417	0.0521	0.0625	0.0729				

EXAMPLE: 5 ft 7⅜ in. = 5.0 + 0.5833 + 0.0313 = 5.6146 ft

* From Marks, *Mechanical Engineers' Handbook*, 4th ed., McGraw-Hill, 1941.

TABLE 7.4(b) Circle Diameters (inches and fractions of an in.) and areas (decimals of a ft)

Trigonometric Functions

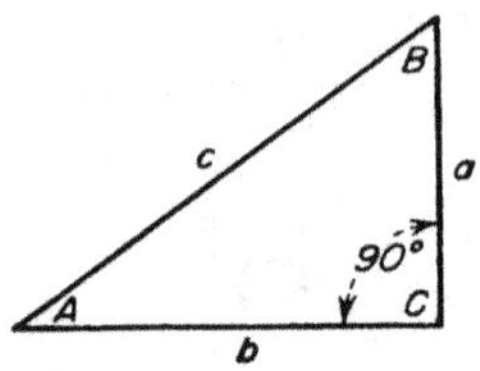

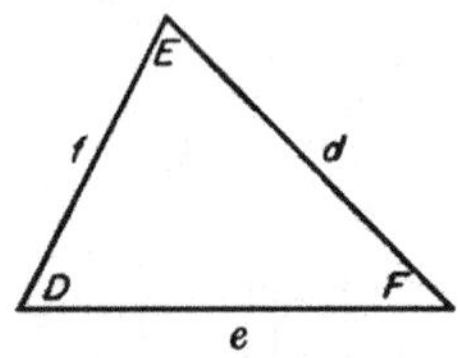

Given	Required	Formulas		
a, c	A, B, b	$\sin A = \frac{a}{c};$	$\cos B = \frac{a}{c};$	$b = \sqrt{c^2 - a^2}$
a, c	Area	$\frac{a}{2}\sqrt{c^2 - a^2}$		
a, b	A, B, c	$\tan A = \frac{a}{b};$	$\tan B = \frac{b}{a};$	$c = \sqrt{a^2 + b^2}$
a, b	Area	$\frac{ab}{2}$		
A, a	B, b, c	$B = 90° - A;$	$b = a \cot A;$	$c = \frac{a}{\sin A}$
A, a	Area	$\frac{a^2 \cot A}{2}$		
A, b	B, a, c	$B = 90° - A;$	$a = b \tan A;$	$c = \frac{b}{\cos A}$
A, b	Area	$\frac{b^2 \tan A}{2}$		
A, c	B, a, b	$B = 90° - A;$	$a = c \sin A;$	$b = c \cos A$
A, c	Area	$\frac{c^2 \sin 2A}{4}$		
d, e, f	D	$\sin \frac{D}{2} = \sqrt{\frac{(s-e)(s-f)}{ef}};$	$s = \frac{d+e+f}{2}$	
d, e, f	E	$\sin \frac{E}{2} = \sqrt{\frac{(s-d)(s-f)}{df}}$		
d, e, f	F	$\sin \frac{F}{2} = \sqrt{\frac{(s-d)(s-e)}{de}}$		

TABLE 7.5 Trigonometric Formulas

Given	Required	Formulas
d, e, f	Area	$\sqrt{s(s-d)(s-e)(s-f)}$
d, D, E	e, f	$e = \frac{d \text{ s n } E}{\sin D}$; $f = \frac{d \sin F}{\sin D}$
d, D, E	Area	$\frac{de \sin F}{2}$ (e from above formula)
d, e, D	E	$\sin E = \frac{e \sin D}{d}$
d, e, D	f	$f = \frac{e \sin F}{\sin E}$
d, e, F	D	$\tan D = \frac{d \sin F}{e - d \cos F}$
d, e, F	f	$f = \frac{d \sin F}{\sin D}$ (D from above formula)
d, e, F	Area	$\frac{de \sin F}{2}$

Table 7.5 Trigonometric Formulas (*Continued*)

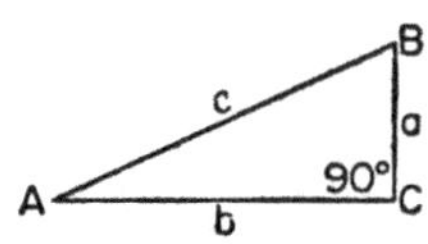

Known	Required				
	A	B	a	b	c
a, b	$\tan A = \frac{a}{b}$	$\cot B = \frac{a}{b}$			$\sqrt{a^2 + b^2}$
a, c	$\sin A = \frac{a}{c}$	$\cos B = \frac{a}{c}$		$\sqrt{c^2 - a^2}$	
b, c	$\cos A = \frac{b}{c}$	$\sin B = \frac{b}{c}$	$\sqrt{c^2 - b^2}$		
A, a		$90° - A$		$a \cot A$	$\frac{a}{\sin A}$
A, b		$90° - A$	$b \tan A$		$\frac{b}{\cos A}$
A, c		$90° - A$	$c \sin A$	$c \cos A$	

Table 7.6 Trigonometric Solutions

Natural Sines at intervals of 0°.1, or 6′.

Degrees	°.0 =(0′)	°.1 (6′)	°.2 (12′)	°.3 (18′)	°.4 (24′)	°.5 (30′)	°.6 (36′)	°.7 (42′)	°.8 (48′)	°.9 (54′)			Average differ-ence
											0.0000	90°	
0°	0.0000	0017	0035	0052	0070	0087	0105	0122	0140	0157	0175	89	17
1	0175	0192	0209	0227	0244	0262	0279	0297	0314	0332	0349	88	17
2	0349	0366	0684	0401	0419	0436	0454	0471	0488	0506	0523	87	17
3	0523	0541	0558	0576	0593	0610	0628	0645	0663	0680	0698	86	17
4	0698	0715	0732	0750	0767	0785	0802	0819	0837	0854	0.0872	85	17
5	0.0872	0889	0906	0924	0941	0958	0976	0993	1011	1028	1045	84	17
6	1045	1063	1080	1097	1115	1132	1149	1167	1184	1201	1219	83	17
7	1219	1236	1253	1271	1288	1305	1323	1340	1357	1374	1392	82	17
8	1392	1409	1426	1444	1461	1478	1495	1513	1530	1547	1564	81	17
9	1564	1582	1599	1616	1633	1650	1668	1685	1702	1719	0.1736	80°	17
10°	0.1736	1754	1771	1788	1805	1822	1840	1857	1874	1891	1908	79	17
11	1908	1925	1942	1959	1977	1994	2011	2028	2045	2062	2079	78	17
12	2079	2096	2113	2130	2147	2164	2181	2198	2215	2233	2250	77	17
13	2250	2267	2284	2300	2317	2334	2351	2368	2385	2402	2419	76	17
14	2419	2436	2453	2470	2487	2504	2521	2538	2554	2571	0.2588	75	17
15	0.2588	2605	2622	2639	2656	2672	2689	2706	2723	2740	2756	74	17
16	2756	2773	2790	2807	2823	2840	2857	2874	2890	2907	2924	73	17
17	2924	2940	2957	2974	2990	3007	3024	3040	3057	3074	3090	72	17
18	3090	3107	3123	3140	3156	3173	3190	3206	3223	3239	3256	71	17
19	3256	3272	3289	3305	3322	3338	3355	3371	3387	3404	0.3420	70°	16
20°	0.3420	3437	3453	3469	3486	3502	3518	3535	3551	3567	3584	69	16
21	3584	3600	3616	3633	3649	3665	3681	3697	3714	3730	3746	68	16
22	3746	3762	3778	3795	3811	3827	3843	3859	3875	3891	3907	67	16
23	3907	3923	3939	3955	3971	3987	4003	4019	4035	4051	4067	66	16
24	4067	4083	4099	4115	4131	4147	4163	4179	4195	4210	0.4226	65	16
25	0.4226	4242	4258	4274	4289	4305	4321	4337	4352	4368	4384	64	16
26	4384	4399	4415	4431	4446	4462	4478	4493	4509	4524	4540	63	16
27	4540	4555	4571	4586	4602	4617	4633	4648	4664	4679	4695	62	16
28	4695	4710	4726	4741	4756	4772	4787	4802	4818	4833	4848	61	15
29	4848	4863	4879	4894	4909	4924	4939	4955	4970	4985	0.5000	60°	15
30°	0.5000	5015	5030	5045	5060	5075	5090	5105	5120	5135	5150	59	15
31	5150	5165	5180	5195	5210	5225	5240	5255	5270	5284	5299	58	15
32	5299	5314	5329	5344	5358	5373	5388	5402	5417	5432	5446	57	15
33	5446	5461	5476	5490	5505	5519	5534	5548	5563	5577	5592	56	15
34	5592	5606	5621	5635	5650	5664	5678	5693	5707	5721	0.5736	55	14
35	0.5736	5750	5764	5779	5793	5807	5821	5835	5850	5864	5878	54	14
36	5878	5892	5906	5920	5934	5948	5962	5976	5990	6004	6018	53	14
37	6018	6032	6046	6060	6074	6088	6101	6115	6129	6143	6157	52	14
38	6157	6170	6184	6198	6211	6225	6239	6252	6266	6280	6293	51	14
39	6293	6307	6320	6334	6347	6361	6374	6388	6401	6414	0.6428	50°	13
40°	0.6428	6441	6455	6468	6481	6494	6508	6521	6534	6547	6561	49	13
41	6561	6574	6587	6600	6613	6626	6639	6652	6665	6678	6691	48	13
42	6691	6704	6717	6730	6743	6756	6769	6782	6794	6807	6820	47	13
43	6820	6833	6845	6858	6871	6884	6896	6909	6921	6934	6947	46	13
44	6947	6959	6972	6984	6997	7009	7022	7034	7046	7059	0.7071	45°	12
45°	0.7071												
		°.9 = (54′)	°.8 (48′)	°.7 (42′)	°.6 (36′)	°.5 (30′)	°.4 (24′)	°.3 (18′)	°.2 (12′)	°.1 (6′)	°.0 (0′)	Degrees	

Natural Cosines

* From Marks, *Mechanical Engineers' Handbook*, 4th ed., McGraw-Hill, 1941.

TABLE 7.7 Natural Sines and Cosines*

Natural Sines at intervals of 0°.1, or 6′.

Degrees	°.0 =(0′)	°.1 (6′)	°.2 (12′)	°.3 (18′)	°.4 (24′)	°.5 (30′)	°.6 (36′)	°.7 (42′)	°.8 (48′)	°.9 (54′)			Average difference
											0.7071	45°	
45°	0.7071	7083	7096	7108	7120	7133	7145	7157	7169	7181	7193	44	12
46	7193	7206	7218	7230	7242	7254	7266	7278	7290	7302	7314	43	12
47	7314	7325	7337	7349	7361	7373	7385	7396	7408	7420	7431	42	12
48	7431	7443	7455	7466	7478	7490	7501	7513	7524	7536	7547	41	12
49	7547	7559	7570	7581	7593	7604	7615	7627	7638	7649	0.7660	40°	11
50°	0.7660	7672	7683	7694	7705	7716	7727	7738	7749	7760	7771	39	11
51	7771	7782	7793	7804	7815	7826	7837	7848	7859	7869	7880	38	11
52	7880	7891	7902	7912	7923	7934	7944	7955	7965	7976	7986	37	11
53	7986	7997	8007	8018	8028	8039	8049	8059	8070	8080	8090	36	10
54	8090	8100	8111	8121	8131	8141	8151	8161	8171	8181	0.8192	35	10
55	0.8192	8202	8211	8221	8231	8241	8251	8261	8271	8281	8290	34	10
56	8290	8300	8310	8320	8329	8339	8348	8358	8368	8377	8387	33	10
57	8387	8396	8406	8415	8425	8434	8443	8453	8462	8471	8480	32	9
58	8480	8490	8499	8508	8517	8526	8536	8545	8554	8563	8572	31	9
59	8572	8581	8590	8599	8607	8616	8625	8634	8643	8652	0.8660	30°	9
60°	0.8660	8669	8678	8686	8695	8704	8712	8721	8729	8738	8746	29	9
61	8746	8755	8763	8771	8780	8788	8796	8805	8813	8821	8829	28	8
62	8829	8838	8846	8854	8862	8870	8878	8886	8894	8902	8910	27	8
63	8910	8918	8926	8934	8942	8949	8957	8965	8973	8980	8988	26	8
64	8988	8996	9003	9011	9018	9026	9033	9041	9048	9056	0.9063	25	7
65	0.9063	9070	9078	9085	9092	9100	9107	9114	9121	9128	9135	24	7
66	9135	9143	9150	9157	9164	9171	9178	9184	9191	9198	9205	23	7
67	9205	9212	9219	9225	9232	9239	9245	9252	9259	9265	9272	22	7
68	9272	9278	9285	9291	9298	9304	9311	9317	9323	9330	9336	21	6
69	9336	9342	9348	9354	9361	9367	9373	9379	9385	9391	0.9397	20°	6
70°	0.9397	9403	9409	9415	9421	9426	9432	9438	9444	9449	9455	19	6
71	9455	9461	9466	9472	9478	9483	9489	9494	9500	9505	9511	18	6
72	9511	9516	9521	9527	9532	9537	9542	9548	9553	9558	9563	17	5
73	8563	9568	9573	9578	9583	9588	9593	9598	9603	9608	9613	16	5
74	9613	9617	9622	9627	9632	9636	9641	9646	9650	9655	0.9659	15	5
75	0.9659	9664	9668	9673	9677	9681	9686	9690	9694	9699	9703	14	4
76	9703	9707	9711	9715	9720	9724	9728	9732	9736	9740	9744	13	4
77	9744	9748	9751	9755	9759	9763	9767	9770	9774	9778	9781	12	4
78	9781	9785	9789	9792	9796	9799	9803	9806	9810	9813	9816	11	3
79	9816	9820	9823	9826	9829	9833	9836	9839	9842	9845	0.9848	10°	3
80°	0.9848	9851	9854	9857	9860	9863	9866	9869	9871	9874	9877	9	3
81	9877	9880	9882	9885	9888	9890	9893	9895	9898	9900	9903	8	3
82	9903	9905	9907	9910	9912	9914	9917	9919	9921	9923	9925	7	2
83	9925	9928	9930	9932	9934	9936	9938	9940	9942	9943	9945	6	2
84	9945	9947	9949	9951	9952	9954	9956	9957	9959	9960	0.9962	5	2
85	0.9962	9963	9965	9966	9968	9969	9971	9972	9973	9974	9976	4	1
86	9976	9977	9978	9979	9980	9981	9982	9983	9984	9985	9986	3	1
87	9986	9987	9988	9989	9990	9990	9991	9992	9993	9993	9994	2	1
88	9994	9995	9995	9996	9996	9997	9997	9997	9998	9998	0.9998	1	0
89	0.9998	9999	9999	9999	9999	0000	0000	0000	0000	0000	1.0000	0°	0
90°	1.0000												
		°.9 = (54′)	°.8 (48′)	°.7 (42′)	°.6 (36′)	°.5 (30′)	°.4 (24′)	°.3 (18′)	°.2 (12′)	°.1 (6′)	°.0 (0′)	Degrees	

Natural Cosines

Table 7.7 Natural Sines and Cosines*(*Continued*)

Natural Tangents at intervals of 0°.1, or 6′.

Degrees	°.0 =(0′)	°.1 (6′)	°.2 (12′)	°.3 (18′)	°.4 (24′)	°.5 (30′)	°.6 (36′)	°.7 (42′)	°.8 (48′)	°.9 (54′)			Average difference
											0.0000	90°	
0°	0.0000	0017	0035	0052	0070	0087	0105	0122	0140	0157	0175	89	17
1	0175	0192	0209	0227	0244	0262	0279	0297	0314	0332	0349	88	17
2	0349	0367	0384	0402	0419	0437	0454	0472	0489	0507	0524	87	17
3	0524	0542	0559	0577	0594	0612	0629	0647	0664	0682	0699	86	18
4	0699	0717	0734	0752	0769	0787	0805	0822	0840	0857	0.0875	85	18
5	0.0875	0892	0910	0928	0945	0963	0981	0998	1016	1033	1051	84	18
6	1051	1069	1086	1104	1122	1139	1157	1175	1192	1210	1228	83	18
7	1228	1246	1263	1281	1299	1317	1334	1352	1370	1388	1405	82	18
8	1405	1423	1441	1459	1477	1495	1512	1530	1548	1566	1584	81	18
9	1584	1602	1620	1638	1655	1673	1691	1709	1727	1745	0.1763	80°	18
10°	0.1763	1781	1799	1817	1835	1853	1871	1890	1908	1926	1944	79	18
11	1944	1962	1980	1998	2016	2035	2053	2071	2089	2107	2126	78	18
12	2126	2144	2162	2180	2199	2217	2235	2254	2272	2290	2309	77	18
13	2309	2327	2345	2364	2382	2401	2419	2438	2456	2475	2493	76	18
14	2493	2512	2530	2549	2568	2586	2605	2623	2642	2661	0.2679	75	19
15	0.2679	2698	2717	2736	2754	2773	2792	2811	2830	2849	2867	74	19
16	2867	2886	2905	2924	2943	2962	2981	3000	3019	3038	3057	73	19
17	3057	3076	3096	3115	3134	3153	3172	3191	3211	3230	3249	72	19
18	3249	3269	3288	3307	3327	3346	3365	3385	3404	3424	3443	71	19
19	3443	3463	3482	3502	3522	3541	3561	3581	3600	3620	0.3640	70°	20
20°	0.3640	3659	3679	3699	3719	3739	3759	3779	3799	3819	3839	69	20
21	3839	3859	3879	3899	3919	3939	3959	3979	4000	4020	4040	68	20
22	4040	4061	4081	4101	4122	4142	4163	4183	4204	4224	4245	67	21
23	4245	4265	4286	4307	4327	4348	4369	4390	4411	4431	4452	66	21
24	4452	4473	4494	4515	4536	4557	4578	4599	4621	4642	0.4663	65	21
25	0.4663	4684	4706	4727	4748	4770	4791	4813	4834	4856	4877	64	21
26	4877	4899	4921	4942	4964	4986	5008	5029	5051	5073	5095	63	22
27	5095	5117	5139	5161	5184	5206	5228	5250	5272	5295	5317	62	22
28	5317	5340	5362	5384	5407	5430	5452	5475	5498	5520	5543	61	23
29	5543	5566	5589	5612	5635	5658	5681	5704	5727	5750	0.5774	60°	23
30°	0.5774	5797	5820	5844	5867	5890	5914	5938	5961	5985	6009	59	24
31	6009	6032	6056	6080	6104	6128	6152	6176	6200	6224	6249	58	24
32	6249	6273	6297	6322	6346	6371	6395	6420	6445	6469	6494	57	25
33	6494	6519	6544	6569	6594	6619	6644	6669	6694	6720	6745	56	25
34	6745	6771	6796	6822	6847	6873	6899	6924	6950	6976	0.7002	55	26
35	0.7002	7028	7054	7080	7107	7133	7159	7186	7212	7239	7265	54	26
36	7265	7292	7319	7346	7373	7400	7427	7454	7481	7508	7536	53	27
37	7536	7563	7590	7618	7646	7673	7701	7729	7757	7785	7813	52	28
38	7813	7841	7869	7898	7926	7954	7983	8012	8040	8069	8098	51	28
39	8098	8127	8156	8185	8214	8243	8273	8302	8332	8361	0.8391	50°	29
40°	0.8391	8421	8451	8481	8511	8541	8571	8601	8632	8662	8693	49	30
41	8693	8724	8754	8785	8816	8847	8878	8910	8941	8972	9004	48	31
42	9004	9036	9067	9099	9131	9163	9195	9228	9260	9293	9325	47	32
43	9325	9358	9391	9424	9457	9490	9523	9556	9590	9623	0.9657	46	33
44	0.9657	9691	9725	9759	9793	9827	9861	9896	9930	9965	1.0000	45°	34
45°	1.0000												
		°.9 = (54′)	°.8 (48′)	°.7 (42′)	°.6 (36′)	°.5 (30′)	°.4 (24′)	°.3 (18′)	°.2 (12′)	°.1 (6′)	°.0 (0′)	Degrees	

Natural Cotangents

* From Marks, *Mechanical Engineers' Handbook*, 4th ed., McGraw-Hill, 1941.

Table 7.8 Natural Tangents and Cotangents*

Natural Tangents at intervals of 0°.1, or 6′.

Degrees	°.0 =(0′)	°.1 (6′)	°.2 (12′)	°.3 (18′)	°.4 (24′)	°.5 (30′)	°.6 (36′)	°.7 (42′)	°.8 (48′)	°.9 (54′)			Average difference
											1.0000	45°	
45°	1.0000	0035	0070	0105	0141	0176	0212	0247	0283	0319	0355	44	35
46	0355	0392	0428	0464	0501	0538	0575	0612	0649	0686	0724	43	37
47	0724	0761	0799	0837	0875	0913	0951	0990	1028	1067	1106	42	38
48	1106	1145	1184	1224	1263	1303	1343	1383	1423	1463	1504	41	40
49	1504	1544	1585	1626	1667	1708	1750	1792	1833	1875	1.1918	40°	41
50°	1.1918	1960	2002	2045	2088	2131	2174	2218	2261	2305	2349	39	43
51	2349	2393	2437	2482	2527	2572	2617	2662	2708	2753	2799	38	45
52	2799	2846	2892	2938	2985	3032	3079	3127	3175	3222	3270	37	47
53	3270	3319	3367	3416	3465	3514	3564	3613	3663	3713	3764	36	49
54	3764	3814	3865	3916	3968	4019	4071	4124	4176	4229	1.4281	35	52
55	1.4281	4335	4388	4442	4496	4550	4605	4659	4715	4770	4826	34	55
56	4826	4882	4938	4994	5051	5108	5166	5224	5282	5340	5399	33	57
57	5399	5458	5517	5577	5637	5697	5757	5818	5880	5941	6003	32	60
58	6003	6066	6128	6191	6255	6319	6383	6447	6512	6577	6643	31	64
59	1.6643	6709	6775	6842	6909	6977	7045	7113	7182	7251	1.7321	30°	67
60°	1.732	1.739	1.746	1.753	1.760	1.767	1.775	1.782	1.789	1.797	1.804	29	7
61	1.804	1.811	1.819	1.827	1.834	1.842	1.849	1.857	1.865	1.873	1.881	28	8
62	1.881	1.889	1.897	1.905	1.913	1.921	1.929	1.937	1.946	1.954	1.963	27	8
63	1.963	1.971	1.980	1.988	1.997	2.006	2.014	2.023	2.032	2.041	2.050	26	9
64	2.050	2.059	2.069	2.078	2.087	2.097	2.106	2.116	2.125	2.135	2.145	25	9
65	2.145	2.154	2.164	2.174	2.184	2.194	2.204	2.215	2.225	2.236	2.246	24	10
66	2.246	2.257	2.267	2.278	2.289	2.300	2.311	2.322	2.333	2.344	2.356	23	11
67	2.356	2.367	2.379	2.391	2.402	2.414	2.426	2.438	2.450	2.463	2.475	22	12
68	2.475	2.488	2.500	2.513	2.526	2.539	2.552	2.565	2.578	2.592	2.605	21	13
69	2.605	2.619	2.633	2.646	2.660	2.675	2.689	2.703	2.718	2.733	2.747	20°	14
70°	2.747	2.762	2.778	2.793	2.888	2.824	2.840	2.856	2.872	2.888	2.904	19	16
71	2.904	2.921	2.937	2.954	2.971	2.989	3.006	3.024	3.042	3.860	2.078	18	17
72	3.078	3.096	3.115	3.133	3.152	3.172	3.191	3.211	3.230	3.251	3.271	17	19
73	3.271	3.291	3.312	3.333	3.354	3.376	3.398	3.420	3.442	3.465	3.487	16	22
74	3.487	3.511	3.534	3.558	3.582	3.606	3.630	3.655	3.681	3.706	3.732	15	24
75	3.732	3.758	3.785	3.812	3.839	3.867	3.895	3.923	3.952	3.981	4.011	14	28
76	4.011	4.041	4.071	4.102	4.134	4.165	4.198	4.230	4.264	4.297	4.331	13	32
77	4.331	4.366	4.402	4.437	4.474	4.511	4.548	4.586	4.625	4.665	4.705	12	37
78	4.705	4.745	4.787	4.829	4.872	4.915	4.959	5.005	5.050	5.097	5.145	11	44
79	5.145	5.193	5.242	5.292	5.343	5.396	5.449	5.503	5.558	5.614	5.671	10°	53
80°	5.671	5.730	5.789	5.850	5.912	5.976	6.041	6.107	6.174	6.243	6.314	9	
81	6.314	6.386	6.460	6.535	6.612	6.691	6.772	6.855	6.940	7.026	7.115	8	
82	7.115	7.207	7.300	7.396	7.495	7.596	7.700	7.806	7.916	8.028	8.144	7	
83	8.144	8.264	8.386	8.513	8.643	8.777	8.915	9.058	9.205	9.357	9.514	6	
84	9.514	9.677	9.845	10.02	10.20	10.39	10.58	10.78	10.99	11.20	11.43	5	
85	11.43	11.66	11.91	12.16	12.43	12.71	13.00	13.30	13.62	13.95	14.30	4	
86	14.30	14.67	15.06	15.46	15.90	16.35	16.83	17.34	17.89	18.46	19.08	3	
87	19.08	19.74	20.45	21.20	22.02	22.90	23.86	24.90	26.03	27.27	28.64	2	
88	28.64	30.14	31.82	33.69	35.80	38.19	40.92	44.07	47.74	52.08	57.29	1	
89	57.29	63.66	71.62	81.85	95.49	114.6	143.2	191.0	286.5	573.0	∞	0°	
90°	∞												
		°.9 = (54′)	°.8 (48′)	°.7 (42′)	°.6 (36′)	°.5 (30′)	°.4 (24′)	°.3 (18′)	°.2 (12′)	°.1 (6′)	°.0 (0′)	Degrees	

Natural Cotangents

TABLE 7.8 Natural Tangents and Cotangents*(*Continued*)

Conversion Factors

Linear Measure		
1 millimeter	=	0.03937 inch
1 centimeter	=	0.3937 inch
1 decimeter	=	3.937 inches
1 meter	=	39.37 inches
		3.28083 feet
		1.09361 yards
1 kilometer	=	3280.83 feet
		1093.61 yards
		0.62137 mile
1 inch	=	25.4 millimeters
		2.54 centimeters
		0.254 decimeter
		0.0254 meter
1 foot	=	0.3048 meter
1 yard	=	0.9144 meter
1 mile	=	1.60935 kilometers
Square Measure		
1 square millimeter	=	0.00155 square inch
		1973.5 circular mils
1 square centimeter	=	00.155 square inch
1 square decimeter	=	15.5 square inches
1 square meter	=	1550 square inches
		10.7639 square feet
		1.196 square yards
1 square kilometer	=	0.386109 square mile
		247.11 acres
1 square myriameter	=	38.6109 square miles
1 square inch	=	645.2 square millimeters
		6.452 square centimeters
		0.06452 square decimeter
1 square foot	=	0.0929 square meter
1 square yard	=	0.836 square meter
Cubic Measure		
1 cubic centimeter	=	0.061 cubic inch
1 cubic decimeter	=	61.0234 cubic inches
		0.035314 cubic foot
1 cubic meter	=	35.314 cubic feet
		1.308 cubic yards

Table 7.9 Metric System of Measures and Weights

	Weight	
1 gram	=	15.432 grains
1 kilogram	=	2.204622 pounds
1 metric ton, 1000 kilograms	=	2204.6 pounds 0.9842 gross ton 1.1023 net tons
1 grain	=	0.0648 gram
1 pound	=	0.4536 kilogram
	Capacity	
1 liter	=	61.0234 cubic inches 0.03531 cubic foot 0.2642 gallon
1 cubic foot	=	28.317 liters
1 gallon	=	3.785 liters
	Compound Units	
1 gram per square millimeter	=	1.422 pounds per square inch
1 kilogram per square millimeter	=	1422.32 pounds per square inch
1 kilogram per square centimeter	=	14.2232 pounds per square inch
1 kilogram per square meter	=	0.2048 pound per square foot 1.8433 pounds per square yard
1 kilogram-meter	=	7.2330 foot-pounds
1 kilogram per meter	=	0.6720 pound per foot
1 pound per square inch	=	0.07031 kilogram per square centimeter
1 pound per square foot	=	0.0004882 kilogram per square centimeter 0.006944 pound per square inch
1 pound per cubic inch	=	27679.7 kilograms per cubic meter
1 pound per cubic foot	=	16.0184 kilograms per cubic meter
1 kilogram per cubic meter	=	0.06243 pound per cubic foot
1 foot per second	=	0.30480 meter per second
1 meter per second	=	3.28083 feet per second 2.23693 miles per hour

TABLE 7.9 Metric System of Measures and Weights (*Continued*)

Farenheit	Celsius	Farenheit	Celsius	Farenheit	Celsius	Farenheit	Celsius
−35	−35	32	0	95	35	158	70
−24	−30	41	5	104	40	167	75
−13	−25	50	10	113	45	176	80
−4	−20	59	15	122	50	185	85
5	−15	68	20	131	55	194	90
14	−10	77	25	140	60	203	95
23	−5	86	30	149	65	212	100

Table 7.10 Temperature Equivalents

Beaufort number	Miles per hour	Knots	Wind effects observed on land	Terms used in USWB forecasts
0	Less than 1	Less than 1	Calm; smoke rises vertically	
1	1–3	1–3	Direction of wind shown by smoke drift; but not by wind vanes	Light
2	4–7	4–6	Wind felt on face; leaves rustle; ordinary vane moved by wind	
3	8–12	7–10	Leaves and small twigs in constant motion; wind extends light flag	Gentle
4	13–18	11–16	Raises dust, loose paper; small branches are moved	Moderate
5	19–24	17–21	Small trees in leaf begin to sway; created wavelets form on inland waters	Fresh
6	25–31	22–27	Large branches in motion; whistling heard in telegraph wires; umbrellas used with difficulty	Strong
7	32–38	28–33	Whole trees in motion; inconvenience felt walking against wind	
8	39–46	34–40	Breaks twigs off trees; generally impedes progress	Gale
9	47–54	41–47	Slight structural damage occurs; chimney pots, slates removed	
10	55–63	48–55	Seldom experienced inland; trees uprooted; considerable structural damage occurs	Whole gale
11	64–72	56–63	Very rarely experienced; accompanied by widespread damage	Hurricane
12 or more	73 or more	64 or more	Very rarely experienced; accompanied by widespread damage	

(Compiled by U.S. Weather Bureau 1955)

Table 7.11 Wind Forces (Beaufort Scale)

8ths	16ths	32nds	64ths	Exact decimal values	8ths	16ths	32nds	64ths	Exact decimal values
					4	8	16	32	0.50
			1	0.01 5625				33	.51 5625
		1	2	.03 125			17	34	.53 125
			3	.04 6875				35	.54 6875
	1	2	4	.06 25		9	18	36	.56 25
			5	.07 8125				37	.57 8125
		3	6	.09 375			19	38	.59 375
			7	.10 9375				39	.60 9375
1	2	4	8	.12 5	5	10	20	40	.62 5
			9	.14 0625				41	.64 0625
		5	10	.15 625			21	42	.65 625
			11	.17 1875				43	.67 1875
	3	6	12	.18 75		11	22	44	.68 75
			13	.20 3125				45	.70 3125
		7	14	.21 875			23	46	.71 875
			15	.23 4375				47	.73 4375
2	4	8	16	.25	6	12	24	48	.75
			17	.26 5625				49	.76 5625
		9	18	.28 125			25	50	.78 125
			19	.29 6875				51	.79 6875
	5	10	20	.31 25		13	26	52	.81 25
			21	.32 8125				53	.82 8125
		11	22	.34 375			27	54	.84 375
			23	.35 9375				55	.85 9375
3	6	12	24	.37 5	7	14	28	56	.87 5
			25	.39 0625				57	.89 0625
		13	26	.40 625			29	58	.90 625
			27	.42 1875				59	.92 1875
	7	14	28	.43 75		15	30	60	.93 75
			29	.45 3125				61	.95 3125
		15	30	.46 875			31	62	.96 875
			31	.48 4375				63	.98 4375

*From Marks, Mechanical Engineer's Handbook, 4th ed., McGraw-Hill, 1941.

Table 7.12 Decimal Equivalents of Common Fractions*

Material Weights

Substance	Weight, lb/ft^3
Metals, Alloys, Ores	
Aluminum, cast-hammered	165
Aluminum, bronze	481
Brass, cast-rolled	534
Bronze, 7.9 to 14%Sn	509
Bronze, phosphor	554
Copper, cast-rolled	556
Copper ore, pyrites	262
Gold coin (U.S.)	1073
Iron, gray cast	442
Iron, cast, pig	450
Iron, wrought	485
Iron ore, hematite	325
Iron ore, limonite	237
Iron ore, magnetite	315
Iron slag	172
Lead	710
Lead ore, galena	465
Manganese	475
Manganese ore, pyrolusite	259
Mercury	849
Nickel	537
Platinum, cast-hammered	1330
Silver, cast-hammered	656
Steel, cold-drawn	489
Tin, cast-hammered	459
Tin ore, cassiterite	418
Tungsten	1200
Zinc, cast-rolled	440
Zinc ore, blende	253
Timber, U.S. Seasoned (Moisture 15 to 20%)	
Ash, white-red	40
Birch	32
Cedar, white-red	22
Cypress	30
Fir, Douglas spruce	32
Fir, eastern	25
Elm, white	45
Hemlock	29
Hickory	48
Mahogany	44
Maple, hard	43
Maple, white	33
Oak, chestnut	54
Oak, live	59
Oak, red, black	41
Oak, white	46
Pine, Oregon	32
Pine, red	30
Pine, white	26
Pine, yellow, long-leaf	44
Pine, yellow, short-leaf	38

Table 7.13 Weights of Various Materials

Substance	Weight, lb/ft^3
Timber (Continued)	
Poplar	27
Redwood, California	26
Spruce, white, black	27
Walnut, black	42
Walnut, white	26
Water, 4°C, maximum density	62.428
Water, 100C	59.830
Water, ice	56
Water, snow, fresh fallen	8
Water, seawater	64
Various Building Materials	
Ashes, cinders	40–45
Cement, portland, loose	90
Cement, portland, set	183
Lime, gypsum, loose	53–64
Mortar, set	103
Slags, bank slag	67–72
Slags, bank screenings	98–117
Slags, machine slag	96
Slags, slag sand	49–55
Earth, etc. Excavated	
Clay, dry	63
Clay, damp, plastic	110
Clay and gravel, dry	100
Earth, dry, loose	76
Earth, dry, packed	95
Earth, mud, packed	115
Riprap, limestone	80–85
Riprap, sandstone	90
Riprap, shale	105
Sand, gravel, dry, loose	90–105
Sand, gravel, dry, packed	100–120
Sand, gravel, wet	118–120
Excavations in Water	
Sand or gravel	60
Sand or gravel and clay	65
Clay	80
River mud	90
Soil	70
Stone riprap	65
Stone, Quarried, Piled	
Basalt, granite, gneiss	96
Limestone, marble, quartz	95
Sandstone	82
Shale	92

TABLE 7.13 Weights of Various Materials (*Continued*)

Substance	Weight, lb/ft³
Coal and Coke, Piled	
Coal, anthracite	47–58
Coal, bituminous, lignite	40–54
Coal, peat, turf	20–26
Coal, charcoal	10–14
Coal, coke	23–32

TABLE 7.13 Weights of Various Materials (*Continued*)

Size, in.	Weight, lb per lin ft		Size, in.	Weight, lb per lin ft		Size, in.	Weight, lb per lin ft		Size, in.	Weight, lb per lin ft	
	Square	Round		Square	Round		Square	Round		Square	Round
0	.....	.....	3	30.60	24.03	6	122.4	96.1	9	275.4	216.3
1/16	0.013	0.010	1/16	31.89	25.05	1/16	125.0	98.2	1/16	279.2	219.3
1/8	0.053	0.042	1/8	33.20	26.08	1/8	127.6	100.2	1/8	283.1	222.4
3/16	0.120	0.094	3/16	34.54	27.13	3/16	130.2	102.2	3/16	287.0	225.4
1/4	0.213	0.167	1/4	35.91	28.21	1/4	132.8	104.3	1/4	290.9	228.5
5/16	0.332	0.261	5/16	37.31	29.30	5/16	135.5	106.4	5/16	294.9	231.6
3/8	0.478	0.376	3/8	38.73	30.42	3/8	138.2	108.5	3/8	298.8	234.7
7/16	0.651	0.511	7/16	40.18	31.55	7/16	140.9	110.7	7/16	302.8	237.8
1/2	0.850	0.668	1/2	41.65	32.71	1/2	143.7	112.8	1/2	306.9	241.0
9/16	1.076	0.845	9/16	43.15	33.89	9/16	146.4	115.0	9/16	310.9	244.2
5/8	1.328	1.043	5/8	44.68	35.09	5/8	149.2	117.2	5/8	315.0	247.4
11/16	1.607	1.262	11/16	46.23	36.31	11/16	152.1	119.4	11/16	319.1	250.6
3/4	1.913	1.502	3/4	47.81	37.55	3/4	154.9	121.7	3/4	323.2	253.9
13/16	2.245	1.763	13/16	49.42	38.81	13/16	157.8	123.9	13/16	327.4	257.1
7/8	2.603	2.044	7/8	51.05	40.10	7/8	160.7	126.2	7/8	331.6	260.4
15/16	2.988	2.347	15/16	52.71	41.40	15/16	163.6	128.5	15/16	335.8	263.7
1	3.400	2.670	4	54.40	42.73	7	166.6	130.9	10	340.0	267.0
1/16	3.838	3.015	1/16	56.11	44.07	1/16	169.6	133.2	1/16	344.3	270.4
1/8	4.303	3.380	1/8	57.85	45.44	1/8	172.6	135.6	1/8	348.6	273.8
3/16	4.795	3.766	3/16	59.62	46.83	3/16	175.6	137.9	3/16	352.9	277.1
1/4	5.313	4.172	1/4	61.41	48.23	1/4	178.7	140.4	1/4	357.2	280.6
5/16	5.857	4.600	5/16	63.23	49.66	5/16	181.8	142.8	5/16	361.6	284.0
3/8	6.428	5.049	3/8	65.08	51.11	3/8	184.9	145.2	3/8	366.0	287.4
7/16	7.026	5.518	7/16	66.95	52.58	7/16	188.1	147.7	7/16	370.4	290.9
1/2	7.650	6.008	1/2	68.85	54.07	1/2	191.3	150.2	1/2	374.9	294.4
9/16	8.301	6.519	9/16	70.78	55.59	9/16	194.5	152.7	9/16	379.3	297.9
5/8	8.978	7.051	5/8	72.73	57.12	5/8	197.7	155.3	5/8	383.8	301.5
11/16	9.682	7.604	11/16	74.71	58.67	11/16	200.9	157.8	11/16	388.4	305.0
3/4	10.413	8.178	3/4	76.71	60.25	3/4	204.2	160.4	3/4	392.9	308.6
13/16	11.170	8.773	13/16	78.74	61.85	13/16	207.5	163.0	13/16	397.5	312.2
7/8	11.953	9.388	7/8	80.80	63.46	7/8	210.9	165.6	7/8	402.1	315.8
15/16	12.763	10.024	15/16	82.89	65.10	15/16	214.2	168.2	15/16	406.7	319.5
2	13.600	10.681	5	85.00	66.76	8	217.6	170.9	11	411.4	323.1
1/16	14.463	11.359	1/16	87.14	68.44	1/16	221.0	173.6	1/16	416.1	326.8
1/8	15.353	12.058	1/8	89.30	70.14	1/8	224.5	176.3	1/8	420.8	330.5
3/16	16.270	12.778	3/16	91.49	71.86	3/16	227.9	179.0	3/16	425.5	334.2
1/4	17.213	13.519	1/4	93.71	73.60	1/4	231.4	181.8	1/4	430.3	338.0
5/16	18.182	14.280	5/16	95.96	75.36	5/16	234.9	184.5	5/16	435.1	341.7
3/8	19.178	15.062	3/8	98.23	77.15	3/8	238.5	187.3	3/8	439.9	345.5
7/16	20.201	15.866	7/16	100.53	78.95	7/16	242.1	190.1	7/16	444.8	349.3
1/2	21.250	16.690	1/2	102.85	80.78	1/2	245.7	192.9	1/2	449.7	353.2
9/16	22.326	17.534	9/16	105.20	82.62	9/16	249.3	195.8	9/16	454.6	357.0
5/8	23.428	18.400	5/8	107.58	84.49	5/8	252.9	198.7	5/8	459.5	360.9
11/16	24.557	19.287	11/16	109.98	86.38	11/16	256.6	201.5	11/16	464.4	364.8
3/4	25.713	20.195	3/4	112.41	88.29	3/4	260.3	204.5	3/4	469.4	368.7
13/16	26.895	21.123	13/16	114.87	90.22	13/16	264.0	207.4	13/16	474.4	372.6
7/8	28.103	22.072	7/8	117.35	92.17	3/8	267.8	210.3	3/8	479.5	376.6
15/16	29.338	23.042	15/16	119.86	94.14	15/16	271.6	213.3	15/16	484.5	380.5

* From Marks, *Mechanical Engineers' Handbook*, 4th ed., McGraw-Hill, 1941.

TABLE 7.14 Weights of Square and Bound Steel Bars* (For iron, subtract 2%)

Diameters of Standard Bolts and Pipes

Bolt		Areas		Tensile strength, lb			Shearing strength, lb			
							Full bolt		Bottom of thread	
Diameter of bolt, in.	Number of threads per in.	Full bolt, sq in.	Bottom of thread, sq in.	At 10,000 lb per sq in.	At 12,500 lb per sq in.	At 17,500 lb per sq in.	At 7,500 lb per sq in.	At 10,000 lb per sq in.	At 7,500 lb per sq in.	At 10,000 lb per sq in.
1/4	20	0.049	0.027	270	340	470	380	490	200	270
5/16	18	0.077	0.045	450	570	790	580	770	340	450
3/8	16	0.110	0.068	680	850	1,190	830	1,100	510	680
7/16	14	0.150	0.093	930	1,170	1,630	1,130	1,500	700	930
1/2	13	0.196	0.126	1,260	1,570	2,200	1,470	1,960	940	1,260
9/16	12	0.248	0.162	1,620	2,030	2,840	1,860	2,480	1,220	1,620
5/8	11	0.307	0.202	2,020	2,520	3,530	2,300	3,070	1,510	2,020
3/4	10	0.442	0.302	3,020	3,770	5,290	3,310	4,420	2,270	3,020
7/8	9	0.601	0.419	4,190	5,240	7,340	4,510	6,010	3,150	4,190
1	8	0.785	0.551	5,510	6,890	9,640	5,890	7,850	4,130	5,510
1 1/8	7	0.994	0.693	6,990	8,660	12,130	7,450	9,940	5,200	6,930
1 1/4	7	1.227	0.890	8,890	11,120	15,570	9,200	12,270	6,670	8,900
1 3/8	6	1,435	1.054	10,540	13,180	18,450	11,140	14,850	7,910	10,540
1 1/2	6	1.767	1.294	12,940	16,170	22,640	13,250	17,670	9,700	12,940
1 5/8	5 1/2	2.074	1.515	15,150	18,940	26,510	15,550	20,740	11,360	15,150
1 3/4	5	2.405	1.745	17,450	21,800	30,520	18,040	24,050	13,080	17,440
1 7/8	5	2.761	2.049	20,490	25,610	35,860	20,710	27,610	15,370	20,490
2	4 1/2	3.142	2.300	23,000	28,750	40,500	23,560	31,420	17,250	23,000
2 1/4	4 1/2	3.976	3.021	30,210	37,770	52,870	29,820	39,760	22,660	30,210
2 1/2	4	4.909	3.716	37,160	46,450	65,040	36,820	49,090	27,870	37,160
2 3/4	4	5.940	4.620	46,200	57,750	80,840	44,580	59,400	34,650	46,200
3	3 1/2	7.069	5.428	54,280	67,850	94,990	53,020	70,690	40,710	54,280

* From Marks, *Mechanical Engineers' Handbook*, 4th ed., McGraw-Hill, 1941.

TABLE 7.15 U.S. Standard Bolts from $^1/_4$ to 3 in. Diameter*

(All dimensions in inches.)

Bolt diameter	Rough and semifinished					Finished				
	Width across flats		Min width across corners		Height	Width across flats		Min width across corners		Height
	Max	Min	Hex	Square		Max	Min	Hex	Square	
1/4	3/8	0.363	0.414	0.498	11/64	7/16	0.428	0.488	0.588	3/16
5/16	1/2	0.484	0.552	0.665	13/64	9/16	0.552	0.629	0.758	15/64
3/8	9/16	0.544	0.620	0.747	1/4	5/8	0.613	0.699	0.842	9/32
7/16	5/8	0.603	0.687	0.828	19/64	3/4	0.737	0.840	1.012	21/64
1/2	3/4	0.725	0.827	0.995	21/64	13/16	0.799	0.911	1.097	3/8
9/16	7/8	0.847	0.966	1.163	3/8	7/8	0.861	0.982	1.182	27/64
5/8	15/16	0.906	1.033	1.244	27/64	15/16	0.922	1.051	1.266	15/32
3/4	1 1/8	1.088	1.240	1.494	1/2	1 1/8	1.108	1.263	1.521	9/16
7/8	1 5/16	1.269	1.447	1.742	19/32	1 5/16	1.293	1.474	1.775	21/32
1	1 1/2	1.450	1.653	1.991	21/32	1 1/2	1.479	1.686	2.031	3/4
1 1/8	1 11/16	1.631	1.859	2.239	3/4	1 11/16	1.665	1.898	2.286	27/32
1 1/4	1 7/8	1.813	2.067	2.489	27/32	1 7/8	1.850	2.109	2.540	15/16
1 1/2	2 1/4	2.175	2.480	2.986	1	2 1/4	2.222	2.533	3.051	1 1/8
1 3/4	2 5/8	2.538	2.893	3.485	1 5/32	2 5/8	2.593	2.956	3.560	1 5/16
2	3	2.900	3.306	3.982	1 11/32	3	2.964	3.379	4.070	1 1/2
2 1/4	3 3/8	3.263	3.720	4.480	1 1/2	3 3/8	3.335	3.802	4.579	1 11/16
2 1/2	3 3/4	3.625	4.133	4.977	1 21/32	3 3/4	3.707	4.226	5.090	1 7/8
2 3/4	4 1/8	3.988	4.546	5.476	1 53/64	4 1/8	4.078	4.649	5.599	2 1/16
3	4 1/2	4.350	4.959	5.973	2	4 1/2	4.449	5.072	6.108	2 1/4

Regular nuts (rough, semifinished, and finished) have a maximum width across flats of $1\frac{1}{2}D$ except for $D = \frac{1}{4}$ to $\frac{9}{16}$ when the width $= 1\frac{1}{2}D + \frac{1}{16}$. D is bolt diameter. Tolerance for width is $-0.050D$. Thickness is $\frac{7}{8}D$.

* From Marks, *Mechanical Engineers' Handbook*, 4th ed., McGraw-Hill, 1941.

TABLE 7.16 Square and Hexagonal Regular Bolt Heads*

Diameter			Nominal thickness, in.	Circumference		Transverse areas			Length of pipe per sq ft of		Length of pipe containing 1 cu ft, ft	Nominal weight per ft, lb	No. of threads per in. of screw
Nominal internal, in.	Actual external, in.	Approx internal diam, in.		External, in.	Internal, in.	External, sq in.	Internal, sq in.	Metal, sq in.	External surface, ft	Internal surface, ft			
1/8	0.405	0.27	0.068	1.27	0.85	0.13	0.06	0.07	9.44	14.15	2513.00	0.24	27
1/4	0.540	0.36	0.088	1.70	1.14	0.23	0.10	0.12	7.08	10.49	1383.30	0.42	18
3/8	0.675	0.49	0.091	2.12	1.55	0.36	0.19	0.17	5.66	7.76	751.20	0.57	18
1/2	0.840	0.62	0.109	2.63	1.95	0.55	0.30	0.25	4.55	6.15	472.40	0.85	14
3/4	1.050	0.82	0.113	3.30	2.59	0.87	0.53	0.33	3.64	4.64	270.00	1.13	14
1	1.315	1.05	0.134	4.13	3.29	1.36	0.86	0.50	2.90	3.65	166.90	1.68	11½
1¼	1.660	1.38	0.140	5.22	4.34	2.16	1.50	0.67	2.30	2.77	96.25	2.27	11½
1½	1.900	1.61	0.145	5.97	5.06	2.84	2.04	0.80	2.01	2.37	70.66	2.72	11½
2	2.375	2.07	0.154	7.46	6.49	4.43	3.36	1.07	1.61	1.85	42.91	3.65	11½
2½	2.875	2.47	0.204	9.03	7.75	6.49	4.78	1.71	1.33	1.55	30.10	5.79	8
3	3.500	3.07	0.217	11.00	9.63	9.62	7.39	2.24	1.09	1.25	19.50	7.57	8
3½	4.000	3.55	0.226	12.57	11.15	12.57	9.89	2.68	0.96	1.08	14.57	9.11	8
4	4.500	4.03	0.237	14.14	12.65	15.90	12.73	3.18	0.85	0.95	11.31	10.79	8
5	5.563	5.05	0.259	17.48	15.85	24.31	19.99	4.32	0.69	0.76	7.20	14.62	8
6	6.625	6.07	0.280	20.81	19.05	34.47	28.89	5.59	0.58	0.63	4.98	18.97	8
8	8.625	8.07	0.276	27.10	25.35	58.43	51.15	7.28	0.44	0.47	2.82	24.69	8
8	8.625	7.98	0.322	27.10	25.07	58.43	50.02	8.41	0.44	0.48	2.88	28.55	8
9	9.625	8.94	0.344	30.24	28.08	72.76	62.72	10.04	0.40	0.43	2.29	33.91	8
10	10.750	10.19	0.278	33.77	32.01	90.76	81.55	9.21	0.36	0.37	1.76	31.20	8
10	10.750	10.14	0.306	33.77	31.86	90.76	80.75	10.01	0.36	0.38	1.78	34.24	8
10	10.750	10.02	0.366	33.77	31.47	90.76	78.82	11.94	0.36	0.38	1.82	40.48	8
12	12.750	12.09	0.328	40.06	37.98	127.68	114.80	12.88	0.30	0.32	1.25	43.77	8
12	12.750	12.00	0.375	40.06	37.70	127.68	113.10	14.59	0.30	0.32	1.27	49.56	8

Table 7.17 Standard Pipe and Line Pipe*

Diameter			Nominal thickness, in.	Circumference		Transverse areas			Length of pipe per sq ft of		Nominal weight per ft, lb
Nominal internal, in.	Actual external, in.	Approx internal diam, in.		External, in.	Internal, in.	External, sq in.	Internal, sq in.	Metal, sq in.	External surface, ft	Internal surface, ft	
1/8	0.405	0.21	0.100	1.27	0.64	0.13	0.03	0.10	9.43	18.63	0.31
1/4	0.540	0.29	0.123	1.70	0.92	0.23	0.07	0.16	7.08	12.99	0.54
3/8	0.675	0.42	0.127	2.12	1.32	0.36	0.14	0.22	5.66	9.07	0.74
1/2	0.840	0.54	0.149	2.64	1.70	0.55	0.23	0.32	4.55	7.05	1.09
3/4	1.050	0.74	0.157	3.30	2.31	0.87	0.43	0.44	3.64	5.11	1.47
1	1.315	0.95	0.182	4.13	2.99	1.36	0.71	0.65	2.90	4.02	2.17
1¼	1.660	1.27	0.194	5.22	4.00	2.16	1.27	0.89	2.30	3.00	2.99
1½	1.900	1.49	0.203	5.97	4.69	2.84	1.75	1.08	2.01	2.56	3.63
2	2.375	1.93	0.221	7.46	6.07	4.43	2.94	1.50	1.61	1.98	5.02
2½	2.875	2.32	0.280	9.03	7.27	6.49	4.21	2.28	1.33	1.65	7.66
3	3.500	2.89	0.304	11.00	9.09	9.62	6.57	3.05	1.09	1.33	10.25
3½	4.000	3.36	0.321	12.57	10.55	12.57	8.86	3.71	0.96	1.14	12.50
4	4.500	3.82	0.341	14.14	12.00	15.90	11.45	4.46	0.85	1.00	14.98
5	5.563	4.81	0.375	17.48	15.12	24.31	18.19	6.11	0.69	0.79	20.78
6	6.625	5.75	0.437	20.81	18.07	34.47	25.98	8.50	0.58	0.66	28.57
8	8.625	7.63	0.500	27.10	23.96	58.43	45.66	12.76	0.44	0.50	43.34
10	10.750	9.75	0.500	33.77	30.63	90.76	74.66	16.10	0.36	0.40	54.73
12	12.750	11.75	0.500	40.06	36.91	127.68	108.43	19.25	0.30	0.33	65.41

* From Marks, *Mechanical Engineers' Handbook*, 4th ed., McGraw-Hill, 1941.

Table 7.18 Extra-Strong Pipe*

Nominal pipe size, in.	Outside diameter, in.	Nominal wall thickness, in.		Wt per ft, lb, plain ends	Nominal pipe size, in.	Outside diameter, in.	Nominal wall thickness, in.		Wt per ft, lb, plain ends
		Wrought iron	Steel				Wrought iron	Steel	
½	0.840	0.307	0.294	1.714	2½	2.875	0.565	0.552	13.695
¾	1.050	0.318	0.308	2.440	3	3.500	0.615	0.600	18.583
1	1.315	0.369	0.358	3.659	4	4.500	0.690	0.674	27.541
1¼	1.660	0.393	0.382	5.214	5	5.563	0.768	0.750	38.552
1½	1.900	0.411	0.400	6.408	6	6.625	0.884	0.864	53.160
2	2.375	0.447	0.436	9.029	8	8.625	0.895	0.875	72.424

* From Marks, *Mechanical Engineers' Handbook*, 4th ed., McGraw-Hill, 1941.

Table 7.19 Double-Extra-Strong Pipe*

Nominal Size (OD)	Wall thickness inches	Weight lb/ft	Schedule
3/4 (1.050)	0.113	1.13	40
	0.154	1.47	80
	0.219	1.94	160
	0.308	2.44	–
1 (1.315)	0.133	1.68	40
	0.179	2.17	80
	0.250	2.84	160
	0.358	3.66	–
1-1/4 (1.660)	0.140	2.27	40
	0.191	3.00	80
	0.250	3.76	160
	0.382	5.21	–
1-1/2 (1.900)	0.145	2.72	40
	0.200	3.63	80
	0.281	4.86	160
	0.400	6.41	–
2 (2.375)	0.083	2.03	–
	0.109	2.64	–
	0.125	3.00	–
	0.141	3.36	–
	0.154	3.65	40
	0.172	4.05	–
	0.188	4.39	–
	0.218	5.02	80
	0.250	5.67	–
	0.281	6.28	–
	0.344	7.46	160
	0.436	9.03	–

Table 7.20 Pipe Weight (lb/ft)

Gauge No.	American wire gauge, or Brown & Sharpe (for nonferrous sheet and wire)	Steel wire gauge or Washburn & Moen or Roebling (for steel wire)	Birmingham wire gauge (B.W.G.) or Stubs iron wire (for steel rods or sheets)	Stubs steel wire gauge	British Imperial standard wire gauge (S.W.G.)	U.S. standard gauge for wrought iron sheet (480 lb per cu ft)	U.S. standard gauge for steel and open-hearth iron sheet (489.6 lb per cu ft)	British standard gauge for iron and steel sheets and hoops, 1914 (B.G.)
0000000	......	0.4900	......	.....	0.500	0.500	0.4902	0.6666
000000	......	0.4615	......	.....	0.464	0.469	0.4596	0.6250
00000	......	0.4305	......	.....	0.432	0.438	0.4289	0.5883
0000	0.460	0.3938	0.454	.....	0.400	0.406	0.3983	0.5416
000	0.410	0.3625	0.425	.....	0.372	0.375	0.3676	0.5000
00	0.365	0.3310	0.380	.....	0.348	0.344	0.3370	0.4452
0	0.325	0.3065	0.340	.....	0.324	0.312	0.3064	0.3964
1	0.289	0.2830	0.300	0.227	0.300	0.281	0.2757	0.3532
2	0.258	0.2625	0.284	0.219	0.276	0.266	0.2604	0.3147
3	0.229	0.2437	0.259	0.212	0.252	0.250	0.2451	0.2804
4	0.204	0.2253	0.238	0.207	0.232	0.234	0.2298	0.2500
5	0.182	0.2070	0.220	0.204	0.212	0.219	0.2145	0.2225
6	0.162	0.1920	0.203	0.201	0.192	0.203	0.1991	0.1981
7	0.144	0.1770	0.180	0.199	0.176	0.188	0.1838	0.1764
8	0.128	0.1620	0.165	0.197	0.160	0.172	0.1685	0.1570
9	0.114	0.1483	0.148	0.194	0.144	0.156	0.1532	0.1398
10	0.102	0.1350	0.134	0.191	0.128	0.141	0.1379	0.1250
11	0.091	0.1205	0.120	0.188	0.116	0.125	0.1225	0.1113
12	0.081	0.1055	0.109	0.185	0.104	0.109	0.1072	0.0991
13	0.072	0.0915	0.095	0.182	0.092	0.094	0.0919	0.0882
14	0.064	0.0800	0.083	0.180	0.080	0.078	0.0766	0.0785
15	0.057	0.0720	0.072	0.178	0.072	0.070	0.0689	0.0699
16	0.051	0.0625	0.065	0.175	0.064	0.062	0.0613	0.0625
17	0.045	0.0540	0.058	0.172	0.056	0.056	0.0551	0.0556
18	0.040	0.0475	0.049	0.168	0.048	0.050	0.0490	0.0495
19	0.036	0.0410	0.042	0.164	0.040	0.0438	0.0429	0.0440
20	0.032	0.0348	0.035	0.161	0.036	0.0375	0.0368	0.0392
21	0.0285	0.0317	0.032	0.157	0.032	0.0344	0.0337	0.0349
22	0.0253	0.0286	0.028	0.155	0.028	0.0312	0.0306	0.0313
23	0.0226	0.0258	0.025	0.153	0.024	0.0281	0.0276	0.0278
24	0.0201	0.0230	0.022	0.151	0.022	0.0250	0.0245	0.0248
25	0.0179	0.0204	0.020	0.148	0.020	0.0219	0.0214	0.0220
26	0.0159	0.0181	0.018	0.146	0.018	0.0188	0.0184	0.0196
27	0.0142	0.0173	0.016	0.143	0.0164	0.0172	0.0169	0.0175
28	0.0126	0.0162	0.014	0.139	0.0148	0.0156	0.0153	0.0156
29	0.0113	0.0150	0.013	0.134	0.0136	0.0141	0.0138	0.0139
30	0.0100	0.0140	0.012	0.127	0.0124	0.0125	0.0123	0.0123
31	0.0089	0.0132	0.010	0.120	0.0116	0.0109	0.0107	0.0110
32	0.0080	0.0128	0.009	0.115	0.0108	0.0102	0.0100	0.0098
33	0.0071	0.0118	0.008	0.112	0.0100	0.0094	0.0092	0.0087
34	0.0063	0.0104	0.007	0.110	0.0092	0.0086	0.0084	0.0077
35	0.0056	0.0095	0.005	0.108	0.0084	0.0078	0.0077	0.0069
36	0.0050	0.0090	0.004	0.106	0.0076	0.0070	0.0069	0.0061
37	0.0045	0.0085	......	0.103	0.0068	0.0066	0.0065	0.0054
38	0.0040	0.0080	......	0.101	0.0060	0.0062	0.0061	0.0048
39	0.0035	0.0075	......	0.099	0.0052	0.0059	0.0057	0.0043
40	0.0031	0.0070	......	0.097	0.0048	0.0055	0.0054	0.0039
41	......	0.0066	......	0.095	0.0044	0.0053	0.0052	0.0034
42	......	0.0062	......	0.092	0.0040	0.0051	0.0050	0.0031
43	......	0.0060	......	0.088	0.0036	0.0049	0.0048	0.0027
44	......	0.0058	......	0.085	0.0032	0.0047	0.0046	0.0024
45	......	0.0055	......	0.081	0.0028	......	......	0.0022
46	......	0.0052	......	0.079	0.0024	......	......	0.0019
47	......	0.0050	......	0.077	0.0020	......	......	0.0017
48	......	0.0048	......	0.075	0.0016	......	......	0.0015
49	......	0.0046	......	0.072	0.0012	......	......	0.0014
50	......	0.0044	......	0.069	0.0010	......	......	0.0012

* From Marks, *Mechanical Engineers' Handbook*, 4th ed., McGraw-Hill, 1941.

TABLE 7.21 Wire and Sheet-Metal Gauges* (Diameters and thicknesses in decimel parts of an inch)

Rope Sizes

Never base selection of ropes of any material on size alone. Always consider size (diameter) and weight together. The amount, or weight, of material in a standard length of rope determines rope characteristics, such as breaking strength, working load, the cyclic and shock loading forces it can withstand. The weight of a standard length of rope is the rope's weight: length ratio, pounds per 100 ft (lb/ft) or kilograms per 100 meters (kg/100 m).

Diameter		Circumference		Diameter		Circumference	
Inches	Millimeters	Inches	Millimeters	Inches	Millimeters	Inches	Millimeter
1/8	3.2	3/8	10.0	15/16	23.8	3	76.2
5/32	4.0	15/32	12	1	25.4	3-1/8	79.4
3/16	4.8	9/16	15	1-1/16	27.0	3-3/8	85.7
7/32	5.6	11/16	17	1-1/8	28.6	3-1/2	88.9
1/4	6.3	3/4	20	1-3/16	30.2	3-3/4	95.2
5/16	7.9	1	25.4	1-1/4	31.7	3-7/8	98.4
3/8	9.5	1-1/8	28.6	1-3/8	34.9	4-3/8	111
7/16	11.1	1-3/8	34.9	1-7/16	36.5	4-1/2	114
1/2	12.7	1-5/8	41.3	1-1/2	38.1	4-3/4	121
9/16	14.3	1-3/4	44.4	1-5/8	41.3	5-1/8	130
5/8	15.9	2	50.8	1-11/16	42.9	5-1/4	133
11/16	17.5	2-1/4	57.1	1-3/4	44.4	5-1/2	140
3/4	19.0	2-3/8	60.3	1-13/16	46.0	5-3/4	146
13/16	20.6	2-1/2	63.5	1-7/8	47.6	5-7/8	149
7/8	22.2	2-3/4	69.8	1-15/16	49.2	6-1/8	156
				2	50.8	6-1/4	159

*U.S. sizes designed by diameters. Circumferences shown are for comparison only.

TABLE 7.22 Metric Equivalents of Standard Rope Sizes*

Wire Rope Size (in.)	Web thickness		Rim width		Groove depth	
	Light weight (in.)	Heavy weight (in.)	Light weight (in.)	Heavy weight (in.)	Light weight (in.)	Heavy weight (in.)
1/2	3/8	1/2	1-1/8	1-1/4	3/4	3/4
5/8	1/2	3/4	1-7/16	1-3/4	15/16	15/16
3/4	1/2	3/4	1-1/2	1-3/4	1	1-1/8
7/8	3/4	3/4	2	2	1-5/16	1-5/16
1	3/4	3/4	2-1/4	2-1/4	1-1/2	1-1/2
1-1/8	3/4	1	2-1/4	2-1/2	1-5/8	1-11/16
1-1/4	1	1-1/4	2-1/2	2-3/4	1-3/4	1-7/8
1-3/8	1-1/4	1-1/2	3	3-1/4	1-15/16	2-1/16
1-1/2	1-1/4	1-1/2	3-1/4	3-1/2	2-1/8	2-1/4

TABLE 7.23 Sheave Rim Dimensions

Blocks and Tackles

Adopted symbol	Description	Formula
M	Mechanical advantage	$M = N$
N	Number of parts of line supporting load	Known quantity
W	Load being hoisted, pounds	$W = P \times N \times E$
P	Lead-line pull, pounds	$P = W / (N \times E)$
E	Efficiency of tackle	Refer to chart below
F	Load on standing block, pounds	$F = W + P$
L	Length of lead line (hauled in), feet	$L = D \times N$
D	Distance load is hoisted, feet	Known distance
T	Time required to hoist load, minutes	Known time
A	Accomplishment (work put in), foot pounds	$A = P \times L$
H	Horsepower (work expressed mathematically) (1 hp =33,000 lb. of work done in 1 min)	$H = (P \times L) / (33{,}000 \times T)$

TABLE 7.24 Basic Formulas for computing Loads on Blocks and Tackles

Block and tackle makes work easier. Like any other machine, however, it can not decrease the total amount of work to be done.

Thus: Work input = work output

$$F_1 \times S_1 = F_2 \times S_2$$

Where:

F_1 = effort applied, in pounds
S_1 = distance through which effort moves, in feet
F_2 = resistance overcome, in pounds
S_2 = distance resistance is moved, in feet

A small force operating through a large distance, therefore, can move a big load through a small distance

Example: To raise a 600-lb load 20 ft, a pull of only one-fifth of 600—or 120 lb is needed—but actually requires a 160-lb pull through 100 ft of rope.

To calculate the efficiency of tackle:

$$\text{Efficiency} = \frac{\text{Output}}{\text{Input}} = \frac{F_2 \times S_2}{F_1 \times S_1}$$

$$\text{Efficiency} = \frac{600 \times 20}{160 \times 100} = 0.75 \quad (75\%)$$

TABLE 7.25 Computing Efficiency of Tackle

SECTION 3

Worksite Preparation

CHAPTER 8

Protection of Worksite, Adjacent Areas, and Utilities

> The problems of protection at construction sites are immense. Protection clearly implies a liability, and the bulk of litigation derived from construction sites alleges improper protection of lives and property . . . Most construction damage claims involve more than one contractor, therefore, if an award is made, it is often apportioned among the defendants on the basis of a degree of liability.
>
> Since this is often a very subjective area . . . a claim will often turn against you on seemingly unimportant evidence—such as having used the wrong color danger sign. The important thing to remember, is that form may be as important as substance in a legal sense. It is not enough to put up a required danger sign—it must be the prescribed color and size.
>
> Robert T. Ratay, *Handbook of Temporary Structures in Construction*

8.1 Work Area Safety

Construction work safety rules are promulgated and enforced by the federal government and by many states and municipalities. The contractor faces substantial civil penalties for failure to comply with the federal code as administered by the Occupational Health & Safety Administration (OSHA). Direct disobedience can expose the contractor to criminal penalties.

Safety requirements for construction activities tend to vary greatly. For example, state requirements generally are considerably less extensive than OSHA requirements; however, some state regulations are much broader, while municipal codes generally are somewhat between the state and federal codes in length, dealing with most local problems.

When the codes supplement each other, it is best that the contractor follow the more strict requirements. The federal codes should be followed when state and federal or municipal codes conflict; the sequence to follow is: federal, followed by state, followed by municipal regulations.

NOTE: *Every construction site should have a copy of OSHA manual* CFR 29 Part 1926—Safety and Health Regulations for Construction *to use as a reference over the duration of the project.*

In addition to OSHA requirements, which apply uniformly across the country, each state and locality may have differing building codes that will regulate construction projects in their jurisdictions. The contractor is also required to obtain permits for any work done, including sidewalk crossings and for storage of materials in the street.

8.2 Signage and Protective Barriers

Proper signs are a contractor's legal obligation to protect the public and workers from the hazards that may exist around the construction site. The two basic OSHA-approved DANGER and CAUTION signs are available at most contractor supply houses (Fig. 8.1). Traffic control signs are more complicated in complying with regulations for directing moving traffic. Those signs have been standardized nationally, as detailed in *ANSI D6.1-1971, Manual on Uniform Traffic Control Devices for Street and Highways.*

When conditions exist that require a flag-person to direct traffic, OSHA requires that the flag-person *must* be properly equipped using a red flag of at least 18 in. x 18 in. (45 cm x 45 cm) during the day and red lights at night, and wearing a red or orange vest by day and a reflectorized vest by night.

Traffic signs *must* be mounted at right angles to traffic and about 6 ft (2 m) from the roadway. Advance warning signs *must* begin at least 1500 ft (450 m) before the construction site and be followed by repeat signs at 1000 ft (300 m) and 500 ft (150 m) from the site. A typical sign would be diamond-shaped with an orange background and black lettering. The suggested speed limits should be readable for existing conditions—if they are too low, they will be ignored.

Use only standardized orange and black, nonregulatory signs for construction warning signs. *Avoid* using standardized regulatory signs of red and yellow, and black and white types normally placed and enforced by police and highway departments. If they are to be used, it should be with the approval and supervision of the police agency having jurisdiction. Such signs are *not* just warning signs—they impose legal obligation on the public.

Figure 8.1 OSHA approved basic warning signs. (*Source:* FREE Printable OSHA and Statutory Signs—*www.freesinage.com/osha_caution*.)

Barricades and Fences

Many forms of barricades are used (*ANSI D6.1-1971*), including:

- *Drums.* 30-gal (0.11 m^3) and 55-gal (0.21 m^3) are used, closely spaced; they should be marked with 6-in. (15-cm) orange stripes and 4-in. (10-cm) white strips using reflectorized paint. Plastic drum shapes are available for traffic control by building supply houses. Drums must never be used in the road without advance warning signs.

- *Cones.* Used for short-duration jobs of less than a day; should be 18 in. (45 cm) high and preferably orange in color. Cones must never be used in the road without advance warning signs.
- *Fences.* Erected to protect equipment and materials, and to reduce contractor exposure to public liability claims, especially trespassing children. Chain link is commonly used on large open sites while a plywood fence is more practical for city construction, reducing the invitation to thefts and offering better protection to pedestrians.

Because gates for truck entry are typically 16 ft (5 m) wide, swinging gates are not suitable for an opening this size, posing a maintenance problem and difficulty in securing them well enough to prevent entry by children. The safest and most secure gate is a chain link sliding gate.

Avoid providing spectator viewing openings in the fence. Where viewing windows are unavoidable, they should be covered with heavy shatterproof plastic and located away from compressor hoses, which should be kept as far from the fence as possible.

The construction superintendent is directly responsible for making the job site safe at night by properly closing the fence and placing barricades at the end of the day.

8.3 Site Lighting and Perimeter Fencing

The process of providing proper illumination for the work area is an ill-defined area of the contract and specifications. Yet, the responsibility to provide temporary lighting falls in the scope of work of the general contractor—or is included in the scope of work of the electrical contractor.

Studies have shown that at least parts of some construction projects do not have the minimum illumination required by accepted safety standards, even when the contract requirements are met. An examination of construction standards and requirements indicates that the contract language defining the extent of the temporary lighting often is vague and the requirements minimal.

Construction worksite lighting OSHA regulations, based on *ANSI A11.1-1965; R 1970–Standard UL 1572,* apply to luminaires for use in nonhazardous locations and intended for installation on branch circuits of 600 V or less between conductors in accordance with the *US National Electrical Code NEC, ANSI/NFPA 70,* and the *Canadian Electrical Code Part 1 (CEC), CSA C22.1.*

Lighting and Illumination Fundamentals

Temporary lighting is the process and the equipment used to provide artificial illumination in a work area. Illumination is the amount of

Location	Light meter readings (FC = footcandles)*
Outside on a sunny day	4000 FC
Parking deck	3 FC
Shopping mall common area	56 FC

*Readings were taken 36 in. off the floor to reflect the common task height.

Table 8.1 Miscellaneous Light Readings (3 ft above the floor)

light falling on a surface and is measured in footcandles (FC), the measurement of the light falling on a square foot of surface.

Measuring illumination at a specific point only describes the illumination environment in one dimension—that is, the quantity of illumination. Other aspects, such as brightness, contrast, and glare help define the quality of illumination. (Brightness refers to the perception of the human observer.)

The overall lighting in an area may be adequate for the work process, but as workers move from one area to another, or look from one area to another, the amount of illumination is different, and the eye must keep adjusting (Table 8.1).

Temporary Lighting Requirements

Although some level of lighting is necessary to provide workers with a safe environment, safety is only one aspect of illumination of a construction project. The quality of work and productivity may be related to illumination.

OSHA standard *29 CFR 1929.56 Illumination* and *29 CFR 1926.56(a) General* specify that construction areas, runways, corridors, offices, shops, and storage areas *must* be lighted to not less than the minimum illumination while any work is in progress (Table 8.2). For areas or operations *not* covered in Table 8.2 refer to *ANSI A11.1-1965; R 1970 - Practice for Industrial Lighting for Recommended Values of Illumination.*

Project Specifications

The construction manager or general contractor can place the duty to provide temporary lighting under the scope of work for the electrical contractor. Although construction design documents define the contractor's responsibilities during construction, they focus on the design elements of the finished product, but rarely setting environmental guidelines to be followed during the construction phase.

The design team should work with the construction contractor to adopt environmental guidelines to be followed during construction (U.S. DOE – *Energy and Environmental Guidelines for Construction*).

Intensities (footcandles)	Areas of operation
5	General construction area lighting
3	General construction areas, concrete placement, excavation, waste areas, accessways, active storage areas, loading platforms, refueling, and field maintenance areas
5	Indoor warehouses, corridors, hallways, and exitways
5	Tunnels, shafts, and general underground work areas. Exception: minimum of 10 footcandles is *REQUIRED* at tunnel and shaft heading during drilling, mucking, and scaling. Bureau of Mines–approved cap lights shall be acceptable for use in a tunnel heading.
10	General construction plant and shops (such as batch plants, screening plants, mechanical and electrical equipment rooms, carpenter shops, rigging lofts, and active store rooms, barracks or living quarters, locker or dressing rooms mess halls, indoor toilets, and workrooms).
30	First-aid stations, infirmaries, and offices

Source: 29 CFR 1926.56—Table D-3, OSHA

TABLE 8.2 Minimum Illumination Intensities (footcandles)

Task Lighting

Illumination on the construction site is based on improving task lighting for the workers. Task lighting requirements, however, are generally difficult to identify on the construction project since there are many workers performing different tasks with different lighting requirements. (Masons have different requirements than electricians, and electricians putting in conduit have different requirements than electricians installing a control panel on an HVAC unit.)

There is also the problem of where the workers are performing their tasks. Construction projects require constant movement of many workers, performing tasks at different levels at a location. The larger problem, however, is that construction workers are asked to work in various locations within and outside a building with various light levels at each location. Furthermore, the contrast of light at many locations can be drastic. And the light conditions are constantly changing with the time of day, outside conditions, and the constant installation of more building components. These factors all contribute to the visual working conditions of the typical construction project.

Use efficient work lighting, exceeding 40 lumens per watt, such as *T5* or *T8* fluorescent lamps with electronic ballasts and low mercury content for general work area lighting. Metal halide lamps are an energy-efficient choice for high-bay overhead installations, as are high-pressure sodium lamps if color rendition is not critical.

Minimize the use of quartz/halogen and conventional incandescent lamps. Compact fluorescent lamps are a better choice with high-efficiency lighting paying back quickly through reduced operation and maintenance/replacement costs.

Temporary construction lighting is a complex problem that needs attention. The safety, productivity, and quality of construction project are influenced by the quantity and quality of the lighting. The construction environment is constantly changing, and the dynamic quality of the environment requires solutions that are just as dynamic.

There is no simple solution but to look at one possible solution at a time and analyze the costs and benefits, which then may open the door to combinations of solutions that are cost effective, provide a safer work environment, promote higher productivity, and help to insure higher quality.

Perimeter Fencing

Perimeter fencing should always be erected before construction activities start, particularly if the site is located near a school or in a residential area, and is unattended outside work hours. At no time shall a temporary/security fence be installed in a manner that would prohibit the safe and continued operation of an occupied building in accordance with the local building code. All fences *must* be maintained in a structurally sound and attractive manner.

The standard of fencing required for a particular construction site will depend on the hazards and environmental conditions at the worksite, as well as location. Some factors to consider include building height (more than one story); location near stores, offices, schools, residential areas; whether there are any trenches and trench depth; whether the site is vacant outside work hours.

8.4 Worksite Housekeeping

Keeping a construction site relatively clean of debris can reduce hazards to employees during construction work (*CFR 29 1926.25.OSHA Safety and Health Standards for Construction Industry*).

Waste Material Removal

Most of the waste generated through construction activities is nonhazardous solid waste, including concrete, steel wood, rubber, asphalt, soil, and organic matter (such as tree stumps and roots). Disposal of these wastes may be regulated under a variety of federal, state, and local laws, with many states and local governments specifying special construction waste landfills for disposing nonhazardous construction and demolition debris.

OSHA 1926.252 (c) requires contractors to *remove* all scrap lumber, waste material, and rubbish from the immediate work area as the work progresses. Disposal of waste material or debris by burning *must* comply with local fire regulations.

All solvent waste, oily rags, and flammable liquids *must* be kept in fire resistant covered containers until removed from the worksite. Hazardous construction wastes, if generated, are regulated under the federal RCRA hazardous waste regulations.

The benefits of good housekeeping practices far exceeds the small additional effort required to establish good housekeeping practices at a construction site (Table 8.3).

- Store waste materials in covered areas; cover exposed piles of soil, construction materials, and waste with a plastic sheeting or temporary roof.
- Keep covers on all trash cans and dumpsters.
- Keep the area around trash receptacles free of debris.
- Keep trash receptacles in a covered area to prevent wind or rain from getting into the trash.
- Keep all stored scrap material from entering the storm drain system.
- Prevent liquids from leaking out of the trash area and entering the storm drain system.
- Protect all storm drain inlets using berms, screens, or filter fabric.
- Maximize trash cans to prevent littering.
- Waste being held for recycling or disposal should be held separate and covered.
- Reduce waste by ordering only the amount needed to finish the job at hand.
- Recycle leftover material whenever possible.
- Make sure that trailers carrying materials are covered during transit.
- Use a vacuum or other effective capture and container system to collect all of the water and sediment from saw cutting asphalt or concrete.
- Pick up and sweep dirt and debris from all work areas—including streets, sidewalks, and alleys—when a project is complete,
- *Do not* wash down work areas, tools, and equipment on surfaces that drain to the storm drains or creeks.

Table 8.3 Waste Management Practices

Water/Wastewater Treatment/Disposal

The contractor is responsible for making arrangements for obtaining water for construction use. Proof of such arrangement is subject to review and approval by the local water authority. If required, the contractor *must* obtain a permit from the authority before drawing any water from the supply system. The contractor *must* adhere to all stormwater pollution control requirements, to prevent sediment from entering the stormwater collection and conveyance system.

Managing the proper use of water on the construction site includes, among other practices:

- Implementing a water conservation awareness program within the worksite
- Monitoring daily water consumption by taking water meter readings daily
- Including discussion of water consumption, water bill, and water conservation in the monthly project meeting
- Installing pressure-reducing valves after the meter position and service stop valve where pressure exceeds 2.5 bar
- Installing throttle control valve along pipes leading to toilets, wash basins, etc. to reduce excessive flow rate at the taps
- Installing sub-meters at constructional and nonconstructional activities areas to monitor water usage and determine the norm of water usage for those activities areas; identifying the reasons and taking appropriate action if the amount of water uses in any operation is above its norm
- Installing service pipes that are more durable at worksite to minimize leakage due to damage; lay exposed pipe where feasible
- Performing regular leak detection tests and checks for water leakage in taps, hoses, pipes, water tanks, etc. so that leaks can be discovered as early as possible; and taking immediate action to repair the leak(s)
- Ensuring water saving devices are installed and in good working condition, such as:
    - Low-capacity flushing toilets
    - Self-closing delayed-action taps at all washing facilities
- Using nonpotable water for vehicle washing at the vehicle washing area whenever possible; and recycling the used water
- Installing an earth removal facility for the vehicles to remove the earth from the ties before commencing washing.

Best Water Management Practices[1]

- *Natural features and soil exposure minimization.* Leave vegetation in place when possible and disturb the smallest area of land possible. Control access to surface water with fencing or bugger zones. Divert runon/runoff away from exposed areas. Phase construction to minimize the amount of exposed soil at one time.
- *Erosion control (soil stabilization).* Install temporary or permanent seeding and planting; mulching or cover using geotextiles, mats, and erosion control blankets.
- *Groundwater protection.* Protect drinking water supply wellheads from construction activity and stormwater by sloping, fencing, or berming. If infiltration devices are used, do not excavate to or through the water table. Drywells and other infiltration devices are subject to underground injection control regulations.
- *Good housekeeping.* Dedicate covered contained areas for any activities involving toxic contaminants, or move such activities off-site, exposed and spread by stormwater.
- *Inspection and maintenance.* Inspect construction site regularly and train staff and subcontractors to recognize and discuss potential problems at the construction site so they can be prevented.
- *Postconstruction stormwater management.* Provide detention ponds, infiltration basins, and trenches; pervious pavements, "green" parking and alternative pavers, grassed swales, grassed filter strip and buffer zones.

8.5 Jobsite Safety Checklist

Program administrator (1926.20.35)	**A**	**B**	**Comments**
1. OSHA Postings, Emergency Numbers			
2. Hazcom (MSDS, Labeling)			
3. Job Specific Safety Orientation			
4. Safety Meeting (Last Meeting Topic)			
5. Job Site Injury Records			
6. Medical Services (Kits, Training)			

[1]*Source:* Environmental Protection Agency—*www.epa.gov/safewater.*

Housekeeping/sanitation (1926.20.35)	A	B	Comments
1. Work Area Orderly (Nails Removed)			
2. Passageways and Walkways Clear			
3. Lighting			
4. Waste Containers Provided and Used			
5. Sanitary Facilities Clean			
6. Adequate Eating and Drinking Areas			
Fire prevention (1926.24, .25, .97)	**A**	**B**	**Comments**
1. Hazards ID and Control Fueling/Welding			
2. Extinguishers Identified and Charged			
3. Correct Type for the Job			
4. "No Smoking" Posted and Enforced			
5. Hydrants Clear			
6. Flammable Liquids Storage			
Electric utility (1926.20, .21)	**A**	**B**	**Comments**
1. Electrical Dangers Posted			
2. GFCI and Overcurrent Protection			
3. Access to Breaker Boxes Clear			
4. Lockout Procedures Used			
5. Drop Cords Protected			
6. Utility Lines Located and Market			
7. Overhead Lines Protection			
Hand and power tools (1926.300, .400)	**A**	**B**	**Comments**
1. Tools in Good Working Condition			
2. Damaged Tools Repaired/Replaced			
3. Proper Tool Used for Each Job			
4. Power Tools Grounded/Double Insulated			
5. Cords in Good Condition			
6. Mechanical Safeguards in Use			
7. Power Actuated Tools Safety/Certification			
Ladders (1926.1053)	**A**	**B**	**Comments**
1. Ladders Inspected/Good Condition			
2. Job Made Ladders Construction			
3. Ladders Proper for Job and Secured			
4. Siderail Extend 36 in. above Landings			

5. Stepladders Fully Open When in Use			
6. Overhead Electrical Exposure			
Scaffolding (1926.451)	**A**	**B**	**Comments**
1. Erection Properly Supervised			
2. Scaffolding Plumb and Square			
3. Connections Sound and Secure			
4. Footings and Mudsills			
5. Scaffold Tied to Structure			
6. Guardrails, Midrails, and Toeboards			
7. Worker Protection (Falling Objects)			
8. Scaffold Access			
9. Planks Properly secured/Overlapped			
10. Planking Free from Debris, Ice, Snow			
11. Overhead Electrical Exposure			
Heavy equipment, motor vehicles (1926.600)	**A**	**B**	**Comments**
1. Equipment Maintenance and Inspection			
2. Operating Manuals Available			
3. Operators Qualified			
4. Haul Roads Maintained			
5. Speed Limits Posted			
6. Seat Belts Operative and Used/ROPs			
7. Horns, Brakes, Lights, Back-up Alarms			
8. Wheels Cocked When Required			
9. Glass in Good Condition			
10. Weight Limits and Load Capacities			
11. Personnel Transported Safely			
Welding and cutting (1926.350)	**A**	**B**	**Comments**
1. Extinguishers Available/Firewatch			
2. Screens and Shields in Place/Personal Protective Equipment			
3. Cylinders Secured/Stored Properly			
4. Electrical Ground/Cables Sound			
Materials handling/storage (1926.953)	**A**	**B**	**Comments**
1. Materials Properly Stored and Stacked			
2. Dust Protection			

3. Proper Number of Employees for Operations			
4. Employees Lifting Loads Correctly			
Barricades and fencing (1926.502)		**A**	**B**
1. Site Fenced			
2. Floor Openings Planked/Protected			
3. Roadways and Sidewalks Protected			
4. Proper Access/Traffic Control			
Hoists, cranes, and derricks (1926.550)	**A**	**B**	**Comments**
1. Equipment Maintenance/Inspections			
2. Equipment Support/Outriggers			
3. Load Capacities Posted and Observed			
4. Signalman Use as Appropriate			
5. Overhead Electrical Exposure			
Personal protective equipment (1926.100)	**A**	**B**	**Comments**
1. Eye/Face Protection			
2. Hard Hats Mandatory			
3. Safety Shoes/Boots			
4. Fall Protection Equipment			
5. Respiration/Ventilation Required			
6. Noise Exposures			
Roadway construction (1926.600)	**A**	**B**	**Comments**
1. Laws and Ordinances Observed			
2. Flagmen Certified/Positioning			
3. Warning Signs and Markers			
4. Traffic Control/Devices			
5. Lighting Equipment Exposure/Traffic			

8.6 Site Features

Before beginning any construction rigging operations, it is important to conduct a careful inspection of the site and its boundaries to determine all possible conditions that must be considered in planning the way work on the project will be carried out as well as to calculate the possible additional costs of removing obstacles, or providing temporary construction to circumvent them. Inspection should follow a predetermined checklist to assure nothing is overlooked.

Temporary Access

On-site

If the local authority imposes weight restrictions on approach roads or bridges, reductions in the size of equipment used on the worksite may be necessary that, in turn, may mean an increased time frame due to lower output. If a temporary access approach system is necessary, the system implemented *must* be capable of being readily removed at the end of the contract to minimize reinstatement costs.

Off-site

The most common overhead obstructions are high-tension electricity cables. Before starting work on a site, contractors *must* contact the local utility company to determine the safety requirements in relation to clearances to ensure that flash-over from cables to site, vehicles, and personnel does not occur, especially in wet conditions, when the flash-over distance is less than when it is dry.

Determine safety requirements in terms of clearances to prevent discharge from overhead lines to construction facility, equipment, and workers. Provide minimum horizontal distances and travel clearances.

OSHA requires contractors to ensure that when cranes and boomed vehicles are operated near overhead power lines, the power lines are de-energized or separated from cranes and their loads by implementing one or more procedures:

- De-energize and visibly ground electrical distribution and transmission lines.
- Use independent insulated barriers to prevent physical contact with the power line.
- Maintain minimum clearance between energized power lines and cranes and their loads.

ANSI standard B30.5-199 specifies guidelines to prevent contact between cranes and electrical energy. For mobile cranes operating near overhead power lines:

- Consider any overhead wire to be energized unless and until the person owning the line or the utility authorities verify that the line is not energized.
- De-energize power lines before work begins, erecting insulated barriers to prevent physical contact with the energized lines of maintaining safe clearances between the energized lines and boomed equipment.
- Limit the use of cage-type boom guards, insulating links, and proximity warning devices.
- Notify line owners before work is performed near power lines.

- Post warnings on cranes cautioning operators to maintain safe clearance between energized power lines and their equipment.

Once the minimum working clearances have been established, traveling under cables *must* be restricted to defined points, where *goal posts* are erected on each side of the overhead line to define the maximum height that can pass under the lines. Elsewhere, barriers *must* be erected at safe distances away from the lines to prevent vehicles traveling inadvertently under the lines at that point.

Ground-level and Underground Obstructions

Other than open excavations, the most likely ground-level obstructions on a construction site are surface drainage pipes, industrial mains, and similar installations. These obstructions generally require some form of bridging to permit travel over them. Barriers may also be needed to prevent damage from construction or machinery adjacent to any service pipes or other installations.

Water, gas, and electric services normally present underground obstructions that require temporary protection only if they are sited close to the surface and need protection from traffic passing over them, or from other forms of vibration. As with ground obstructions, bridging across will usually provide the best solution.

Whether or not true bridges are needed for ground-level or underground obstructions will depend on the circumstances. Often the use of standard steel sections or timber sections will readily provide adequate protection.

Space Allocation

Construction projects often occupy the entire site, or a very high percentage of the area, to provide construction work area for operation at or near ground level. In planning the project, the contractor *must* consider the sequence of operation to determine the method of creating space for contractor and subcontractor offices; workers sheds; trailers and materials storage; unloading area; first-aid station; and sanitary facilities.

Planning also should aim to restore space, at or near ground level, upon completion of construction work.

8.7 Boundary Considerations

Conduct a preconstruction survey using video and still photographs to identify the construction worksite boundaries and mark adjacent properties and structures for protection. In addition, provide each property owner with a copy of the survey and have each owner sign an affidavit acknowledging the results of the survey.

Adjacent Buildings

The contractor *must* provide protection for adjacent buildings and structures at all times for the duration of construction—especially from adjacent excavations. Of particular importance is the protection of historic facades, with new interior construction.

Protective measures and temporary construction *must* be mutually agreed upon, in writing, by both property owners, and include shoring, temporary walls, and permanent common walls.

Consider the height of adjacent buildings for crane swing clearances. Crane overswing on adjoining property is considered trespassing, and as such requires obtaining the adjoining owner's consent. Installation of protective fans or nets may be required.

Trees

Tall trees, adjacent to the site boundary, create obstacles that may inhibit construction operations and impinge on the location of work facilities. In particular, the height of trees may limit crane overswing.

If access is needed past trees at the boundaries, or just outside, protection against damage is a likely requirement. Careful trimming and fencing around the base may be all that is required, with the adjacent property owner's agreement.

Provide minimum protective distances for trees to be retained, using fencing at suitable distance around the trees. Make plans for removing trees that are not to be saved.

Railway Lines

Special requirements normally govern the use of cranes adjacent to railway property, as well as protecting the railway from falling objects. Railway companies or local transportation authorities have a legal duty to protect the traveling public, and thus make and enforce rules about what goes on land adjacent to their operational property.

Contractors *must* protect the tracks and property of a railroad, and are responsible for any damage to a railroad's roadbed, tracks, trains and equipment, catenary wires and supports, and any other railroad property. The contractor *must* take protective measures necessary to prevent any material, equipment, or debris from falling on railroad tracks or equipment operating on the tracks.

The contractor *must* pay the railroad for its replacing or repairing any damage to railroad's facilities or trains resulting from the contractor's operations.

Drainage and Watercourses

Maintenance of a well-drained site is of prime importance. In the initial site inspection carefully examine the state of the ground and, in particular, any existing watercourses:

- If water courses exist, is there any local knowledge in relation to liability or flood at certain times of the year? If so, what has been the worst extent?
- If no permanent diversions are shown on the drawings, will any temporary diversions be needed? If so, what will be the likely size and extent?
- If no diversions are specified, can access be maintained to all parts of the worksite? Or will temporary bridging of any kind be necessary?
- Will water runoff ditches be needed to avoid washout danger to falsework or other temporary foundations, or to protect excavations from flooding?

Trespass

Possible encroachments on adjacent properties include:

- Crane overswings
- Ground anchors for boundary retaining walls on excavations
- Right-of-way passage to inaccessible areas of the construction site
- Roadways with sewer line beneath them.

8.8 Public Safety

Safety, health, and noise exposure during construction not only includes the protection of workers and the general public, but also the adjacent public and private properties. Make sure to review EPA, OSHA, and local ordinances to protect the public outside the construction site from risks to safety and health.

Health

Proper control of dust and disposal of debris, garbage, waste materials, and water *MUST* be provided on the construction site.

Noise

Control of and protection from noise generated by construction activities *must* be provided not only for those working on the site, but also the public outside the site area. Meeting any noise level restrictions covered by OSHA requirements or imposed by local authorities may require temporary works in the shape of baffles, shelters, or other noise-reducing measures.

Several steps can be taken at the construction site to reduce the level of noise emanating from the site. A construction fence made with plywood is the single most effective noise barrier, providing excellent noise attenuation. Noise attenuation is a function of material and mass, therefore the heavier the plywood the more effective the noise barrier. The height of the fence is also important in reducing the amount of noise that leaves the site. As a practical matter, fences over 12 ft (3.8 m) in height pose stability and wind problems (although with proper engineering anything is possible).

Special enclosures around excessively noisy machines are often appropriate, particularly during alterations inside buildings where noise levels can reach dangerous levels very quickly. The use of plywood panels sandwiched with insulation is an effective enclosure; however, because plywood is combustible, where fireproof materials are required, prefabricated fireproof shelters and enclosures *must* be used that provide noise reductions of 25 dB to 40 dB.

Sound absorbers may also be used to temporarily reduce the construction noise leaving an enclosed area. Two types include: a cylinder of absorbent material, the size of a 50-gal (0.2 m^3) drum designed to be hung from the ceiling; and a sound-absorbent panel for wall and ceiling mounting.

Pedestrian Protection

The contractor must perform construction operations so as *not* to expose the public to risks of safety or health, including installing of warning signs, barriers, and protective structures to prevent objects from falling outside site boundary onto passers-by, adjacent school yards, parking facilities, or other areas where large numbers of people are likely to congregate at regular intervals.

Required pedestrian protection *must* be maintained in place and kept in good condition for the entire length of time pedestrians may be endangered. Upon completion of construction activity, walkways, debris, and other obstructions *must* be removed, leaving public property in as good a condition as it was before such work was commenced.

- *Signs* to direct pedestrian traffic safely around the construction site and for protection *must* be provided during construction and demolition activities, as required by OSHA 1926:.
- *Walkways* for pedestrian travel *must* be provided in front of every construction and demolition site (unless the appropriate authority authorizes the sidewalk to be fenced or closed). Walkways *must* be of sufficient width to accommodate pedestrian traffic, but *not less* than 4 ft (1.2 m) wide; *must* be provided with a durable walking surface; and *must* be designed to support all imposed loads, having a live load design *not less* than 150 psi (7.2 kN/m^2).
- *Construction railings* at least 42 in. (1.1 m) in height *must* be sufficient to direct pedestrians around construction areas.
- *Covered walkways must* have a minimum height of 8 ft (2.44 m), as measured from the floor surface to the canopy overhead;

have adequate lighting at all times; and be designed to support all imposed loads with live load design *not* less than 150 psi (7.2 kN/m^2). Openings in the overhead structure are prohibited.

- *Barriers* at least 8 ft (2.44 m) in height *must* be placed on the side of the walkway nearest the construction and extend the entire length of the construction site, with openings in the barriers protected by doors that are normally kept closed.
- *Directional barricades must* be used to protect pedestrian traffic where the walkway extends into the street; barricade *must* be of sufficient size and construction to direct vehicular traffic away from the pedestrian path.
- *Catch platforms* or netting, installed below the construction deck, *must* be used to protect the public from material inadvertently falling from high-rise structures during construction; with the platform or netting raised as construction progresses, so as not to be more than three or four floors below the construction deck.
- *Scaffolds must* be covered with wire mesh, and toe boards and handrails *must* always be in place when working over the sidewalk to prevent materials, tools, and hardware from falling through the scaffold planking; scaffolds on the side of a building *must* be secured to the building at every 30 ft (9 m) horizontally and 26 ft (8 m) vertically and should be checked daily to make sure that any ties, removed to facilitate work, have been replaced.

8.9 Property Protection

Adjoining public and private property *must* be protected from damage—and water runoff and erosion *must* be controlled—during construction and demolition work.

- *Streets, alleys, and public property.* Storing and handling of materials or of equipment on streets or public property *must* comply with the local building regulations; and meet the satisfaction of appropriate authorities.
- *Storage and placement.* Construction equipment and materials *must* be stored and placed so as not to endanger the public, the workers, or adjoining property for the duration of the construction project. Materials, equipment, fences, sheds, or any other structure:
  - *must not* be placed or stored so as to obstruct access to fire hydrants, standpipes, fire or police alarm boxes, utility poles, catch basins, or manholes; or so as to interfere with

the passage of water in the gutter; such fixtures *must* be protected during the progress of work, but sight of them *must not* be obstructed.

- *must not* be located within 20 ft (6.1 m) of a street intersection, or placed so as to obstruct normal observations of traffic signals or to hinder the use of public transit loading platforms.

- *Unobstructed areas.* Maintain, at all times, required exits, existing structural elements, fire protection devices, and sanitary safeguards during alterations, repairs, or additions to any building or structure.
- *Fire extinguishers.* All structures under construction, alteration, or demolition *must* be provided with not less than one approved portable fire extinguisher at each stairway, of all floor levels where combustible materials have accumulated; and in every storage and construction shed. Additional approved portable fire extinguishers *must* be provided where special hazards exists, such as flammable or combustible liquid storage areas.
- *Water supply.* Temporary or permanent water supply *must* be made for fire protection; additional fire protection safeguards *must* be made as outlined in Chapter 29, *NFPA Fire Prevention Code 1* during construction.
- *Exit maintenance.* Required exits *must* be maintained at all times during alterations and additions to any building; where an existing building exceeds 50 ft (1520 m) in height is altered, at least one temporary lighted stairway *must* be provided, unless one or more of the permanent stairways are available for egress as construction progresses.
- *Waste disposal.* Removal of waste materials from the construction site *must* be made in a manner that prevents injury or damage to persons, adjoining properties, and public right-of-way.
- *Sanitary facilities.* Sanitary facilities *must* be provided during construction or demolition activities in accordance with the *International Plumbing Code.*

CHAPTER 9

Support Systems

9.1 Sanitary Facilities

The contractor *must* provide an adequate supply of potable water on the construction site and portable containers to dispense drinking water that are capable of being tightly closed and equipped with a tap (*OSHA 1926.51 Sanitation*).

- Water *must not* be dipped from containers.
- Common drinking cup is *prohibited;* where single service cups (to be used but once) are supplied, both a sanitary container for the unused cups and a receptacle for disposing of the used cups *must* be provided.
- Any container used to distribute drinking water *must* be clearly marked as to the nature of its contents and *not* used for any other purpose.
- Nonpotable water, such as water used for construction work or firefighting purposes only *must* be identified by signs that meet OSHA requirements to indicate clearly that the water is unsafe and is not to be used for drinking, washing, or cooking purposes.
- Toilets *must* be provided for employees, according to *OSHA 1926.51 Sanitation Regulations* (Table 9.1).

9.2 Medical Services and First Aid

For enabling first aid to be rendered to construction workers, if they are injured or become ill at work, the employer:

- *Must* provide, or ensure, that such equipment and facilities are available—as are adequate and appropriate in the circumstances.
- *Must* ensure first-aid supplies, approved by the consulting physician, are readily available on the worksite.

Number of employees	Minimum number of facilities
20 or less	1
20 or more	1 toilet seat and 1 urinal per 40 workers
200 or more	1 toilet seat and 1 urinal per 50 workers

TABLE 9.1 Minimum Number of Toilet Facilities per Employee

- *Must* provide a suitable number of persons, for rendering first aid to the employees—as are adequate and appropriate in the circumstances. A person shall not be suitable unless he or she has undergone such training and has such qualifications as the Health and Safety Executive may approve for the time being in respect of that case; with such additional training as may be appropriate in the circumstances of that case.
- *Must* make an assessment of first-aid needs appropriate to the circumstances of each workplace. OSHA, however, does not fix the level for the provision of first aid; each employer *must* assess what facilities and personnel are appropriate.
- *Must* conspicuously POST the telephone numbers of the local physicians, hospitals, or ambulances.

9.3 Equipment and Materials Storage

Where possible, appropriate equipment and materials storage areas (and equipment maintenance areas) should be identified before any construction work is done on the site. Most important is to provide diversions to keep runoff from entering storage or maintenance areas.

Equipment storage areas that properly control runoff will prevent oil, grease, solvents, hydraulic fluids, sediment, wash water, and other pollutants from being carried off the area and entering surface waters; and prevent pollutants from filtering into the ground.

Equipment

The maintenance, repair, cleaning, and storage of construction machinery, vehicles, and equipment should be confined to areas specifically designed and designated for that purpose. This practice includes both open and covered equipment maintenance and storage areas.

Where possible, maintenance areas should be placed on flat areas to prevent surface runoff from entering or leaving the area. Most important is controlling runoff from both kinds of storage, not only from construction sites but also existing permanent storage facilities.

If equipment is to be maintained and stored in an open area (temporary storage), consider the following:

- Site should *not* be within the drip line of trees.
- Site should *not* be within 100 ft of a watercourse or wetland; runoff should be diverted away from watercourses and wetlands.

If equipment is to be maintained and stored in a permanent structure (building), consider the following:

- Building should *not* be located within 100 ft of a watercourse.
- Building should *not* be constructed, when possible, within 100 ft of a wetland.

Under *no* circumstance should buildings or equipment be located in floodplains, stream beds, or the channel of any watercourse.

Outside equipment/maintenance storage areas should be inspected daily to ensure equipment is not being stored within the drip line of trees, and to ensure the vehicles and equipment are not leaking. Also, make sure waste materials are being properly disposed of. Periodic checks of the equipment wash area should also be done to ensure it is not failing.

Ongoing maintenance of structural equipment maintenance/storage areas should include periodic inspections of the structure to check for cracks in the floor, and for other structural flaws. In existing building make sure to inspect the floor drains to make sure they are not discharging to storm drains.

Building Storage

All floors of an equipment storage building should be constructed of concrete or other impervious materials to prevent contaminants from leaching into the soils or goundwater.

All floor drains that discharge to storm sewers *must* be sealed, or plugged; new floor drains should discharge to a sanitary sewer. Equipment wash areas *must* discharge into a sanitary sewer line.

Depending on the amount of oil, grease, and other pollutants, pretreatment of wastewater may be needed before it enters the sanitary sewer.

Trucks and other equipment with large quantities of mud should be washed outside on a designated wash area, so as not to clog sanitary sewer lines.

Open Area Storage

All maintenance, ideally, should be done on impervious areas surrounded with impervious berms. Where this is not possible, use pads designed to contain the pollutants that may leak or spill during maintenance operations—especially on sandy and other coarse soils where spilled materials can easily leach into the groundwater.

Wash areas should be constructed out of 2-in. to 3-in. stone or other approved material, with a minimum 6-in. base. They should be underlain with geotextile materials, and protected using berms or diversions to prevent the runoff water from leaving the site.

Materials

All construction materials stored in tiers *must* be stacked, racked, blocked, interlocked, or otherwise secured to prevent sliding, falling, or collapse.

Stored materials *must not* create a hazard. Storage areas *must* be kept free from accumulated materials that may cause tripping, fires, or explosions, or that may contribute to the harboring of rats and other pests.

When stacking and piling materials, it is important to be aware of factors such as materials' height and weight, how accessible the stored materials are to the user, and the condition of any containers where the materials are being stored. Height limitations *must* be observed when stacking material.

Lumber. Lumber *must* be stacked and leveled on solidly supported bracing. If handled manually, lumber *must* not be stacked higher than 16 ft; if a forklift is used, 20 ft is the maximum stacking height. For quick reference, paint stripes on walls or posts to indicate maximum stacking heights.

Used lumber. All nails *must* be removed before stacking.

Bricks. Stacks of loose bricks *must* be tapered back 2 in. for every foot of height above the 4-ft level.

Masonry blocks. When stacked higher than 6 ft, masonry blocks should be tapered back one-half block for each tier above the 6-ft level.

Bags and bundles. These *must* be stacked in interlocking rows to remain secure. Bagged material *must* be stacked by stepping back the layers and cross-keying the bags at least every ten layers. To remove bags from the stack, start from the top row first.

Bound material. Bound material should be stacked, placed on racks, blocked, interlocked, or otherwise secured to prevent it from sliding, falling, or collapsing. Loads greater than that approved by a building official *must not* be placed on any floor of a building or other structure. Where applicable, load limits approved by the building inspector should be conspicuously posted in all storage areas.

Drums, barrels, and kegs. These *must* be stacked symmetrically. If stored on their sides, the bottom tier *must* be blocked to keep them from rolling. When stacked on end, put planks, sheets of plywood dunnage, or pallets between each tier to make a firm, flat stacking surface.

Structural steel, bar stock, poles, and other cylindrical materials. Unless on racks, these materials *must* be stacked and blocked to prevent spreading or tilting. Pipes and bars *must not* be stored in racks that face main aisles, this could create a hazard to passers-by when supplies are being removed. When stacking any material two or more tiers high, the bottom tier *must* be chocked on each side to prevent shifting in either direction. Material that cannot be stacked due to size, shape, or fragility can be safely stored on shelves or bins.

Flammable/combustible materials. Almost every construction site will store a certain amount of flammable (ignition below 140°F [60°C]) or combustible (ignition above 140°F [60°C]) liquids, and liquid petroleum gas. Flammable and combustible liquids *must never* be stored in hallways, on stairs, or at exits. OSHA regulations permit storage of up to 25 gal (0.1 m^3) of flammable or combustible liquids within a building without special storage cabinets. Quantities above 25 gal (0.1 m^3) must be stored in an approved metal cabinet. A given storage cabinet *cannot* have over 60 gal (0.22 m^3) of flammable liquids or 120 gal (0.45 m^3) of combustible liquids. Containers or cylinders of liquified petroleum gases, such as butane and propane, *must never* be stored inside the building (Table 9.2).

Proper use of temporary heat at the construction site is regulated by considerable state and local law as well as by detailed OSHA regulations. Strict compliance is important for safety reasons as well as legal reasons. Serious fires due in part to violations of fire regulations fall in the criminal negligence area. Inspections for compliance with fire regulations by various agencies are quite frequent, and noncompliance tends to be quite obvious.

Approved fire extinguishers *must always* be available within 50 ft (15 m) of stored flammables. (Local fire departments have specific regulations on the approved number, type, and size of extinguishers for their jurisdictions.) *Never* use water-type extinguishers on flammable or combustible liquids and electrical fires; only foam, carbon dioxide, and dry chemicals are suitable and effective.

Quantity of LP gas stored		Distance from nearest building	
lb	**kg**	**ft**	**m**
500 or less	228	0	0
501 to 6000	229 to 2736	10	3
6001 to 10,000	2737 to 4560	20	6
Over 10,000	Over 4560	25	7.6

Table 9.2 Storage Regulations for LP Gas in the Vicinity of Buildings

Do not use temporary LP gas heaters without first consulting the local fire department on the various regulations for such heaters, due to the ever-present danger of explosion. All temporary heaters *must* be kept at least 10 ft (3 m) from tarpaulins to avoid being turned over by a thrashing tarpaulin in high winds.

OSHA standards require proper ventilation for solid fuel salamanders using coke or coal only.

9.4 Storage Area Checklist[1]

- Direct runoff from equipment maintenance/storage areas to stabilized outlets designed to assimilate the volume and type of pollutants discharged to them.
- Keep heavy equipment well maintained to prevent leaks.
- *Do not* wash vehicles and other equipment at locations where the runoff will flow directly into a watercourse of storm sewer.
- Store, cover, and isolate construction materials, including topsoil and chemicals, to prevent runoff of pollutants and contamination of groundwater.
- Develop a spill response plan, including procedures that *must* be taken in case of a spill.
- Develop a waste management plan that includes daily disposal of canisters or other chemical containers (such as from hydraulic fluids, etc.), scrap wood, scrap metal, and all other waste materials; or keep in sealed waste containers until they can be disposed of off-site in a landfill.
- *Do not* bury waste materials on-site.
- Designate and maintain specific areas for employee parking.

9.5 Communications Systems

A proper communications system *must* be made up of strategically placed equipment and properly defined responsibilities. The emergency response plan *must* posted in a conspicuous place on the project identifying the designated equipment and the people to operate it.

Administrating and organizing the plan is vital to its effectiveness. Thus, the person who has this task will normally be the person in charge of the emergency response operation and will be responsible to ensure that:

- Everyone clearly understands their roles and responsibilities within the emergency response plan.

---

[1]*Source:* Michigan Department of Natural Resources.

- Emergency resources, whether people or equipment, are kept at adequate levels in step with the progress of the project.

An important key to effective emergency response is a communications system that can relay accurate information quickly. This requires using reliable communications equipment, developing procedures, and training personnel as well as providing a backup system in case the system is rendered useless by the emergency, such as telephone lines that may be cut.

Post the type and location of emergency communication systems on the project, including location of telephones, a list of site personnel with cellular phones or two-way radios, and any other equipment available.

Clearly mark emergency phone numbers, as well as the site address/location, for:

- Fire
- Police
- Ambulance
- Utilities (electric and water)

Data Exchange

The main communications issue in construction is to provide a method to exchange data between the field operation and the field office, normally located in a trailer within close proximity of the field operation. The foreman or supervisor in the field needs to be free to roam within the boundaries of field operations. The information needs under consideration are time-critical to assist in maintaining or improving the efficiency at the jobsite.

Construction site communications can take place through different media and technologies, including satellite, terrestrial, mobile, wireless, and wired systems (Table 9.3). When choosing an effective communication system, consider:

- Benefits, in terms of the nature of information that needs to be transmitted, including voice, data, video, web collaboration, etc.
- Access to all members of the project teams, including workers, foremen, and superintendent
- Reliability, availability, and quality of service
- Cost of service, including network, administration, maintenance, and upgrade
- Availability of terminals and users' devices, such as mobile handsets

Technology	Advantages	Drawbacks
Cellular/Mobile *GSM:* Global Systems for mobile communications	• Voice/Data transfer • Familiar technologies • Instant availability • Roaming • Miniaturized terminals • Well-established	• Full dependence on the service provider • Communications costs are difficult to control
DEC – Digital Enhanced Cordless Telecommunications standard that provides a general radio access for wireless communications operating in the 1880 to 1900 MHz band	• Voice/data transfer • No license required • High-speed data up to 552 kbits/s with migration to up to 2 Mbits/s • High reliability through reserved frequency band resulting in no interference problems for industrial applications	• Limited coverage
VPN – Virtual Private Network Consists of registering the mobile phone within the same service provider that provides the VNP at a pre-agreed rate	• Voice/Data transfer • Cost control • Communications with other networks	• Full dependence on service provider for coverage and quality of service • Costs are governed by length of calls
WLAN – Wireless Local Area Network in which a mobile user can connect to a LAN through a wireless (radio) connection	• Voice/data transfer • Unlimited *free* voice calls between devices • External communications cost-control • Well-established technology • High bandwidth data transfer	• VoIP quality of service issues • WLAN IP phones are mainly made for indoor use • Short coverage perimeters may involve higher number Network Access Points

PMR – Private Mobile Radio Broadly covers all forms of two-way radio system	• Well-established technology • Coverage up to 15 miles in ideal conditions	• No data transfer • Short-length conversations • Internal communications only • Bulky handsets
Advanced PMR Interfaces to the PSTN to provide access to external voice communications	• Coverage up to 15 miles in ideal conditions • Voice/data supported	• Limited data transfer • Bulky handsets • Number of simultaneous calls limited by the number of channels available
LEO – Low Earth Orbit Satellite	• Universal coverage • Voice/data supported • Quick deployment for services	• Very expensive per minute calls and subscriptions • Bulky handsets
GEO – Geostationary Earth Orbit Satellite Requires a satellite receiver to provide a direct connection to the Geostationary Satellite and enable information to be passed onwards; WLAN is also required to connect to personnel on construction sites	• Good coverage with the satellite footprint • Voice/data supported • Quick deployment • Supports high level encryption • True broadband is achievable delivering high speed "Always On" access to the Internet	• Voice transmission delay • Coverage issues inside • Communications is not always reliable; however, some providers to the geostationary offer a Service Level Agreement

Source: Sebastian Beyh & Mike Kangioglou, University of Salford, U.K.

TABLE 9.3 Various Communications Systems Used in Construction

9.6 Emergency Response Plan

Planning for emergency response *must* begin before any work commences on the project. A good emergency response plan can be generic and, with some minor changes, can be easily adapted to specific sites and readily implemented. Development of the plan should include:

- Hazard identification/assessment of emergency resources
- Communications systems
- Administration response procedure
- Communication of the procedure
- Debriefing and posttraumatic stress procedure

The plan should be used to set emergency procedures, implement and communicate the procedures, and ensure that any required training has been completed. The plan should also be evaluated regularly to ensure that it conforms to current operations and conditions.

It is very important to review the emergency plan on a regular basis, especially after an emergency has occurred. Changes may be necessary where deficiencies became apparent as the plan went into operation.

To be effective, an Emergency Response Procedure must be clearly communicated to all site personnel, and the procedure *must* be posted in a conspicuous location.

Review the procedure with:

- New site subcontractors and new workers to ensure that it covers their activities adequately
- Suppliers to ensure that it covers any hazards that the storage or delivery of their materials might create
- Owner/client, on any new work areas in operating plants, to ensure that new hazards are identical and covered in the procedure
- Construction team members on a regular basis to address new hazards or significant changes in the conditions

Response Procedure[2]

The emergency response procedure for a construction project must continually undergo review and revision to meet changing conditions. Recovery process, or what happens after the emergency response has been completed, is a critical step in the plan.

Many emergency tasks may be handled by people who are not accustomed to dealing with emergencies; they may have seen their

[2] *Source:* Construction Safety Association of Ontario.

work partners and friends badly injured and suffering great pain. Once the emergency is over, some of the people involved may need assistance in order to recover. And, in some cases, professional counseling may be needed.

As part of the site emergency planning, construction companies should have measures in place to deal with posttraumatic stress. For more information, refer to local hospitals, ambulance services, and medical practitioners that may be able to help.

Debriefing is necessary to review how well the plan worked in the emergency and to correct any deficiencies that were identified. Debriefing is critical to the success of future emergency response planning.

9.7 Security Systems

Construction sites are vulnerable to theft and vandalism; theft of machinery, pilfering or removal of materials; and damage to property. Before the project begins, create a site security plan. Consider the following:

- How to secure jobsite perimeter
- What type of lighting is needed
- Where to place video surveillance
- How to secure valuable equipment and materials
- Whether to employ on-site security

To combat equipment theft, Liberty Mutual Insurance's Loss Prevention Department recommends permanently etching an identifying mark on the equipment (spray paint and initials on a piece of equipment, however, *does not* qualify as being "positively" identified). Liberty also suggests asking for identification from drivers before loading equipment onto low boys, as well as removing keys and securing the jobsite at the end of the day.

Equipment should also be immobilized when it is not in use by removing rotors, lowering blades and buckets, and disabling batteries and electric starting systems. If mobilization is not possible, then equipment should be parked in a "wagon train" formation, using larger equipment to protect small equipment. Using front-end loaders and forklifts to block trailer entrances can also help protect equipment on site.

To help prevent theft of office equipment from jobsite offices, position the site or trailer office in an area with limited access to the opportunist thief, yet highly visible to the public. Expensive pieces of office equipment should be well marked with indelible markings.

Valuable business information such as payroll figures, work schedules and future ventures, or anything that could undermine a company's strategy if it were lost or put in the hands of competition should also be recognized as a potential target, and protected accordingly.

Site Lighting

Use high-efficiency security lighting and motion sensor controls on construction sites according to energy conservation standards specified by U.S. Department of Energy–*ANSI/ASHRAE/IESNA Standard 90-1-2004 – Energy Standard for Buildings (Except Low-Rise Residential Building)* – which includes specific lighting power limits for a variety of exterior lighting applications.

Outdoor lighting should average 65 lumens per watt, and be designed so that only lighting necessary for security, such as at site gates and trailers, operates through the night using light-sensitive controls. Control other site security lighting using motion detectors. Motion detectors, however, *must* be carefully placed and tested for best security and to avoid disturbing the neighbors.

Construction areas, ramps, runways, corridors, offices, shops, and storage areas *must* be lighted to *not* less than minimum illumination intensity standards while any work is in progress. OSHA regulations, derived from *ANSI A11.1-1965; R1970,* contain very specific requirements for illumination of construction site work areas (Table 9.4).

Footcandles	Area of operation
5	General construction area lighting
3	General construction areas, concrete placement, excavation, waste areas, accessways, active storage areas, loading platforms, refueling, and field maintenance areas
5	Indoor warehouses, corridors, hallways, and exitways
5	Tunnels, shafts, and general underground work areas. (Exception: minimum of 10 footcandles required at tunnel and shaft headings during drilling, mucking, and scaling. Bureau of Mines–approved cap lights shall be acceptable for use in the tunnel lining.)
10	General construction plant and shops such as batch plants, screening plants, mechanical and electrical equipment rooms, carpenters shops, rigging lofts, and active store rooms, barracks, or living quarters, locker or dressing rooms, mess halls, indoor toilets, and workrooms
30	First-aid stations, infirmaries, and offices

TABLE 9.4 Minimum Illumination Intensities on Construction Worksites

For areas of operation not covered by these rules, refer to *ANSI A11.1-1965; R 1970 – Practice for Industrial Lighting for Recommended Values of Illumination.*

Although use of high-intensity discharge (HID) lamps provides temporary increased illumination on construction sites to perform work tasks, it also increases the risk of eye and facial injuries from exposure to ultraviolet (UV) radiation. NECA has issued a warning about eye burn hazard from exposure to HID lighting, and certain precautions must be taken when using such lamps.

It is generally recommended to mount HID fixtures at a high level in the work area to minimize employee exposure, and to turn off HID fixtures when people are working in an elevated position near them.

- Use only self-extinguishing HID lamps or luminaires that protect the lamp with a glass or plastic lens.
- Provide additional guarding such as metal screens under the HID lamp to protect it against breakage.
- Use HID luminaires only in high ceiling areas.
- If elevated work is to be performed, extinguish the HID luminaire and substitute alternative lighting sources (such as higher wattage fluorescent light bulbs).
- *Do not* field-modify HID luminaires; doing so increases the contractor's liability in the event of an incident.
- Use frosted lamps rather than clear ones; frosted lamps reduce the glare but don't eliminate it. Looking at the lamp could result in possible eye damage.

Fencing

Temporary/security fencing *must* be provided when:

- Construction or demolition (or both) occurs on any site visible from a public or private street
- Required in accordance with OSHA and local building codes to prevent unauthorized persons gaining access to the construction site, where they could be exposed to hazards such as building debris, exposed steel reinforcing, trenching, and excavation

During nonworking hours, the main groups exposed to hazards on construction sites are children and adolescents. Although a fence will not always stop those who are determined to gain access to a construction site (children and thieves), it will discourage most people and may demonstrate that reasonably practical steps have been taken to meet contractor responsibilities under OSHA regulations.

Levels of risk may be categorized as:

- *High risk.* Areas that the public would normally access or pass through, such as inner city areas, main business areas, and education institutions. For such sites, it is necessary to prevent public access to the worksite.
- *Medium risk.* Areas having moderate public traffic, such as suburban residential areas. For such sites, it is necessary to restrict access to the worksite, and provide warning of the hazard.
- *Low risk.* Areas having minimal public access, such as rural areas or new subdivisions.

A building permit is required to erect a security fence. The contractor should submit, with the permit application, a site plan (drawn to scale) showing abutting streets, lot lines, and their dimensions; existing building and their dimensions; the proposed fence location; and proposed setbacks. A drawing or picture of the fence indicating the style and height should also be provided.

The standard *Guide for Design and Construction of Chain Link Fencing (ASTM 2166)* provides design and security engineers all the proper ASTM chain link fence material and system installation and layout options to design a security fence that will comply with their specific security requirements—and substantially increase the difficulty of penetration of a construction site.

Alarms

Construction sites are easy targets for thieves. Constant staff turnover means it is difficult to keep good security records, and the high value of plant and equipment can lead to quick and easy profit for successful thieves. A proportion of construction site crime is committed by people working in the industry. This means that great responsibility rests with site supervisors. Make it a company policy to inform everyone that they can go a long way toward reducing the problem.

It is important to install adequate security alarm systems, including boundary alarms, motion detectors, and area sensors. In addition, consider using remotely monitored IP CCTV systems, and covert CCTV, which can serve to:

Replace static security guards (up to 70% more cost effective)

Provide security evenings and weekends

Provide security throughout the day while staff and contractors are on site

Record activity on site for health and safety

Record general site progress

Allow on-site activity to be viewed in real-time through any laptop/PC that has an Internet connection and web browser.

Make sure to post warning signs prominently that clearly state both visible and covert security systems are in place throughout the worksite.

9.8 Security Checklist

Security	Check
Employees, contractors, security guards	
• Conduct background checks on employees, and check references	☐
• Check out performance and references of subcontractor and security guards prior to hiring	☐
Perimeter fencing, gates, and locks	
• Limit number of access points onto construction site; where feasible, situate in locations highly visible from adjacent streets	☐
• Install fencing around site perimeter adjacent to streets • *Do not* cover fencing for 20 ft on either side of gates	☐
• Install, where possible, barbed wire on top of and through the fencing	☐

• Secure vehicle and pedestrian gates and openings after hours of operation
  • Use padlocks for securing gates designed to prohibit cutting of shackle
  • Secure locking gates with minimum 3/8-in. thick, 30 grade coil chain, or 5/16-in. cable
	☐
	
	☐

Signage	
• Post address signs, prominently, at all vehicle entrances listing street name and number using letters and numbers at least 6 in. in height	☐
• Install **NO TRESPASSING** signs listing state penal code at all perimeter access points	☐
• Post signs, at the main entrance gate and on the exterior of on-site office trailer or building, with a 24-hr emergency number for a person who can respond to site	☐
• Post all signs a minimum of 5 ft from ground level	☐
Site lighting	

• Illuminate construction worksite during hours of darkness, specifically:
  • Vehicle gate locations
  • All open centralized storage areas for construction materials or fixtures
  • Trailers, temporary buildings, or containers used as an office or storage for construction materials or fixtures
	
	☐
	☐
	
	☐

• Maintain minimum light level at required locations according to OSHA regulations (developed from *ANSI A11.1-1965; R 1970*)	☐

• Install light fixtures a minimum 18 ft high, with tempered or poly carbonate lens approved for wet locations *(UL Bulletin 1572)*	☐
• Use motion sensors to control required lighting that would affect occupied residential properties	☐
Office trailers and temporary buildings	
• Install on doors deadbolt locks and auxiliary locking devices that use slide bolt or hasp, with a protective device to prohibit cutting the padlock • Bolt or hasp attached with nonremovable bolts from the exterior and padlock with minimum 1/2-in. thick shackle having heel and toe locking	☐ ☐
• Secure windows with either • Steel bars, minimum 1/2-in. round or 1-in by 1/4-in. flat, spaced not more than 5-in.apart, secured with bolts that are nonremovable from the exterior; or • Steel gate mesh, minimum 1/8-in. thick, securely attached using a means resistant to removal from the exterior	 ☐ ☐
Storage containers (with at least 64 sq. ft. storage area)	
• Secure doors using a hasp or slide bolt with a protective device to prohibit cutting the padlock, and attached with nonremovable bolts from the exterior	☐
• Render exterior hinge pins nonremovable by design or welding	☐
Materials	
• Protect likely targets, such as copper wire, aluminum, and fittings	☐
• Time deliveries for the day of installation, when possible (deliveries late in the week give thieves the entire weekend to steal them	☐
• Hire temporary security if unable to install or secure valuable deliveries	☐
• Have suppliers include serial numbers on sales invoices to increase chances of tracking stolen goods	☐
Tools	
• Make tools easy to identify by engraving them with visible identification numbers; paint them bright, and unique colors	☐
• Lock gang boxes and supply sheds at times with sturdy locks and case-hardened chain	☐
• Make sure all tools are checked-out and checked -in each day; make an individual responsible for the tools	☐
• Require employees to have their personal properly engraved with a unique identification number (such as, driver's license number)	☐

Heavy equipment	
• Make heavy equipment easy to identify; use bright colored paint, engrave with a large company logo, or make heavy equipment with unique color schemes (or both)	☐
• Render moving equipment inoperable when the hours of operation are ceased, making them unable to start, operate, or move	☐
• Remove keys and lock cabs when equipment is not in use	☐
• Disable equipment in any way possible: remove the battery, remove wheels	☐
• Maintain on-site record for all motorized equipment with a minimum 15-in. diameter, listing the equipment manufacturer, model, license plate number, vehicle identification number (VIN), and product identification number (PIN)	☐
Police report	
• File a police report, providing as much information as possible; description of stolen equipment, including brand name, serial number, and markings; who might have been on the site when the theft occurred; and any possible suspect information	☐

Regulations and Standards

OSHA Regulations

1926 Subpart D—*Occupational Health and Environmental Controls*

1926.59 *Hazard Communication*

1926 Subpart G—*Signs, Signals, and Barricades*

1926.200 *Accident Prevention Signs and Tags*
1926.201 *Signaling*
1926.202 *Barricades*
1926.203 *Definitions Applicable to this Subpart*

1926 Subpart H—*Materials Handling, Storage, Use, and Disposal*

1926.250 *General Requirements for Storage*
1926.251 *Rigging Equipment for Material Handling*
1926.252 *Disposal of Waste Materials*

1926 Subpart P—*Excavations*

1926.650 *Scope, Application, and Definitions Applicable to this Subpart*

1926.651 *Specific Excavation Requirements*

1926.652 *Requirements for Protective Systems*

Appendix B *Sloping and Benching*

Appendix C *Timber Shoring for Trenches*

Appendix D *Aluminum Hydraulic Shoring for Trenches*

Appendix E *Alternatives to Timber Shoring*

Appendix F *Selection of Protective Systems*

U.S. Consensus Standards

ANSI A11 1-1965; R 1970 *Practice for Industrial Lighting for Recommended Values of Illumination*

ASTM F 2611 *Guide for Design and Construction of Chain Link Security Fencing*

UL 1572 *Luminaires for Use in Non-hazardous Locations*

Industry Consensus Standards and Guidelines

American Society of Civil Engineers, Reston, VA *(www.asce.org)*

Associated General Contractors of America, Arlington, VA *(www.agc.org)*

American Lighting Association, Dallas, TX *(www.americanlightingassoc.com)*

National Electrical Contractors Association, Bethesda, MD *(www.necanet.org)*

National Alarm Association of America, Dayton, OH *(www.naaa.org)*

National Burglar and Fire Alarm Association, Irving, TX *(www.alarm.org)*

American Water Works Association, Denver, CO *(www.awwa.org)*

Water Environment Federation, Alexandria, VA *(www.wef.org)*

SECTION 4

Rigging Systems

CHAPTER 10

Fiber Rope (Cordage)

Steelworkers require tools to hoist and maneuver the steel members into place to erect a structure of any magnitude. These hoisting tools range from uncomplicated devices, such as tripods and gin poles, to more complex mechanisms, such as cranes and motor-powered derricks. Whatever the case, one of the most important components of these hoisting mechanisms is the fiber rope (cordage) or wire rope that must be attached to and hold the load to be hoisted and maneuvered.

Before becoming skilled in the supervision of hoisting devices, a steelworker must first understand the use and maintenance of both fiber ropes and wire ropes.

10.1 Natural Fiber Ropes

These ropes are made with short, overlapping fibers that come from plants and include Manila, hemp, sisal, jute, coil, and cotton. (Hemp and jute, today, are used for industrial purposes, such as textiles, packaging material, burlap, sacks, carpet yarn, felts, padding, and miscellaneous items.)

Each type of natural fiber rope has its own advantages and disadvantages, and varying grades of a particular type or rope are used for different operating conditions. Strength among natural fiber ropes varies considerably depending on such diverse factors as climate, curing, manufacturing, and maintenance.

Manila. Manila is made from the leaf stems of the abaca plant stalk, which belongs to the banana family. Manila is strong and durable; and should be used for operations requiring a dependable rope that stands up under severe use and resists weathering (Table 10.1).

A good grade of Manila rope is light yellow (cream) in color, with a smooth waxy surface, clean, and pliable. Poorer grades of Manila are characterized by varying shades of brown. Only number 1

Nominal size		Manila			
Diameter	Circumference	Linear density,* lb/100 ft	New rope tensile strength,† lb	Safety factor	Working load,‡ lb
3/16	5/8	1.50	406	10	41
1/4	3/4	2.00	540	10	54
5/16	1	2.90	900	10	90
3/8	1 1/8	4.10	1,220	10	122
7/16	1 1/4	5.25	1,580	9	176
1/2	1 1/2	7.50	2,380	9	264
9/16	1 3/4	10.4	3,100	8	388
5/8	2	13.3	3,960	8	496
3/4	2 1/4	16.7	4,860	7	695
13/16	2 1/2	19.5	5,850	7	835
7/8	2 3/4	22.4	6,950	7	995
1	3	27.0	8,100	7	1,160
1 1/16	3 1/4	31.2	9,450	7	1,350
1 1/8	3 1/2	36.0	10,800	7	1,540
1 1/4	3 3/4	41.6	12,200	7	1,740
1 5/16	4	47.8	13,500	7	1,930
1 1/2	4 1/2	60.0	16,700	7	2,380
1 5/8	5	74.5	20,200	7	2,880
1 3/4	5 1/2	89.5	23,800	7	3,400
2	6	108.	28,000	7	4,000
2 1/8	6 1/2	125.	32,400	7	4,620
2 1/4	7	146.	37,000	7	5,300
2 1/2	7 1/2	167.	41,800	7	5,950
2 5/8	8	191.	46,800	7	6,700
2 7/8	8 1/2	215.	52,000	7	7,450
3	9	242.	57,500	7	8,200
3 1/4	10	298.	69,500	7	9,950
3 1/2	11	366.	82,000	7	11,700
4	12	434.	94,500	7	13,500

*Linear density (lb/100 ft) shown is average. Maximum is 5% higher.

†New rope tensile strengths are based on tests of new and unused rope of standard construction in accordance with Cordage Institute standard test methods.

‡Working loads are for rope in good condition with appropriate splices, in noncritical applications, and under normal service conditions. Working loads should be exceeded only with expert knowledge of conditions and professional estimates of risk. Working load should be reduced where life, limb, or valuable property is involved, or for exceptional service conditions such as shock loads or sustained loads. Working loads are not necessarily intended to apply in those applications where a thorough engineering analysis of all conditions of use has been made and these conditions will not be exceeded in service. In such cases tensile strength, elongation, energy absorption, behavior under long-term or cyclic loading, and other pertinent properties of a rope may be evaluated to allow the selection of the rope best suited to the requirements of the application and an analysis of any risk involved in its use.

Source: The Cordage Institute.

TABLE 10.1 Manila Rope Strength (Three-Strand Laid and Eight-Strand Plaited, Standard Construction)

Yacht	Extremely light color, high-grade Manila rope of the highest quality; costly and thus used on jobs where appearance is most important
Bolt	Very light-color, high-grade Manila rope; about 10 percent to 15 percent stronger than number 1 grade Manila
Number 1	Very light-color, standard high-grade Manila hoisting rope; specified for most rigging applications
Number 2	Slightly darker Manila rope with about the same initial strength as number 1 grade; loses strength very rapidly through use
Hardware store	Darkest of all Manila ropes; very poor grade with many fiber ends protruding from the strands; has low strength and short life

Table 10.2 Manila Rope Grades

grade Manila fiber rope should be specified for rigging operations (Table 10.2).

The quality of the fiber and its length give Manila line relatively high elasticity, strength, and resistance to wear and deterioration. In many instances, the manufacturer treats the line with chemicals to make it more mildew resistant and thereby increases its quality.

Sisal. The next best fiber rope is a whitish, rather coarse rope made from agave sisalana and henequen. The fiber is similar to Manila, but lighter in color, less strong, and durable. It should be used where rope requirements are less demanding and cost is a major consideration (Table 10.3).

Coir. Coir is made from the fiber of coconut husks and is light enough to float on water. A resilient rough line, it has about one fourth the strength of hemp; therefore, the use of coir is restricted to small lines.

Cotton. Smooth white line that withstands much bending and running.

When purchasing hoisting rope *be sure* to specify that new Manila rope be made of first-grade Manila fibers. Most rope manufacturers identify their first-grade Manila rope by twisting one or more colored yarns, or a colored string, in a strand. Rope without such a fiber marking *should not* be used for hoisting.

Note: *The size of a rope larger than 1¾ in. (44.5 mm) in circumference is generally designated by its circumference in inches. Fiber rope is available up to 16 in. (40 cm) in circumference, but 12 in. (30 cm) is normally the largest size carried in stock. Anything larger is used only on special jobs.*

Nominal size		Sisal			
Diameter	Circumference	Linear density,* lb/100 ft	New rope tensile strength,† lb	Safety factor	Working load,‡ lb
3/16	5/8	1.50	360	10	36
1/4	3/4	2.00	480	10	48
5/16	1	2.90	800	10	80
3/8	1 1/8	4.10	1,080	10	108
7/16	1 1/4	5.26	1,400	9	156
1/2	1 1/2	7.52	2,120	9	236
9/16	1 3/4	10.4	2,760	8	354
5/8	2	13.3	3,520	8	440
3/4	2 1/4	16.7	4,320	7	617
13/16	2 1/2	19.5	5,200	7	743
7/8	2 3/4	22.5	6,160	7	880
1	3	27.0	7,200	7	1,030
1 1/16	3 1/4	31.3	8,400	7	1,200
1 1/8	3 1/2	36.0	9,600	7	1,370
1 1/4	3 3/4	41.7	10,800	7	1,540
1 5/16	4	47.8	12,000	7	1,710
1 1/2	4 1/2	59.9	14,800	7	2,110
1 5/8	5	74.6	18,000	7	2,570
1 3/4	5 1/2	89.3	21,200	7	3,030
2	6	108.	24,800	7	3,540
2 1/8	6 1/2	—	—	7	—
2 1/4	7	146.	32,800	7	4,690
2 1/2	7 1/2	—	—	7	—
2 5/8	8	191.	41,600	7	5,940
2 7/8	8 1/2	—	—	7	—
3	9	242.	51,200	7	7,300
3 1/4	10	299.	61,600	7	8,800
3 1/2	11	—	—	7	—
4	12	435.	84,000	7	13,000

*Linear density (lb/100 ft) shown is average. Maximum is 5% higher.

†New rope tensile strengths are based on tests of new and unused rope of standard construction in accordance with Cordage Institute standard test methods.

‡Working loads are for rope in good condition with appropriate splices, in noncritical applications, and under normal service conditions. Working loads should be exceeded only with expert knowledge of conditions and professional estimates of risk. Working load should be reduced where life, limb, or valuable property is involved, or for exceptional service conditions such as shock loads or sustained loads. Working loads are not necessarily intended to apply in those applications where a thorough engineering analysis of all conditions of use has been made and these conditions will not be exceeded in service. In such cases tensile strength, elongation, energy absorption, behavior under long-term or cyclic loading, and other pertinent properties of a rope may be evaluated to allow the selection of the rope best suited to the requirements of the application and an analysis of any risk involved in its use.

Source: The Cordage Institute.

TABLE 10.3 Sisal Rope Strength (Three-Strand Laid and Eight-Strand Plaited, Standard Construction)

REMEMBER: *It is up to the purchaser to specify the type, size, and construction of rope best suited to the specific job requirements.*

Construction

For a natural fiber rope to function effectively, it must have a closely packed structural form that will remain compact and maintain its cross-sectional dimension throughout its serviceable life. The degree and effectiveness of this compactness will depend on the amount of twist applied, the tension in various components during twisting operations, and the compression that the tubes and dies impart to a rope's components while being twisted (Fig. 10.1).

The construction most commonly encountered in fiber ropes is one in which the fibers are first compacted by a right-hand twist into yarns; the yarns are then compacted by a left-hand twist into strands; and finally the strands are compacted by a right-hand twist

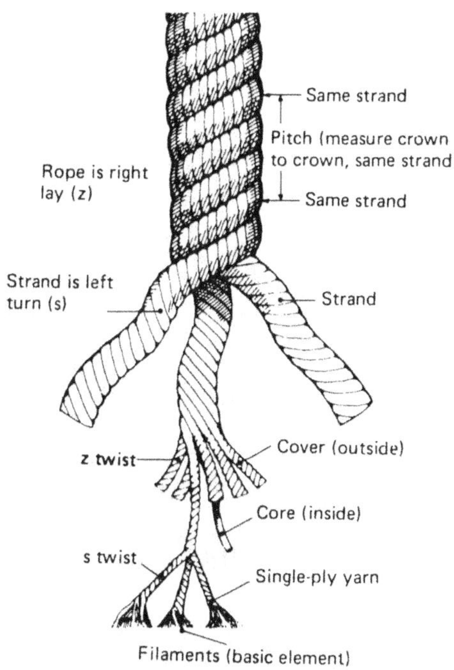

FIGURE 10.1 Fiber rope construction. (a) Three-strand rope; (b) eight-strand plaited rope; (c) double-braided rope. (*Source:* The Cordage Group, Auburn, NY.) (*Continued*)

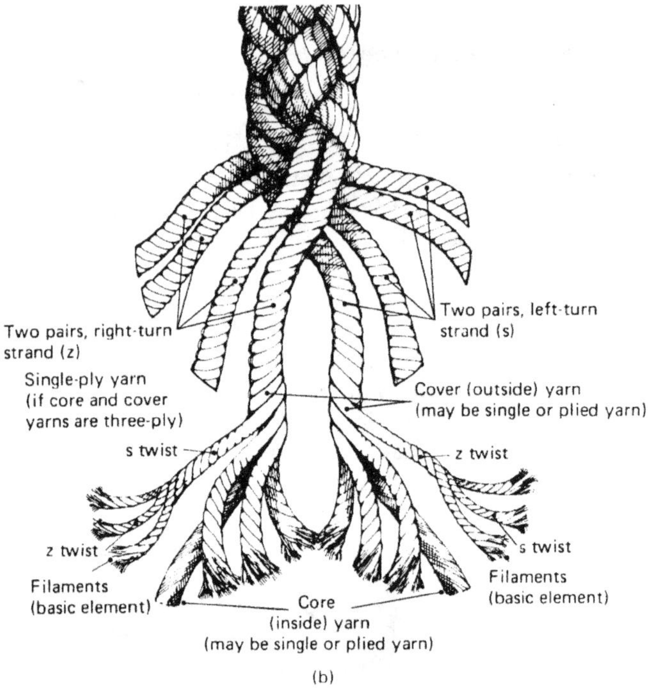

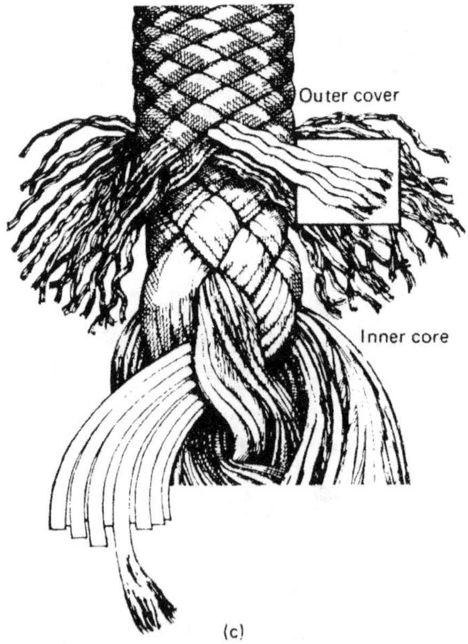

Figure 10.1 Fiber rope construction. (a) Three-strand rope; (b) eight-strand plaited rope; (c) double-braided rope. (*Source:* The Cordage Group, Auburn, NY.) (*Continued*)

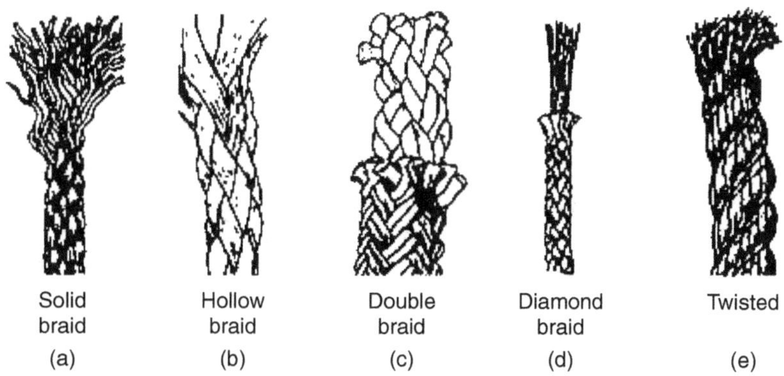

FIGURE 10.2 Fiber rope comparisons. (a) Solid braid; (b) hollow braid; (c) double braid; (d) diamond braid; (e) twisted.

into ropes. Cables are constructed by twisting these ropes in a left-hand direction (Fig. 10.2).

With its components twisted in opposing directions, a fiber rope will maintain its compactness during use, but only as long as the opposing twists remain in equilibrium. Any use of a rope that disrupts this equilibrium will alter the rope's properties. For example, improper coiling and handling of rope removes the turns, eventually inducing strand kinking and thus will reduce a fiber rope's serviceability.

Repeated stretching may alter the rope and its strand-twist relationship; the rope's swelling from wetting has a strong effect on the equilibrium of twists.

Most Manila rope used in rigging operations is of three-strand construction. Where abrasion is the key factor, however, a four-strand rope with a rope core is recommended for work (Fig. 10.3). Such a rope is rounder in cross section and only slightly less strong than the

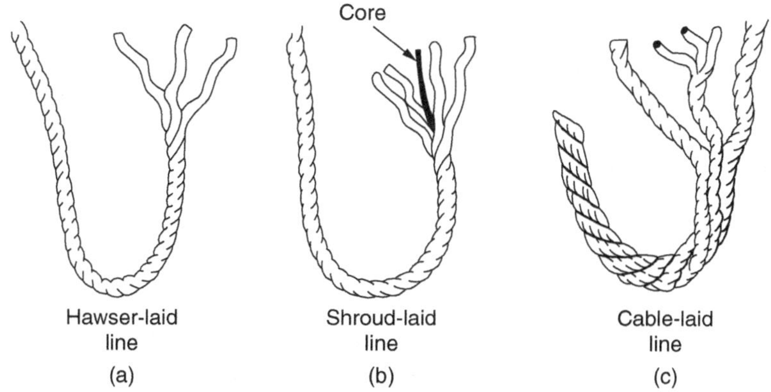

FIGURE 10.3 Types of fiber rope lays.

more standard three-strand rope. It is used almost exclusively for power hoisting.

- *Hawser-laid* consists of three strands twisted together, usually in a right-hand direction.
- *Shroud-laid* is composed of four strands twisted together in a right-hand direction around a center strand or core—usually of the same material, but smaller in diameter than the four strands. A shroud-laid rope is more pliable and stronger than hawser-laid rope. It not only prevents kinking, but also makes use of its pliability and strength when used on sheaves and drums.
- *Cable-laid* usually consists of three right-hand hawser-laid lines twisted together in a left-hand direction. It is epecially useful in heavy construction work because if it tends to untwist, it will tighten any regular right-hand screw connection on which it may be attached, hence its use provides an added safety feature.

10.2 Synthetic Fiber Ropes

Although natural fiber rope is normally used in rigging operations, a number of synthetic fibers are also used. Unlike those constructed of natural fibers, synthetic fiber ropes consist of individual fibers running the entire length of a rope. These longer fibers impart a greater strength to the rope and, as a result, ropes of synthetic fibers such as nylon, polyester, polyethylene, and polypropylene have replaced Manila rope for many uses.

Synthetic fiber ropes cost considerably more than their natural fiber counterparts, but numerous advantages justify the expense. Synthetic fiber ropes are stable under the most adverse conditions; are impervious to rot, mildew, and fungus; and have a high resistance to chemicals.

These ropes also are lighter than natural fiber ropes, more flexible, easier to handle, and have excellent impact and fatigue characteristics. However, they tend to soften at high temperatures. Hence, synthetic fiber ropes should *never* be used under conditions of extreme heat, such as near welding or other open fires, or where high and concentrated friction might cause the fibers to melt.

Synthetic fibers used to fabricate rigging rope are described in the following sections.

Nylon

Nylon has a chalky white texture with a smooth surface is soft and pliant, and feels elastic to the touch. Continuous filaments, running the full length, provide nylon rope with high strength (three times that of Manila rope) and high resistance to creep under sustained loading.

Nylon rope stretches permanently upon initial hoisting. Once this permanent stretch occurs, nylon rope recovers completely from subsequent stretching under load. To compensate for this, all new nylon ropes should be broken in and allowed to stretch before being put into use.

Nylon rope's high resistance to abrasion results from a protective fuzz coating that develops after its outer fibers are initially abraded. Although highly impervious to alkalis, however, nylon fibers are easily weakened by chemicals, most acids, paints, and linseed oil. If contact with chemicals is suspected, wash the nylon rope thoroughly in cold water and carefully examine its fibers to detect any possible damage.

Because nylon fibers absorb moisture readily, nylon rope when wet becomes slippery and loses about 10 percent of its strength. When dry again, nylon rope regains full strength. Among its other characteristics, nylon rope is not susceptible to mildew or rot and thus can be stored wet or dry. Nylon rope retains its physical properties almost to its melting point, permitting use at temperatures up to 300° F without loss of strength (see Table 10.4).

Braided Nylon

Cover over a braided nylon construction provides a soft flexible rope that will not twist or kink, and displays the highest possible strength. With this type of construction, stress from loading is divided equally between the rope's cover and core. Thus, regardless of cover damage, 50 percent of the rope's initial strength still remains in the core.

Polyester (Dacron® and Terylene®)

Although almost the same as nylon in appearance, polyester rope does not nave the elastic feeling of nylon. In comparable sizes, polyester rope is much heavier than nylon, but not as strong.

The same as nylon, polyester rope stretches permanently under initial loading and recovers fully from subsequent loading stretch. Hence, it too must be broken in to allow initial stretching before use. Under normal loading, polyester fiber rope stretches approximately the same as Manila rope. Polyester fiber rope's ability to absorb shock loading is only about two-thirds that of nylon; however, it is considerably more than that of Manila rope.

Polyester has an abrasion resistance similar to that of nylon, and can be used under similar hot working condition—up to 330°F (148.9°C) for long periods—without loss of strength. Unlike nylon, however, polyester fiber rope absorbs almost no water and thus does not lose its strength when wet.

It is highly resistant to damage from sun and weather, mildew and dry rot, acids, and alkalines, but its fiber can be damaged by contact with chemicals. Polyester fiber slings should be washed frequently in cold water (see Table 10.5).

Nominal size		Nylon			
Diameter	Circumference	Linear density,* lb/100 ft	New rope tensile strength,† lb	Safety factor	Working load,‡ lb
3/16	5/8	1.00	900	12	75
1/4	3/4	1.50	1,490	12	124
5/16	1	2.50	2,300	12	192
3/8	1 1/8	3.50	3,340	12	278
7/16	1 1/4	5.00	4,500	11	410
1/2	1 1/2	6.50	5,750	11	525
9/16	1 3/4	8.15	7,200	10	720
5/8	2	10.5	9,350	10	935
3/4	2 1/4	14.5	12,800	9	1,420
13/16	2 1/2	17.0	15,300	9	1,700
7/8	2 3/4	20.0	18,000	9	2,000
1	3	26.4	22,600	9	2,520
1 1/16	3 1/4	29.0	26,000	9	2,880
1 1/8	3 1/2	34.0	29,800	9	3,320
1 1/4	3 3/4	40.0	33,800	9	3,760
1 5/16	4	45.0	38,800	9	4,320
1 1/2	4 1/2	55.0	47,800	9	5,320
1 5/8	5	66.5	58,500	9	6,500
1 3/4	5 1/2	83.0	70,000	9	7,800
2	6	95.0	83,000	9	9,200
2 1/8	6 1/2	109.	95,500	9	10,600
2 1/4	7	129.	113,000	9	12,600
2 1/2	7 1/2	149.	126,000	9	14,000
2 5/8	8	168.	146,000	9	16,200
2 7/8	8 1/2	189.	162,000	9	18,000
3	9	210.	180,000	9	20,000
3 1/4	10	264.	226,000	9	25,200
3 1/2	11	312.	270,000	9	30,000
4	12	380.	324,000	9	36,000

*Linear density (lb/100 ft) shown is average. Maximum is 5% higher.

†New rope tensile strengths are based on tests of new and unused rope of standard construction in accordance with Cordage Institute standard test methods.

‡Working loads are for rope in good condition with appropriate splices, in noncritical applications, and under normal service conditions. Working loads should be exceeded only with expert knowledge of conditions and professional estimates of risk. Working load should be reduced where life, limb, or valuable property is involved, or for exceptional service conditions such as shock loads or sustained loads. Working loads are not necessarily intended to apply in those applications where a thorough engineering analysis of all conditions of use has been made and these conditions will not be exceeded in service. In such cases tensile strength, elongation, energy absorption, behavior under long-term or cyclic loading, and other pertinent properties of a rope may be evaluated to allow the selection of the rope best suited to the requirements of the application and an analysis of any risk involved in its use.

Source: The Cordage Institute.

Table 10.4 Nylon Fiber Rope Material and Construction (Three-Strand Laid and Eight-Strand Plaited, Standard Construction)

Nominal size		Polyester			
Diameter	Circumference	Linear density,* lb/100 ft	New rope tensile strength,† lb	Safety factor	Working load,‡ lb
3/16	5/8	1.20	900	10	90
1/4	3/4	2.00	1,490	10	149
5/16	1	3.10	2,300	10	230
3/8	1 1/8	4.50	3,340	10	334
7/16	1 1/4	6.20	4,500	9	500
1/2	1 1/2	8.00	5,750	9	640
9/16	1 3/4	10.2	7,200	8	900
5/8	2	13.0	9,000	8	1,130
3/4	2 1/4	17.5	11,300	7	1,610
13/16	2 1/2	21.0	14,000	7	2,000
7/8	2 3/4	25.0	16,200	7	2,320
1	3	30.4	19,800	7	2,820
1 1/16	3 1/4	34.4	23,000	7	3,280
1 1/8	3 1/2	40.0	26,600	7	3,800
1 1/4	3 3/4	46.2	29,800	7	4,260
1 5/16	4	52.5	33,800	7	4,820
1 1/2	4 1/2	67.0	42,200	7	6,050
1 5/8	5	82.0	51,500	7	7,350
1 3/4	5 1/2	98.0	61,000	7	8,700
2	6	118.	72,000	7	10,300
2 1/8	6 1/2	135.	83,000	7	11,900
2 1/4	7	157.	96,500	7	13,800
2 1/2	7 1/2	181.	110,000	7	15,700
2 5/8	8	204.	123,000	7	17,600
2 7/8	8 1/2	230.	139,000	7	19,900
3	9	258.	157,000	7	22,400
3 1/4	10	318.	189,000	7	27,000
3 1/2	11	384.	228,000	7	32,600
4	12	454.	270,000	7	38,600

*Linear density (lb/100 ft) shown is average. Maximum is 5% higher.

†New rope tensile strengths are based on tests of new and unused rope of standard construction in accordance with Cordage Institute standard test methods.

‡Working loads are for rope in good condition with appropriate splices, in noncritical applications, and under normal service conditions. Working loads should be exceeded only with expert knowledge of conditions and professional estimates of risk. Working load should be reduced where life, limb, or valuable property is involved, or for exceptional service conditions such as shock loads or sustained loads. Working loads are not necessarily intended to apply in those applications where a thorough engineering analysis of all conditions of use has been made and these conditions will not be exceeded in service. In such cases tensile strength, elongation, energy absorption, behavior under long-term or cyclic loading, and other pertinent properties of a rope may be evaluated to allow the selection of the rope best suited to the requirements of the application and an analysis of any risk involved in its use.

Source: The Cordage Institute.

Table 10.5 Polyester Fiber Rope Material and Construction (Three-Strand Laid and Eight-Strand Plaited, Standard Construction)

Polypropylene and Polyethylene

Nearly identical in appearance, these two synthetic fiber ropes are pliable and have a smooth, almost slippery feel. Both ropes have considerably lower strength than nylon or polyester.

Polypropylene ropes stretch considerably more than polyester ropes, while polyethylene ropes stretch about the same as polyester ropes. Shock loading properties are about half that of nylon ropes.

Although not as tough as nylon or polyester ropes, polypropylene rope does not lose its strength when wet. Polyethylene rope's strength, however, is much lower compared with that of other synthetic fiber ropes.

Both ropes soften progressively with rising temperatures and thus are unsuitable for use in high-temperature operations. Both are highly resistant to attack by acids and alkalies, mildew, and rot, but are weakened by prolonged exposure to sunlight (see Tables 10.6 and 10.7).

10.3 Characteristics of Fiber Rope

The function of any rope, basically, is to transmit a tensile force from its point of origin to a point of applications. Comparing properties and characteristics of different fiber ropes is necessary in choosing the right rope for a particular task–including strength, weight, durability, and stretch. To choose the rope with the combination of properties that best fit a specific job of application see Table 10.9.

A rope in use is subject to three basic mechanical stresses: tensile pull, structural friction (associated with bending and flexing), and surface friction that contributes to abrasion. These forces act on a rope under dynamic operating conditions and repeated loadings. Secondary stresses resulting from acceleration and heat are usually also involved.

Tensile Pull or Load

Acceleration stresses that result from reversing a load's motion momentarily increase the tensile load on a rope by as much as 100 percent. When a rope bends on sheaves, the tensile load will increase by as much as 10 percent for each sheave. Tensile stress applied and maintained during a pull will stretch a rope continually, reducing the diameter until the rope breaks. Of course, the amount of stretch will depend on the fiber, amount of twist, and size of rope, with the shortest components—the rope's core or inner fibers—stretching to their limit and breaking first.

Bending Flexing

When a rope is bent, its individual strands roll over one another, contracting for the inner part of the bend and pulling away from each other to form its outer part. The strands rub against each other and produce powdering and chafing on the surface. If the rope is excessively overloaded, internal chafing will result where strands are in contact.

Nominal size		Propylene			
Diameter	Circum-ference	Linear density,* lb/100 ft	New rope tensile strength,† lb	Safety factor	Working load,‡ lb
3/16	5/8	.70	720	10	72
1/4	3/4	1.20	1,130	10	113
5/16	1	1.80	1,710	10	171
3/8	1 1/8	2.80	2,440	10	244
7/16	1 1/4	3.80	3,160	9	352
1/2	1 1/2	4.70	3,780	9	420
9/16	1 3/4	6.10	4,600	8	575
5/8	2	7.50	5,600	8	700
3/4	2 1/4	10.7	7,650	7	1,090
13/16	2 1/2	12.7	8,900	7	1,270
7/8	2 3/4	15.0	10,400	7	1,490
1	3	18.0	12,600	7	1,800
1 1/16	3 1/4	20.4	14,400	7	2,060
1 1/8	3 1/2	23.8	16,500	7	2,360
1 1/4	3 3/4	27.0	18,900	7	2,700
1 5/16	4	30.4	21,200	7	3,020
1 1/2	4 1/2	38.4	26,800	7	3,820
1 5/8	5	47.6	32,400	7	4,620
1 3/4	5 1/2	59.0	38,800	7	5,550
2	6	69.0	46,800	7	6,700
2 1/8	6 1/2	80.0	55,000	7	7,850
2 1/4	7	92.0	62,000	7	8,850
2 1/2	7 1/2	107.	72,000	7	10,300
2 5/8	8	120.	81,000	7	11,600
2 7/8	8 1/2	137.	91,000	7	13,000
3	9	153.	103,000	7	14,700
3 1/4	10	190.	123,000	7	17,600
3 1/2	11	232.	146,000	7	20,800
4	12	276.	171,000	7	24,400

*Linear density (lb/100 ft) shown is average. Maximum is 5% higher.

†New rope tensile strengths are based on tests of new and unused rope of standard construction in accordance with Cordage Institute standard test methods.

‡Working loads are for rope in good condition with appropriate splices, in noncritical applications, and under normal service conditions. Working loads should be exceeded only with expert knowledge of conditions and professional estimates of risk. Working load should be reduced where life, limb, or valuable property is involved, or for exceptional service conditions such as shock loads or sustained loads. Working loads are not necessarily intended to apply in those applications where a thorough engineering analysis of all conditions of use has been made and these conditions will not be exceeded in service. In such cases tensile strength, elongation, energy absorption, behavior under long-term or cyclic loading, and other pertinent properties of a rope may be evaluated to allow the selection of the rope best suited to the requirements of the application and an analysis of any risk involved in its use.

Source: The Cordage Institute.

Table 10.6 Propylene Fiber Rope Material and Construction (Three-Strand Laid and Eight-Strand Plaited, Standard Construction)

Nominal size		Composite‖			
Diameter	Circumference	Linear density,* lb/100 ft	New rope tensile strength,† lb	Safety factor	Working load,‡ lb
3/16	5/8	.94	720	10	72
1/4	3/4	1.61	1,130	10	113
5/16	1	2.48	1,710	10	171
3/8	1 1/8	3.60	2,430	10	243
7/16	1 1/4	5.00	3,150	9	350
1/2	1 1/2	6.50	3,960	9	440
9/16	1 3/4	8.00	4,860	8	610
5/8	2	9.50	5,760	8	720
3/4	2 1/4	12.5	7,560	7	1,080
13/16	2 1/2	15.2	9,180	7	1,310
7/8	2 3/4	18.0	10,800	7	1,540
1	3	21.8	13,100	7	1,870
1 1/16	3 1/4	25.6	15,200	7	2,170
1 1/8	3 1/2	29.0	17,400	7	2,490
1 1/4	3 3/4	33.4	19,800	7	2,830
1 5/16	4	35.6	21,200	7	3,020
1 1/2	4 1/2	45.0	26,800	7	3,820
1 5/8	5	55.5	32,400	7	4,620
1 3/4	5 1/2	66.5	38,800	7	5,550
2	6	78.0	46,800	7	6,700
2 1/8	6 1/2	92.0	55,000	7	7,850
2 1/4	7	105.	62,000	7	8,850
2 1/2	7 1/2	122.	72,000	7	10,300
2 5/8	8	138.	81,000	7	11,600
2 7/8	8 1/2	155.	91,000	7	13,000
3	9	174.	103,000	7	14,700
3 1/4	10	210.	123,000	7	17,600
3 1/2	11	256.	146,000	7	20,800
4	12	300.	171,000	7	24,400

*Linear density (lb/100 ft) shown is average. Maximum is 5% higher.

†New rope tensile strengths are based on tests of new and unused rope of standard construction in accordance with Cordage Institute standard test methods.

‡Working loads are for rope in good condition with appropriate splices, in noncritical applications, and under normal service conditions. Working loads should be exceeded only with expert knowledge of conditions and professional estimates of risk. Working load should be reduced where life, limb, or valuable property is involved, or for exceptional service conditions such as shock loads or sustained loads. Working loads are not necessarily intended to apply in those applications where a thorough engineering analysis of all conditions of use has been made and these conditions will not be exceeded in service. In such cases tensile strength, elongation, energy absorption, behavior under long-term or cyclic loading, and other pertinent properties of a rope may be evaluated to allow the selection of the rope best suited to the requirements of the application and an analysis of any risk involved in its use.

‖Materials and construction of this polyester/polypropylene composite rope conform to MIL-R-43942 and MIL-R-43952. For other composite ropes, consult the manufacturer.

Source: The Cordage Institute.

Table 10.7 Composite (Polypropylene and Polyethylene) fiber rope material and construction (Three-Strand Laid and Eight-Strand Plaited, Standard Construction)

Because strand movement depends on the bend radius, a larger sheave will minimize the wear. Sheave diameters should always be five times a rope's diameter—or more—depending on the speed and frequency of strand-bending adjustments. No frictional damage occurs to a rope under tensile stress when subjected to static bending, such as in a knot. Load failure, instead, results from a shearing force applied to the rope by the knot's loop.

Surface Friction

A rope in use normally is subjected to surface friction as it rubs against itself and against sharp edges, both in straight tension and in bending moments. It is also subject to internal and external wear from imbedded grit, resulting in the major cause of most fiber rope failure—surface abrasion.

10.4 Breaking Strengths and Safe Working Loads of Fiber Rope

The breaking strength of rope refers to the tension at which the rope will break apart when an additional load is applied. Breaking strength is a measure of a rope's serviceability. Strength as a property of the rope, however, is significant only in relation to its size. Rope strengths are not measured directly in terms of stress per unit area—as with most materials—but rather on the basis of strength and length-to-weight ratio.

The degree of twist on a rope and its strands is the primary factor influencing a rope's strength. The greater the twist, the lower a rope's strength; a tightly twisted hard-lay rope will be as much as 20 percent lower in strength than the corresponding standard-lay rope (see Table 10.8).

Wetting a rope has a profound effect on its strength, with wet natural fiber ropes tending to swell in size and shrink in length. Swelling increases the angle of twist, thus making the rope harder and reducing its strength. (Synthetic fibers, in particular, lose considerable strength when wet while Manila and cotton ropes tend to increase in strength.)

Repeated tensile loading will not adversely affect fiber rope strength. In fact, if a rope's internal yarns are not broken, repeated loading can actually increase breaking strengths as much as 20 percent, because yarn and strands subject to increased tensions become better adjusted as the rope is stretched repeatedly.

A rope's strength is only about two-thirds the potential strength of its combined components. Because rope efficiency depends on the degree of twist, a slacker twist results in higher strength. But, as the number of strands increases for a given size, a rope's strength will decrease.

Weather also affects a rope's serviceable life by degrading the fibers and weakening them. Alternate wetting, drying, and stretching structurally loosens a rope, and this in turn increases elongation, even during minor loading.

New Three-Strand Fiber Ropes					
Nominal rope diameter, in.	Manila	Nylon	Poly-propy-lene	Polyes-ter	Poly-ethyl-ene
3/16	100	200	150	200	150
1/4	120	300	250	300	250
5/16	200	500	400	500	350
3/8	270	700	500	700	500
1/2	530	1,250	830	1,200	800
5/8	880	2,000	1,300	1,900	1,050
3/4	1,080	2,800	1,700	2,400	1,500
7/8	1,540	3,800	2,200	3,400	2,100
1	1,800	4,800	2,900	4,200	2,500
1 1/8	2,400	6,300	3,750	5,600	3,300
1 1/4	2,700	7,200	4,200	6,300	3,700
1 1/2	3,700	10,200	6,000	8,900	5,300
1 5/8	4,500	12,400	7,300	10,800	6,500
1 3/4	5,300	15,000	8,700	12,900	7,900
2	6,200	17,900	10,400	15,200	9,500

Braided Synthetic Fiber Ropes			
Nominal rope diameter (inches)	Nylon cover, nylon core	Nylon cover, polypropy-lene core	Polyester cover, poly-propylene core
1/4	420	—	380
5/16	640	—	540
3/8	880	680	740
7/16	1,200	1,000	1,060
1/2	1,500	1,480	1,380
9/16	2,100	1,720	—
5/8	2,400	2,100	2,400
3/4	3,500	3,200	2,860
7/8	4,800	4,150	3,800
1	5,700	4,800	5,600
1 1/8	8,000	7,000	—
1 1/4	8,800	8,000	—
1 1/2	12,800	12,400	—
1 5/8	16,000	14,000	—
1 3/4	19,400	18,000	—
2	23,600	20,000	—

*Safety factor = 5.

TABLE 10.8 Approximate Safe Working Strength of Manila and Nylon Ropes

Elongation, Elasticity, and Creep

The degree to which a rope elongates depends on fiber type and rope size. Elongation also depends on a rope's construction. Hard fiber ropes do not stretch as much as some made of soft fibers, and small diameter ropes stretch less than larger-sized ones when loaded to the breaking point. In addition, the more the twist (the harder the lay), the more a rope will stretch.

After initial loading, each time a rope is used within a given loading range it will behave more like a rigid material. Unlike rigid materials, however, ropes do not display true elasticity. Once stretched by an appreciable load, a rope remains more or less permanently elongated.

Some contraction of the rope begins to take place immediately upon release of a load. More than half the total recovery that will ever be attained occurs in a matter of minutes. The extent of a rope's recovery from stretch will depend on the magnitude of the applied load, rope size, nature of fiber, and structure of the rope.

- *Rope loaded lightly*, in comparison to its breaking strength, will recover faster than when loaded heavily.
- *Larger-sized rope*, when stretched, tends to readjust internally, more or less permanently, and thereby has less of a chance to recover from stretch.
- *Four-strand rope* shows less recovery than a three-strand rope.
- *Hard-lay rope*, having more twist, will recover less than a soft-lay rope.
- *Rope loaded repeatedly*, thereby compacted to its maximum tensile load, will ultimately show complete recovery from stretch provided the applied load does not exceed the load used in prestretching the rope.
- *Wet rope* will not show as marked recovery as a dry rope, because the wet one has a higher initial stretch, and wetting internally adjusts the rope's yarn and strand tension relationship.

Rope loaded for any sustained time will continue to stretch beyond its initial elongation when the load was first applied. This slow, continual stretch under load is called *creep*. A rope lulled taut and secured to maintain tension will stretch until it breaks, even though the tension is well be low its normal breaking strength.

Impact Loading

Although a rope's ultimate elongation remains about the same, regardless of the rate of loading, its breaking strength will increase with a faster loading rate. With impact loading, however, a rope can absorb considerably more energy that it can during a slow steadily applied load. Tests indicate that under extreme impact loading, a new hard-fiber rope will absorb as much as twice the amount of energy it would absorb from a slow, steady loading.

Flexibility

A rope's flexibility depends on the degree with which strands that are normally compressed against each other can overcome the frictional forces binding them together, and thus adjust to a new position.

As the rope flexes (bends), lubricant is forced to the outer surface of the strands, which tends to restrict yarn and strand movement and thus results in poor rope performance. Improving both the type and amount of lubrication will improve a rope's flexing endurance. Hard fiber rope requires a minimum lubricant content of about 10 percent to 15 percent; however, sticky ingredients in lubricants will have an adverse effect on a rope's flexing endurance.

A twisted fiber rope will have better flexing endurance for a comparable loading than a braided one of the same size and material. Increasing the twist will help improve a rope's flexing endurance. An increase in the number of strands, however, generally will have no effect.

REMEMBER: *All the forces acting on a rope when it is pulled in tension are also present when the rope is bent. As this tension increases, the compacting action of the rope's strand surface tends to reduce rope flexibility. In addition, the outer surface compression that develops in a bent rope, in relation to its internal compressive forces, tends to create shearing stresses that further damage the rope.*

Strand surface condition and fiber rigidity will also influence knot strength. Where strand surfaces are smooth, the frictional forces involved in bending are reduced and a knot will break more readily. When a rope is wet, knot strength efficiency ordinarily will increase; this is particularly true of Manila rope.

10.5 Calculating Breaking Strength of Fiber Rope

Rope manufacturers have determined the breaking strength of various ropes through tests, and provide published tables with this information. The fiber from which a particular rope is constructed has a definite bearing on its breaking strength.

Maximum safe working load for high-grade Manila rope, common-lay three-strand *must not* exceed values listed in Table 10.1.

Manila Rope

The breaking strength of Manila rope is higher than that of an equal-size sisal line because of the difference in strength of the two fibers.

$$\frac{\text{rope tensile strength}}{\text{design factor}} = \text{safe working load}$$

The rule of thumb for estimating the breaking strength of Manila rope is:

$$C \text{ squared (in.)} \times 900 = \text{BS (lb)}$$

$$C \text{ squared (cm)} \times 64.8 = \text{BS (kg)}$$

Nylon Rope

The breaking strength of nylon rope is almost three times that of Manila line of the same size. The rule of thumb for estimating the breaking strength of nylon rope is:

$$C \text{ squared (in.)} \times 2400 = \text{BS (lb)}$$

$$C \text{ squared (cm)} \times 172.8 = \text{BS (kg)}$$

where C is the circumference of rope.

Nylon rope can withstand repeated stretching to the breaking point with no serious effects. When nylon line is under load, it thins out. Under normal safe working loads, nylon line will stretch about one-third of its length. When free of tension, it returns to its normal size. When nylon line is stretched more than 40 percent, it is likely to part—and the stretch immediately recovered with a snapback that sounds like a pistol shot.

WARNING: *The snapback of a nylon line can be as deadly as a bullet. While this feature is also true for other types of rope, overconfidence in the strength of nylon may lead one to underestimate its backlash; therefore, ensure that no one stands in the direct line of pull when a heavy strain is applied to a line.*

The critical point of loading is 40 percent extension of length; for example, a 10-ft length of nylon rope would stretch to 14 ft when under load. Should the stretch exceed 40 percent, the rope will be in danger of parting.

Nylon rope, however, will hold a load even though a considerable number of strands are abraded. Ordinarily, when abrasion is localized the rope may be made satisfactory for reuse by cutting away the chafed section and splicing the ends.

10.6 Calculating Safe Working Load of Fiber Rope

Briefly defined, the safe working load of a fiber rope is the load that can be applied without causing damage to the rope.

REMEMBER: *The safe working load of a rope is considerably less than the breaking strength. A wide margin of difference between breaking strength and safe working load is necessary to allow for such factors as additional strain imposed on the rope by jerky movements in hoisting or bending over sheaves in a pulley block.*

In the absence of a Safe Working Load chart for a particular size fiber rope, a rule of thumb for determining the safe working load (SWL) is:

$$SWL = \text{C squared (in.)} \times 150 \text{ SWL (lb)}$$

$$SWL = \text{C squared (cm)} \times 10.8 \text{ SWL (kg)}$$

where C is the circumference of rope.

If a fiber rope is in

- Good shape—add 30 percent to the calculated SWL
- Bad shape—subtract 30 percent from the SWL

REMEMBER: *The strength of a line will decrease with age, use, and exposure to excessive heat, boiling water, or sharp bends. Especially with used line, these and other factors affecting strength should be given careful consideration, making proper adjustment in the breaking strength and safe working load capacity of the line.*

Although rope manufacturers provide tables that show the breaking strength and safe working load capacity of rope, the values given in those tables apply to *new rope* used under favorable conditions. Therefore, as the rope ages or deteriorates with use, the values given in the manufacturers' tables *must* be progressively reduced.

10.7 Safety Factor of Fiber Rope

The safety factor of a rope is the ratio between its breaking strength and the safe working load. Usually a safety factor of 4 is acceptable, but this is not always the case. In other words, the safety factor will vary, depending on such things as the condition of the line and circumstances under which it is to be used. While the safety factor should never be less than 3, it often must be well above 4 (possibly as high as 8 or 10).

For best average or unfavorable conditions, the following safety factors are usually suitable:

4—best condition (new rope)

6—average condition (rope used but in good condition)

8—unfavorable condition (frequently used rope, such as running rigging)

Where uses of rope, however, involve a serious risk of injury to personnel or damage to valuable property, or if there is any question about the loads involved or the conditions of use, the working load should be reduced substantially and the rope properly inspected.

Because of the wide range of rope use, rope condition, exposure to the different factors that affect rope behavior, and the degree of risk to life and property involved, it is impossible to make blanket recommendations for safe working loads.

CAUTION: *Working loads are only guidelines for rope in good condition with appropriate splices, in noncritical applications, and under normal service conditions.*

- Working loads should be reduced where life, limb, or valuable property is involved, or for exceptional service conditions such as shock loads.
- Working loads are *not* applicable when rope is subjected to significant dynamic loading resulting from picking up, stopping, moving, or swinging a load rapidly or suddenly, such as when rope is used as towing lines, lifelines, safety lines, and climbing ropes.

Overloading a rope poses a serious safety threat to personnel. It is also likely to result in heavy losses through damage to material and equipment. To avoid overloading, the strength of the rope being used *must* be known and involves three factors: breaking strength, safe working load, and safety factor.

To provide the rigger with some guidelines for using rope under normal service conditions in noncritical applications, the Cordage Institute publishes strength tables for various types of rope, covering a wide range of diameters, tensile strengths, safety factors, and working loads for rope in good condition with appropriate splices (*Rope Standards CI 1201-CI 131*).

10.8 Handling and Care of Fiber Rope

Manila ropes are chemically treated to resist mildew and dry rot, thus permitting their use for prolonged periods under wet and dirty conditions. Nevertheless, all Manila ropes are subject to mildew and dry rot and therefore must be cleaned and dried at regular intervals.

Proper maintenance of Manila rope will extend serviceability. To slow down rotting, wash rope in cold water after use and hang in loosely formed coils over wooden pegs to dry. Allow air to circulate around stored ropes. Carefully check slings and safety lines frequently for signs of mildew or rot.

Keep Manila rope protected from prolonged exposure to sunlight, which also causes deterioration. If a rope stiffens with storage, it can be made pliable again with a thin coating of warm lubricating oil.

When nylon rope is properly handled and maintained, it should last more than five times longer than Manila rope subjected to the same use. *Do not* store nylon rope in strong sunlight; cover it with tarpaulins. If a nylon rope becomes slippery because of grease, it should be cleaned with light oils, such as kerosene or diesel oil.

To keep fiber ropes in good condition to provide safe and dependable service and extend their working life, *make sure* they are handled and cared for properly. Procedures for handling and caring of fiber rope include the following:

- Keep them away from all sources of heat and exhaust gases that will cause rapid deterioration of fibers.
- *Never* drag fiber rope over the ground or over rough or dirty surfaces. The rope can easily pick up sand and grit that can work into the strands and wear the fibers.
- Clean dirty ropes with water; cleanliness is a key part of the care of fiber rope.
- Do not use soap—it takes oil out of the rope.
- Avoid pulling a rope over sharp edges because the strands may break. When rope must pass over a sharp edge, place chafing gear such as a board, folded cardboard or canvas, or part of a rubber tire between the rope and the sharp edge to prevent damaging the rope.
- *Never* cut a rope unless absolutely necessary. When possible, always use knots that can be untied easily.
- Protect fiber rope from undue exposure to weather, dampness, and sunlight. Fiber rope contracts (shrinks) if it gets wet. If there is not enough slack in a wet rope to perrmit shrinkage, the rope is likely to overstrain and weaken. If a taut rope is exposed to rain or dampness, make sure that the rope, while still dry, is slackened to allow for the shrinkage.
- *Always* thaw out a frozen rope before using it to prevent breaking the frozen fibers. But allow wet and frozen ropes to dry naturally, as too much heat will cause the fibers to become brittle.
- *Always* clean and dry wet ropes before storing them.
- Make sure to air natural fiber ropes whenever they have sour odor until the odor disappears.
- Do not store fiber ropes on the floor, in boxes, or on shelves where air circulation is restricted. To permit proper air circulation, place the rope in loose coils on a wood grating platform about 6 in. (15 cm) above the floor; or hang the line in loose coils on a wooden peg.
- Store fiber ropes in a dry, cool room having good air circulation, at a temperature of 50° to 70°F, and at a humidity of 40 to 60 percent.
- Avoid continuous exposure of rope to sunlight because excessive sunlight can damage the rope.
- *Never* expose fiber rope to lime, acids, or other chemicals, or store it in a room containing chemicals—even the fumes may severely damage the line.

Uncoiling Fiber Rope

New fiber rope is coiled, bound, and wrapped in burlap. This protective covering should not be removed until the rope is to be used because it protects the line during storage and prevents tangling.

When opening a coil of new fiber rope, carefully follow the instructions printed on the tag. Fiber rope, like wire rope, can be damaged if it is removed from the shipping coil incorrectly.

After opening and removing the burlap wrapping, lay the coil of rope flat on the floor with the inside rope end at the bottom of the core. If the rope's end is not there, turn the coil over so that the end will be at the bottom. Cut the lashings that bind the coil together, then reach down inside the core and pull the end of the rope up through the core (Fig. 10.4).

As the rope comes up through the coil, it will unwind in a counterclockwise direction. Even though the rope is unwound properly, loops and kinks may form in it. Remove them immediately and carefully to prevent their being pulled taught, thereby damaging the rope.

If, for any reason, a rope must be uncoiled from the outside, place the coil so that as the rope pays out it will unwind in a counterclockwise direction. If a rope has a large number of kinks, coil it counterclockwise on the floor, then pass the end through the coil and proceed to uncoil.

After being used, rope should always be coiled down in a clockwise direction (assuming it is a right-hand lay). Should the rope be kinked from excessive turns, remove the kinks by coiling the rope down counterclockwise and then pulling the bottom end of the coil up and out the middle of the coil.

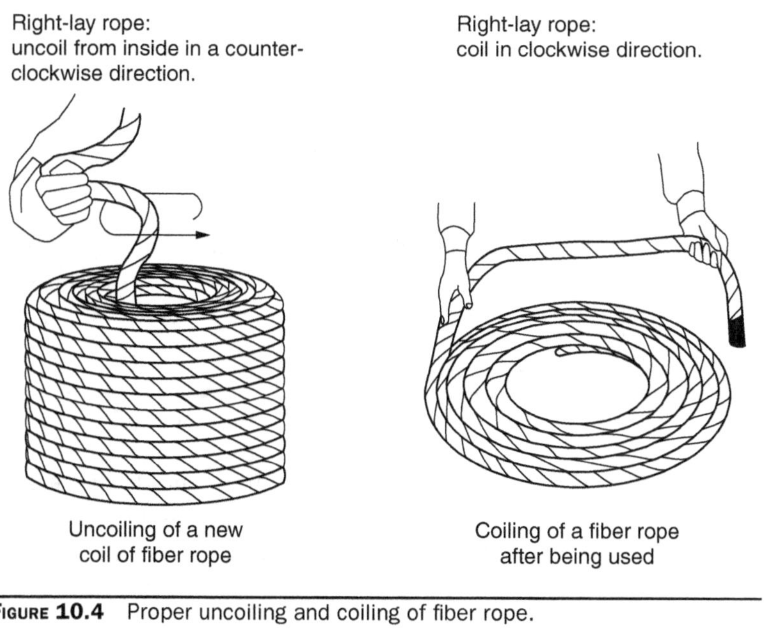

FIGURE 10.4 Proper uncoiling and coiling of fiber rope.

- If the rope is free of kinks as it leaves the coil, make it up in the correct manner.
- If the rope is still kinked, repeat the process before making up the line for storage.

Do not uncoil new nylon line, however, by pulling the end up through the eye of the coil. *Avoid* coiling nylon in the same direction all the time, or you could unbalance the lay.

Making Up Rope

After removing the rope from the manufacturer's coil, it may be made up (i.e., prepared for storage or for use) by winding it on a reel or by cooling it down, that is, laying it in circles, roughly one on top of the other. Rope should always be coiled in the same direction as the lay—clockwise for right lay and counterclockwise for left lay. When a line has been coiled down, one end is ready to run off. This is the end that went down last and is now on top. If, for some reason, the bottom end must go out first, turn the coil over to free it for running.

When rope is cut to a required length, the rope ends *must* be bound or whipped with yarn to prevent the strands from untwisting and the end from fraying. If the ends are not properly whipped, the strands will slip in relation to each other, causing one of them to assume more or less than its share of the load, thus shortening the rope's useful life.

The process of securing the ends of a rope will not increase the size of the rope enough to prevent the fitting of the blocks or openings through which it must pass.

Whippings are made with yarn, or fine twine. In applying a whipping, make a loop in the end of the yarn and place the loop at the end of the rope. Wind the standing part around the rope covering all but a portion of the loop. Pass the remainder of the standing end up through the small loop and pull the dead end of the yarn, thus pulling the standing end and the small loop (through which it is threaded) back underneath the whipping until they reach a point midway underneath the whipping. Trim both ends of the yarn close up against the loops to finish off the whipping.

Before cutting a rope, however, place two whippings on the rope 1 or 2 in. apart and make the cut between the whippings. This procedure prevents the ends from untwisting after they are cut.

10.9 Inspection of Fiber Rope

The exterior appearance of fiber line is not always a good indication of its internal condition. Fiber rope softens with use and dampness, heavy loads, fraying, breaking or broken strands, and dragging over rough surfaces all contribute to rope weakening and failure. Additonally, overloading a rope can cause it to part, resulting in heavy damage to material, equipment, and serious injury to personnel.

For these reasons, every foot of a rope *must* be inspected carefully at regular intervals to determine whether it is safe for use, its life expectancy, and its load-carrying ability.

- Frequency of sling use
- Severity of service conditions
- Nature of lifts being made
- Experience gained during the service life of slings used in similar circumstances and for special and infrequent service use as recommended by a qualified person.

Designate a qualified person to inspect slings and all fastenings and attachments *each day* before use for damage or defects, thoroughly covering slings and attachments. When inspecting a fiber rope sling make sure to check:

- *Exterior*. Look for surface cuts, gouges, or worn surface area; dry, brittle, scorched, or discolored fibers; or melting or charring of any part of the sling. Check for signs of abrasion, cuts, broken fibers and yarns, burns, any unlaying of the twist, or a reduction the rope's diameter. If any of these conditions are found, notify a supervisor to determine the safety of the sling. If the sling is found to be unsafe, it must be discarded immediately.
- *Interior*. Should be as clean as when the rope was new. A buildup of powder-like sawdust on the inside of the fiber rope indicates excessive internal wear and that the sling is unsafe.
- *Scratch*. Fibers with a finger nail; if the fibers separate easily, the fiber sling has suffered some kind of chemical damage and *must* be discarded.

Look for external wear and cutting, internal wear between the strands, and deterioration of the fibers. The section of the rope showing the most deterioration should be the determining factor when estimating a rope's condition. During inspection, however, be careful not to distort a rope's lay during inspection.

Visible Defects

Remove all synthetic rope slings from service if visible defects include any of the following:

- Acid or caustic burns
- Melting or charring of any surface part
- Snags, punctures, tears, or cuts
- Broken or worn stitches

- Wear or elongation exceeding the amount recommended by manufacturer
- Distortion of end fittings
- Missing or illegible sling identifications
- Cuts, gouges, areas of extensive fiber breakage along the length and abraded areas on the rope
- Damage to 10 percent or more of the ropes diameter
- Uniform fiber breakage along the major part of the length of the rope in the sling such that the entire rope appears covered with fuzz or whiskers
- Fiber breakage or melted fiber inside the rope that appears along the length at the same relative position and involves damage estimated at 10 percent of the fiber in the strand at that point
- Discoloration and brittle or stiff areas on any part of the sling
- Excessive dirt and grit in the interior of the rope structure
- Foreign matter that has permeated the rope and attracts and holds grit
- Kinks, distortion, or other damage in the rope structure
- Melted or charred areas that affect more than 10 percent of the diameter of the rope or affect several adjacent strands along the length to more than 10 percent of their individual diameters
- Poor condition of thimbles or other fittings manifested by corrosion, cracks, distortion, or localized wear
- Other conditions that cause doubt as to continued use of the sling.

Remove the sling or attachment from service immediately where any such defect or deterioration is present; *do not* use worn or damaged slings or attachments.

Do not use repaired or reconditioned fiber rope slings. Do not use old or used rope to make up a fiber rope sling; modifications or alterations to end attachments or fittings are considered a repair.

Checking Rope Strength

Broken fibers or yarns or any other failures represent a loss of rope strength. But *do not* be misled by dirt on the rope's surface. Any rope that has been used will be dirty on the outside. Carefully open up the rope by untwisting the strands to observe the condition of the rope's inside. Try not to kink the strands. The rope's interior should be as bright and clean as when the rope was new. Rope that is mildewed gives off a musty odor.

Check for broken yarns inside, excessively loose strands and yarns, or any accumulation of powder such as dust that results from excessive strand wear.

- If the fiber rope is large enough to permit it, open up a strand and, with a pencil or other blunt instrument, try to pull out one of the inside yarns. Keep in mind that if a rope has been overloaded, it is the interior yarns that will have failed first. Excessive oil on the outside of new rope also indicates that the rope has been overloaded.

- If the fiber rope is four-strand construction with a core, try to gently pull out the core. If it readily comes out in short pieces, the rope has been overloaded and should not be used for hoisting.
- If possible, pull out a couple of long fibers from the end of the rope and try to break them. The finer fibers are relatively stronger than the coarser ones, and all should be broken only with difficulty. Some fibers have a tensile strength as high as 30,000 psi.
- If the inside of the rope is dirty, if the strands have begun to unlay, or if the rope has lost its life and elasticity, it *must not* be used for hosting purposes.
- If the rope is high stranded and presents a spiral appearance, or if the core protrudes, the load will not be equally distributed on the strands and a very short rope life may be expected.
- If a rope feels dry and brittle, has sections that are glazed or fused, or is discolored from exposure to acid fumes, it *must* always be discarded.
- If a rope is weak in just one spot, cut out that proportion and splice the rope. Make sure that all splices are properly served or taped. *Do not* allow any tuck to become undone, since every tuck is necessary for optimum splice efficiency.
- If thimbles are loose in eyes because of rope stretch they *must* be retightened in the eye by seizing the eye. *Never* allow a thimble to become so loose that it rocks in the eye, resulting in chafing of the rope inside the eye, with ultimate breakdown.
- If there is any doubt as to whether the rope is fit for use, *replace* it at once. *Never* risk danger to life or damage to property by taking a chance.

NOTE: *When a rope has been condemned, it should be destroyed at once, or cut up into short hand lines so that it cannot be used again for hoisting purposes.*

10.10 Comparative Characteristics of Cordage Fibers

When selecting natural or synthetic fiber rope for rigging operations, properties important enough to be taken into account include strength (dry and wet), weight, elongation, surface characteristics, water absorption, rot and mildew resistance, abrasion resistance, chemical resistance, sunlight degradation, and high and low temperature properties (Table 10.9).

	3- and 8-strand, plaited		Double braid	
	Manila		Core and cover nylon	
Rope diameter	1 in.	2 in.	1 in.	2 in.
Strength characteristics				
Tensile strength, dry (approximate average), lb	9,000	31,000	31,300	117,000
Recommended factor of safety	5	5	5 to 8*	5 to 8*
Working strength, lb	1,800	6,200	To 6,260*	To 23,400*
Wet strength compared to dry strength, %	To 120	To 120	88	88
Strength per unit of weight or "breaking length" (breaking strength), lb/ft	33,000	29,000	118,000	110,000
Cyclic loading characteristics	Poor	Poor	Excellent	Excellent
Individual filament or fiber strength, grams per denier	6.0 to 7.5	6.0 to 7.5	7.5 to 8.3	7.5 to 8.3
Weight and density characteristics				
Weight per 100 ft, lb	27.0	108.0	26.0	106.0
Specific gravity of fiber	1.38	1.38	1.14	1.14
Ability to float	No	No	No	No
Elasticity, stretch (approximate)				
Permanent elongation at working load (20% of breaking strength), %	4.8	4.8	6.5 (11.0 wet)	
Working elasticity (temporary stretch under load) at working load (20% of breaking strength), %	5.0	5.0	6.5 (11.0 wet)	
Elongation at 100% load (at break) for broken-in ropes, %	13	13	23.5	
Individual filament or fiber elongation, %	2 to 3	2 to 3	16 to 20	
Surface characteristics				
Rendering qualities, ability to ease out smoothly under load over bitts	Poor		Good, requires fewer wraps	
Hand (feeling of rope to the touch)	Some harshness due to hairs; after use considerable harshness due to broken fiber ends		Smooth; after use becomes fuzzy with a softer feel	

Water absorbed into fiber (Some water will be held between fibers of all ropes)	Up to 100% of weight of rope	5 to 6%
Resistance to rot, mildew, and attack by marine organisms (Some marine organisms will attach themselves to any submerged object, including synthetic ropes)	Poor	100% resistant
Deterioration		
Due to aging (stored ropes, ideal conditions)	About 1% per year	Zero
Due to exposure to sunlight	Some slight	Some slight
Resistance to chemicals		
Acids	Very poor	Fair, except to concentrated sulfuric and hydrochloric acids
Alkalis	Very poor	Excellent
Solvents	Good	Good
Wear		
Resistance to surface abrasion	Good	Excellent
Resistance to internal wear from flexing	Good	Excellent
Resistance to cutting (toughness)	Good	Excellent
High- and low-temperature properties		
Melting point	Loses strength rapidly over 180°F	480°F; progressive strength loss above 350°F
Low-temperature properties	No change	No change
Flammability	Burns like wood	Burns with difficulty

*Depending on application.

Source: Sampson Rope Manual, Sampson Ocean Systems, Inc.

TABLE 10.9 Comparative Characteristics of Various Fibers Used in Cordage (*Continued*)

	3- and 8-strand, plaited			
	Nylon		Polyester	
Rope diameter	1 in.	2 in.	1 in.	2 in.
Strength characteristics				
Tensile strength, dry (approximate average), lb	25,000	92,000	20,000 to 22,000	65,000 to 80,000
Recommended factor of safety	9	9	9	9
Working strength, lb	2,890	10,000	2,220 to 2,450†	7,200 to 8,900†
Wet strength compared to dry strength, %	90 to 95	90 to 95	100	100
Strength per unit of weight or "breaking length" (breaking strength), lb/ft	96,000	97,000	65,000 to 72,000	55,000 to 64,000
Cyclic loading characteristics	Good	Good	Good to excellent	Good to excellent
Individual filament or fiber strength, grams per denier	7.5 to 8.3	7.5 to 8.3	6.0 to 7.0	6.0 to 7.0
Weight and density characteristics				
Weight per 100 ft, lb	26.0	95.0	30.5	118.0
Specific gravity of fiber	1.14	1.14	1.38	1.38
Ability to float	No	No	No	No
Elasticity, stretch (approximate)				
Permanent elongation at working load (20% of breaking strength), %	8.0	8.0	6.0 to 6.2	6.0 to 6.2
Working elasticity (temporary stretch under load) at working load (20% of breaking strength), %	16.0	16.0	5.9 to 6.5	5.9 to 6.5
Elongation at 100% load (at break) for broken-in ropes, %	35	35	20 to 22	20 to 22
Individual filament or fiber elongation, %	16 to 20		11 to 13	11 to 13
Surface characteristics				
Rendering qualities, ability to ease out smoothly under load over bitts	Poor		Good	
Hand (feeling of rope to the touch)	Smooth; after use becomes fuzzy with a softer feel		Smooth and hard, not slippery; after use becomes fuzzy with a softer feel	

Water absorbed into fiber (Some water will be held between fibers of all ropes)	5 to 6%	Less than 1%
Resistance to rot, mildew, and attack by marine organisms (Some marine organisms will attach themselves to any submerged object, including synthetic ropes)	100% resistant	100% resistant
Deterioration		
Due to aging (stored ropes, ideal conditions)	Zero	Zero
Due to exposure to sunlight	Some slight	Almost none
Resistance to chemicals		
Acids	Fair, except to concentrated sulfuric and hydrochloric acids	Very good to excellent
Alkalis	Excellent	Very good, except to concentrated sodium hydroxide at high temperatures
Solvents	Good	Very good to excellent
Wear		
Resistance to surface abrasion	Very good	Excellent
Resistance to internal wear from flexing	Excellent	Very good to excellent
Resistance to cutting (toughness)	Excellent	Very good to excellent
High- and low-temperature properties		
Melting point	480°F; progressive strength loss above 350°F	480°F; progressive strength loss above 350°F
Low-temperature properties	No change	No change
Flammability	Burns with difficulty	Burns with difficulty

*Depending on application.
†Depending on grade of polyester used.

Source: Sampson Rope Manual, Sampson Ocean Systems, Inc.

TABLE 10.9 Comparative Characteristics of Various Fibers Used in Cordage (*Continued*)

	3-strand		12-strand single braid	
	Polyethlene		Polyester and polypropylene	
Rope diameter	1 in.	2 in.	1 in.	2 in.
Strength characteristics				
Tensile strength dry, (approximate average), lb	12,600	47,700	23,500	88,000
Recommended factor of safety	6	6	5 to 8*	5 to 8*
Working strength, lb	2,100	7,900	To 4,700*	To 17,600*
Wet strength compared to dry strength, %	100	100	100	100
Strength per unit of weight or "breaking length" (breaking strength), lb/ft	66,000	75,500	94,000	89,000
Cyclic loading characteristics	Fair	Fair	Excellent	
Individual filament or fiber strength, grams per denier	3.0 to 7.0	3.0 to 7.0	Polyester 7.5 to 9.0; monopoly 3.0 to 7.0	
Weight and density characteristics				
Weight per 100 ft, lb	18.5	72.5	25.0	99.0
Specific gravity of fiber	0.95	0.95	1.14	1.14
Ability to float	Yes	Yes	No	No
Elasticity, stretch (approximate)				
Permanent elongation at working load (20% of breaking strength), %	5.8	5.8	3.9	3.9
Working elasticity (temporary stretch under load) at working load (20% of breaking strength), %	5.9	5.9	3.0	3.0
Elongation at 100% load (at break) for broken-in ropes, %	22	22	16	16
Individual filament or fiber elonation, %	10 to 12	10 to 12	11 to 20	
Surface characteristics				
Rendering qualities, ability to ease out smoothly under load over bitts	Good but requires extra wraps		Good	
Hand (feeling of rope to the touch)	Smooth and very slippery; after use becomes slightly harsh due to broken fiber ends		Smooth, not slippery; after use, fuzz and a few broken fiber ends but little harshness	

Water absorbed into fiber (Some water will be held between fibers of all ropes)	Zero	Less than 1%
Resistance to rot, mildew, and attack by marine organisms (Some marine organisms will attach themselves to any submerged object, including synthetic ropes)	100% resistant	100% resistant
Deterioration		
Due to aging (stored ropes, ideal conditions)	Zero	Zero
Due to exposure to sunlight	Some; black resists best	Some slight
Resistance to chemicals		
Acids	Excellent except to concentrated sulfuric acid	Very good to excellent
Alkalis	Good	Good, except to concentrated sodium hydroxide at high temperatures
Solvents	Good	Good to excellent
Wear		
Resistance to surface abrasion	Good	Very good
Resistance to internal wear from flexing	Very good	Very good
Resistance to cutting (toughness)	Good	Very good
High- and low-temperature properties		
Melting point	280°F; softens above 250°F	330°F; softens above 300°F
Low-temperature properties	Brittle below −150°F	No change
Flammability	Burns with difficulty	Burns with difficulty

*Depending on application.

Source: Sampson Rope Manual, Sampson Ocean Systems, Inc.

TABLE 10.9 Comparative Characteristics of Various Fibers Used in Cordage (*Continued*)

	3- and 8-strand, plaited			
	Polypropylene monofilament		Polypropylene multifilament	
Rope diameter	1 in.	2 in.	1 in.	2 in.
Strength characteristics				
Tensile strength, dry (approximate average), lb	14,000	52,000	15,700	55,500
Recommended factor of safety	6	6	6	6
Working strength, lb	2,330	8,700	2,620*	9,250*
Wet strength compared to dry strength, %	102 to 105	102 to 105	102 to 105	102 to 105
Strength per unit of weight or "breaking length" (breaking strength), lb/ft	77,800	75,500	82,000	75,000
Cyclic loading characteristics	Excellent	Excellent	Excellent	Excellent
Individual filament or fiber strength, grams per denier	3.0 to 7.0	3.0 to 7.0	3.0 to 7.0	3.0 to 7.0
Weight and density characteristics				
Weight per 100 ft, lb	18.0	69.0	19.2	74.0
Specific gravity of fiber	0.91	0.91	0.91	0.91
Ability to float	Yes	Yes	Yes	Yes
Elasticity, stretch (approximate)				
Permanent elongation at working load (20% of breaking strength), %	3.8	3.8	7.5	7.5
Working elasticity (temporary stretch under load) at working load (20% of breaking strength), %	8.9	8.9	10.5	10.5
Elongation at 100% load (at break) for broken-in ropes, %	24	24	36	36
Individual filament or fiber elongation, %	16 to 20	10 to 12	22 to 28	22 to 28
Surface characteristics				
Rendering qualities, ability to ease out smoothly under load over bitts	Poor		Fair	
Hand (feeling of rope to the touch)	Smooth but not slippery; after use becomes harsh due to broken fiber ends		Smooth and soft with some natural fuzziness; remains same after use	

Water absorbed into fiber (Some water will be held between fibers of all ropes)	Zero	Zero
Resistance to rot, mildew, and attack by marine organisms (Some marine organisms will attach themselves to any submerged object, including synthetic ropes)	100% resistant	100% resistant
Deterioration		
Due to aging (stored ropes, ideal conditions)	Zero	Zero
Due to exposure to sunlight	White some; black resists best	White some; black resists best
Resistance to chemicals		
Acids	Excellent	Excellent
Alkalis	Good	Good
Solvents	Good	Good
Wear		
Resistance to surface abrasion	Good	Good
Resistance to internal wear from flexing	Very good	Very good
Resistance to cutting (toughness)	Good	Very good
High- and low-temperature properties		
Melting point	330°F, softens above 300°F	330°F, softens above 300°F
Low-temperature properties	No change	No change
Flammability	Burns with difficulty	Burns with difficulty

*Depending on application.

Source: Sampson Rope Manual, Sampson Ocean Systems, Inc.

TABLE 10.9 Comparative Characteristics of Various Fibers Used in Cordage (*Continued*)

	3- and 8-strand, plaited			
	Core polypropylene; cover polyester and polyethlene		Core polyethlene; cover veneer of polyethylene and dacron	
Rope diameter	1 in.	2 in.	1 in.	2 in.
Strength characteristics				
Tensile strength, dry (approximate average), lb	15,000	56,500	14,000	60,000
Recommended factor of safety	6¾	6¾	6¾	6¾
Working strength, lb	2,500	9,400	2,300	10,000
Wet strength compared to dry strength, %	100	100	100	100
Strength per unit of weight or "breaking length" (breaking strength), lb/ft	69,800	68,000	53,000	63,000
Cyclic loading characteristics	Very good	Very good	Good	Good
Individual filament or fiber strength, grams per denier	—	—	—	—
Weight and density characteristics				
Weight per 100 ft, lb	21.5	83.0	26.5	95.0
Specific gravity of fiber	Varies with rope size	Varies with rope size	Varies with rope size	Varies with rope size
Ability to float	Varies with rope size	Varies with rope size	No	No
Elasticity, stretch (approximate)				
Permanent elongation at working load (20% of breaking strength), %	5.9	5.9	4.7	4.7
Working elasticity (temporary stretch under load) at working load (20% of breaking strength), %	7.1	7.1	5.1	5.1
Elongation at 100% load (at break) for broken-in ropes, %	27	27	21	21
Individual filament or fiber elongation, %	—	—	—	—

Surface characteristics		
Rendering qualities, ability to ease out smoothly under load over bitts	Good	Good
Hand (feeling of rope to the touch)	Smooth and hard; after use bristles and fuzz appear but little harshness	Smooth and hard, not slippery; after use becomes fuzzy but not harsh
Water absorbed into fiber (Some water will be held between fibers of all ropes)	Almost zero	Almost zero
Resistance to rot, mildew, and attack by marine organisms (Some marine organisms will attach themselves to any submerged object, including synthetic ropes)	100% resistant	100% resistant
Deterioration		
Due to aging (stored ropes, ideal conditions)	Zero	Zero
Due to exposure to sunlight	Slight	Some slight
Resistance to chemicals		
Acids	Good to excellent	Excellent, except to concentrated nitric acid
Alkalis	Excellent	Excellent
Solvents	Good to excellent	Good to excellent
Wear		
Resistance to surface abrasion	Very good	Very good
Resistance to internal wear from flexing	Very good	Very good
Resistance to cutting (toughness)	Very good	Very good
High- and low-temperature properties		
Melting point	Progressive strength loss above 250°F	Progressive strength loss above 250°F
Low-temperature properties	Some brittleness below −150°F	Some brittleness below −150°F
Flammability	Burns with difficulty	Burns with difficulty

Source: Sampson Rope Manual, Sampson Ocean Systems, Inc.

TABLE 10.9 Comparative Characteristics of Various Fibers Used in Cordage (*Continued*)

	Double braid			
	Core polypropylene; cover nylon		Core and cover polyester	
Rope diameter	1 in.	2 in.	1 in.	2 in.
Strength characteristics				
Tensile strength, dry (approximate average), lb	28,400	106,000	28,400	106,000
Recommended factor of safety	5 to 8*	5 to 8*	5 to 8*	5 to 8*
Working strength, lb	To 5,700*	To 21,200*	To 5,700*	To 21,200*
Wet strength compared to dry strength, %	95	95	100	100
Strength per unit of weight or “breaking length” (breaking strength), lb/ft	120,000	112,000	88,700	82,800
Cyclic loading characteristics	Excellent		Excellent	
Individual filament or fiber strength, grams per denier	Cover 7.5 to 8.3; core 3.0 to 7.0		Cover 6.0 to 7.0; core 3.0 to 7.0	
Weight and density characteristics				
Weight per 100 ft, lb	24.0	95.0	32.0	126.0
Specific gravity of fiber	1.02	1.02	1.38	1.38
Ability to float	Yes	Yes	No	No
Elasticity, stretch (approximate)				
Permanent elongation at working load (20% of breaking strength), %	6 (11.0 wet)		3 (5 wet)	
Working elasticity (temporary stretch under load) at working load (20% of breaking strength), %	6 (11.0 wet)		3 (5 wet)	
Elongation at 100% load (at break) for broken-in ropes, %	30		17	
Individual filament or fiber elonation, %	16 to 30		11 to 13	
Surface characteristics				
Rendering qualities, ability to ease out smoothly under load over bitts	Good		Good	
Hand (feeling of rope to the touch)	Smooth; after use becomes fuzzy with a softer feel		Smooth, not slippery; after use becomes fuzzy with a softer feel	

Water absorbed into fiber (Some water will be held between fibers of all ropes)	3.0%	Less than 1%
Resistance to rot, mildew, and attack by marine organisms (Some marine organisms will attach themselves to any submerged object, including synthetic ropes)	100% resistant	100% resistant
Deterioration		
Due to aging (stored ropes, ideal conditions)	Zero	Zero
Due to exposure to sunlight	Some slight	Almost none
Resistance to chemicals		
Acids	Fair, except to concentrated sulfuric and hydrochloric acids	Very good to excellent
Alkalis	Excellent	Very good, except to concentrated sodium hydroxide at high temperatures
Solvents	Good	Very good to excellent
Wear		
Resistance to surface abrasion	Very good to excellent	Excellent
Resistance to internal wear from flexing	Good	Very good to excellent
Resistance to cutting (toughness)	Very good to excellent	Very good to excellent
High- and low-temperature properties		
Melting point	Good	480°F; progressive strength loss above 350°F
Low-temperature properties	Very good	No change
Flammability	Burns with difficulty	Burns with difficulty

*Depending on application.

Source: Sampson Rope Manual, Sampson Ocean Systems, Inc.

TABLE 10.9 Comparative Characteristics of Various Fibers Used in Cordage (*Continued*)

Regulations and Standards

OSHA Regulations

1910 Subpart N—*Material Handling and Storage*

1910.184 *Slings*

1926 Subpart H—*Materials Handling, Storage, use and Disposal*

1926.25/ *Rigging Equipment for Material Handling*

1926.25/(d) *Natural Fiber and Synthetic Web Slings*

1926.25/(d)(4) *Splicing*

U.S. Consensus Standards

ANSI/ASME B30.23 *Personnel Lifting Systems*

ASME B30.9–2006 *Contains Provisions on Maintaining Inspection Records*

Industry Concensus Standards and Guidelines

Cordage Institute

CI-1401* 2006 *Safer Use of Fiber Rope*

CIS-1 2006 *Rope Standards CI 1201 - CI 1312*
Physical Properties, Design Factors, and Safe Working Loads

CIS-2 2006 *Solid Braid Standards, Complete Set CI 1300 Through CI 322 Includes Standards CI 1201 and CI 1320 Through CI 1321*

Web Sling and Tiedown Association

WSTDA-WS-1-04 *Web Sling Standard*

WSTDA-RS-1-01 *Roundsling Standard*

WSTDA-T-3-07 *Winches Standard*

WSTDA-T-4-06 *Webbing Standard*

WSTDA-WB-1-05 *Synthetic Webbing for Slings*

CHAPTER 11

Knots, Bends, Hitches, and Splices

There are hundreds, if not thousands, of different rope knots, bends, and hitches, but the average riggers can get along with the knowledge of a comparatively few. The term *knot* is usually applied to any tie or fastening formed with a rope. In a general sense, however, knot also includes the words *bends* and *hitches*.

The distinction between knots, bends, and hitches is generally accepted as:

- *Knot*. An intertwining of the end of a rope within a portion of the rope.
- *Bend*. Intertwining of the ends or two ropes, or of the same rope, to make one continuous rope or endless rope, or to fasten a line to a ring or loop.
- *Hitch*. Attachment of a rope to a post, pole, ring, hook, or other object so it will hold temporarily but can be readily untied.
- *Loops*. Made to be dropped over an object, unlike hitches that are made directly around the object and follow its shape. (When synthetic rope is used to tie this knot, it might be less reliable, thus the end should be secured with an extra half hitch, or tucked and wrapped beneath one of the rope's strands.)

NOTE: *A good knot, bend, or hitch is one that can be tied rapidly and easily, holds fast when pulled tight, and is easily untied. Their prime requirements are suitability, strength, and security against slippage.*

REMEMBER: *A rope fastening is never as strong as the original rope—knots, bends, and hitches reduce the rope's strength by 5 to 10 percent. The proper fastening must be selected for the job to be done, and tied correctly to obtain the maximum strength.*

11.1 Knots, Bends and Hitches

Knots, bends, and hitches are made from three fundamental elements: a *bight*, a *loop*, and a *round turn*. The free or working end of a rope is known as the *running end*. The remainder of the rope is called the *standing part*.

One of the main reason for selecting one knot, rather than another, is the relative strength of that particular knot.

Knots

Knots are various methods for fastening or securing a length of one or more segments of rope by tying or interweaving, or webbing, and include:

- *Stoppers*. Most often used to prevent the ends of a length of rope from slipping through an eye or a hole, and can be used to bind the end of a line so that it will not unravel. These include:
    - *Overhand knot*—forms the basic knot of most other knots. It is used as a simple stopper knot in the end of a rope; difficult to untie when wet (Fig. 11.1a).
    - *Figure-eight*—forms a larger knot than would be formed by an overhand knot in the end of a rope to prevent the end from slipping through a fastening or loop in another line. To make the figure-eight knot, make a loop in the standing part, then pass the running end around the standing part, back over one side of the loop and down through the loop, and pull tight (Fig. 11.1b).
    - *Overhand loop*—extremely useful where a bulky stopper knot is required; however, difficult to untie (Fig. 11.1c).
- *Bends*. These are used to join two lengths of rope:
    - *Square (reef) knot*—ties two lines of the same size together so they will not slip. To tie a square knot, first bring the two ends of the line together and make an overhand knot. Then form another overhand knot in the opposite direction (Fig. 11.2a).
- *Sheet bend*—joins ropes of unequal diameters; *not* 100 percent secure, and should *not* be used where it will be subject to great strain (Fig. 11.2b).
- *Double sheet bend*—a strong knot used to tie two ropes (usually of different thicknesses or rigidity) together. It is a doubled version of the sheet bend (Fig. 11.2c).

(a) (b)

(c)

FIGURE 11.1 Stopper knots. (a) Overhand; (b) figure eight; (c) overhand loop.

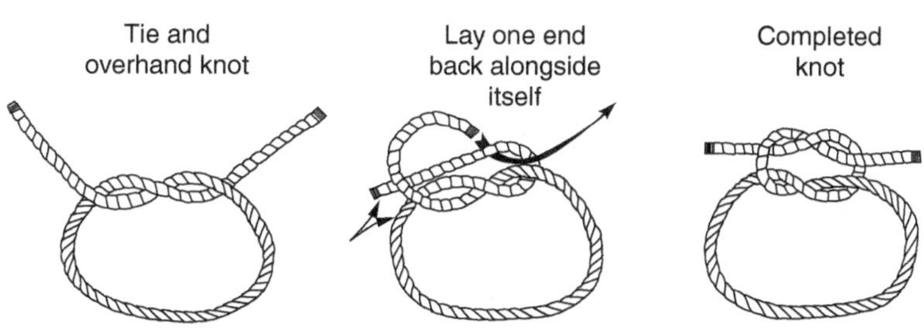

(a) Square knot

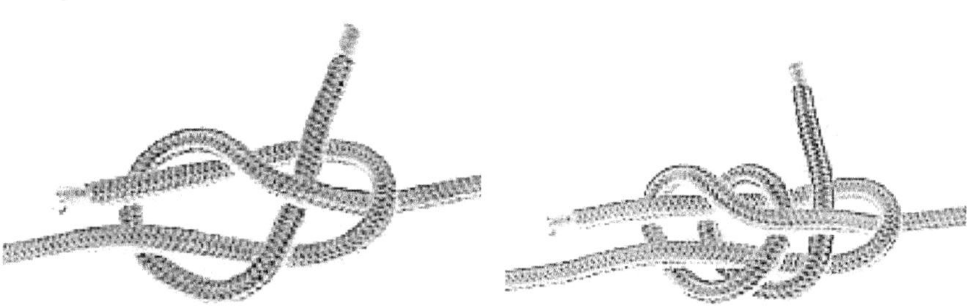

(b) Sheet bend (c) Double sheet bend

FIGURE 11.2 Bends. (a) Square knot; (b) sheet bend; (c) double sheet bend.

NOTE: *A good rule to follow for a square knot is left over right and right over left. When tying a square knot, make sure the two overhand knots are parallel; this means that each running end must come out parallel to the standing part of its own rope. If the knot fails to meet this test, it is known as a "granny" knot, which should never be used; it is unsafe because it will slip under strain. A true square knot instead of slipping under strain will only draw tighter.*

Hitches

These are used for binding a rope to an object:

- *Barrel Hitch.* Lifts a barrel or other rounded object that is either in a horizontal or a vertical position.
- *Cat's paw.* Slings heavy loads, best for rope of medium diameter because the strain is equal on both sides. Even if a single part of a loaded rope is hung over a hook, the rope is weakened by a about one-half. This knot, however, gives the extra assurance that should one leg break the other will last long enough to allow the load to be safely lowered to the ground.
- *Clove hitch.* Tied around a post with one hand; *not* totally secure, however, if the strain is intermittent and at an inconsistent angle. Adding a stopper knot or making one or two half hitches around the standing part will make a clove hitch more secure.
- *Half hitch.* Temporary knot of a single hitch made around the standing part of another hitch; *not* meant to take any strain, but rather used to complete and strengthen other knots.
- *Scaffold hitch.* Useful for suspending timber struts such as those used in scaffolding; can also be used as a clamp knot to hold two objects together while nailing or gluing takes place. When used as a lashing, the hitch should be pulled tight and then sealed off with a reef knot.
- *Timber hitch.* Temporary noose, formed by twisting the working end of a rope around its own part and *not* around the standing part. Three twists are usually sufficient to secure the rope around planks, poles, or tree trunks so that they may be raised or lowered, dragged, or pulled. More twists may be needed if the object to be moved is especially thick (Fig. 11.3).

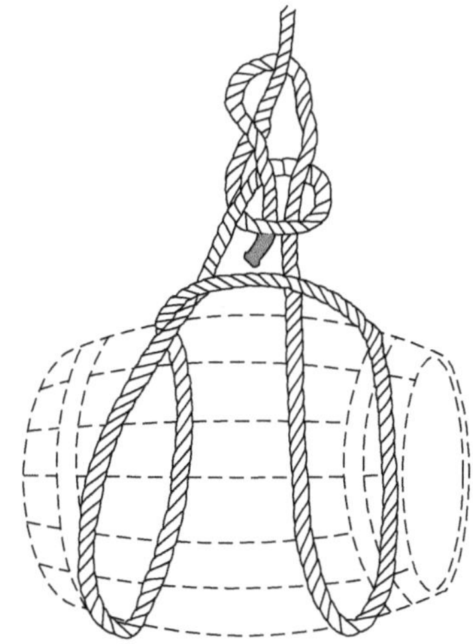

(a) Barrel hitch

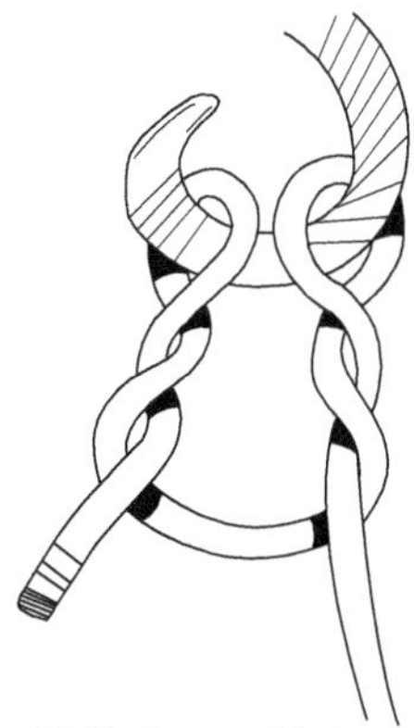

(b) Cat's paw hitch

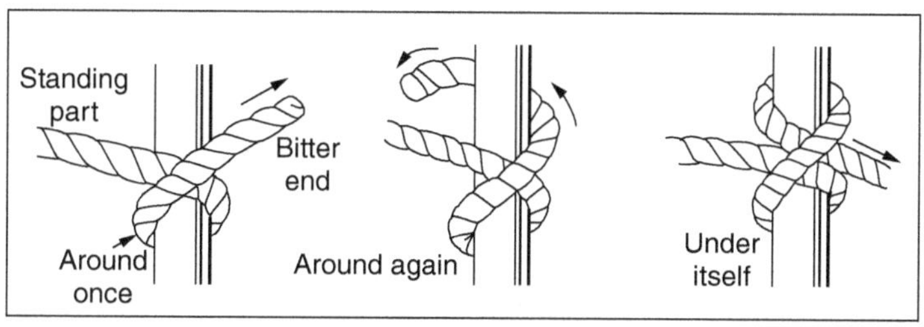

(c) Clove hitch

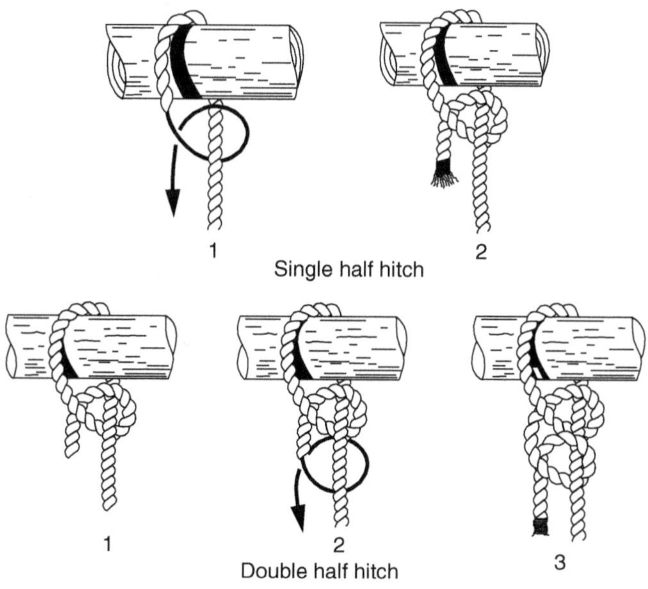

(d) Half hitches

FIGURE 11.3 Hitches. (a) Barrel; (b) cat's paw; (c) clove hitch; (d) half hitches; (e) scaffold hitch; (f) rolling hitch; (g) timber hitch; (h) tautline hitch. (*Continued*)

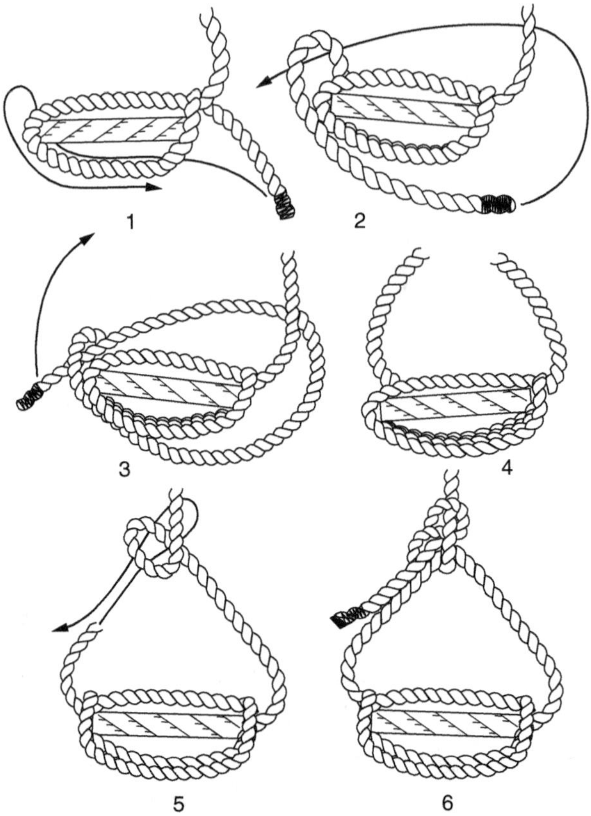

(e) Scaffold hitch

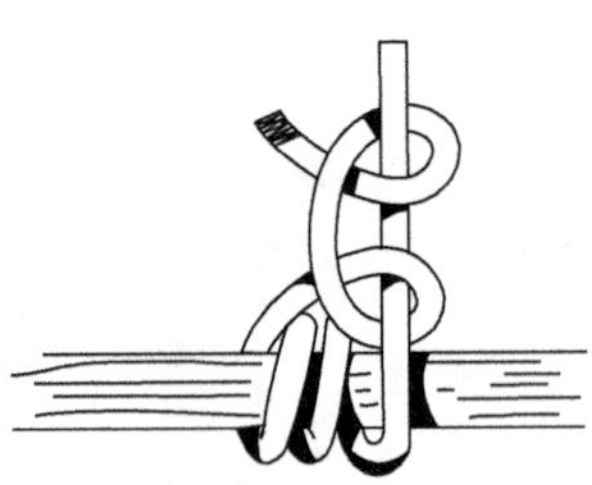

(f) Rolling hitch

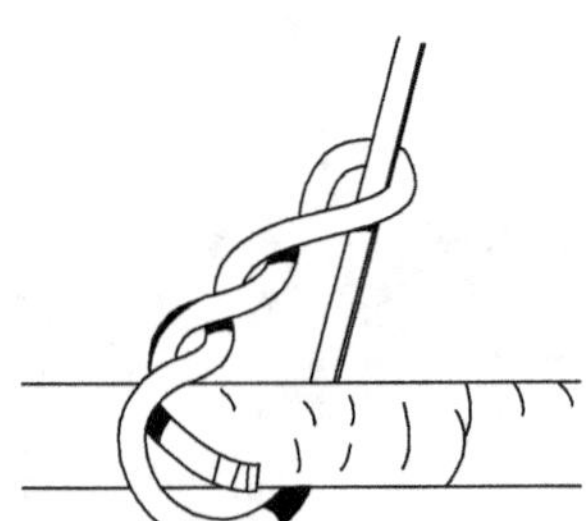

(g) Timber hitch

FIGURE 11.3 Hitches. (a) Barrel; (b) cat's paw; (c) clove hitch; (d) half hitches; (e) scaffold hitch; (f) rolling hitch; (g) timber hitch; (h) tautline hitch. (*Continued*)

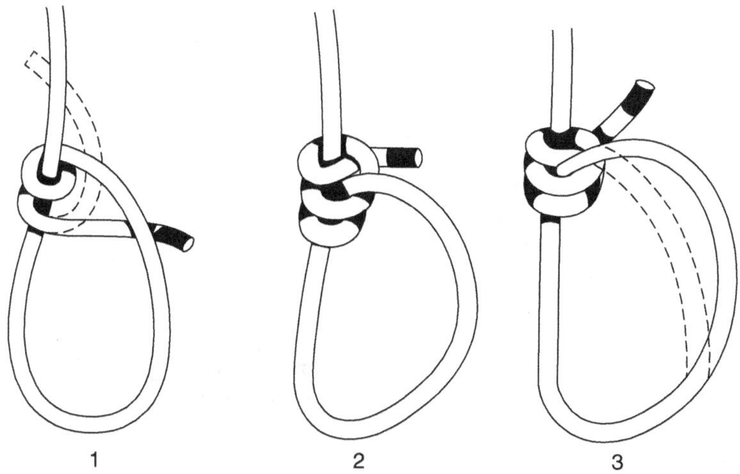

(h) Tautline hitch

FIGURE 11.3 Hitches. (a) Barrel; (b) cat's paw; (c) clove hitch; (d) half hitches; (e) scaffold hitch; (f) rolling hitch; (g) timber hitch; (h) tautline hitch. *(Continued)*

Loops

These are used to create a closed circle in a rope.

- *Bowline.* Especially useful, a temporary eye is needed in the end of a rope; it will neither slip nor jam and can be untied easily (Fig. 11.4).
- *Bowline on a bight.* Used especially in sea rescues, with the victim placing one leg in each loop and hanging on the standing part of the rope.
- *Figure-eight loop knot*
- *Loop knot dressed*

Shortenings

These are used to shorten a rope or take up a weakened part of a rope.

- *Loop knot.* The best way of shortening a damaged rope, taking up the weakened part of the rope in the center of the knot; used for towing equipment on wheels.
- *Sheepshank.* Generally thought of as merely a means to shorten a line, but in an emergency it can also be used to take the load off a weak spot in the line. To make a sheepshank, form two bights. Then take a half hitch around each bight. When using a sheepshank to take the load off a weak spot, *make sure* the spot is in the part of the rope not under strain (Fig. 11.5).

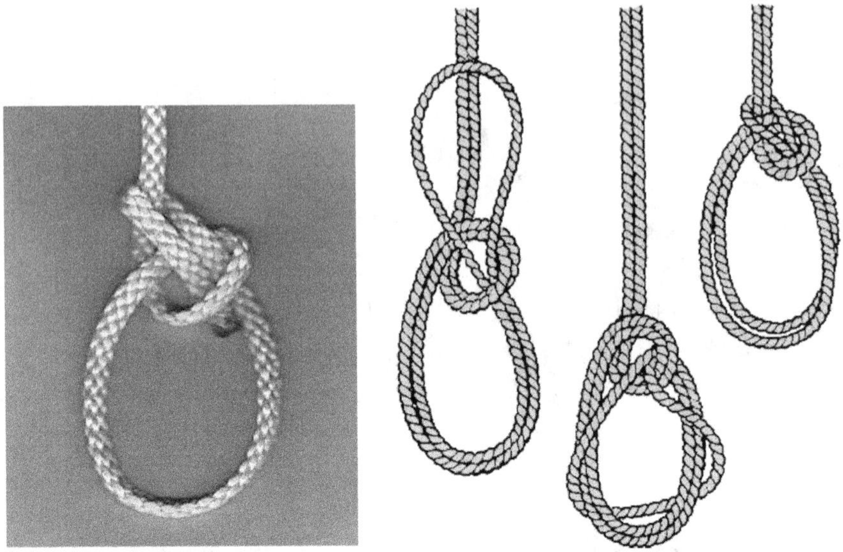

(a) Bowline (b) Bowline on a bight

FIGURE 11.4 Loops. (a) Bowline; (b) Bowline on a bight.

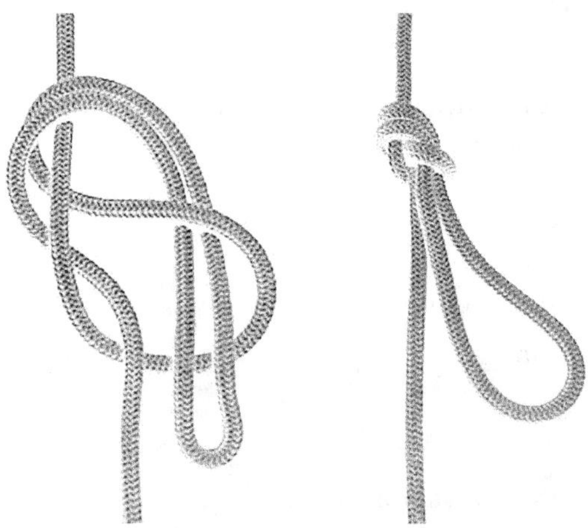

(a) Figure-8 loop knot (b) Loop knot dressed

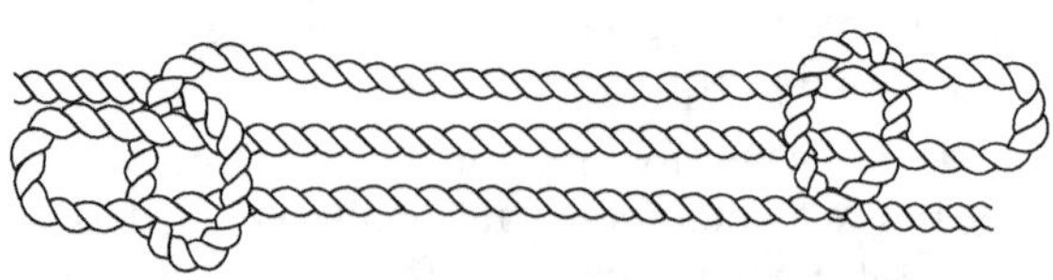

(c) Sheepshank

FIGURE 11.5 Shortenings. (a) Figure-8 loop knot; (b) loop knot dressed; (c) sheepshank.

11.2 Splicing Natural Fiber Rope

Splicing in rope work is the forming of a semipermanent join between two ropes or two parts of the same rope by partly untwisting and then interweaving their strands. Splices can be used to form a stopper at the end of a line, to form a loop or an eye in a rope, or to join two ropes together. When properly installed, splices *do not* seriously affect rope strength.

Splices are preferred to knotted rope, because a knot reduces the strength of a rope by as much as 40 percent or more, while a splice retains up to 95 percent of the rope's strength. However, splicing usually results in a thickening of the line and, if subsequently removed, a distortion of the rope. Most types of splices are used on three strand rope, but some can be done on twelve or greater strand braided rope.

Four general types of fiber rope splices commonly used for rigging operations are discussed below.

Short Splice

Used to join the ends of two ropes, the short splice is more similar to the technique used in other splices and results in the spliced part being about twice as thick as the nonspliced part. The short splice, however, has greater strength than the long splice, and retains more of the rope strength than any knots that join rope ends.

Even though it is the strongest of all splices, the short splice is not suitable for a rope that must run over sheaves or pass through blocks because in a short splice, the ends of a rope are joined together or the ends of two different ropes are joined, causing an increase in the diameter of the line for a short distance.

One purpose for which the short splice is especially useful is in making endless slings. It is also used for making straps. Slings and straps are made of pieces of rope with their own ends short-spliced together.

Where possible, a short splice, rather than a long splice, should be used because the short splice requires less line and can be fashioned quicker than the long splice. To make a short splice, the ends of the two ropes are unlaid for a sufficient distance and placed together with the strands of one rope passing alternately between the strands of the other. Then the two ropes are jammed closely together. Holding the end of one rope and the strand of the other rope firmly in the left hand, pass the middle strand over the strand of the second rope, which goes down to the left of it; then tuck it under, and haul taut (Fig. 11.6). Do the same to each of the other strands in succession, putting them over the strand next to them and under the next one beyond. Turn the rope around and do the same to the other set of strands. Repeat on both ropes.

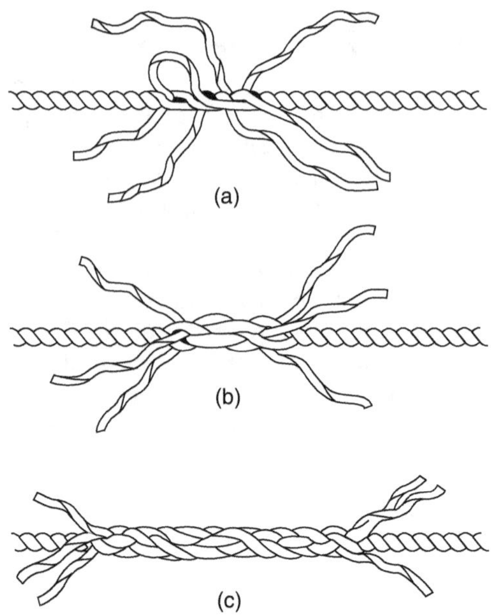

FIGURE 11.6 Making a short splice. (*Source:* U.S. Army Field Manual, Army Department, Washington, DC.)

Take care not to bring two strands up through the same interval in the rope. Each strand should come up separately between the two strands of the rope into which they are passed. To finish the splice, tape it by dividing the yarns after the ends have been interwoven, passing one-half as before and then cutting off the other half. To bring the rope to shape again after splicing, roll it under the foot. For small cord, use a piece of flat wood.

Long Splice

A long splice is used to join two rope ends forming one rope—the length of the two ropes. Unlike most splice types, the long splice results in a splice that is only very slightly thicker than the rope without the splice; however, a long splice sacrifices some of the strength of the short splice by replacing two of the strands of each rope end with those from the other, and cutting off some of the extra strands that result. The long splice allows the spliced rope to still fit through the same pulleys, which is necessary in some applications.

When properly made, the long splice strength will be equal to that of the rope itself.

Although slightly weaker than the short splice, the long splice has definite advantages for specific applications. Because it increases the rope diameter only slightly, it allows the rope to run through a sheave or block easily. A long splice also lessens wear and chafing of

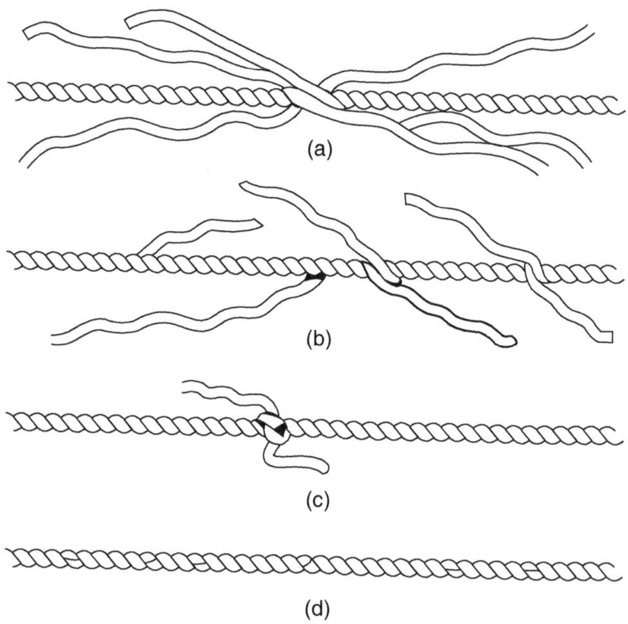

FIGURE 11.7 Making a long splice. (*Source:* U.S. Army Field Manual, Army Department, Washington, DC.)

the rope's fibers at the point of splicing, and has a neater appearance than a short splice.

When making a long splice, it is important to choose the correct pairs of strands for opposites. To determine the correct pair, try laying one of the tucking ends into the opposite standing line. The strand that this tucking end tends to push out and replace will be the correct opposing strand.

To make a long splice, unlay the end of two ropes for a much greater distance (about 15 turns) than with the short splice, and put the ends together. Unlay one strand for some length and fill up the space left by its removal with the opposite strand from the other rope. Do the same with two more strands. Roll and pound the rope so the tucks will settle in tight (Fig. 11.7). Make an overhand knot with the two remaining strands. Divide both strands into halves and pass one half over the next strand and under the following one. Do this, two or three times and cut all the ends off close. Work the remaining two pairs of strands the same way, and the splice is finished.

The rope should be well stretched before the ends of the strands are cut off.

Back Splice (End Splice)

Strands in the end of the rope are spliced directly back into the end without forming a loop. This splice is used to finish off the end of the rope to keep it from fraying.

Making a back splice follows the same principle as with the eye and short splice—over one and under one. It starts from a crown knot (Fig. 11.8). After the crown knot is hauled down tight by heaving on each of the three strands, proceed to lay up the back splice. This merely requires splicing the three loose strands back into the rope, following the same principle as with the eye and short splice—over one and under one. Because the back splice leaves a lump in the rope, it should not be used where there is a possibility of the enlarged end hanging up, such as trying to run it through hoisting blocks.

The end of the rope with the splice is about twice the thickness of the rest of the rope. With nylon and other plastic materials, the back splice is often no longer used; the rope strands are simply fused together with heat to prevent fraying. The strands at the end of rope are spliced back into its own strands. This splice is used to prevent a rope from unlaying when an enlargement at the end of the rope is not objectionable.

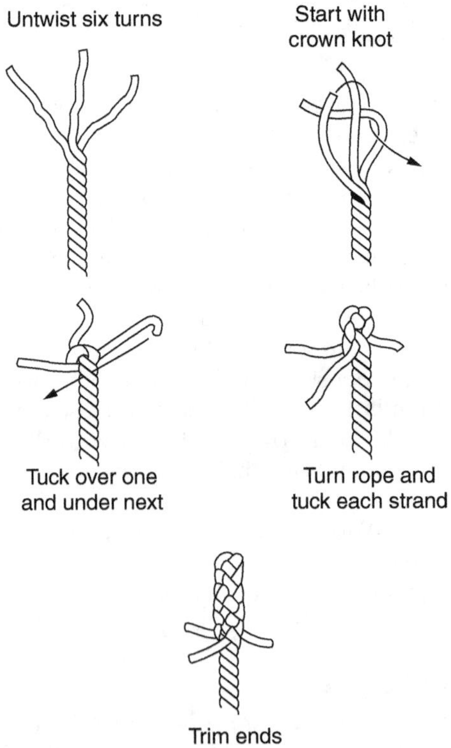

FIGURE 11.8 Making a back splice.

Eye Splice

The eye splice is normally used to form an eye in the end of a fiber rope to fasten the rope to a ring or hook. It is made like the short splice, except that only one rope is used, with the end spliced back into its own side. Metal or nylon thimbles should be set in all eye splices used for lifting to prevent chafing of the rope. The eye splice is estimated to be 90 percent as strong as the rope itself (Fig. 11.9).

Make an eye splice on plain-laid fiber rope by unlaying (untwisting) a portion of the strands, in the end of the rope—about five turns from right to left through the intact portion of the rope in a manner similar to weaving. Place the strands on the standing part to form the eye and splice them into the standing part of the line by tucking the unlaid strands from the end into the standing part.

Each free strand passes between the different rope strands to form the first series of tucks. Thereafter, each free strand passes over one rope strand and under the next rope strand. Repeat this operation until three complete tucks are inserted into natural ropes, or until four complete tucks are inserted into synthetic fiber ropes. The last strand must go through the third strand on the other side of the rope.

An original round of tucks, plus two more complete rounds, is enough for an ordinary eye splice. For synthetic fiber ropes, maintain strand turns for the first two tucks to taper the splice.

Taper an eye splice similarly to a long splice by halving the strands and tucking them securely. With large lines, whip the ends of the strands before starting the splice; otherwise they will frazzle out and

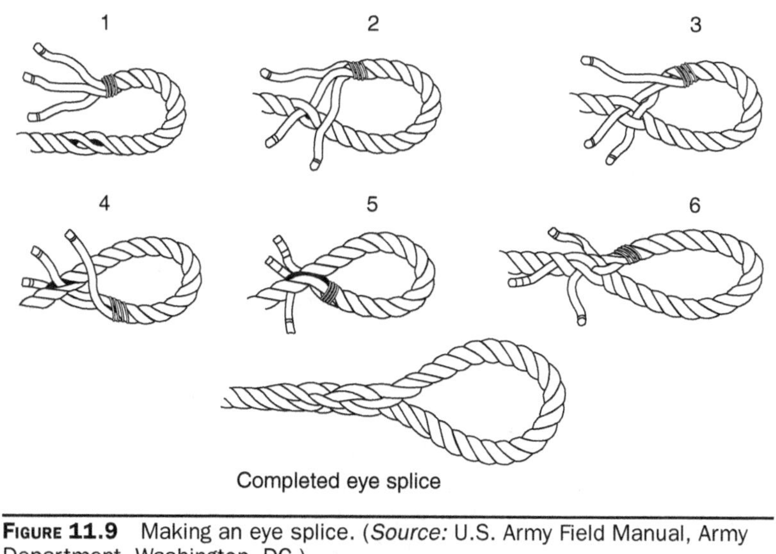

FIGURE 11.9 Making an eye splice. (*Source:* U.S. Army Field Manual, Army Department, Washington, DC.)

result in problems. Large lines must also be seized at the point where unlaying stops or they will cause trouble working them.

Tapering is accomplished by cutting approximately 1/3 of the fibers of each strand, tucking each strand over and under the rope strands, cutting approximately half the remaining fibers of each strand, and completing the final tuck before cutting the remaining loose strands (approximately 2/3 and 1/3 of the fibers remain in the last two tucks).

With any rope up to about 2 in. (50 mm) diameter, pull the strands apart in the rope's standing with fingers. For larger diameter ropes, use a *fid*—a tapered and pointed tool made from maple, hickory, or other hardwood.

Tucks Efficiency

With short and eye splices, four full tucks tapered to finish with one-half and one-quarter tucks will give full efficiency. Medium-sized and large ropes always require tapering the splice to obtain maximum strength. Where very heavy and rapidly fluctuating loads are involved, increase the number of tucks as a safety measure.

To tuck any rope up to 2 in. (50 mm) diameter, open the strands in a standing part with fingers; for larger ropes, use a fid. In addition, with large rope it is necessary to whip the ends of the strands before starting a splice, otherwise they will frazzle out and cause trouble. Large lines must also be seized at the point where unlaying stops or it will be difficult to work them.

Normal splices can be made on synthetic ropes, but because of the smooth surface of the yarns, it is essential to insert extra tucks to maintain full efficiency and prevent slippage. Care must also be taken when splicing synthetic fiber rope to avoid losing the twist and to maintain the form and lay of yarns and strands that will separate easily because of their smooth nature.

When tapering a splice, make sure to leave the yarn ends of the strands with long tails, preferably not less than 1 in. or 2 in. (25 mm or 50 mm), and simply seized to the finished rope.

Whipping Natural Fiber Rope

Fiber rope ends should be whipped (seized) to stop the rope from unraveling. A common whipping requires no tools and the rope does not need to be unlayed. The problem is that it will slide off the end of the rope with little provocation and thus should not be used for rope ends that will be handled. This is avoided in other whippings by interleaving the whipping with the strands of the rope and creating friction with the strands to avoid slipping (Fig. 11.10).

Natural fiber rope normally is whipped with twine—its size dictated by the size of the rope. Artificial fiber ropes should have their ends fused by heat rather than whipped to prevent unraveling.

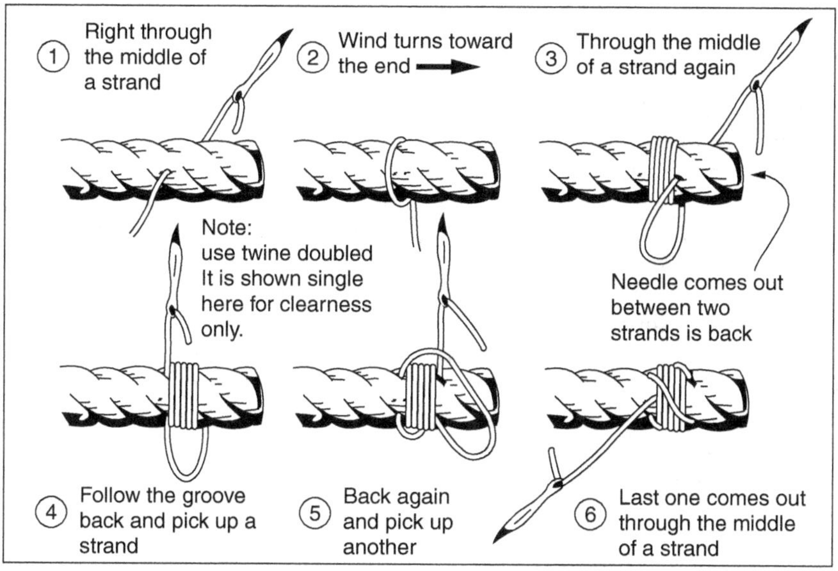

FIGURE 11.10 Whipping (seizing) natural fiber rope. (*Source:* U.S. Army Field Manual, Army Department, Washington, DC.)

A rope should be whipped a short distance (1½ times the diameter) from its end. Begin the whipping by laying the head of the twine along the rope and make a bight back along the rope. Wrap the twine around the rope and bight of twine securely. Continue wrapping until the whipping is 1½ times wider than the rope is thick.

Slip the working end of the twine through the bight and carefully pull on the standing end of the twine until the bight and working end are pulled under the whipping. Normally it is not necessary to maintain tension on the working end to prevent the bight from being dragged completely through and so destroying the whipping.

Cut the twine flush with the edges of the whipping and the rope end not less than half its width from the whipping to give the rope end a finished look.

11.3 Splicing Synthetic Fiber Rope

As with natural fiber rope, when fabricating a synthetic fiber rope sling use splicing. *Do not* use knots, clips, or clamps. If thimbles do not have ears, lash the thimbles to the rope to prevent rotation.

Splicing synthetic fiber rope is similar to that of Manila rope; however, when splicing synthetic fiber rope friction tape should be used instead of seizing stuff for whipping the strands and rope as whipping tends to slip on the smooth synthetic fibers. And, because it is smooth and elastic, nylon rope requires at least one tuck more than does manila rope. For heavy loads, a back tuck should be taken with each strand.

For tuck splices in three- and eight-strand synthetic ropes, no less than four full tucks *must* be used. Short splices contain at least eight full tucks, four on each side of the center splice. Strand-end tails in all tuck splices *must not* be trimmed short.

Synthetic rope slings have a minimum length of ten times the rope diameter between the last tuck of tuck splices or between the ends of the buried tails or strands of other types of splices. The diameter and width of the bearing surface of the fitting can affect the strength of the sling.

A distinct advantage of a sling made with nylon rope is that it can hold a load even when many strands are abraded. Normally, when abrasion is local, the rope may be restored to use by cutting away the chafed section and splicing the ends. Chafing and stretching do not necessarily affect the load-carrying ability of nylon rope.

References

Knots, Bends, Hitches, Splices and Seizings (*www.2020site.org/knots*)
Marlinspike seamanship (*www. globalsecurity.org/military*)
Rope Management & knots
Military Mountaineering Manual-FM 3-97.61 (TC90-6-1) Department of the Army, Washington DC

Regulations and Standards

OSHA Regulations

1926 Subpart H—*Materials Handling, Storage, Use, and Disposal*
1926.251 *Rigging Equipment for Material Handling - Splices*
1926.251(d)(4) *Splicing*

U.S. Consensus Standards

ANSI B20.23 2005 *Personnel Lifting Systems*

Industry Consensus Standards and Guidelines

Cordage Institute

CHAPTER 12

Slings and Hitches

Rigging in its truest sense is the handling of loads suspended from a crane, hoist, or derrick. But, properly designing a piece of hoisting equipment to safely lift a given load on a hook is not the only problem that the rigger faces. The load must be secured to the hook, and the attachment must be of adequate strength to lift the load safely, thus the reason for using a sling and the need for proper sling design.

The dominant characteristics of a sling are determined by the components of that sling. For example, the strengths and weaknesses of a wire rope sling are essentially the same as the strengths and weaknesses of the wire rope of which it is made.

Slings are generally one of six types: natural fiber rope; synthetic fiber rope; synthetic fiber web; wire strand rope or metal mesh; and chain. In general, use and inspection procedures tend to place these slings into three groups: fiber rope web, wire rope and mesh, and chain.

Each type has its own particular advantages and disadvantages. Factors to consider when choosing the best sling for the job include the size, weight, shape, temperature, and sensitivity of the material to be moved, as well as the environmental conditions under which the sling will be used.

12.1 Natural Fiber Rope Slings

Fiber rope slings are preferred for some applications because they are pliant, grip the load well, and do not mar the surface of the load. They should be used only on light loads, however, and must not be used on objects that have sharp edges capable of cutting the rope or in applications where the sling will be exposed to high temperatures, severe abrasion or acids, and caustics. They are used primarily for temporary work, such as painting jobs, and are the best choice for use on expensive loads, highly finished parts, fragile parts, and delicate equipment. Generally, fiber rope slings are discarded rather than serviced or repaired. Operators must always follow the manufacturer's recommendations.

The choice of rope type and size will depend on the application, the weight to be lifted, and the sling angle. Before lifting any load with a fiber rope sling *make sure* to inspect the sling carefully because these slings deteriorate far more rapidly than wire rope slings and their actual strength is very difficult to estimate.

Manila Rope Slings

Manila rope slings are used for the lightest loads; an endless Manila rope looped into a noose around the object to be lifted is satisfactory. Such a sling is inexpensive, lightweight, flexible, and easy to handle. It can bend around the comparatively sharp edges of boxes or crates, but should be padded when passing over the sharp machined edges of metal parts. Even in some heavier jobs, such as handling of steel shafts, which must not be scratched or burred, Manila rope slings are useful (Table 12.1).

Fiber rope slings *must* be spliced in accordance with OSHA minimum requirements, as well as any recommendations from the manufacturer.

Because fiber rope deteriorates on contact with acids and caustics, however, fiber rope slings *must* not be used around those substances unless the manufacturer specifically recommends them for that use.

12.2 Synthetic Fiber Rope Web Slings

Synthetic web slings are used for rigging loads that must not be marred or scratched, such as finely machined, highly polished, or painted surfaces. These slings offer a number of advantages for rigging purposes, including:

- *Softness and bearing width.* Having less tendency to crush fragile loads.
- *Flexibility.* Tending to mold to the shape of the load, thus gripping a load more securely.
- *Elasticity.* Tending to stretch under load and thus able to absorb heavy shocks and to cushion loads.
- *Long life.* Not affected by moisture and certain chemicals.

The most commonly used synthetic web slings are made of nylon, polyester, and polypropylene. They have a number of advantages over slings made of other materials, including:

- *Strength.* Ability to handle loads up to 300,000 lbs (136,363 kg); having a design factor of 5.
- *Convenience.* Conform to any shape.

Nominal rope diameter, in.	Nominal weight per 100 ft, lb	Minimum breaking strength, lb	Eye and eye sling						Endless sling					
					Basket hitch						Basket hitch			
					Angle of rope to horizontal						Angle of rope to horizontal			
					90°	60°	45°	30°			90°	60°	45°	30°
					Angle of rope to vertical						Angle of rope to vertical			
			Vertical hitch	Choker hitch	0°	30°	45°	60°	Vertical hitch	Choker hitch	0°	30°	45°	60°
½	7.5	2,650	550	250	1,100	900	750	550	950	500	1,900	1,700	1,400	950
9/16	10.4	3,450	700	350	1,400	1,200	1,000	700	1,200	600	2,500	2,200	1,800	1,200
⅝	13.3	4,400	900	450	1,800	1,500	1,200	900	1,600	800	3,200	2,700	2,200	1,600
¾	16.7	5,400	1,100	550	2,200	1,900	1,500	1,100	2,000	950	3,900	3,400	2,800	2,000
13/16	19.5	6,500	1,300	650	2,600	2,300	1,800	1,300	2,300	1,200	4,700	4,100	3,300	2,300
⅞	22.5	7,700	1,500	750	3,100	2,700	2,200	1,500	2,800	1,400	5,600	4,800	3,900	2,800
1	27.0	9,000	1,800	900	3,600	3,100	2,600	1,800	3,200	1,600	6,500	5,600	4,600	3,200
1 1/16	31.3	10,500	2,100	1,100	4,200	3,600	3,000	2,100	3,800	1,900	7,600	6,600	5,400	3,800
1⅛	36.0	12,000	2,400	1,200	4,800	4,200	3,400	2,400	4,300	2,200	8,600	7,600	6,100	4,300
1¼	41.7	13,500	2,700	1,400	5,400	4,700	3,800	2,700	4,900	2,400	9,700	8,400	6,900	4,900
1 5/16	47.9	15,000	3,000	1,500	6,000	5,200	4,300	3,000	5,400	2,700	11,000	9,400	7,700	5,400
1½	59.9	18,500	3,700	1,850	7,400	6,400	5,200	3,700	6,700	3,300	13,500	11,500	9,400	6,700
1⅝	74.6	22,500	4,500	2,300	9,000	7,800	6,400	4,500	8,100	4,100	16,000	14,000	11,500	8,000
1¾	89.3	26,500	5,300	2,700	10,500	9,200	7,500	5,300	9,500	4,800	19,000	16,500	13,500	9,500
2	107.5	31,000	6,200	3,100	12,500	10,500	8,800	6,200	11,000	5,600	22,500	19,500	16,000	11,000
2⅛	125.0	36,000	7,200	3,600	14,500	12,500	10,000	7,200	13,000	6,500	26,000	22,500	18,500	13,000
2¼	146.0	41,000	8,200	4,100	16,500	14,000	11,500	8,200	15,000	7,400	29,500	25,500	21,000	15,000
2½	166.7	46,500	9,300	4,700	18,500	16,000	13,000	9,300	16,500	8,400	33,500	29,000	23,500	16,500
2⅝	190.8	52,000	10,500	5,200	21,000	18,000	14,500	10,500	18,500	9,500	37,500	32,500	26,500	18,500

TABLE 12.1 Rated Capacity of Manila Rope Slings, in Pounds (Safety Factor = 5)

- *Safety*. Adjust to the load contour and hold it with a tight, nonslip grip.
- *Load protection*. Will not mar, deface, or scratch highly polished or delicate surfaces.
- *Long life*. Unaffected by mildew, rot, or bacteria and resist some chemical action.
- *Economy*. Low initial cost, long service life.
- *Shock absorbency*. Absorb heavy shocks without damage.
- *Temperature resistance*. Unaffected by temperatures up to 180°F (82.2°C).

Because each synthetic material has unique properties, however, synthetic fiber web slings should be used according to the manufacturer's instructions, especially when dealing with chemically active environments.

- Nylon. *Must* be used wherever alkaline or greasy conditions exist. It is also preferable when neutral conditions prevail and when resistance to chemicals and solvents is important.
- Dacron. *Must* be used where high concentrations of acid solutions—such as sulfuric, hydrochloric, nitric, and formic acids—and where high-temperature bleach solutions are prevalent. (Nylon will deteriorate under these conditions.) *Do not* use Dacron in alkaline conditions because it will deteriorate; use nylon or polypropylene instead.
- Polyester. *Must* be used where acids or bleaching agents are present; is also ideal for applications where a minimum of stretching is important.

Synthetic fiber web slings *must* always be removed from service if any defects exist, such as:

- Acid or caustic burns
- Melting or charring of any part of the surface
- Snags, punctures, tears, or cuts
- Broken or worn stitches
- Wear or elongation exceeding the manufacturer's recommended amount
- Distortion of fittings

12.3 Wire Rope Slings

Most slings used today are made of wire rope. Although they have become the workhorses of the sling field, with exceptionally sound safety features, wire rope slings must be chosen carefully for the services in which they are to be placed since they also have certain limitations.

Like manila rope slings, however, wire rope slings should be padded where they bend over sharp edges of the load to be lifted. Although load edges made of soft material such as wood crate or skid will not cut the sling, individual wires are bent sharply and highly stressed—possibly beyond their elastic limit, even before the live load is applied.

Wire rope is composed of individual wires that have been twisted to form strands. The strands are then twisted to form a wire rope. When wire rope has a fiber core, it is usually more flexible but less resistant to environmental damage. Conversely, a core that is made of a wire rope strand tends to have greater strength and is more resistant to heat damage.

Wire rope slings are a basic material handling tool offering a strong, dependable, and economical option for most lifting applications. Relative to other types of slings, however, wire rope slings are average in strength/weight ratio, average in abrasion and cut resistance, poorest in elongation and flexibility, and average for high temperatures.

Even so, their popularity is enhanced by the numerous sling configurations available to support a broad range of applications, and include:

- *Single, double, triple, and quad leg slings.* These can be designed and fabricated within close tolerances to desired specifications. For safety purposes, it is essential that all fittings and attachments carry a rated capacity equal to or greater than that of the wire rope sling. Where this is not possible, the rated capacity of the entire assembly must be downgraded to the weakest component.
- *Cable laid.* Fabricated from a rope comprised of seven small wire ropes; used where flexibility and resistance to kinking are more important than resistance to abrasion. Since the rope is made up of many smaller wire ropes, the sling can bend around smaller diameters without taking a permanent set or kink. The small wires, however, are susceptible to abrasion.
- *Eight-part braided.* Fabricated from eight pieces of smaller diameter wire ropes, braided together to form one large sling; commonly used for high capacity lifts; can be either flat or round, snug tightly to a load in a choker hitch, resist kinking, and offer the ultimate flexibility and versatility.
- *Hand-tucked splice.* Created by forming an eye and "tucking" and "locking" a strand of the wire rope under other adjacent strands. Although not as strong as mechanically spliced slings, the absence of steel sleeves allows this sling to be easily removed from underneath loads. Caution, however, must be exercised as hand-tucked splices may unravel if the sling rotates during use.

12.4 Sling Construction

Three configurations of fiber and wire rope slings are commonly used for lifting a load.

Endless sling (grommet). Usually referred to simply by the term *sling*, it can be made by splicing the ends of a piece of fiber line or wire rope to form an endless loop.

Single-leg sling. Commonly referred to as a *strap*, it can be made by forming a spliced eye in each end of a piece of fiber or wire rope. Sometimes the ends of a piece of wire rope are spliced into eyes around thimbles, and one eye is fastened to a hook with a shackle. With this arrangement, the shackle and hook are removable.

The single-leg sling maybe used as a choker hitch in hoisting by passing one eye through the other eye and over the hoisting hook; also useful as a double-anchor hitch that works well for hoisting drums or other cylindrical objects where a sling must tighten itself under strain and lift by friction against the sides of the object. Single-leg slings can be used to make various types of *bridle hitches*.

Bridle hitch. Consists of two, three, or four single hitches used together for hoisting an object. The hitch provides excellent load stability when the load is distributed equally among each sling leg, the load hook is directly over the center of gravity of the load, and the load is raised. Using bridle slings requires that the sling angles be carefully determined to ensure that the individual legs are not overloaded.

NOTE: *It is wrong to conclude that a three- or four-leg bridle will safely lift a load equal to the safe load on one leg multiplied by the number of legs. This is because there is no way of knowing that each leg is carrying its share of the load. With a four-legged bride sling lifting a rigid load, it is possible for two of the legs to support practically the full load while the other two legs only balance it.*

Use manufacturer's rated capacity for two-leg bridle slings as the safe working load for three- or four-leg bridle hitches.

Wire Rope Sling Elements

Although wire rope slings are manufactured in a wide variety of configurations, all can be identified in terms of just three elements, described below.

- *Type of sling*—eye and eye, or endless
- *Form of hitch*—vertical, bridle, choker, or basket
- *Angle of legs*—vertical, included, or horizontal

Types of Slings

Eye and eye. Made by forming an eye in each end of a length of wire rope, using either a mechanical or a hand splice.

Endless (grommet). Made of either strand-laid or cable-laid construction. Strand-laid grommets are formed from a single length of wire strand that is laid or twisted helically around itself on each of six successive loops, with the end tucked into the space where a short length of the hem center has been removed. Cable laid grommets are formed of six wire ropes laid around a wire core, with the ends joined together by hand, or mechanically, forming an endless body (Fig. 12.1).

Forms of Hitches

Vertical (straight). Permits the use of full-rated capacity of a sling in a different link between load and lifting device. No load rotation, however, can be permitted with this type of sling, because there is a danger of the rope losing its twist or lay.

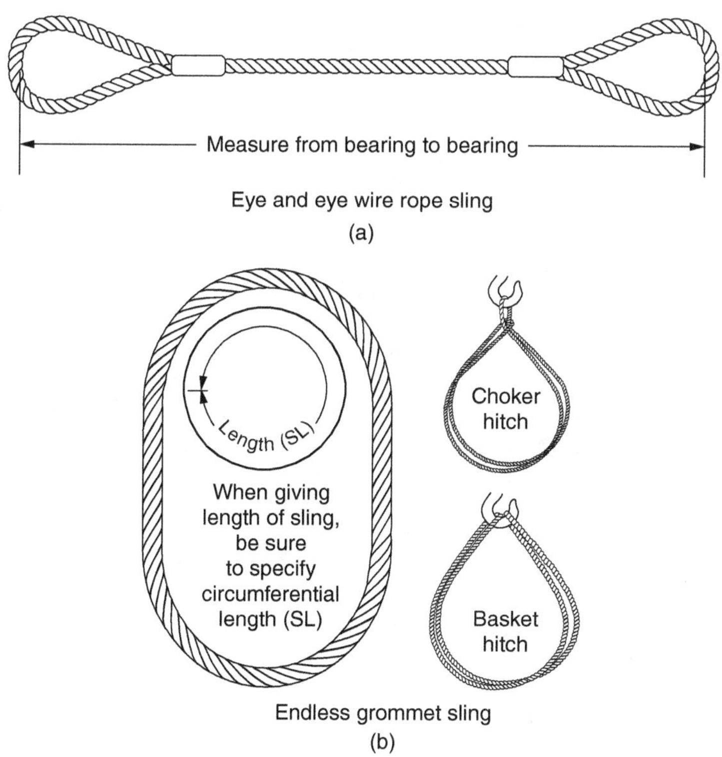

FIGURE 12.1 Basic types of fiber and wire rope slings. (a) Eye and eye (*Source*: American Sling Company, Inc., Hurst, TX.); (b) endless grommet (*Source*: Ashley Sling Inc., Atlanta, GA.)

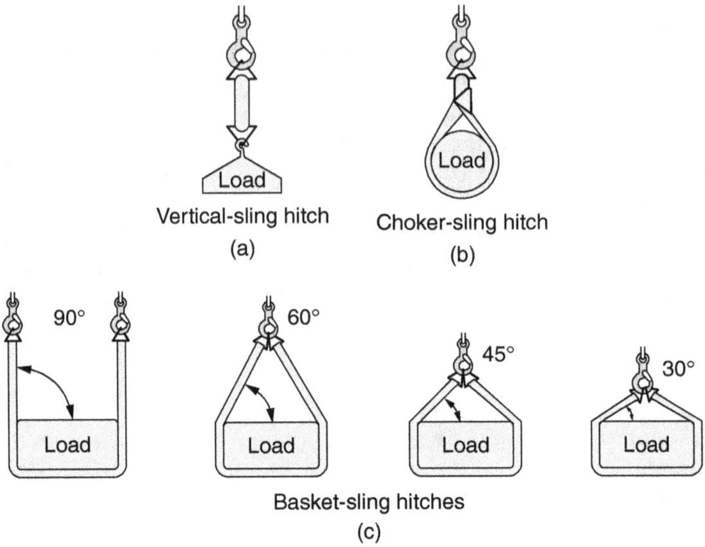

FIGURE 12.2 Basic forms of fiber rope sling hitches. (a) Vertical; (b) choker; (c) basket. (*Source*: F.D. Lake, Grand Rapids MI.)

Choker. Used to hold irregular or unbalanced loads securely. Rated capacities, however, are reduced by 25 percent—as compared to a straight attachment—since the noose configuration affects the ability of the component wires and strands in the rope to adjust during the lift, and thereby can actually damage the body at the point of choke.

Basket. Used in pairs for better balance and greatest strength, distributing load evenly between the two ends of a sling. Since the sling's two ends are not vertical, however, lifting capacity may not be double the capacity of a vertical hitch with the same sling (Fig. 12.2).

Bridle. Combination of two, three, or four single hitches, used together for hoisting an object that has the necessary lifting attachments (Fig. 12.3). A bridle hitch provides excellent load stability when the load is distributed equally among the legs, with the hook located directly over the center of gravity of the load. Its use, however, requires that sling angles be determined carefully to ensure that individual legs are not overloaded.

Angle of Hitch Legs

Unless the sling angle is stated, the lifting capacities are misleading because the capacity of a sling leg is reduced as the rigging angle is reduced (Table 12.2).

Whenever possible, sling angles should be greater than 45° from the horizontal; use of any sling with legs at angles less than 30° from

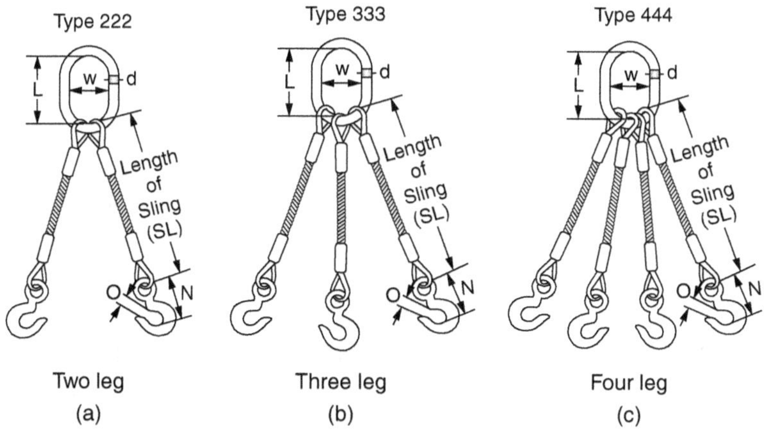

FIGURE 12.3 Wire rope bridle slings. (a) Two leg; (b) three leg; (c) four leg. (*Source*: Safety Sling, Pittsburgh, PA.)

the horizontal is extremely dangerous and not recommended because of effect on the load that would result from an error in sling angle measurement of as little as 5° (Fig. 12.4).

Comparing the rated capacities of a sling on the manufacturer's charts readily indicates the loss of capacity between a straight sling with one vertical leg and a sling with angled legs.

Rigging the lifting device close to the load, thereby increasing the angle at which the sling legs are spread (the *included* angle),

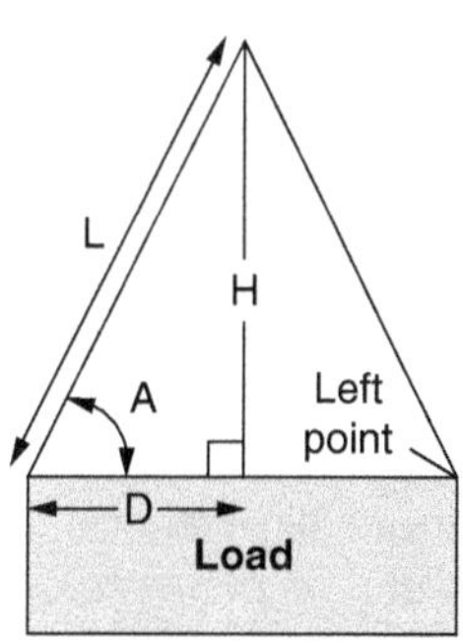

Sling angle (A)	Sling angle factor	Sling angle (A)	Sling angle factor
90°	1.000	55°	0.819
85°	0.996	50°	0.766
80°	0.985	45°	0.707
75°	0.966	40°	0.643
70°	0.940	35°	0.574
65°	0.906	30°	0.500
60°	0.866		

To calculate the reach for equal legs needed to rig at a 45° or 60° angle

Reach at 45°: L = 1.4 × D Reach at 60°: L = 2 × D

Source: F.D. Lake, Grand Rapids, MI.

TABLE 12.2 Estimating Lift Angle of Basket and Bridle Hitches

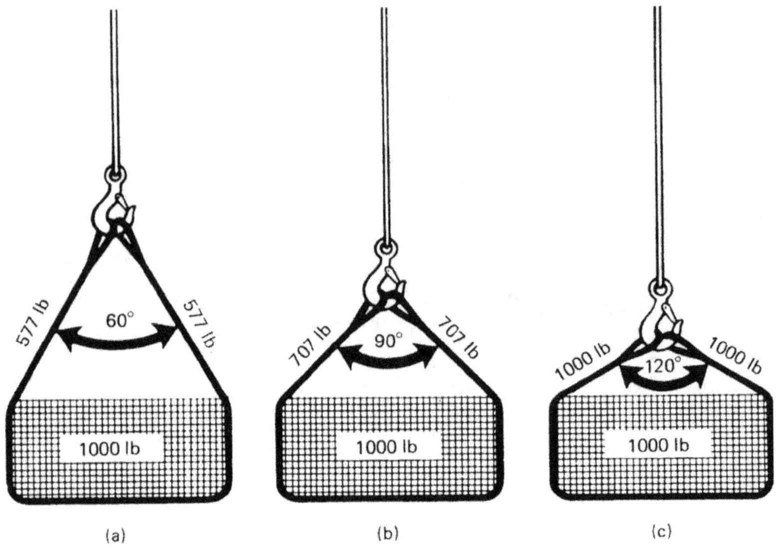

FIGURE 12.4 Stresses induced in multilegged slings due to angle between legs. (*Source*: Wire Rope Works Inc. (formerly Paulsen Wire Rope Corp.), Williamsport, PA, *www.wireropeworks.com*.)

sharply increases the tension placed on the legs of a sling. Although the length and width of the load, the sling length, and the available headroom will determine the sling angle that can be used to lift a particular load, it is imperative that the sling-leg angle be kept as large as possible.

A sling-leg angle less than 30° is neither economical nor good practice. Such angles not only build up the tension in the sling legs out of all proportion to the weight of the load, they also create a much greater "in pull" on the ends of the load, thus producing an eccentrically loaded column effect (long, slender objects have a tendency to buckle).

> **Rule of thumb**: When the sling leg angle (as measured from the horizontal) is:
>
> 60°—the lifting capacity is reduced by about 15 percent
> 45°—the lifting capacity is reduced by about 30 percent
> 30°—the lifting capacity is reduced by about 50 percent

Never assume that a three- or four-leg hitch will safely lift a load equal to the safe load on one leg multiplied by the number of legs. There is no way of knowing that each leg is carrying its share of the load—two loads may take the full load, while the other legs serve only to balance the load.

12.5 Alloy Steel Chain Slings

Chain slings can be manufactured using permanent, welded coupling links, or if preferred, mechanical coupling links for quicker "in the field" assemblies. Either way, the sling must have an attached identification tag providing the grade, size, reach, type of sling, working load limit at a specific angle of lift, and serial number.

Relative to other types of slings, chain slings have the poorest strength/weight ratio, best abrasion and cut resistance, average elongation and shock resistance, best flexibility, and best resistance to high temperature. They are commonly used because of their strength and ability to adapt to the shape of the load. They should be used to lift heavy objects with sharp edges that could cut wire or are hot.

Only alloy steel chains *must* be used for overhead lifting operations. They are flexible, durable and long lasting, ductile, easy to inspect, collapsible for convenient storage, and will adhere securely to the contours of a load. Although chain slings have little flexibility, they grip the load well.

Despite giving no warning of failure, chain still is better suited for some jobs because it is more resistant to abrasion, corrosion, and heat. In particular, chain slings are the best choice for lifting very hot materials. They can be heated to temperatures up to 1000°F (537.78°C). When alloy chain slings are consistently exposed, however, to service temperatures in excess of 600°F (315.6°C), operators must reduce the working load limits in accordance with the manufacturer's recommendations.

Care should be taken when using alloy chain slings because they are subject to damage by sudden shocks. Misuse of chain slings could damage the sling, resulting in sling failure and possible injury to an employee.

12.6 Selecting Proper Slings

Slings must be selected in accordance with their intended use, based on the size and type of load, and the environmental conditions of the workplace.[1]

- *Determine weight.* Make sure that the load to be lifted is known. If not, take proper steps to ensure the sling has more than adequate rated capacity.
- *Decide on hitch.* Make sure that the hitch accommodates the load's shape and size as well as its weight. Selection must

[1]*Source:* Wire Rope Producers Committee of the American Iron and Steel Institute, Washington, DC.

take into consideration any possible physical damage to the load as well as providing a positive attachment. The hitch chosen may affect the choice of sling construction.

- *Check lifting device.* Make sure that the lifting device has sufficient capacity, is in proper working condition, and provides the maneuverability required once the load is hoisted, and that the sling fits on this device.
- *Consider room to lift.* Make sure that the lifting device has sufficient head room to pick up the load and handle it when the length of the sling is added to the hook.
- *Determine sling length.* Make sure to use the longest sling possible for completing the lift, since the longest sling will provide the smallest angle of spread between the legs for minimum stress on the sling.
- *Check rated capacity.* Ascertain the sling's safe working load capacity. *Do not guess.* Use the manufacturer's rated capacity chart. Double-check that the sling length, type, and diameter chosen, when rigged at the selected angle, will accommodate the load to be lifted.
- *Determine leg configuration.* Make sure to use the correct leg configuration. Whether to use a single-part sling (one rope) or a multipart sling (several ropes) involves the handling characteristics of the sling more than any other factor. Considering the capacity alone, multipart slings will be more flexible, more easily handled, than single-part slings, and will often provide the only practical means for handling extremely heavy lifts.
  Estimate the proper angle between the leg of the sling and the load.
  Select the proper fittings (hooks, master rings, etc.).
  Determine the overall reach (measured from the bearing point on the master link to the bearing point of fitting).
  Choose chain size that meets the required work load, angles, and reduction factor.
  Choose the grade, type, and finish of steel that meets job requirements.

12.7 Safe Working Loads for Slings

Formulas for estimating loads in most sling configurations are based on the safe working loads (SWL) of the single-vertical hitch of a particular sling. The efficiencies of the end fittings used also have to be considered when determining the capacity of the combination.

Bridle hitch—with two, three, or four legs:

$$SWL = SWL_1 \times H/L \times 2$$

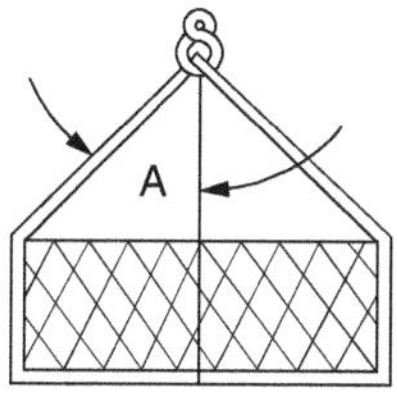

Leg angle A	Reduction
30°	14%
45°	30%
60°	50%
75°	75%

TABLE 12.3 Reduction in Lifting Capacity When Lifting with Basket Hitch

where SWL_1 = for single-vertical hitch
H = height
L = length

When the sling legs are not of equal length, use the smallest H/L measurement. This formula is for a two-leg bridle hitch, but its use is strongly recommended also for the three- and four-leg hitches (Table 12.3).

NOTE: *Never assume that a three- or four-leg hitch can safely lift a load equal to the safe load on one leg multiplied by the number of legs.*[1]

Single-basket hitch:

Vertical legs: $SWL = SWL_1 \times 2$

Inclined legs: $SWL = SWL_1 \times H/4$

Double-basket hitch—angles of 45° or more:

$$SWL = SWL_1 \times 3/4 \text{ (or .75)} \times H/L \times 2$$

Angles of less than 45° generally are not recommended; however, if they are used:

$$SWL = SWL_1 \times A/B \times H/L \times 2$$

NOTE: *Make sure when lifting heavy loads that the bottoms of the sling legs are fastened to the load to prevent damage to the load.*

Many pieces of equipment have eyes fastened to them during the process of manufacture to aid in lifting. With some loads, though, fastening a hook to the eye on one end of each sling leg suffices to secure the sling to the load. Use a protective pad when a fiber line or

[1]*Source*: Safe Working Loads for single vertical hitch (SWLv) listed under OSHA—Guidance on Safe Sling Use – Tables 1 through 26.

Size of rope	Slip noose	Two end bridle sling			Three-end bridle sling		
	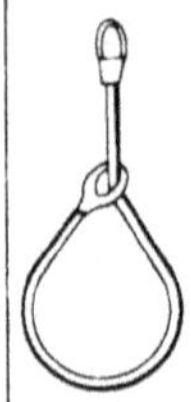	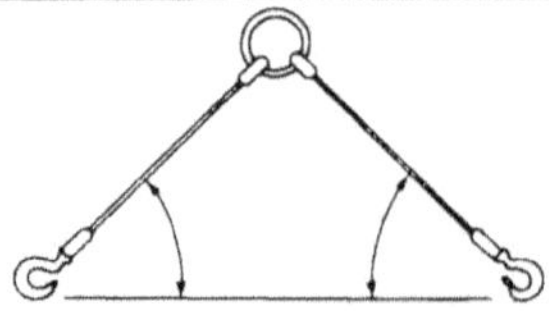 When angle is			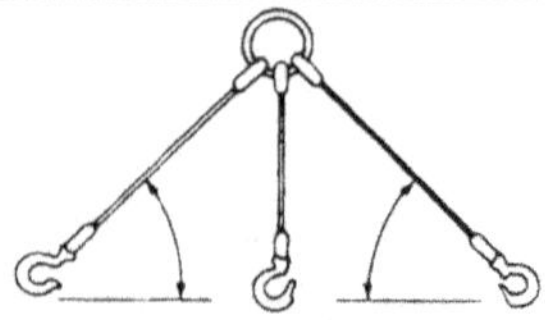 When angle is		
		75°	60°	45°	75°	60°	45°
3/8	3,200	3,000	2,800	2,200	4,600	4,200	3,400
1/2	5,400	5,200	4,800	4,000	7,800	7,200	6,000
5/8	8,400	8,000	7,200	6,000	12,000	10,800	9,000
3/4	12,000	11,400	10,200	8,400	17,000	15,400	12,600
3/8	16,000	16,000	14,000	11,400	24,000	21,000	17,200
1	22,000	20,000	18,000	15,000	30,000	26,000	22,000
1 1/8	26,000	26,000	23,000	19,000	40,000	36,000	28,000
1 1/4	32,000	32,000	28,000	23,000	48,000	42,000	36,000
1 3/8	40,000	38,000	34,000	28,000	58,000	52,000	42,000
1 1/2	46,000	44,000	40,000	32,000	66,000	60,000	48,000

Source: AMHOIST—American Hoist & Derrick Co.

TABLE 12.4 Safe Working Loads for Typical Slings, in Pounds

wire rope sling is exposed to sharp edges at the corners of a load. Pieces of wood or old rubber tires are fine for padding (Table 12.4).

To ensure safe use of slings between scheduled inspections, each operator should keep a close watch on those slings being used. If any accident involving the movement of materials occurs, the operator *must* immediately shut down the equipment and *report* the accident to a supervisor.

The cause of the accident *must* be determined and corrected before resuming operations.

Sling Angle

The rated capacity of any sling depends on the size, the configuration, and the angles formed by the legs of the sling and the horizontal. A sling with two legs, used to lift a 1000-lb object, will have 500 lbs of the load on each leg when the sling angle is 90°. The load stress on each leg increases, however, as the angle decreases.

For example, if the sling angle is 30° when lifting the same 1000-lb object, the load is 1000 lbs on each leg. Try to keep all sling angles greater than 45°. *Avoid* sling angles approaching 30°—they are considered hazardous (Table 12.5).

REMEMBER: *When using slings, the greater the angle from the vertical the greater the stress on the sling legs.*

Largest angle between any leg and vertical*	Capacity reduction factor (cosine)
0°	1.000
5°	0.996
10°	0.985
15°	0.966
20°	0.940
25°	0.906
30°	0.866
35°	0.819
40°	0.766
45°	0.707
50°	0.643
55°	0.574
60°	0.500
65°	0.423
70°	0.342
75°	0.259
80°	0.174
85°	0.087
90°	0.000

*Normally a sling should not be used when the vertical angle exceeds 45°.

Source: Paulsen Wire Rope Corporation.

Table 12.5 Sling Capacity Reduction Factors

12.8 Safe Lifting Practices with Slings

Each lifting job varies in size, shape, weight, and the location of lift points for sling attachment. A good rigger *must* learn to recognize the difference between a good hitch and an unsafe one, know where and how to attach a sling, and be able to anticipate how a load will react when lifted onto the sling. Improper rigging is not only hazardous it can be costly and damaging to both the load and the sling.

The rigger *must* exercise intelligence, care, and common sense in selecting a sling in accordance with the sling's intended use, based on the size and type of load and the environmental conditions of the workplace. All slings *must* be visually inspected before use to ensure that there is no obvious damage.

A well-trained rigger can prolong the service life of equipment and reduce costs by avoiding the potentially hazardous effects of overloading equipment, operating it at excessive speeds, taking up slack with a sudden jerk, and suddenly accelerating or decelerating equipment.

The operator *must* look for causes and seek corrections whenever a danger exists, and should cooperate with coworkers and supervisors and become a leader in carrying out safety measures—not merely for the good of the equipment and the production schedule, but more importantly, for the safety of everyone concerned.

There are four primary factors the operator *must* consider to lift a load safely:

1. *Size, weight, and center of gravity of the load.* To make a level lift, the crane must be directly above the load's center of gravity, that point at which the entire weight may be considered as concentrated. While slight variations are usually permissible, if the crane hook is too far to one side of the center of gravity, dangerous tilting will result causing unequal stresses in the different sling legs. This imbalance must be compensated immediately.
2. *Number of legs and angle the sling makes with the horizontal line.* As the angle formed by the sling leg and the horizontal line decreases, the rated capacity of the sling also decreases—that is, the smaller the angle between the sling leg and the horizontal, the greater the stress on the sling leg and the smaller (lighter) the load the sling can safely support; larger (heavier) loads can be safely moved if the weight of the load is distributed among more sling legs.
3. *Rated capacity of the sling.* This varies depending on the type of sling, the size of the sling, and the type of hitch. Operators must know the capacity of the sling. Sling manufacturers generally make available charts or tables that contain this information. The values given are for new slings. Older slings must be used with additional caution. Under NO circumstances shall a sling's rated capacity be exceeded.
4. *History of sling's care and use.* Mishandling and misuse of slings are the leading causes of sling-related accidents. The majority of injuries and accidents, however, can be avoided by operators becoming familiar with the essentials of proper sling care and use.

Proper care and use are essential for maximum service and safety. Slings must be protected with cover saddles, burlap padding, or wood blocking as well as from unsafe lifting procedures such as overloading to prevent sharp bends and cutting edges.

Before making a lift, check to be certain that the sling is properly secured around the load and that the weight and balance of the load have been accurately determined. If the load is on the ground, *do not* drag the load along the ground. This could damage the sling. If the load is already resting on the sling, ensure that there is no sling damage prior to making the lift.

Next, position the hook directly over the load and seat the sling squarely within the hook bowl. This gives the operator maximum lifting efficiency without bending the hook or overstressing the sling.

Wire rope slings also are subjected to damage resulting from contact with sharp edges of the loads being lifted. These edges can be blocked of padded to minimize damage to the sling.

After the sling is properly attached to the load, a number of good lifting techniques common to all slings include the following:

- Make sure that the load is not lagged, clamped, or bolted to the floor.
- Guard against shock loading by taking up the slack in the sling slowly; apply power cautiously to prevent jerking at the beginning of the lift, and slowly accelerate or decelerate.
- Check the tension on the sling, raise the load a few inches, stop, then check for proper balance and that all items are clear of the path of travel.
- *Never* allow anyone to ride on the hood or load.
- Keep all personnel clear while the load is being raised, moved, or lowered; crane or hoist operators should watch the load at all times when it is in motion.
- *Never* allow more than one person to control a lift or give signals to a crane or hoist operator except to warn of a hazardous situation.
- *Never* raise the load more than necessary.
- *Never* leave the load suspended in the air.
- *Never* allow a worker (or anyone else) to ride a suspended load.

Once the lift has been completed, clean the sling; check it for damage, and store it in a clean, dry airy place—on a rack or wall.

REMEMBER: *Damaged slings cannot lift as much weight as new or older well-cared slings. Proper and safe use and storage of slings will increase their service life.*

Center of Gravity

It is extremely important that everyone involved in the lifting and transporting of loads be familiar with the fundamentals of statics and the effect of the center of gravity of a body in relation to the lifting point, or the distribution of forces in the sling and body (Fig. 12.5).

For example, in a uniform, rectangular load the center of gravity is directly beneath the intersection of the diagonals. As a rule of thumb, when an irregular-shaped object is to be lifted, it is helpful to visualize the load as "enclosed" by an imaginary rectangle, and the center of gravity can be perceived as a point where diagonals of this imaginary rectangle intersect.

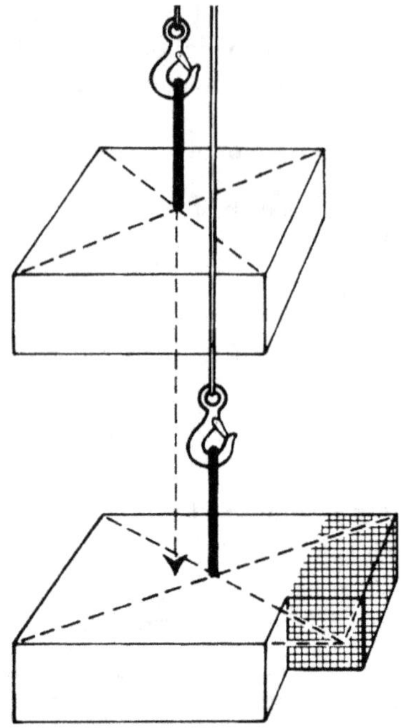

Figure 12.5 Compensating for center of gravity in irregularly shaped objects. (*Source*: Wire Rope Works Inc. (formerly Paulsen Wire Rope Corp.), Williamsport, PA, *www.wireropeworks.com*.)

Keep in mind, however, that such a general rule cannot apply to objects that have weight concentrated at one end or side. The effects of such forces in equilibrium include the following:

- *Center of gravity is under the crane hook.* The angles of lift are equal and each sling leg is carrying the same load.
- *Center of gravity is not in line with the point of lift.* This requires the use of unequal leg lengths to put the center of gravity under the point of lift and balance the load. The rated capacity of the sling must be based on the smallest horizontal angle.
- *Center of gravity of the load is above the points of attachment.* When lifted, the load will rotate to bring the center of gravity to the lowest point. Whenever possible, have the points of attachment above the center of gravity of the load.
- *With a two-legged sling, the load will tilt and possibly slip from hooks unless precisely balanced.*
- *With a three-legged sling the third leg provides stability.* One leg, however, is carrying half the weight of the load, and the capacity should be based on a two-legged sling at some horizontal lift angle.
- *With a four-legged sling there is good stability with the load if equally distributed on each leg.*

- *When the load is lifted by a sling, the legs of the sling will exert lifting forces on the load to raise it.* The legs will also exert other forces, which are horizontal components of the loads on the sling legs. The forces, however, can crush or deform the load if it cannot withstand these. Therefore, when fragile loads are involved, a spreader beam is recommended (Fig. 12.6).

The center of gravity is the point in a load at which all the weight can be said to be concentrated during a lift. It acts downward to bring the load to a position of equilibrium directly below the lifting hook, even though the load may not be level.

Before attempting to lift a load that may be much heavier on one end than on the other, or where eyebolts or lifting lugs may not have been located with respect to the center of gravity of the load, the rigger must estimate the location of the load's center of gravity and spot the crane hook directly over this theoretical point (Fig. 12.7).

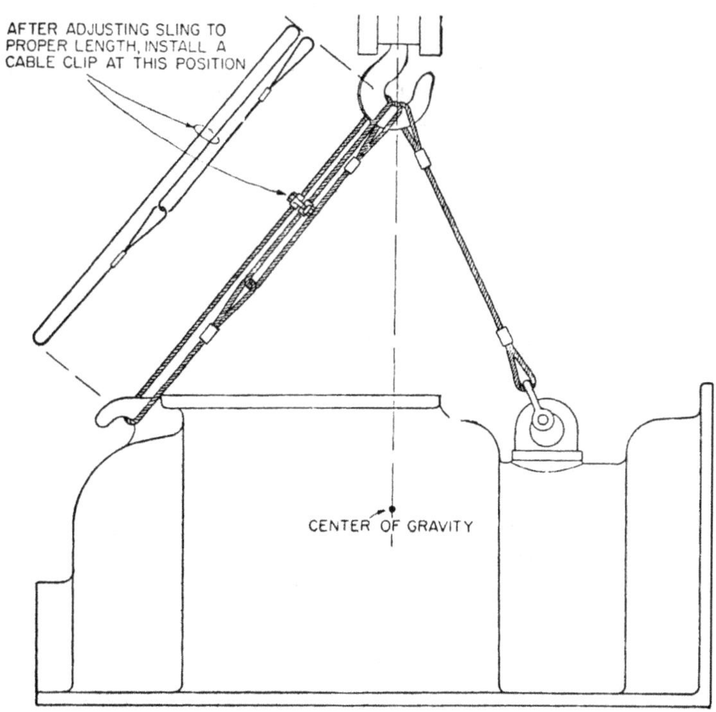

FIGURE 12.6 Center of gravity not midway between hitch points; unequal legs.

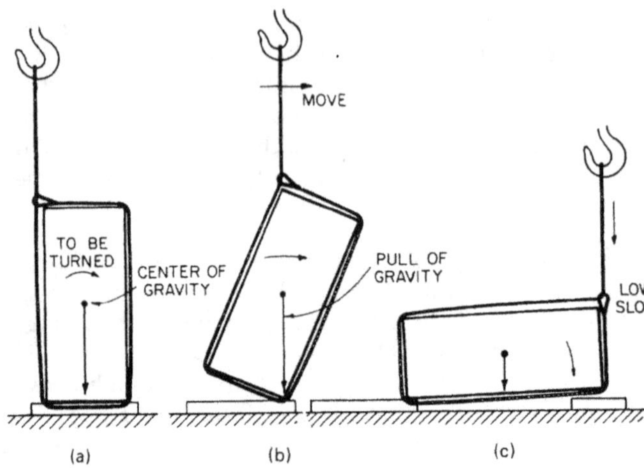

FIGURE 12.7 Proper hitch for turning a load on its side.

12.9 Care and Maintenance of Slings[1]

When not in use, all slings should be hung up in an orderly manner on special hooks or brackets to keep them as straight as practicable.

All slings should be inspected periodically and condemned when found to be in an unsafe condition. Inspect all slings in service daily, and remove from service a sling that shows defect, including:

- Abnormal wear
- Powdered fiber between strands
- Broken or cut fibers or strands
- Variation in size or roundness of strands
- Discoloration or rotting (fiber rope)
- Distortion of sling hardware

Fiber ropes and synthetic webs are generally discarded rather than serviced or repaired. Operators must always follow the manufacturer's recommendations.

Protect all slings from sharp corners by using wooden blocks and pads at lift contact points. (A protective pad should be used any time a sling passes around a sharp corner.) Examine every sling visually from end to end before every lift. And examine them carefully to determine that the slings are in as new condition as practicable before each lift. Remember, the manufacturer's capacity applies only to a new sling in unused condition.

---

[1]*Source:* LiftAll Co. Inc., Landsville, PA.

When slings and rigging are initially broken out for field use, they *must* be proof-tested and tagged before being returned for storage. Care in the use and handling of slings will prolong their operational life significantly. Factors that affect the strength of slings include:

- *Abuse.* Avoid abusing slings by actions such as dragging them on the ground through oil, chemicals, or water; running over them with vehicles; contact with flying welding or burning slag; or unnecessary exposure to ultraviolet rays from sunlight.
- *External damages.* Primary causes of sling failure are the cutting of synthetic slings and the nicking or gouging of steel and chain slings. These damages are usually caused by sharp or small diameter load edge against the sling.
- *Improper loading.* Shock loading, unbalanced loading, overloading, and inadequate consideration for the effect of angle factors can adversely affect safety. Make sure the load weight is within the rated capacity of the sling(s) being used for both type of hitch and angle of lift.
- *Temperature.* Avoid loads and environments where temperatures exceed the limits of the slings being used. All slings can be damaged by excessive heat.
- *Punctures and abrasions.* Sling strength is seriously degraded from punctures (synthetic fiber ropes) and abrasion (fiber and wire ropes). Rough load surfaces and dragging slings on the ground will damage all slings—fiber as well as wire ropes. Use proper padding between slings and rough loads; and *never* drag slings on the ground or over concrete floors.
- *Foreign matter.* Material such as metal strips and heavy grit can damage web slings, externally and internally. And, both synthetic and steel slings can be damaged by weld spatter and heat from a welding torch. Avoid contact with foreign matter whenever possible.
- *Ultraviolet light.* Nylon and polyester web slings are adversely affected by prolonged exposure to UV light (sunlight or arc welding). Inspect and remove slings that appear bleached and stiff.
- *Improper storage.* Even in storage, synthetic and steel slings can degrade if not kept in clean, dry conditions.

    Wire rope slings and associated hardware should be stored either in coils or on reels, hung in the rigging loft, or laid on racks indoors to protect them from corrosive weather and other types of damage, such as kinking or being backed over.

Web slings should be stored properly, when not in use, hanging on pegs or on a rack in a dark area to avoid unnecessary sunlight UV degradation. *Never* leave slings out at the end of the workday.

- *Chemical environment.* Slings exposed to certain chemicals, or to vapors of chemicals and paints, can lose some or all of their strength. Check with the sling manufacturer before using slings in a chemical environment to assure sling compatibility.

12.10 Inspection of Slings

The operator or person in charge (or both) *must* visually inspect slings selected for the job for obvious unsafe conditions—before, during, and after each use. Determining to remove slings from service, however, requires experience and good judgment, especially when evaluating the remaining strength in a sling—after allowing for normal wear. The safety of the sling depends primarily on the remaining strength.

Proper usage and knowledge of slings is an essential safety factor for all sling usage. *Never* use a repaired or reconditioned fiber rope lifting sling; and *never* use a sling without inspecting it prior to *each* use. Frequency of fiber sling inspections should be based on frequency of use, severity of service conditions, and experience with service life of slings used in similar applications.

- *Initial inspection.* Before any new sling is placed in service, it should be inspected to insure that the correct sling is being used, as well as to determine that the sling meets the requirements of the manufacturer's specifications.
- *Frequent inspection.* The person handling the sling each time it is used should make this inspection.
- *Periodic inspection.* Should be conducted at least monthly by a designated person.

Possible Defects

A fiber sling should be removed from service if any defects are visible, such as:

- Acid of alkali burns
- Melting, charring, or weld splatter on any part of the sling
- Holes, tears, cuts, snags, or embedded particles
- Broken or worn stitching in load bearing splices
- Excessive abrasive wear
- Knots in any part of the sling

- Distortion and excessive pitting or corrosion or broken fittings
- Other apparent defects that cause doubt as to the sling's strength
- Missing sling-rated capacity, or not readable sling material identification.

12.11 Safe Working Load Formulas

Always refer to the sling manufacturer's SWL tables when selecting a sling. Formulas, providing approximate usable values for estimating the loads on the most common configuration, are based on the SWL of the single vertical hitch of that sling and include:

Choker hitches and bridle hitches—For two-, three-, and four-leg hitches

$$SWL = SWL_0 \times H/L \times 2$$

Choker hitches—Single, angles greater than 45°

$$SWL = SWL_0 \times 3/4$$

Choker hitches—Double, angles greater than 45°

$$SWL = SWL_0 \times 3/4 \times H/L \times 2$$

Basket hitches—Single, vertical legs

$$SWL = SWL_0 \times 2$$

Basket hitches—Single, inclined legs

$$SWL = SWL_0 \times 3/4 \times H/L \times 2$$

Basket hitches—Double, vertical legs

$$SWL = SWL_0 \times 4$$

Basket hitches—Double, inclined legs

$$SWL = SWL_0 \times H/L \times 4$$

Endless slings and grommets

$$SWL = 2 \times \text{values for previous configurations}$$

where SWL = safe working load
SWL_0 = safe working load, single vertical hitch
L = length of sling legs
H = headroom between hook and load

Each formula for a given sling configuration, material, and size is based on the SWL of the single vertical hitch of that sling. Make sure the efficiencies of end fitting used are also taken into account when determining the capacity of the combination.

Regulations and Standards

OSHA Regulations

1910 Subpart N *Materials Handling and Storage*
1910.176 *Handling Materials, General Regulations*
1910.184 *Slings*

U.S. Consensus Standards

ANSI/ASME B30.9 – 2006
Safety Standards for Synthetic Rope Slings
Safety Standards for Round Slings
ASTM A906/A906M *Standard Specifications for Grade 80 and Grade 100 Alloy Steel Chain Slings*

Industry Consensus Standards and Guidelines

Fiber Rope Technical Information and Application Manual – The Cordage Institute

CHAPTER 13

Fiber Rope Slings and Synthetic Fiber Web Slings

Although today wire rope slings are the most commonly used type of slings in rigging, fiber rope slings have their proper place on rigging jobs—for lifting comparatively light loads and on temporary work. Natural and synthetic fiber rope slings, except wet frozen slings, may be used in temperatures that range from –20°F to + 180°F (–28.9°C to + 82.2°C) without decreasing the working load limit.

- Follow the sling manufacturer's recommendations for operations outside this temperature range, and for wet frozen slings.
- Store fiber and synthetic rope slings by hanging on hooks or brackets.
- *Never* splice these slings; only the manufacturer must make alterations to slings.

13.1 Fiber Rope Slings

Fiber rope slings are preferred for some applications because they are pliant, they grip the load well, and they do not mar the surface of the load (Table 13.1). Manila rope slings, however, should be used only on light loads and must not be used on objects that have sharp edges capable of cutting the rope or in applications where the sling will be exposed to high temperatures, severe abrasion, or acids.

Sling Selection

The choice of rope type and size will depend on the application, the weight to be lifted, and the sling angle. Before lifting any load with a

Nominal rope diameter, in.	Nominal weight per 100 ft, lb	Minimum breaking strength, lb	Eye and eye sling						Endless sling					
					Basket hitch						Basket hitch			
					Angle of rope to horizontal						Angle of rope to horizontal			
					90°	60°	45°	30°			90°	60°	45°	30°
					Angle of rope to vertical						Angle of rope to vertical			
			Vertical hitch	Choker hitch	0°	30°	45°	60°	Vertical hitch	Choker hitch	0°	30°	45°	60°
½	7.5	2,650	550	250	1,100	900	750	550	950	500	1,900	1,700	1,400	950
9/16	10.4	3,450	700	350	1,400	1,200	1,000	700	1,200	600	2,500	2,200	1,800	1,200
⅝	13.3	4,400	900	450	1,800	1,500	1,200	900	1,600	800	3,200	2,700	2,200	1,600
¾	16.7	5,400	1,100	550	2,200	1,900	1,500	1,100	2,000	950	3,900	3,400	2,800	2,000
13/16	19.5	6,500	1,300	650	2,600	2,300	1,800	1,300	2,300	1,200	4,700	4,100	3,300	2,300
⅞	22.5	7,700	1,500	750	3,100	2,700	2,200	1,500	2,800	1,400	5,600	4,800	3,900	2,800
1	27.0	9,000	1,800	900	3,600	3,100	2,600	1,800	3,200	1,600	6,500	5,600	4,600	3,200
1 1/16	31.3	10,500	2,100	1,100	4,200	3,600	3,000	2,100	3,800	1,900	7,600	6,600	5,400	3,800
1⅛	36.0	12,000	2,400	1,200	4,800	4,200	3,400	2,400	4,300	2,200	8,600	7,600	6,100	4,300
1¼	41.7	13,500	2,700	1,400	5,400	4,700	3,800	2,700	4,900	2,400	9,700	8,400	6,900	4,900
1 5/16	47.9	15,000	3,000	1,500	6,000	5,200	4,300	3,000	5,400	2,700	11,000	9,400	7,700	5,400
1½	59.9	18,500	3,700	1,850	7,400	6,400	5,200	3,700	6,700	3,300	13,500	11,500	9,400	6,700
1⅝	74.6	22,500	4,500	2,300	9,000	7,800	6,400	4,500	8,100	4,100	16,000	14,000	11,500	8,000
1¾	89.3	26,500	5,300	2,700	10,500	9,200	7,500	5,300	9,500	4,800	19,000	16,500	13,500	9,500
2	107.5	31,000	6,200	3,100	12,500	10,500	8,800	6,200	11,000	5,600	22,500	19,500	16,000	11,000
2⅛	125.0	36,000	7,200	3,600	14,500	12,500	10,000	7,200	13,000	6,500	26,000	22,500	18,500	13,000
2¼	146.0	41,000	8,200	4,100	16,500	14,000	11,500	8,200	15,000	7,400	29,500	25,500	21,000	15,000
2½	166.7	46,500	9,300	4,700	18,500	16,000	13,000	9,300	16,500	8,400	33,500	29,000	23,500	16,500
2⅝	190.8	52,000	10,500	5,200	21,000	18,000	14,500	10,500	18,500	9,500	37,500	32,500	26,500	18,500

TABLE 13.1 Rated capacity of Manila Rope Slings, in Pounds (Safety Factor = 5)

fiber rope sling, be sure to inspect the sling carefully because fiber rope slings deteriorate far more rapidly than wire rope slings and their actual strength is very difficult to estimate.

Always check fiber rope slings before using them:

- Surfaces for dry, brittle, scorched, or discolored fibers—signs of possible damage that requires further analysis.
- Interior of the sling, which should be as clean as when the rope was new.
- Fibers, when scratched should NOT come apart easily, which would indicate possible chemical damage.

13.2 Synthetic Fiber Web Slings

Synthetic webbing slings are widely used for rigging loads that must not be marred or scratched, such as finely machined, highly polished, or painted surfaces. These webbing slings offer a number of advantages over slings made of other materials.

- *Softness and bearing width.* Less tendency to crush fragile loads.
- *Elasticity.* Tends to stretch under load, able to absorb heavy shocks and cushion loads.
- *Strength.* Ability to handle loads up to 300,000 lb (136,363 kg).
- *Flexibility.* Tends to mold to the shape of the load, gripping a load more securely.
- *Safety.* Adjusts to the load contour and holds it with a tight, nonslip grip.
- *Load protection.* Will not mar, deface, or scratch highly polished or delicate surfaces.
- *Long life.* Unaffected by mildew, rot, or bacteria; and resists some chemical action.
- *Economy.* Low initial cost, plus a long service life.
- *Shock absorbency.* Absorbs heavy shocks without damage.
- *Temperature resistance.* Unaffected by temperatures up to 180°F (82.2°C).

Each synthetic material has its own unique properties.

Nylon

The most popular and best general-purpose synthetic webbing slings are made of nylon, which *must* be used wherever alkaline or greasy conditions exist. It is also preferable when neutral conditions prevail and when resistance to chemicals and solvents is important. Nylon slings are unaffected by grease or oil and have good chemical resistance to aldehydes, ethers, and strong alkalies. Nylon slings, however, should not be used with acids and bleaching agents; and they are not suitable for use at temperatures that exceed 250°F (121°C). The stretch-rated capacity is approximately 10 percent (Table 13.2).

Dacron

This synthetic fiber sling *must* be used where high concentrations of acid solutions—such as sulfuric, hydrochloric, nitric, and formic acids—and where high-temperature bleach solutions are prevalent. (Nylon will deteriorate under these conditions.) *Do not* use Dacron in alkaline conditions, however, because it will deteriorate; use nylon or polypropylene instead. Polyester must be used where acids or bleaching agents are present and is also ideal for applications where a minimum of stretching is important.

Polyester

Whenever acid conditions are present and a minimum stretch is desired, use polyester webbing slings. They are unaffected by common acids and hot bleaching solutions, but they should not be used with concentrated sulfuric acid and alkaline conditions. Polyester slings are not suitable for use at temperatures that exceed 250°F (121°C) The stretch-rated capacity is approximately 3 percent (Table 13.3).

Polypropylene

Use polypropylene webbing slings whenever acid or alkaline conditions exist, since the fibers are not affected by these conditions. These fibers, however, are not suitable for use at temperatures exceeding 180°F (100°C). The stretch-rated capacity is approximately 10 percent, as it is for nylon (Table 13.4).

All synthetic webbing slings have a 5:1 design factor. For specific applications, they are available with protective coatings:

- *Latex.* Seals out moisture and dirt; reduces the effects of abrasion.
- *Neoprene.* Increases wear resistance and the sling's coefficient of friction for a firmer grip.
- *Polyurethane.* Increases sling life, by as much as five times that of untreated fabrics, and the coefficient of friction to provide greater gripping power.

Nominal rope diameter, in.	Nominal weight per 100 ft, lb	Minimum breaking strength, lb	Eye and eye sling						Endless sling					
					Basket hitch						Basket hitch			
					Angle of rope to horizontal						Angle of rope to horizontal			
					90°	60°	45°	30°			90°	60°	45°	30°
					Angle of rope to vertical						Angle of rope to vertical			
			Vertical hitch	Choker hitch	0°	30°	45°	60°	Vertical hitch	Choker hitch	0°	30°	45°	60°
½	6.5	6,080	700	350	1,400	1,200	950	700	1,200	600	2,400	2,100	1,700	1,200
9/16	8.3	7,600	850	400	1,700	1,500	1,200	850	1,500	750	3,000	2,600	2,200	1,500
5/8	10.5	9,880	1,100	550	2,200	1,900	1,600	1,100	2,000	1,000	4,000	3,400	2,800	2,000
¾	14.5	13,490	1,500	750	3,000	2,600	2,100	1,500	2,700	1,400	5,400	4,700	3,800	2,700
13/16	17.0	16,150	1,800	900	3,000	3,100	2,600	1,800	3,200	1,600	6,400	5,600	4,600	3,200
7/8	20.0	19,000	2,100	1,100	4,200	3,700	3,000	2,100	3,800	1,900	7,600	6,600	5,400	3,800
1	26.0	23,750	2,600	1,300	5,300	4,600	3,700	2,600	4,800	2,400	9,500	8,200	6,700	4,800
1 1/16	29.0	27,360	3,000	1,300	5,100	5,300	4,300	3,000	5,500	2,700	11,000	9,500	7,700	5,500
1 1/8	34.0	31,350	3,500	1,700	7,000	6,000	5,000	3,500	6,300	3,100	12,500	11,000	8,900	6,300
1¼	40.0	35,625	4,000	2,000	7,900	6,900	5,600	4,000	7,100	3,600	14,500	12,500	10,000	7,100
1 5/16	45.0	40,850	4,500	2,300	9,100	7,900	6,400	4,500	8,200	4,100	16,500	14,000	12,000	8,200
1½	55.0	50,350	5,600	2,800	11,000	9,700	7,900	5,600	10,000	5,000	20,000	17,500	14,000	10,000
1 5/8	68.0	61,750	6,900	3,400	13,500	12,000	9,700	6,900	12,500	6,200	24,500	21,500	17,500	12,500
1¾	83.0	74,100	8,200	4,100	16,500	14,500	11,500	8,200	15,000	7,400	29,500	27,500	21,000	15,000
2	95.0	87,400	9,700	4,900	19,500	17,000	13,500	9,700	17,500	8,700	35,000	30,500	24,500	17,500
2 1/8	109.0	100,700	11,000	5,600	22,500	19,500	16,000	11,000	20,000	10,000	40,500	35,000	28,500	20,000
2¼	129.0	118,750	13,000	5,600	26,500	23,000	18,500	13,000	24,000	12,000	47,500	41,000	33,500	24,000
2½	149.0	132,000	15,000	7,400	29,500	25,500	21,000	15,000	24,500	11,500	53,000	46,000	37,500	26,500
2 5/8	160.0	133,900	17,100	8,600	34,000	29,500	24,000	17,000	31,000	15,500	61,500	53,500	43,500	31,000

Source: OSHA 1926.251, Rigging Equipment for Material Handling.

TABLE 13.2 Rated capacity of Nylon Rope Slings, in Pounds (Safety Factor = 9)

Nominal rope diameter, in.	Nominal weight per 100 ft, lb	Minimum breaking strength, lb	Eye and eye sling						Endless sling					
					Basket hitch						Basket hitch			
					Angle of rope to horizontal						Angle of rope to horizontal			
					90°	60°	45°	30°			90°	60°	45°	30°
					Angle of rope to vertical						Angle of rope to vertical			
			Vertical hitch	Choker hitch	0°	30°	45°	60°	Vertical hitch	Choker hitch	0°	30°	45°	60°
½	8.0	6,080	700	350	1,400	1,200	950	700	1,200	600	2,400	2,100	1,700	1,200
9/16	10.2	7,600	850	400	1,700	1,500	1,200	850	1,500	750	3,000	2,600	2,200	1,500
⅝	13.0	9,500	1,100	550	2,100	1,800	1,500	1,100	1,900	950	3,800	3,300	2,700	1,900
¾	17.5	11,875	1,300	650	2,600	2,300	1,900	1,300	2,400	1,200	4,800	4,100	3,400	2,400
13/16	21.0	14,725	1,600	800	3,300	2,800	2,300	1,600	2,900	1,500	5,900	5,100	4,200	2,900
⅞	25.0	17,100	1,900	950	3,800	3,300	2,700	1,900	3,400	1,700	6,800	5,900	4,800	3,400
1	30.5	20,900	2,300	1,200	4,600	4,000	3,300	2,300	4,200	2,100	8,400	7,200	5,900	4,200
1 1/16	34.5	24,225	2,700	1,300	5,400	4,700	3,800	2,700	4,800	2,400	9,700	8,400	6,900	4,800
1⅛	40.0	28,025	3,100	1,600	6,200	5,400	4,400	3,100	5,600	2,800	11,000	9,700	7,900	5,600
1¼	46.3	31,540	3,500	1,800	7,000	6,100	5,000	3,500	6,300	3,200	12,500	11,000	8,900	6,300
1 5/16	52.5	35,625	4,000	2,000	7,900	6,900	5,600	4,000	7,100	3,600	14,500	12,500	10,000	7,100
1½	66.8	44,460	4,900	2,500	9,900	8,600	7,000	4,900	8,900	4,400	18,000	15,500	12,500	8,900
1⅝	82.0	54,150	6,000	3,000	12,000	10,400	8,500	6,000	11,000	5,400	21,500	19,000	15,500	11,000
1¾	98.0	64,410	7,200	3,600	14,500	12,500	10,000	7,200	13,000	6,400	26,000	22,500	18,000	13,000
2	118.0	76,000	8,400	4,200	17,000	14,500	12,000	8,400	15,000	7,600	30,500	26,500	21,500	15,000
2⅛	135.0	87,400	9,700	4,900	19,500	17,000	13,500	9,700	17,500	8,700	35,000	30,500	24,500	17,500
2¼	157.0	101,650	11,500	5,700	22,500	19,500	16,000	11,500	20,500	10,000	40,500	35,000	29,000	20,500
2½	181.0	115,900	13,000	6,400	26,000	22,500	18,000	13,000	23,000	11,500	46,500	40,000	33,000	23,000
2⅝	205.0	130,150	14,500	7,200	29,000	25,000	20,500	14,500	26,000	13,000	52,000	45,000	37,000	26,000

Source: OSHA 1926.251, Rigging Equipment for Material Handling.

Table 13.3 Rated capacity of Polyester Rope Slings, in Pounds (Safety Factor = 9)

Nominal rope diameter, in.	Nominal weight per 100 ft, lb	Minimum breaking strength, lb	Eye and eye sling						Endless sling					
					Basket hitch						Basket hitch			
					Angle of rope to horizontal						Angle of rope to horizontal			
					90°	60°	45°	30°			90°	60°	45°	30°
					Angle of rope to vertical						Angle of rope to vertical			
			Vertical hitch	Choker hitch	0°	30°	45°	60°	Vertical hitch	Choker hitch	0°	30°	45°	60°
½	4.7	3,990	650	350	1,300	1,200	950	650	1,200	600	2,400	2,100	1,700	1,200
9/16	6.1	4,845	800	400	1,600	1,400	1,100	800	1,500	780	2,900	2,500	2,100	1,500
⅝	7.5	5,890	1,000	500	2,000	1,700	1,400	1,000	1,800	900	3,500	3,100	2,500	1,800
¾	10.7	8,075	1,300	700	2,700	2,300	1,900	1,300	2,400	1,200	4,900	4,200	3,400	2,400
13/16	12.7	9,405	1,600	800	3,100	2,700	2,200	1,600	2,800	1,400	5,600	4,900	4,000	2,800
⅞	15.0	10,925	1,800	900	3,600	3,200	2,600	1,800	3,300	1,600	6,600	5,700	4,600	3,300
1	18.0	13,300	2,200	1,100	4,400	3,800	3,100	2,200	4,000	2,000	8,000	6,900	8,600	4,000
1 1/16	20.4	15,200	2,500	1,300	5,100	4,400	3,600	2,500	4,600	2,300	9,100	7,900	6,500	4,600
1⅛	23.7	17,385	2,900	1,500	5,800	5,000	4,100	2,900	5,200	2,600	10,500	9,000	7,400	5,200
1¼	27.0	19,950	3,300	1,700	6,700	5,800	4,700	3,300	6,000	3,000	12,000	10,500	8,500	6,000
1 5/16	30.5	22,325	3,700	1,900	7,400	6,400	5,300	3,700	6,700	3,400	13,500	11,500	9,500	6,700
1½	38.5	28,215	4,700	2,400	9,400	8,100	6,700	4,700	8,500	4,200	17,000	14,500	12,000	8,500
1⅝	47.5	34,200	5,700	2,900	11,500	9,900	8,100	5,700	10,500	5,100	20,500	18,000	14,500	10,500
1¾	57.0	40,850	6,800	3,400	13,500	12,000	9,600	6,800	12,500	6,100	24,500	21,000	17,500	12,500
2	69.0	49,400	3,200	4,100	16,500	14,500	11,500	8,200	15,000	7,400	29,500	25,500	21,000	15,000
2⅛	80.0	57,950	9,700	4,800	19,500	16,500	13,500	9,700	17,500	8,700	35,000	30,100	24,500	17,500
2¼	92.0	65,550	11,000	5,500	22,000	19,000	15,500	11,000	13,500	9,900	39,500	34,000	28,000	19,500
2½	107.0	76,000	12,500	6,300	25,500	22,000	18,000	12,500	23,000	11,500	45,500	39,500	32,500	23,000
2⅝	120.0	85,500	14,500	7,100	20,500	24,500	20,000	14,300	25,500	13,000	51,500	44,500	36,500	25,000

Source: OSHA 1926.251, Rigging Equipment for Material Handling.

TABLE 13.4 Rated capacity of Polypropylene Rope Slings, in Pounds (Safety Factor = 6)

Because each synthetic material has unique properties, synthetic fiber web slings should be used according to the manufacturer's instructions—especially when dealing with chemically active environments. Synthetic fiber web slings *must* be removed from service if any defects exist, such as:

- Acid or caustic burns
- Melting or charring of any part of the surface
- Snags, punctures, tears, or cuts
- Broken or worn stitches
- Wear or elongation exceeding the manufacturer's recommended amount
- Distortion of fittings

Fiber ropes and synthetic webs are generally discarded rather than serviced or repaired. Operators *must always* follow manufacturer's recommendations.

Types of Synthetic Fiber Web Slings

All synthetic webbing slings have a 5:1 design factor. They are available in one-, two-, three-, or four-ply webbing, with metal end fittings; and include:

- *Triangle and choker triangle.* Fitted on both ends of vertical slings, chokes or basket hitches; available in 2- to 12-in. (5.08- to 30.48-cm) web widths; having 2400- to 38,000-lb (1088.6- to 17,234.5-kg) capacities.
- *Triangles.* Fitted on both ends of vertical slings or basket hitches only; available in 2-to 12-in. (5.08- to 30.48-cm) web widths; having 3200- to 38,400-lb (1431.5- to 17,418-kg) capacities.
- *Sewn eyes.* Either flat tapered eyes or twisted eyes fitted on vertical slings, chokers, or basket hitches; available in 1- to 6-in. (2.54- to 15.24-cm) web widths; having 4600- to 28,600-lb (2086.5- to 12,972.7-kg) capacities.
- *Endless or grommet.* Used on continuous web loops—vertical slings, chokers, or basket hitches; available in 1- to 6-in. (2.54- to 15.24-cm) web widths; having 2600- to 57,200-lb (1179.3- to 29,945.5 kg) capacities.

13.3 Splicing Fiber Rope Slings

Spliced fiber rope slings should *not* be used unless they have been spliced in accordance with the following minimum requirements;

and in accordance with any additional recommendations of the manufacturer.

- Provide, for Manila rope eye splices, at least four full tucks; and for short splices at least six full tucks—three on each side of the splice center line.
- Provide, for synthetic fiber rope eye splices, at least four full tucks, and for short splices at least eight full tucks—four on each side of the splice center line.
- *Do not* trim flush strand end tails of eye and short splices with the surface of the rope immediately adjacent to the full tucks, regardless of the type of fiber. Recommended distances for fiber rope:

  Under 1 in. diameter—tail should project at least six rope diameters beyond last full tuck.

  Over 1 in. diameter—tail should project at least 6 in. beyond last full tuck. Where a projecting tail interferes with use of the sling, the tail should be tapered and spliced into the body of the rope using at least two additional tucks (requiring a tail length of approximately six rope diameters beyond the last full tuck).
- Make sure fiber rope slings have a minimum clear length of rope between eye splices equal to ten times the rope diameter.
- *Never* use knots instead of splices.
- Clamps not designed specifically for fiber ropes should *not* be used for splicing.
- Make sure eye splices are of such size to provide an included angle of not greater than 60° at the splice when the eye is placed over the load or support.
- *Do not* use fiber rope slings if end attachments in contact with the rope have sharp edges or projections.
- Remove from service natural and synthetic fiber rope sling immediately if any of the following conditions are present:

  Abnormal wear
  Powdered fiber between strands
  Broken or cut fibers
  Variations in the size or roundness of strands
  Discoloration or rotting
  Distortion of hardware in the sling
- Only fiber rope slings made from new rope should be used; use of repaired or reconditioned fiber rope slings is *prohibited*.

13.4 Safe Lifting Practices with Slings

Fiber ropes and synthetic webs are generally discarded rather than serviced or repaired. Operators *must always* follow manufacturer's recommendations.

Proper use and knowledge of slings are essential safety factors for all sling usage. Learn as much as possible about the slings, and materials to be lifted; slings come in many different types, one of which is right for the purpose at hand.

Analyze the load to be moved with respect to size, weight, shape, center of gravity of the load, temperature, and sensitivity; then select the sling that best meets those needs (based on the characteristics of the load and the environmental conditions surrounding the lift).

- Consider the number of legs and angle with the horizontal, rated capacity of the sling, and history of sling's care and use.
- Inspect all the equipment before and after a move.
- Always be sure to give the equipment whatever "in service" maintenance it may need.

Maximum Safe Working Loads

Use safe lifting practices, especially the proper lifting technique for the type of sling and the type of load (Table 13.5). Always

	Slip noose	Two-end bridle sling			Three-end bridle sling		
	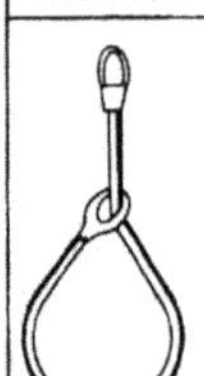	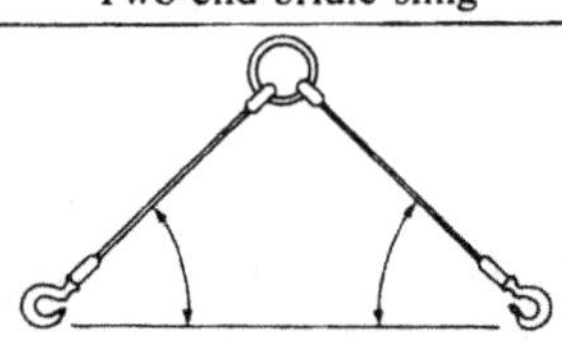 When angle is			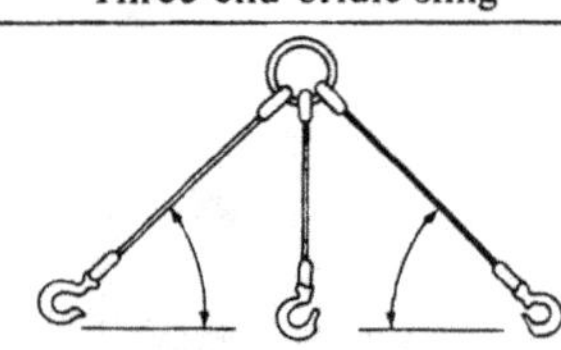 When angle is		
Size of rope		75°	60°	45°	75°	60°	45°
3/8	3,200	3,000	2,800	2,200	4,600	4,200	3,400
1/2	5,400	5,200	4,800	4,000	7,800	7,200	6,000
5/8	8,400	8,000	7,200	6,000	12,000	10,800	9,000
3/4	12,000	11,400	10,200	8,400	17,000	15,400	12,600
3/8	16,000	16,000	14,000	11,400	24,000	21,000	17,200
1	22,000	20,000	18,000	15,000	30,000	26,000	22,000
1 1/8	26,000	26,000	23,000	19,000	40,000	36,000	28,000
1 1/4	32,000	32,000	28,000	23,000	48,000	42,000	36,000
1 3/8	40,000	38,000	34,000	28,000	58,000	52,000	42,000
1 1/2	46,000	44,000	40,000	32,000	66,000	60,000	48,000

Source: AMHOIST—American Hoist & Derrick Co.

TABLE 13.5 Safe Working Loads for Typical Slings, in Pounds follow the manufacturer's recommendations to properly attach the sling to the load.

Good lifting techniques, common to all slings, are listed in Chap. 12.

13.5 Care and Maintenance of Fiber Rope and Synthetic Fiber Web Slings

The mishandling and misuse of slings are the leading cause of sling-related accidents. The majority of injuries and accidents, however, can be avoided by becoming familiar with the essentials of proper sling care and use.

Proper care and use are essential for maximum service and safety. Slings must be protected with cover saddles, burlap padding, or wood blocking as well as from unsafe lifting procedures such as overloading to prevent sharp bends and cutting edges.

Remember, damaged slings cannot lift as much weight as new or older, well-cared for slings. Proper and safe use and storage of slings will increase their service life.

Use fiber slings *only* within a temperatures range of 20°F (29°C) to 180°F (82°C), unless they are wet or frozen. Wet or frozen slings have different limitations, and the manufacturer recommendations must be followed.

Never splice synthetic fiber slings; only the manufacturer must make alterations to slings. Remove natural or synthetic fiber slings from service if:

- There is abnormal wear.
- Powdered fibers appear between strands.
- Fibers are broken or cut.
- Variation in size or roundness of strands occurs.
- Discoloration or rotting is detected.
- Distortion of hardware is detected.

Synthetic Fiber Web Slings

All webbing *must* be uniform in thickness.

- *Do not* use polyester and nylon webs where fumes, vapors, sprays, mists, liquids of acids, phonetics, or caustics are present.

- *Remove* from service any web slings when the following conditions are present
  Acid or caustic burns
  Melting or charring of any part of the sling
  Snags, punctures, tears, or cuts
  Broken or worn stitches
  Distortion of any fitting

Flat Webbing and Round Synthetic Fiber Slings

Made from nylon, polyester, polypropylene, or aramid polyamide. Each sling must be labeled with the SWL.

13.6 Inspection of Fiber Rope and Synthetic Fiber Web Slings

Periodic inspections of natural and synthetic fiber rope and web slings *must* be made at intervals no greater than 12 months: yearly for normal service use; monthly for severe service use; and periodic inspections where service conditions warrant, as determined on the-following basis:

- *Initial*. Before any new sling is placed in service, it should be inspected to insure that the correct sling is being used, as well as to determine that the sling meets the requirements of the manufacturer's specifications.
- *Frequent*. The operator handling the sling, or person in charge (or both) *must* conduct an inspection of the slings selected for the job—each time prior to, during, and after each use.
- *Periodic*. A designated person *must* conduct an inspection of fiber rope sling at least monthly, depending on frequency of use, severity of service conditions, and experience with service life of slings used in similar applications.
- *Recordkeeping*. Inspections (daily, monthly, annual) and proof test certification from manufacturer (check *ASME B30.9 – Slings*).

When inspecting a fiber rope sling, prior to using it, look first at its surface for cuts, gouges, or worn surface areas; dry, brittle, scorched, or discolored fibers; or melting or charring of any part of the sling. If any of these conditions are found, the supervisor must be notified and a determination made regarding the safety of the sling. If the sling is found to be unsafe, it *must* be discarded.

Next, check the sling's interior. It should be as clean as when the rope was new. A buildup of powder-like sawdust on the inside of the fiber rope indicates excessive internal wear and that the sling is unsafe.

Finally, scratch the fibers with a fingernail. If the fibers separate easily, the fiber sling has suffered some kind of chemical damage and must be discarded. When checking fiber line slings for signs of deterioration caused by exposure to the weather:

- *Make sure* none of the fibers have been broken or cut by sharp-edged objects.
- *Never* use a repaired or reconditioned fiber rope lifting sling.
- *Never* use a sling without first inspecting it prior to *each* use.

The inspector should check for possible defects and remove fiber slings from service if any defects are visible, such as:

- Acid of alkali burns
- Melting, charring, or weld splatter on any part of the sling
- Holes, tears, cuts, snags, or embedded particles
- Broken or worn stitching in load bearing splices
- Abnormal wear
- Excessive abrasive wear
- Powdered fibers between strands
- Broken of cut fibers
- Variation in size or roundness of strands
- Knots in any part of the sling
- Discoloration or rotting is detected
- Distortion and excessive pitting or corrosion or broken fittings
- Other apparent defects that cause doubt as to the sling's strength
- Sling rated capacity or sling material identification is missing or not readable

Self-inspection Checklist

The following self-inspection checklist, developed by OSHA under the general industry standard 29 CFR 1910.184 and the construction standard 1926.251, applies to slings used with other equipment to move material by lifting or hoisting (Table 13.6). A yes answer to a question indicates that portion of the inspection complies with the OSHA or U.S. Environmental Protection Agency (EPA) standard, or with a nonregulatory recommendation.

Natural fiber rope	Yes	No	N/A
• Are fiber rope slings that are made from conventional three-strand construction fiber rope used only within their rated capacities and minimum diameter of curvature?	Y	N	N/A
Note: *Consult the tables in the OSHA regulations or the manufacturer's requirements for permitted load ratings and diameter of curvature restrictions for the different rigging situations. Diameter of curvature is important because wrapping a sling around something at a diameter less than that recommended reduces the strength of the sling at the bend and could cause failure of the line.*			
• Are natural and synthetic fiber rope slings used only within a temperature range of –20°F to 180°F?	Y	N	N/A
• If natural and synthetic fiber rope slings are used outside the temperature range of –20°F to 180°F or have been wetted or frozen, are the manufacturer's recommendations followed for continued use?	Y	N	N/A
• Is the use of spliced fiber rope prohibited?	Y	N	N/A
Note: *Consult the OSHA regulations under 29 CFR 1910.184(h)(3) and 1926.251(d)(2) and (4) for additional requirements on splices.*			
• Do fiber rope slings have a minimum clear length of rope between eye splices equal to 10 times the rope diameter?	Y	N	N/A
• Is the use of knots in place of splicing prohibited for fiber rope slings?	Y	N	N/A
• Is the use of fiber rope slings prohibited if the end attachments in contact with the rope have sharp edges or projections?	Y	N	N/A
• Are natural and synthetic fiber rope slings immediately removed from service if any of the following conditions are present? Abnormal wear Powdered fiber between strands Broken or cut fibers Variations in size or roundness of strands Discoloration or rotting Distortion of hardware in the slings	Y	N	N/A
• Are fiber rope slings made only from new rope, i.e., is the use of repaired or reconditioned fiber rope slings prohibited?	Y	N	N/A

TABLE 13.6 Inspection Checklist—Natural and Synthetic Fiber Rope Slings, and Synthetic Fiber Web Slings

Synthetic Web			
• Is each synthetic web slings marked or coated to show the rated capacity for each type of hitch and synthetic web material?	Y	N	N/A
• Is the synthetic webbing of uniform thickness and width?	Y	N	N/A
Note: *Selvage edges must not be split from the webbing's width.*			
• Do fittings have a minimum breaking strength equal to that of the sling?	Y	N	N/A
• Are fittings free of all sharp edges that could damage the webbing?	Y	N	N/A
• Is stitching the only method used to attach end fittings to webbing and to form eyes?	Y	N	N/A
• Are synthetic web slings only used within their rated load capacity?	Y	N	N/A
Note: *Consult the tables in the OSHA regulations or the manufacturer's requirements for permitted load ratings for the different rigging situations.*			
• When synthetic web slings are used, are the following precautions taken?	Y	N	N/A
Nylon web slings are not used where fumes, vapors, sprays, mists, or liquids of acids or phenolics are present.			
Polyester and polypropylene web slings are not used where fumes, vapors sprays, mists, or liquids of caustics are present.			
Web slings with aluminum fittings are not used where fumes, vapors, sprays, mists, or liquids of caustics are present.			
• Is the use of synthetic web slings of polyester and nylon above 180°F prohibited?	Y	N	N/A
• Is the use of polypropylene web slings above 200°F prohibited?	Y	N	N/A
• Are synthetic web slings repaired only by the sling manufacturer or another competent person or organization?	Y	N	N/A

Table 13.6 Inspection Checklist—Natural and Synthetic Fiber Rope Slings, and Synthetic Fiber Web Slings (*Continued*)

• Is each repaired sling proof tested by the manufacturer (or another competent person or organization) to twice the rated capacity before its return to service? Is a certificate of proof test available?	Y	N	N/A
• Is the use of temporarily repaired slings (including webbing and fittings) prohibited?	Y	N	N/A
• Are synthetic web slings immediately removed from service if any of the following conditions are present? Acid or caustic burns Melting or charring of any part of the sling surface Snags, punctures, tears, or cuts Broken or worn stitches Distortion of fittings	Y	N	N/A
• Are shackles and hooks only used within their safe working loads?	Y	N	N/A
Note: *Consult the OSHA regulations for the safe working loads of various sizes of shackles. Use the manufacturer's recommendations to determine the safe working loads for the various sizes and types of hooks. Test all hooks with no applicable manufacturer's recommendations at twice the intended SWL before they are initially put into use. Maintain documentation of the dates and results of the test.*			

Source: Occupational Safety & Health Administration, Washington DC.

TABLE 13.6 Inspection Checklist—Natural and Synthetic Fiber Rope Slings, and Synthetic Fiber Web Slings (*Continued*)

CHAPTER 14

Wire Strand Rope

Wire rope is a machine composed of wires, strands and a core—a number of precise, moving parts, designed and manufactured to bear a very definite relation to one another. This relationship ensures that the wire rope has the flexibility and strength crucial to professional and safe hoisting operations.

14.1 Wire Strand Rope Markings[1]

Wire rope manufactured in the United States normally has some type of colored marker to identify its manufacturer:

- *Strand markers.* Simply a colored lubricant applied externally to one strand during manufacturing; can be seen by looking at the wire rope. (Strand markers are not used in mining rope, elevator rope, galvanized rope, compacted rope, or any rope that is postlubricated.) Strand markers are used in the manufacture of all standard (round) wire ropes but are not a confirmation of the manufacturer in and of itself.
- *Core markers.* Used in most of the wire rope manufactured in the United States; but cannot be seen unless the wire rope is disassembled. Manufacturers add colored threads of filaments to fiber and steel (independent wire rope cores).

Wire rope that is imported is difficult to detect without dismantling the entire rope, and even then there is no clear way to determine if the wire rope is imported. Under federal law, when imported wire rope enters the United States its container must conspicuously state the name of the country of origin in which the wire rope was produced. The "container" for the wire rope may be a reel or shipping cart, or the straps by which a coil is secured.

[1]*Source:* Williamsport Wire Rope Works Inc., Williamsport, PA.

14.2 Types and Sizes

To meet the varied requirements of many applications, wire rope is manufactured from wires made of steel, iron, or other metal in various sizes, and in several tensile strength grades. Each grade provides a different combination of tensile strength, toughness, ability to withstand abrasion, and resistance to fatigue from bending and flexing. The primary grades of wire rope (in increasing degree of strength) include:

- *Mild plow steel.* Tough and pliable; can stand repeated strain and stress and has a tensile strength (resistance to lengthwise stress) of 200,000 to 225,000 psi (14,060 to 15,818 kg/m^2). These characteristics make it desirable for cable tool drilling and other purposes where abrasion is encountered.
- *Plow steel wire.* Unusually tough and strong. This steel has a tensile strength of 195,000 to 258,000 psi (13,709 to 18,138 kg/m^2); plow steel wire rope is suitable for hauling, hoisting, and logging.
- *Improved plow steel.* One of the best grades of rope available. Improved plow steel is stronger, tougher, and more resistant to wear than either mild plow steel or plow steel. Each square inch of improved plow steel can withstand a strain of 220,000 to 300,000 psi (15,466 to 21,090 kg/m^2); therefore, this wire rope is especially useful for heavy-duty service, such as cranes with excavating and weight-handling attachments.
- *Extra improved plow steel.* Has tensile strength of 245,000 to 340,000 psi (15,115 to 23,902 kg/m^2).

Each grade of plow steel is about 15 percent stronger than the preceding grade. Any of these grades are furnished in specially treated or coated versions, where corrosion resistance is of utmost importance, including iron-chromium-nickel composition, zinc-coated, and plastic-jacketed steel ropes.

NOTE: *Although today all these grades are made from steel, the original iron rope steel classification takes its name from the rope originally made from wrought or puddle iron, having a tensile strength of 100,000 psi. High-strength steel rope, developed from the iron grade, was used extensively on steam plows in England in 19th century, thus giving rise to the term plow steel. Iron and traction steel ropes are of lower strength than plow steel and are used primarily for elevators.*

Sizes

The size of wire rope is determined by its diameter—with sizes specified by use:

General hoisting rope	3/16 to 4 in. dia.
Rigging and guy ropes	1/4 to $1^1/_4$ in. dia.
Running ropes	3/4 to $2^1/_2$ in. dia.
Mooring lines	3/8 to $3^1/_2$ in. dia.
Shovel and dragline hoist ropes	3/4 to 5 in. dia.
Elevator ropes	3/15 to $1^1/_{15}$ in. dia.

The actual diameter of a wire rope, when new, is usually slightly larger than its nominal size. Any variation in diameter is always over, never under, and it can range from 1/64 in. (0.4 mm) for smaller-sized ropes to 1/8 in. (3.18 mm) for ropes over 2 in. (50.8 mm) diameter.

The durability and performance of wire rope is determined primarily by the type and size of wire and strands, the type of core, and any special treatment applied to the rope, such as preforming or prestressing. Proper handling and care are essential to maintain wire rope in a durable, safe, and serviceable condition.

14.3 Construction of Wire Strand Rope

Wire rope consists of three parts: wires, strands, and a core. The basic unit is wire, which is carefully processed and drawn from selected grades of steel to predetermined physical properties and sizes. A specific number of finished wires is then helically laid together (not twisted) into a uniform geometric pattern to form a strand.

A number of strands are then either laid, or helically bent around a core of fiber (natural or synthetic), or of wire to form the wire rope (Fig. 14.1). This process must be performed with precision and exactness to form a strand of correct size and characteristics. The required number of suitably fabricated strands are laid symmetrically with a definite length of lay around a core, forming the finished wire rope.

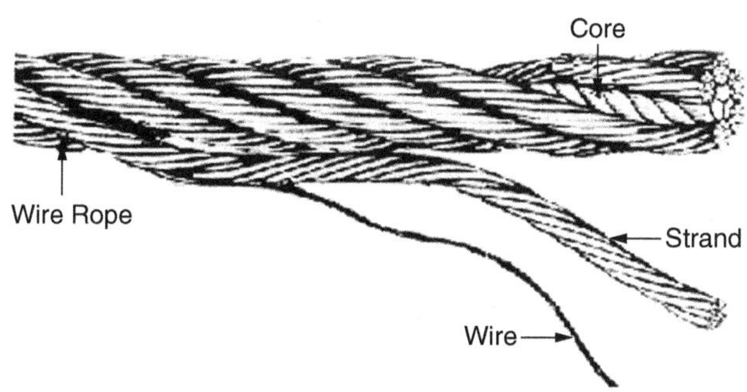

FIGURE 14.1 Basic components of a typical wire strand rope.

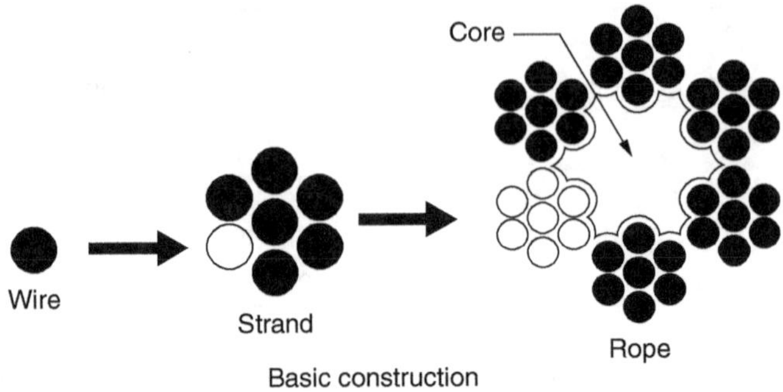

FIGURE 14.2 Construction of a typical wire strand rope.

Strands and Wires

Wire ropes are designated primarily by their type of construction, that is, the number of wires in each strand; and the number and arrangement of strands in the rope (Fig. 14.2). For example, the construction 6 × 25 denotes a six-strand rope, with each strand having 25 wires. Constructions having similar weights and breaking strengths are grouped into a series of basic classifications, including 6 × 7, 6 × 37, and 8 × 19, among others.

Thus a $^1/_2$-in., 6 × 19 wire rope having 6 strands with 19 wires per strand has the same outside diameter as a $^1/_2$-in., 6 × 37 wire rope, which has 6 strands with 37 wires (of smaller size) per strand. The number of wires to a strand varies, depending on the purpose for which the wire rope is intended.

Durability and performance of wire rope is determined primarily by the type and size of wires used, the number of wires and strand arrangement, the lay of wires and strands, the type of core, and any special treatment applied to the wire rope, such as preforming or prestressing. The smaller (and more numerous) the wires used in a rope, the more flexible the rope will be. Such a rope, however, has considerably less abrasion resistance than does a larger-sized rope.

Wires

The wires in a strand may be the same size or a mixture of sizes (Fig. 14.3).

- *Ordinary*. One size of wire is used throughout, with each strand having 4 to 18 outer wires.
- *Seale*. Two sizes of wire are used with the smaller wires on the inside—to provide flexibility—and larger wires on the outside—to resist abrasion; each strand has 8 to 18 wires

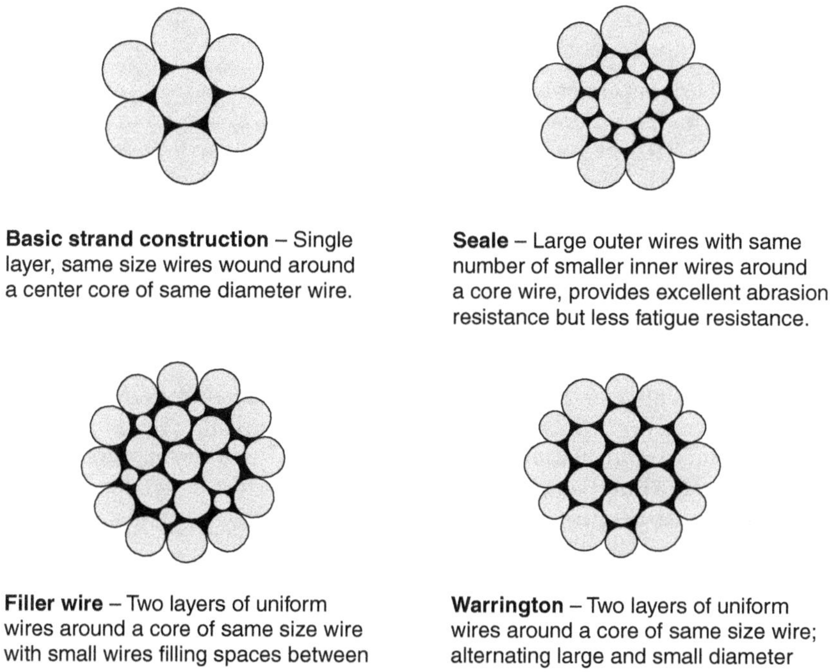

Figure 14.3 Basic strand patterns. (a) Single layer; (b) seale; (c) filler wire; (d) warrington.

- *Warrington.* Two sizes of wire are used, alternately laid, large and small to combine great flexibility with resistance to abrasion; each strand has 10 to 18 outer wires
- *Filler.* Two sizes of wire area used—very small wires filling valleys between outer and inner rope of the same-size wires to provide good abrasion and fatigue resistance; each strand has 8 to 18 wires of the same size.

As the number of wires per strand increases, combinations of Filler, Seale, and Warrington arrangements are used to provide the rope with specific characteristics and produce varying degrees of resistance to bending, fatigue, and abrasion.

Ordinary and Seale-type construction may have an *even* or an *odd* number of outer wires. Warrington and Filler types always have an *even* number in their outer layer. The number of outer wires determines, in a general way, the relative flexibility of the wire rope. By increasing the number of outside wires, the reserve strength of a rope may be increased.

The first number in the classification of wire rope refers to the strands; the second number, to the amount of wires in each strand, thus:

- 6×7 contains 6 strands made up of 3 through 14 wires, of which no more than 9 are outside wires.
- 6×19 contains 6 strands made up of 15 through 26 wires, of which no more than 12 are outside wires; arranged in several different strand patterns and normally recommended where the diameter of rope used is $^1/_8$- through $1^1/_8$-in. (3.175–28.595 mm).
- 6×37 contains 6 strands made up of 27 through 49 wires, of which no more than 18 are outside wires; arranged in several different strand patterns and normally recommended where the diameter range is $1^1/_4$ in. (31.75 mm) and larger.

Wire ropes also are identified by their strand shape—round, flattened, lock coil, and concentric. Most rigging and hoisting ropes are of the round-strand classification, and these are further classified according to the number of wires per strand and the arrangement (or geometric construction) of wires within the strands. The wires in a round strand may be all the same size or a mixture of sizes.

Rope Core

At the center of a wire rope is a core around which strands are laid. The core's principal function is to support the outer strands, maintain their proper position when loads are applied, and prevent them from crushing when the wire is wrapped around a drum or passes through a sheave. Depending on their construction, wire rope cores include:

- *Fiber cores (FC).* Usually made of sisal (occasionally Manila) fibers, offering the advantage of increased flexibility. It also serves as a cushion to reduce the effects of sudden strain and act as an oil reservoir to lubricate the wire and strands (to reduce friction). In the smaller ropes, cotton and jute are sometimes used for the central member. Fiber cores are also manufactured from synthetic fibers such as polypropylene or nylon that are less susceptible to compacting than natural fibers, especially when the rope is wet.
  - *Never* use fiber core wire ropes where temperatures may damage the core.
  - *Never* use fiber core wire ropes when multilayer winding it used, since such ropes are susceptible to crushing.
  - Use fiber-core wire ropes *only* at temperatures below 200°F (93°C).
- *Wire strand cores (WSC).* Consist of a single strand usually (but not always) of the same construction as the main rope strands. They resist more heat and stretch less than fiber cores—actually increasing overall strength by about 15 percent.

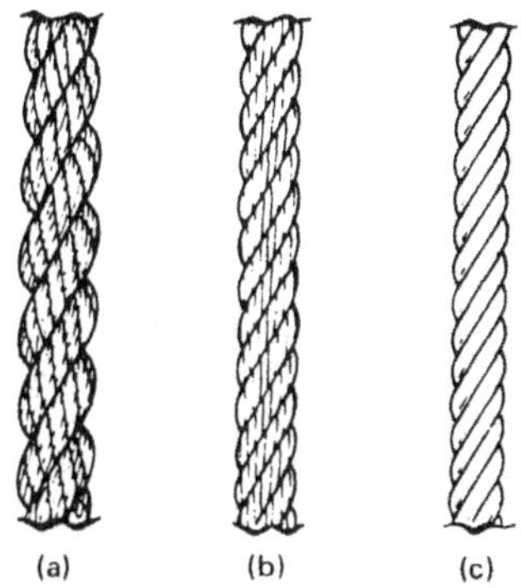

FIGURE 14.4 Wire rope cores. (a) Fiber core, FC; (b) independent wire rope core, IWRC; (c) wire strand core, WSC.

The wire strand core, however, makes the wire rope less flexible than a fiber core, as well as less resilient and less capable of absorbing shock loads than do fiber-core ropes. Strand cores are used only with standing ropes and small-diameter running ropes.

- *Independent wire rope core (IWRC).* Consists of a separate, small wire rope over which the main strands of the rope are laid. This core strengthens the rope, increases the main rope's resistance to crushing by supporting its circular cross section as it bends around sheaves and drums under heavy loading, and supplies maximum resistance to heat (Fig. 14.4).

Depending on for what purpose the wire rope is intended, it may be manufactured as:

- *Preformed strand.* Strands or wires shaped to conform to the curvature of the finished rope before laying up, the rope is termed preformed wire rope.
- *Common strand.* 1×7 construction that includes one straight center strand with six wires wrapped around it. The 1×7 construction is the basis for 1×19, 7×7, and 7×19 wire rope, with 1×7 being the stiffest construction and 7×19 one of the most flexible.

Do not confuse wires and strands—if a strand is broken, the rope *must not be used,* whereas a single broken wire in a sling is not as important.

Rope Lay

In addition to wire and strand size and pattern, the lay of a rope can vary, that is, the direction of the twist of the wires in a strand and the direction that the strands are laid in the rope. The term *lay* can mean:

- One complete wrap of a strand around the core.
- Direction the strands are wound around the core; wire rope is referred to as *right lay* or *left lay.* A right lay rope is one in

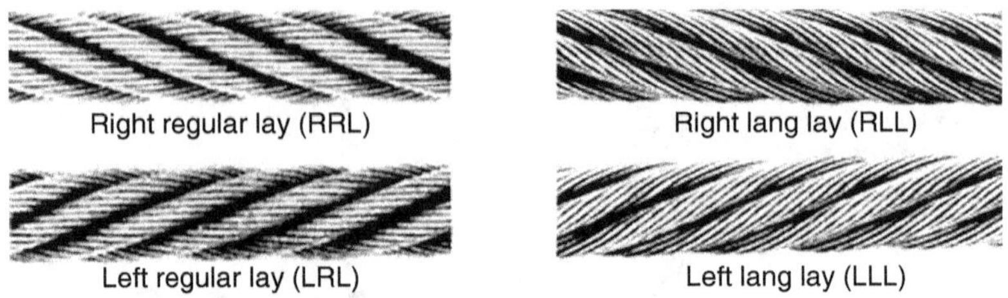

FIGURE 14.5 Wire rope lays. (a) Right, regular lay; (b) right, lang lay; (c) left, regular lay; (d) left, lang lay.

which the strands are wound in a right-hand direction like a conventional screw thread (see Fig. 14.5). A left lay rope is just the opposite

- Direction wires are wound in the strands in relation to the direction of the strands around the core; in regular lay rope, the wires in the strands are laid in one direction while the strands in the rope are laid in the opposite direction.

In some instances, both the wires in the strand and the strands in the rope are laid in the same direction; in other instances, the wires are laid in one direction and the strands are laid in the opposite direction, depending on the intended use of the rope (Fig. 14.5).

Right Lay (or Left Lay)

Right lay ropes rotate to the right (similar to a right-hand screw thread) or to the left while receding from the observer when viewed from above. Right lay is the standard lay of wire rope, with very few types of installations requiring use of left lay rope.

Regular Lay

This is rope lay of either direction, with wires in the individual strands laid in the opposite direction from that of the rope itself, thereby reducing the chances of such ropes kinking and untwisting in use. The result is that the wire crown runs approximately parallel to the longitudinal axis of the rope. These ropes have good resistance to kinking and twisting, and are easy to handle. They are also able to withstand considerable crushing and distortion due to the short length of exposed wires. This type of rope has the widest range of applications.

NOTE: *In a regular lay wire rope, wires in the strands, and strands in the rope are laid in opposite directions; thus, in a right lay rope, the lay of wire in the strands would be "left lay" in making a regular lay wire rope; standard type of construction, unless otherwise specified, is furnished on orders for wire rope.*

Lang Lay

Lang lay is rope in which the wires composing the strands are laid in the same direction as the rope (either right or left lay) forming a left-hand helix, similar to the threads of a left-hand screw thread. Since the outside wires in lang lay ropes lie diagonally across the longitudinal axis of the rope and are exposed for longer lengths than in regular lay, they present a greater wearing surface, have increased resistance to abrasion, and are more flexible. Internal stress due to bending over sheaves and drums is reduced, thus causing lang lay ropes to be more resistant to bending fatigue.

Lang lay rope is recommended for many excavating, construction, and mining applications, including draglines, hoist lines, dredge lines, and other similar lines. Greater care, however, must be exercised when handling lang lay ropes as they are more likely to kink and untwist than regular lay ropes, but the rotation of a left lay rope tightens a standard coupling. Before using lang lay ropes, fasten both ends permanently to prevent untwisting.

NOTE: *In lang lay wire rope, wires in the strands and strands in the rope are laid in the same direction; thus, wires in the strands of a right lay, lang lay rope would be laid "right lay." Lang lay ropes are slightly more susceptible to abuse, must be handled with greater care, and should be used only where specifically recommended.*

Alternate Lay

This is rope in which the wires in one strand are laid to the right, the wires in the nearby strand are laid to the left, the wires in the next strand are laid to the right, and so forth, with alternate directions from one strand to the other. Then all strands are laid to the right.

Herringbone

This rope features two right-hand lang lay strands alternating with two left-hand, regular lay strands.

Most manufacturers specify the types and lays of wire rope to be used on their equipment. Be sure to consult the operator's manual for proper application.

Lay Length

The lay length of wire rope is the distance measured parallel to the center line of a wire rope—in that a strand makes one complete spiral or turn around the rope; and strand lay length is the distance measured parallel to the centerline of the strand—in that one wire makes one complete spiral or turnaround the strand.

The most common type of manufactured wire rope is preformed—wires and strands are shaped in the manufacturing process to fit their position in the finished rope. This removes the tendency of wires and strands to straighten themselves out, leaving them relaxed in their normal position in the rope, thus ensuring their freedom from twisting and their being less likely to kink or foul in use. When preformed wire rope is cut, it tends not to unlay and is more flexible than nonpreformed wire rope.

Although cut or broken ends do not untwist, and thus need not be seized, broken wires tend to lie flat; therefore, preformed wire rope requires careful inspection to detect damages.

Other advantages of preforming wire rope are that each strand in a preformed wire rope carries an equal share of the load, and preformed rope caused less wear on sheaves and drums.

When wires and strands are not shaped before fabrication, the wire rope is termed *nonpreformed wire rope,* in which twisting produces a stress in the wires. As a result, when nonpreformed wire is cut or broken, the stress causes the strands to unlay. The almost instantaneous unlaying of the wires and strands of nonpreformed wire rope can cause serious injury to someone that is careless or not familiar with this rope characteristic.

14.4 Selection Criteria for Wire Rope

Manufacture of a wire rope that can withstand all of the different types of wear and stress it may be subjected to is impossible. Accordingly, selecting a rope is often a matter of compromise—sacrificing one quality to have some other more urgently needed characteristic.

Selecting the proper lay of rope is possible only after having compared the advantages and disadvantages of each. Several factors must be considered when you select a wire rope for use in a particular type of operation.

- *Tensile strength.* Strength necessary to withstand a certain maximum load applied to the rope, including a reserve of strength measured in a factor of safety.
- *Crushing strength.* Strength necessary to resist the compressive and squeezing forces that distort the cross section of a wire rope as it runs over sheaves, rollers, and hoist drums when under a heavy load. Regular lay rope distorts less in these situations than lang lay.
- *Fatigue resistance.* Ability to withstand constant bending and flexing of wire rope that runs continuously on sheaves and hoist drums. Fatigue resistance is important when the wire rope must be run at high speeds. Such constant and rapid bending of the rope can break individual wires in the strands.

- *Abrasion resistance.* Ability to withstand the gradual wearing away of the outer metal, as the rope runs across sheaves and hoist drums. The rate of abrasion depends mainly on the load carried by the rope and the running speed.
- *Corrosion resistance.* Ability to withstand the dissolution of the wire metal that results from chemical attack by moisture in the atmosphere or elsewhere in the working environment.
- *Flexibility.* Individual wires in a rope bend more sharply in regular lay than in lang lay rope because the outside wires in regular lay rope lie parallel with the rope axis and can only take a curvature equal to the radius of the sheave plus one rope diameter.

Fatigue

Lang lay ropes are best for service requiring high fatigue resistance. Ropes with smaller wires around the outside of their strands also have greater fatigue resistance, since these strands are more flexible.

As a load is applied to a rope, each wire in the rope is stretched and drawn tightly together. Then, when the rope bends around a sheave or drum, the tightly compressed wires move, inducing high stresses that continually reverse as the rope moves through a system of sheaves, resulting in failure of individual wires due to fatigue.

The repeated bending and straightening of individual wires passing over sheaves and drums produces small cracks in the wires that increase in number and grow larger with continued bending. This process occurs in all ropes, but is accelerated if drums and sheaves are too small, or if the rope is relatively inflexible.

Lower bending stresses found in lang lay ropes enable such ropes to bend over a sheave for a longer period of time before wires break from fatigue.

Abrasion

Generally, abrasion resistance in a rope depends on the type of metal that the rope is made of and the size of the individual outer wires. Wire rope made of the harder steels, such as improved plow steel, has considerable resistance to abrasion. Ropes that have larger wires forming the outside of their strands are more resistant to wear than ropes having smaller wires that wear away more quickly.

Lang lay rope has more abrasion resistance than regular lay rope, because wear is distributed differently in the two types of rope. In lang lay ropes each wire is worn for a greater length, but there are fewer worn places in a given length of wire; in regular lay ropes the worn places are shorter and there are more of them.

Because the bending stresses are greatest where the cross-section is least, and the reduced cross-section in lang lay rope is about three times that in regular lay rope, these bending stresses are distributed over a greater length of wire and thus are not as great at any one place. Therefore, in lang lay rope more of the metal will be worn away before the outer wires break due to bending stresses.

Although wire rope of Seale construction resists abrasion better than do other types because of its large outer wires, which take longer to wear through, the larger stiffer wires also tend to crack and break. This type of rope, however, should not be used unless the manufacturer specifically recommends using it for a particular application.

Corrosion

Ropes that are put to static work, such as guy wires, may be protected from corrosive elements by paint or other special dressings. Wire rope may also be galvanized for corrosion protection. Most wire ropes used in crane operations must rely on their lubricating dressing to double as corrosion preventive.

Flexibility

With lang lay rope, the outside wires lie at an angle with the axis of the rope and, therefore, the rope's radius of curvature is greater—thus its bending stress is less. Where small diameter sheaves are involved, the difference in the radius of curvature of individual wires can amount to as much as 15 percent in favor of lang lay ropes.

When sheave diameters fall below the economical minimum, however, other conditions make the use of lang lay rope impractical. For instance, the length of wire exposed on the outside of the rope is about twice as long in lang lay as in regular lay. Thus, when a lang lay rope is bent over a very small sheave, there is a tendency for the wires on the underside next to the sheave to loosen and spring away from the rope.

This action is cumulative, so that with repeated use these wires will not return to their proper position in the strand. They become flattened and cut by adjacent wires that deteriorate the rope rapidly. Therefore, with lang lay rope, sheaves should be used that are at least as large as the economical minimum for the type of service involved (Figs. 14.6 and 14.7).

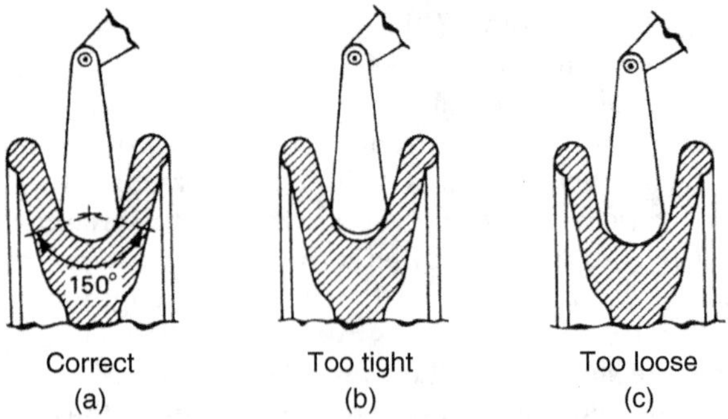

FIGURE 14.6 Sheave groove conditions. (a) Correct; (b) too tight; (c) too loose. (*Source:* American Iron and Steel Institute.)

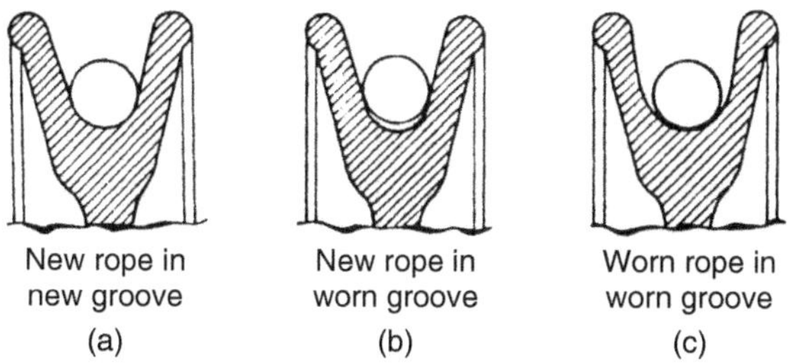

FIGURE 14.7 Sheave-groove cross sections—wire rope seating: (a) New rope in new groove (matched); (b) new rope in worn grove (pinched); (c) worn rope in worn groove (flattened). (*Source*: American Iron and Steel Institute.)

14.5 Rated Capacity of Wire Rope

The rated capacity most frequently chosen for wire rope is the load that a new wire rope may handle under given operating conditions, with an assumed design factor of 5. Operating loads (*not to exceed* 20 percent of catalog breaking strength), however, may have to be reduced when life, limb, or valuable property are at risk or other than new rope is used.

A design factor of 10 is usually chosen when wire rope is used to carry personnel. (Operating loads not to exceed 10 percent of catalog breaking strength.) Choosing a design factor is the responsibility of the user.

Breaking Strength

Breaking strength is an average figure at which wire rope samples have been found to break under laboratory conditions—in straight line with constantly increasing loads—rarely duplicated in actual lifting.

Wire rope breaking strength applies only to new, unused wire rope. *Never* use breaking strength for design or rating purposes. Use only *working load* of wire rope or cable, and *never* apply the maximum load to wire rope under any condition.

The user of wire rope is completely responsible for determining *working load limit* (WWL) for each application. Many factors should be considered, including, but not limited to loads applied, speed of operation, acceleration, deceleration, length of rope or cable, shock loads, abrasion, corrosion, number, condition and location of drums and sheaves, facilities for inspection, and the risk to life and property should a wire rope or cable break.

Attachments

Clips, sockets, thimbles, sleeves, hooks, links, shackles, sheaves, blocks, and all other attachments *must* have at least the same WLL as the wire rope used; and match in size, material, and strength to provide adequate safety protection. Proper installation and use is crucial for maximum efficiency and safety. To provide adequate safety protection,

- Match size, material, and strength of all clips, sockets, thimbles, sleeves, hooks, links, shackles, sheaves, blocks, hitch balls, etc.
- *Keep out* from under a raised load.
- *Do not* operate a load over people.
- *Do not* ride on a load.
- Conduct all lifting operations in such a manner that if equipment were to fall or break, no personnel would be injured.
- *Avoid* shock loads that exceed the static load caused by rapid change of movement such as impacting, jerking, or swinging of a load. WLL will not apply in these cases.
- Replace fall protection gear after being subject to the shock load of a fall.

14.6 Safe Working Loads

The term *safe working load* (*SWL*) of wire rope is used to define the load that can be applied allowing the rope to provide efficient service and also prolong the life of the rope. The formula for computing the SWL of a wire rope is the diameter of the rope squared, multiplied by 8, that is, $D \times D \times 8 = SWL$ (in tons).

The *correct* way to measure the diameter of wire rope is from the top of one strand to the top of the strand directly opposite it; the *wrong* way is to measure across two strands side by side (Fig. 14.8).

To ensure an accurate measurement of the diameter of a wire rope, always measure the rope at three places, at least 5 ft apart. Use the average of the three measurements as the diameter of the rope.

CAUTION: Do not *downgrade the SWL of wire rope because it is old, worn, or in poor condition.* Cut up *and discard wire rope in these conditions.*

Use WLL only, which is based on a load being uniformly applied in a straight line pull. The actual working load of a wire rope includes not only the static or dead load, but also the loading that results from acceleration, retardation, and shock load. This total load, in turn, is influenced by the size of the sheaves and drums, the method of reeving, and the rope's end attachments.

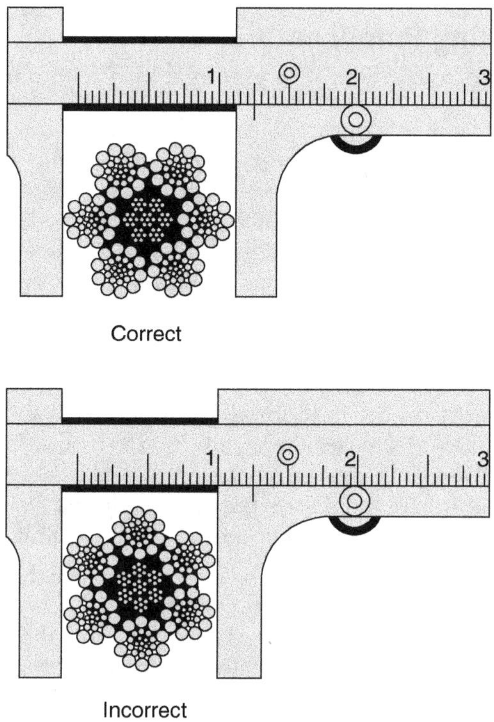

FIGURE 14.8 Correct method to measure wire rope is over any pair of opposite strands. (a) Always measure the larger dimension; (b) *not* the smaller diameter. (*Source:* Hanes Supply Inc., Buffalo, NY.)

The load can vary, within limits, with individual installation conditions. Thus, no fixed or arbitrary values for working loads can be properly applied to all of the numerous classifications of service to which wire rope is subjected.

When the rigger is confronted with a load problem and references for wire rope are not available, either of the following two approximate methods will give safe serviceable results:

Determining the working strength in pounds:

- Change the rope diameter into eighths of an inch
- Square the number of eighths
- Multiply the result by 250 to determine the working strength in pounds

Determining the working strength in tons:

- Square the diameter of rope in inches
- Multiply the result by 8

Avoid shock loads that exceed the static load caused by rapid change of movement, such as jerking, impacting, or swinging of loads. Working load limits will not apply.

14.7 Safe Lifting Practices

Primary factors to consider when lifting a load with wire rope safely include:

- *Load size, weight, and center of gravity.* To make a level lift, the hoist hook must be located directly above the center of gravity (where the entire weight of the object to be lifted may be considered to be concentrated). If the hook is too far to either side of the center of gravity, dangerous tilting will result, causing unequal stress in the sling legs. Immediately correct the load balances.
- *Number of legs at an angle with the horizontal.* The smaller the angle between the sling legs and the horizontal, the greater the stress on the individual sling legs. This increased stress effectively decreases the weight that can be safely lifted with any given sling size. Large (heavy) loads can be safely moved by keeping this angle as large as possible and, when necessary, distributing the weight of the load among more sling legs.
- *Rated capacity of the sling.* The type of material the sling is made of, the size of the sling, and the type of hitch determines the rated capacity of a sling. Workers must know the capacity of the sling, and can obtain this information through charts or tables available from the wire rope manufacturer. The rated capacity of a sling *must not* be exceeded, under any circumstances.
- *History of care and use.* The leading causes of sling failure are their mishandling and misuse. Following the manufacturer's recommendations for proper care and use are essential for maximum sling service life and safety.

Safe working loads of various sizes and classification of improved plow steel wire rope and wire rope slings, with various types of terminals, can be determined from *Tables of Rated Capacities (Working Load Limit) for Alloy Slings—Association for Iron and Steel Technology.*

For sizes, classifications, and grades not included in these tables, identifiable products should be followed, provided that a safety factor of not less than 5 is maintained.

Wire rope *must not* be used in hoisting or lowering, or in pulling loads, if in any length of eight diameters the total number of visible broken wires exceeds 10 percent of the total number of wires; or if the wire rope shows other signs of excessive wear, corrosion, or defect.

Never secure wire ropes with knots. Each wire rope used in hoisting or lowering, or pulling loads, *must* consist of one continuous piece without knots or splices, except for eye splices in the ends of wires and for endless wire rope slings.

NOTE: *Responsibility for choosing a design factor rests entirely with the wire rope user.*

Safety Guidelines

Conduct all lifting operations in such a manner that if equipment were to fall or break, no personnel would be injured.

- *Do not* pass under a raised load.
- *Do not* operate loads over people.
- *Do not* ride on load.
- *Keep out* of line of force of any load.
- *Avoid* shock loads—impacting, jerking, or swinging a load; the WLL will not apply in these circumstances because a shock load is generally significantly greater than the static load.
- Inspect wire rope regularly; use inspection instructions as guidelines only. Two of the most important prerequisites for inspecting wire rope are technical knowledge and experience.
- Check the general condition of the wire rope.
- Look for localized damage and wear, especially at wire rope attachments.
- Inspect all parts that come in contact with the wire rope. (Poor performance of wire rope can often be traced back to worn or wrong-sized sheave, drums, rollers, etc.)
- Look for kinks, broken wires, abrasions, lack of lubrication, rust damage, crushing, reduction of diameter, stretch, or damage to the wire rope. If any of these conditions exists or if there is any other apparent damage to the wire rope, retire the rope.
- Retire a wire rope *immediately* when in doubt about the extent of the damage. (Without laboratory analysis, it is impossible to determine the strength of damaged or used wire, making it impossible to determine the strength of damaged or used wire.)
- Destroy, rather than discard, wire rope to be retired.
- Check various OSHA and ANSI publications for specific inspection procedures.

The user of the rope should have the final responsibility of determining the maximum allowable deterioration before removing the rope from service. Considerations affecting the decision to remove a wire rope from service should be based on conditions of the installation, including size, nature, and frequency of the lifts, the next scheduled inspection, the operating and maintenance practices, as well as the extent of possible or probable injury to people, loss of life, material damage, and the like.

Conditions sufficient to either seriously question wire rope safety or immediately remove the rope from service and replace it include:

- *Severe corrosion.* More dangerous than wear, corrosion can exist internally before it can be detected visually. The remaining strength of the wire rope cannot be calculated with safety. Corrosion can be controlled by proper lubrication. When corrosion is detected, however, the wire rope *must* be removed from service.
- *Broken wires.* This indicates possible fatigue. The effect of broken wires on wire rope strength can be estimated by taking into account the number of broken wires and their distribution among various strands. *Remove* wire rope from service and replace it if one of the following conditions occurs:
  Running ropes—If there are six or more randomly broken wires in one rope lay, or if there are three or more broken wires in one strand within one rope lay.
  Standing ropes—If there are three or more broken wires in one rope lay.
  Any rigging rope—If it has one or more broken wires near an attached fitting.
- Reduced rope diameter. This indicates critical deterioration due to excessive abrasion of outside wire, collapse of core support, or external or internal wire corrosion. Replace any wire rope whose diameter is reduced by more than:
  $^3/_{64}$ in. for wire rope up to and including $^3/_4$-in. diameter
  $^1/_{16}$ in. for rope between $^7/_8$-in. and $1^1/_8$-in. diameter
  $^3/_{32}$ in. for rope of $1^1/_4$- to $1^1/_2$-in. diameter
  Replace any wire rope that stretches beyond these limits.
- *Rope stretch (elongation).* Excessive stretch beyond a wire rope's initial elongation is a deterioration of the rope resulting either from overloading or from loss of strength, and is a reason for replacement. Permanent initial stretch, caused by wires and strands tightening into their respective cores, is about 6 in. (152.4 mm) for every 100 ft (30.48 m) of six-strand wire rope; and about 9 to 10 in. (228.6 to 254 mm) for 100 ft (30.48 m) of eight-strand wire rope.
- *Excessive wire abrasion.* Wear due to friction on sheaves, rollers, or drums flattens out wires and reduces the rope diameter. The smaller cross section reduces the degree of internal lubrication available to a wire rope, which in turn further accelerates deterioration. Replace any wire rope whose diameter has been reduced by more than one-third its nominal size.
- *Internal wire damage.* This is a serious and dangerous condition, which is not visually apparent. Although lubrication may protect external wires in a rope, lack of penetration may

seriously damage internal wires as well as the rope's core. Check for:

Crushed, flattened, or jammed strands
High standing and unlaying
Bird caging, kinks, and bulges
Gaps or excessive clearance between strands
Core protrusion
Unbalanced severely worn areas
Heat damage, torch burns, and electric arc strikes
End connections—corroded, cracked, bent, worn, and improperly applied
Damaged or inadequate splices

Replace the wire rope when wire breaks reach the total number allowable by the ASME standard for the equipment being inspected or other applicable specifications (Table 14.1).

Safety Factors

To guard against possible rope failure, the actual load on wire rope should be only a fraction of the manufacturer's rated breaking strength.

ASME number	Equipment	Number of broken wires in running ropes		Number of broken wires in standing ropes	
		One rope lay	One strand	One rope lay	One strand
B30.4	Portal and tower cranes	6*	3	N/A	N/A
B30.5	Crawler and truck cranes	6*	3	3*	2
B30.6	Derricks	6*	3	3*	2
B0.7	Base-mounted drum hoists	6*	3	3*	2
A10.4	Personnel hoists	6*	3	3*	2
A10.5	Material hoists	6*	3	3*	2

* Also remove *one* valley break.
OSHA requires monthly recordkeeping of wire rope conditions.
Note: Current industry recommendations and OSHA standards are based on steel sheaves.
Consult the manufacturer of plastic and synthetic sheaves or liners for its recommendations on the safe application of its product and for its inspection criteria.
Source: Association for Iron and Steel Technology

TABLE 14.1 Wire Rope Replacement Schedule. Based on Number of Broken Wires

Wire rope	Factor of safety (minimum)
Track cables	3.2
Guys	3.5
Hoisting equipment	5.0
Haulage ropes	6.0
Cranes and derricks	6.0
Electric and air hoists	7.0
Personnel hoists	10.0

Source: Association for Iron and Steel Technology.

TABLE 14.2 Minimum Factors of Safety

Applying a factor of safety to this value provides the margin of strength necessary to handle loads safety and prevent accidents (Table 14.2).

- *Ropes used for rigging.* The minimum acceptable safety factor is 5; and operating loads *must not* exceed 20 percent of catalog breaking strength.
- *Ropes used on equipment that hoists people.* The minimum acceptable safety factor is 10; and operating loads should not exceed 10 percent of catalog breaking strength.

NOTE: *Factor of safety is* not *reserve strength. It* must *never be used to provide additional rope capacity and it* must never *be reduced.*

The maximum safe working load of a rope can be determined by dividing the manufacturer's rated breaking strength of the particular rope by the appropriate factor of safety. For rigging ropes the minimum acceptable safety factor is 5; for ropes used on equipment that hoists people, the factor is 10.

Do not use breaking strength for rating purposes. Breaking strength is an average figure at which samples have been found to break—under laboratory conditions, in straight line with constantly increasing loads. These conditions are rarely duplicated in actual operations.

14.8 Care and Maintenance of Wire Strand Rope

Wire ropes, like the machines and hoists on which they are used, require careful use, handling, care, and maintenance—essential to their satisfactory performance, long life, and adequate safety. To ensure safe, dependable service over a maximum period of time, care and upkeep of the wire rope is necessary to keep it in good condition.

Coiling and Uncoiling

Once a new reel has been opened, it may be coiled or faked down. The proper direction of coiling is counterclockwise for left lay wire rope and clockwise for right lay wire rope. Because of the general toughness and resilience of wire, it often tends to resist being coiled down. When this occurs, it is useless to fight the wire by forcing down the turn because the wire will only spring up again. But if it is thrown in a back turn, it will lie down properly.

When faked down, a wire rope will run right off; but when wound in a coil, it must always be unwound (Fig.14.9, Fig. 14.10).

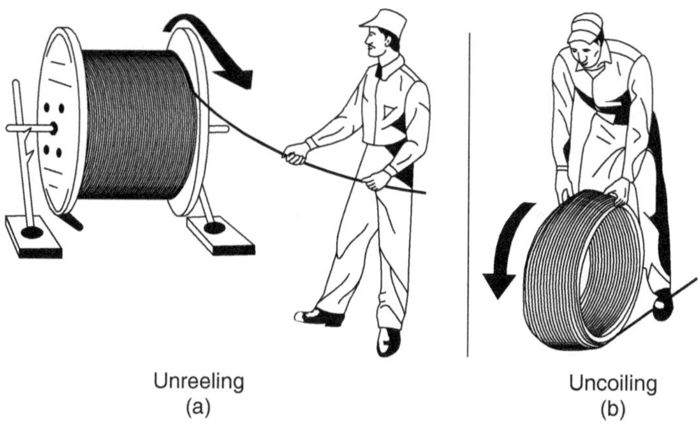

FIGURE 14.9 New wire rope. (a) Unreeling; (b) uncoiling. (*Source:* U.S. Army Field Manual, Army Department, Washington, DC.)

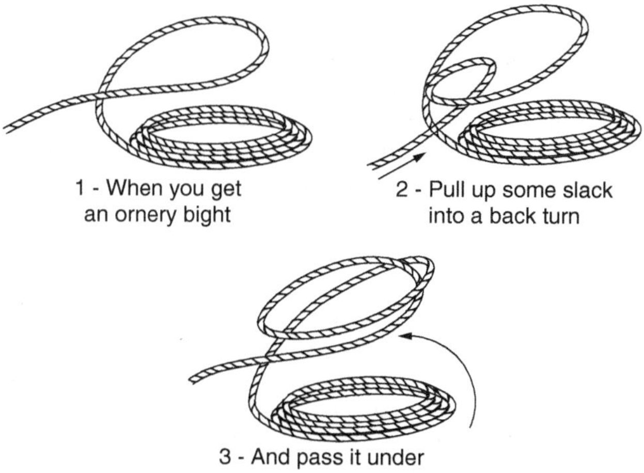

FIGURE 14.10 Throwing down wire rope in a back turn to avoid its springing up. (*Source:* U.S. Army Field Manual, Army Department, Washington, DC.)

Kinks and Reverse Bends

Wire rope tends to kink during uncoiling or unreeling, especially if it has been in service for a long time. A kink can cause a weak spot in the rope that wears out quicker than the rest of the rope.

- *Kinks*. One of the most common types of damage resulting from the improper handling of wire rope is the development of a kink. A kink starts with the formation of a loop. Remove a loop that has not been pulled tight enough to set the wires or strands of the rope into a kink by turning the rope at either end in the proper direction to restore the lay. If this is not done, and the loop is pulled tight enough to cause a kink, it will result in irreparable damage to the rope. Kinking can be easily prevented by proper uncoiling and unreeling methods, and by correctly handling the rope throughout its installation.
- *Reverse bends*. Whenever possible, drums, sheaves, and blocks used with wire rope should be placed to avoid reverse or S-shaped bends. Reverse bends cause the individual wires or strands to shift too much and increase wear and fatigue. For a reverse bend, the drums and blocks affecting the reversal should be of a larger diameter than ordinarily used and should be spaced as far apart as possible

A good method for unreeling wire rope is to run a pipe or rod through the center and mount the reel on drum jacks or other supports, so the reel is off the ground. In this way, the reel will turn as the rope is unwound, and the rotation of the reel helps keep the rope straight. During unreeling, pull the rope straight forward and avoid hurrying the operation. As a safeguard against kinking,

Never unreel wire rope from a reel that is stationary. To uncoil a small coil of wire rope, simply stand the coil on edge and roll it along the ground like a wheel, or a hoop.

Never lay the coil flat on the floor or ground to uncoil it by pulling on the end because such practice can kink or twist the rope.

Wire Rope Failure

Many operating conditions affect wire rope life. They are bending, stresses, loading conditions, speed of load application (jerking), abrasion, corrosion, sling design, materials handled, environmental conditions, and history of previous usage. In addition, the weight, size, and shape of the loads to be handled also affect the service life of a wire rope sling (Fig. 14.11).

Flexibility also is a factor. Generally, more flexible ropes are selected when smaller radius bending is required. Less flexible ropes

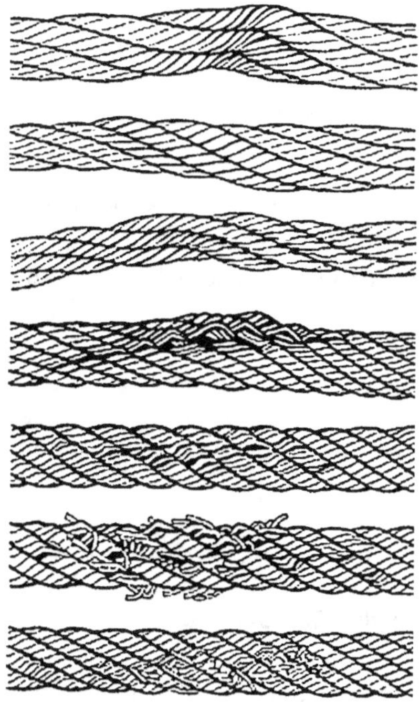

Figure 14.11 Typical wire rope mechanical damages: distortion; kinking, crushing, birdcaging; broken wires; metal loss; heat damage; and corrosion. (*Source:* Unirope Ltd., Missassagua, ON, Canada.)

should be used when the rope must move through or over abrasive materials.

Some of the more common causes of wire rope failure are

- Using incorrect wire rope size, construction, or grade
- Dragging wire rope over obstacles
- Improper lubrication
- Operating over sheaves and drums of inadequate size
- Operating over sheaves and drums with improperly fitted grooves or broken flanges
- Overriding or cross winding on drums
- Jumping off sheaves
- Exposure to acid fumes
- Use of an improperly attached fitting
- Grit being allowed to penetrate between the strands causing internal wear
- Being subjected to severe or continuing overload

Proper Care of Wire Rope

Wire rope needs proper care and maintenance for optimal safety and long service life. In service, wire rope loses strength and work capability.

The strength of a wire rope begins to decrease when the rope is put in use, and continues to decrease with each use.

Abuse and misuse will increase the rate of loss; and wire rope will fail if worn out, overloaded, misused, damaged, or improperly maintained.

- *Never* install or use a wire rope without full knowledge and consideration of the design factor for the application.
- *Never* use the rope where the load applied to it is greater than the working load determined by dividing the nominal strength of the rope by the appropriate design factor.
- *Never* shock-load a wire rope—a sudden application of force or load can cause both visible external damage and internal damage. The sudden release of a load can also damage a wire rope.
- *Clean and lubricate* wire rope properly before placing it in storage. Corrosion of wire rope can be virtually eliminated if lubricant is applied properly and sufficient protection from the weather is provided.
- *Never* store wire rope in an area where acid is or has been kept. The slightest trace of acid or acid fumes coming in contact with wire rope will damage it at the contact spot. Wire that has given way has been found many times to be acid damaged.
- *Remember* that rust, corrosion of wires, and deterioration of the fiber core will significantly reduce the strength of wire rope. Although it is not possible to say exactly the loss due to these effects, it is certainly enough to warrant precautions.

Maintenance of Wire Rope

A wire rope will *fail* if improperly maintained in the field, and proper maintenance of the equipment over which wire ropes operate has an important bearing on the rope's life (Table 14.3).

- Worn grooves, poor alignment of sheaves, and worn parts—resulting in shock loads and excessive vibration—will have a deteriorating effect.
- For wire ropes to work effectively, the wires and strands must be free to move; friction from corrosion or lack of lubrication shortens the service life of wire rope.
- Regular lubrication of wire rope is essential to
    - Reduce friction between individual wires moving over each other—as well as between the rope and sheaves or drums.
    - Provide corrosion protection and lubrication in the core and for the inside wires, as well as on exterior surfaces.

• Inspect wire rope regularly and thoroughly, following manufacturer's guidelines	☐
• Make sure all wire rope ends are properly seized	☐
• Check for local wear in wire rope	☐
• Repair or replace faulty guides and rollers	☐
• Replace sheaves if grooves are deeply worn or scored, if rims are cracked or broken, or if bearings are worn or damaged	☐
• Repair faulty clutches	☐
• *Never* use wire rope that is cut, badly kinked, or crushed	☐
• *Always* spool wire rope on a drum correctly; and make sure the wire rope does not cross wind	☐
• *Never* wind more wire rope on a drum than the drum can properly take	☐
• Use only the correct wire rope for every job	☐
• Use larger-sized rope when the exact load and severity of loading are unknown	☐
• Take special precautions when there is: Possibility of shock loading Lifting operations hazardous to personnel Conditions are abnormal or severe	 ☐ ☐ ☐
• *Never* overload wire rope	☐
• Minimize shock loading to avoid overstressing the rope	☐
• Eliminate slack in the rope at the start of loading	☐
• Make sure that loops formed in slack lines are NOT pulled taught and formed into kinks; once a kink has been made in wire rope, the damage is permanent—a weak spot will always remain no matter how well the kink seems to have been straightened out	☐
• *Never* pull out loops formed in slack lines—always unfold them	☐
• Start loading carefully, and apply power smoothly and steadily	☐
• Avoid sudden loading in cold weather	☐
• *Never* use frozen wire ropes	☐
• Make sure wire ropes do not bind in sheaves	☐
• Check and correct lines from whipping or vibration	☐
• Avoid dropping wire rope from heights	☐
• *Never* roll a load with wire rope passing under the load	☐

TABLE 14.3 Care of Wire Rope Checklist (*Continued*)

• *Do not* pull rope out from under loads or drag it over obstacles	☐
• Unfold any loops that have formed; DO NOT pull them out	☐
• Ensure that drums and sheaves are of sufficient diameter	☐
• *Always* use protective padding where wire rope pass over sharp edges or around sharp corners	☐
• Avoid bending wire ropes in reverse when reeving through blocks or sheaves	☐
• Avoid dragging wire rope from under loads or over obstacles	☐
• Ensure that sheaves are aligned and the fleet angle is correct	☐
• Make sure wire rope spools properly on the drum	☐
• *Never* wind more than the proper amount of wire rope on any drum	☐
• *Never* allow wire rope to cross wind	☐
• Use thimbles in eye fittings at all times	☐
• Lubricate wire rope regularly according to the manufacturer's recommendations	☐
• Watch for local wear—premature wear at one spot is common and can be prevented if the cause is detected	☐
• *Always* store unused rope in a clean, dry place	☐

Source: Wire Rope Users Manual, Wire Rope Technical Board, Alexandria, VA

Table 14.3 Care of Wire Rope Checklist (*Continued*)

Lubrication is applied to the wires and strands of a wire rope when it is manufactured and is adequate for initial storage and the early stages of the rope's working life due to constant bending over sheaves and drums. It is depleted, however, when the rope is in service and *must* be replaced at regular intervals in accordance with the wire rope manufacturer's instructions.

Field Lubrication

Although all wire rope is lubricated during manufacture, to lengthen its useful service life it must also be lubricated "in the field." There is no set rule on how much or how often this should be done. It depends on the conditions under which the rope is used. The heavier the loads, the greater the number of bends; or the more adverse the conditions under which the wire rope operates, the more frequently lubrication will be required.

Wire rope lubricants have two principal functions: to reduce friction as the individual wires move over each other; and to provide corrosion protection and lubrication in the core and inside wires, and on the exterior surfaces.

Both internal and external lubrication protects a wire rope against wear and corrosion. Deterioration from corrosion is more dangerous than that from wear because corrosion ruins the inside wires—a process hard to detect by inspection. Deterioration caused by wear, however, can be detected by examining the outside wires of the wire rope because these wires become flattened and reduced in diameter as the wire rope wears.

Internal lubrication can be properly applied only when the wire rope is being manufactured, and manufacturers customarily coat every wire with a rust-inhibiting lubricant as it is laid into the strand. Lubrication that is applied in the field, however, is designed not only to maintain surface lubrication but also to prevent the loss of the internal lubrication provided by the manufacturer.

Lubricating wire ropes is a difficult proposition, regardless of the construction and composition. Ropes with fiber cores are somewhat easier to lubricate than those made exclusively from steel materials. For this reason, it is important to carefully consider the issue of field relubrication when selecting rope for rigging. Lubricants that may be used include petroleum, asphaltic, grease, petroleum oils, or vegetable oil based.

- **Lubricant types**

  Penetrating lubricants contain a petroleum solvent that carries the lubricant into the core of the wire rope, then evaporates, leaving behind a heavy lubricating film to protect and lubricate each strand.

  Coating lubricants penetrate slightly, sealing the outside of the cable from moisture and reducing wear and fretting corrosion from contact with external bodies.

- **Lubricant characteristics**

  Penetrate the rope core—applied at a place where the rope bends, such as over a sheave, exposing the wires and strands to provide a better opening to the core.

  Adhesive and film strength—enough to stick to wires in the rope and the spaces around the wires.

  Free of acids and alkalis.

  Resist corrosion.

  Retention—should stay put without being easily washed away.

Because most wire rope fails from the inside, it is important to make sure that the center core receives sufficient lubricant. Lubricant engineers recommend using a penetrating lubricant to saturate the wire rope core, and following it with a coating lubricant to seal and protect the outer surface. Because most wire ropes fail from the inside, it is important to make sure that the center core receives sufficient lubricant.

- **Lubricant application**

  Make sure that wire rope is kept properly lubricated, using only manufacturer-specified lubricating designed for specific wire rope operating conditions.

  Never apply heavy grease to the rope—it can trap excessive grit and dust, which can externally damage wire rope or be forced inside, causing hidden damage.

  Never use "engine oil"—it contains materials that can damage wire rope.

  Never use "crankcase oil"—it is acidic and contains metal particles that are highly abrasive and damaging to wire rope.

  Check with a lubricant supplier for the correct field-applied wire rope lubricant; type and amount of lubricant will vary according to type and use of wire rope.

Before lubricating wire rope, make sure it is absolutely clean and dry; remove excess dirt, rock dust, or other materials that can prevent field-applied lubricants from properly penetrating into the ropes. Use a jet of air, steam, or a wire brush to clean the wire rope. Allow the rope to dry thoroughly before lubricating; otherwise the lubricant will not penetrate the rope through a trough filled with warm oil to obtain maximum penetration.

- **Lubricating methods**

  Manual. Use spray or drip systems that apply lubricant directly to the wire rope; swab or paint lube into a moving rope by hand; or pour lube onto the rope as it passes a certain point.

  Automatic. Drip or spray lube onto the rope as it passes over a sheave at preselected intervals.

As a safety precaution, always wipe off any excess when lubricating wire rope, especially with hoisting equipment. Too much lubricant can get into brakes or clutches and cause them to fail.

Never lubricate wire rope that works a dragline or other attachments that normally bring the wire rope in contact with soils—the lubricant will pick up fine particles of material, and the resulting abrasive action will be detrimental to both the wire rope and sheave.

CAUTION: *The motion of machinery in use may sling excess lube around over crane cabs and onto catwalks, making them unsafe.*

Storage

Wire rope slings should be stored in a well-ventilated, dry building or shed. To avoid corrosion and rust, never store wire rope slings on the ground or allow them to be continuously exposed to the elements. And, if it is necessary to store wire rope slings outside, make sure that they are set off the ground and protected.

NOTE: *Using a sling several times a week, even with light loads, is a good practice. Records show that frequently or continuously used slings give useful service far longer than idle ones.*

Discarding Wire Rope Slings

Wire rope slings can provide a margin of safety by showing early signs of failure. Discard wire ropes when there is evidence of:

- Severe corrosion
- Localized wea (shiny worn spots) on the outside
- One-third reduction in outer wire diameter
- Damage, kinking, bird caging, or other evidence of damage to the wire rope structure
- Excessive broken wires

14.9 Inspection of Wire Strand Rope

As with other machines, periodic inspections of wire rope are necessary. As the rope is used, it gradually loses strength. Regularly scheduled inspections will determine whether a wire rope retains sufficient capability to perform the work to be done before the next scheduled inspection (Table 14.4).

Certain governmental regulations require regular inspections. OSHA document *29 CFR 1926-550 (1991)*, ANSI document *B30.5 [1968]*, and SAE document *J959 (1966)* each contain regulatory information on wire rope inspections (Table 14.5).

Inspection Procedures

Appoint a competent person to perform all inspections and to make a complete written report of the wire rope condition, including date of fitting, size, construction, length, defects found during inspections, and length of rope's service.

To ensure safe, economic use of rigging equipment requires a regularly scheduled program of periodic inspections of all load supporting wire rope, chain, and fittings. The frequency of inspection is determined by the use of the rope and the conditions under which it

Inspection type	Equipment type	Frequency	Performed by	Records
Frequent	All	Each day of use	Operator	Not required
Monthly	Overhead and gantry, crawler locomotive, and truck crane	Monthly	Wire rope inspector	Checklist or inspection report; Signed and dated
Periodic	All	To meet conditions but no less than annually. When returning to lift service	Wire rope inspector	Inspection report signed and dated
When equipment has been idle for 1 month or more	All	Before service	Wire rope inspector	Inspection report signed and dated

Table 14.4 Wire Rope Inspection Schedule

• Use only the correct wire rope for every job	☐
• Use larger-sized rope when the exact load and severity of loading are unknown	☐
• Take special precautions when there is: Possibility of shock loading Lifting operations hazardous to personnel	 ☐ ☐
• *Never* overload wire rope	☐
• Minimize shock loading to avoid overstressing the rope	☐
• Eliminate slack in the rope at the start of loading	☐
• Make sure that loops formed in slack lines are *not* pulled taught and formed into kinks; once a kink has been made in wire rope, the damage is permanent—a weak spot will always remain no matter how well the kink seems to have been straightened out	☐
• *Never* pull out loops formed in slack lines—always unfold them	☐
• Start loading carefully; apply power smoothly and steadily	☐
• Avoid sudden loading in cold weather	☐
• *Never* use frozen wire ropes	☐

Table 14.5 Inspection of Wire Rope Checklist

• Make sure that drums and sheaves have large enough diameters for the wire rope being used	☐
• Make sure wire ropes *do not* bind in sheaves	☐
• *Always* spool rope wire on a drum correctly	☐
• Make sure wire rope does not cross wind	☐
• *Never* wind more rope on a drum than the drum can properly take	☐
• Check and correct lines from whipping or vibration	☐
• *Always* use protective padding where wire rope pass over sharp edges or around sharp corners	☐
• Avoid bending wire ropes in reverse when reeving through blocks or sheaves	☐
• Avoid dragging wire rope from under loads or over obstacles	☐
• Avoid dropping wire rope from heights	☐
• *Never* roll a load with wire rope passing under the load	☐
• *Do not* pull rope out from under loads or drag it over obstacles	☐
• *Never* use wire rope that has been cut, badly kinked, or crushed	☐
• Unfold any loops that have formed; *do not* pull them out	☐
• Ensure that drums and sheaves are of sufficient diameter	☐
• Repair or replace faulty guides and rollers	☐
• Ensure that sheaves are aligned and the fleet angle is correct	☐
• Replace sheaves having deeply worn or scored grooves, cracked, or broken rims and worn or damaged bearings	☐
• Repair faulty clutches	☐
• Check for abnormal line whip and vibration	☐
• Make sure wire rope spools properly on the drum	☐
• *Never* wind more than the proper amount of wire rope on any drum	☐
• *Never* allow wire rope to cross wind	☐
• Make sure wire ropes do not bind in sheaves (new wire rope requires a run-in period before operating at full load and full speed)	☐
• Use thimbles in eye fittings at all times	☐
• Lubricate wire rope regularly according to the manufacturer's recommendations	☐
• Watch for local wear—premature wear at one spot is common and can be prevented if the cause is detected	☐
• *Always* store unused rope in a clean, dry place	☐

Source: Wire Rope Users Manual, Wire Rope Technical Board.

Table 14.5 Inspection of Wire Rope Checklist (*Continued*) is used. Make a complete and thorough inspection of all wire ropes in use at least once a month. *Always* thoroughly inspect all wire ropes before putting them back into service; if they have been idle for a period of a month or more.

- *Daily inspections.* User/operator *must* inspect equipment daily before using.
- *Monthly inspections.* Use *only* a person trained to recognize defects and authorized to remove equipment from service to conduct monthly inspections.
- *Annual inspections.* Use *only* the manufacturer or an outside contractor, providing a person trained to inspect equipment for defects and authorized to remove equipment from service (although not required by OSHA).

Regardless of its condition, *retire* wire rope that is consistently in use after a fixed length of service (a predetermined number of hours, weeks, or months) to eliminate the possibility of rope failure due to fatigue.

The primary rule for conducting wire rope inspections on equipment is that each wire rope must be considered individually. This individual treatment is particularly important when inspecting standing ropes, which are considered supporting or structural members. For example, pendant lines that support long crane booms are frequently made up of several sections of individual ropes. Each rope must be inspected and pass on its own merits.

However, because different inspection criteria frequently apply, standing ropes should be inspected separately from the running, or operating ropes on the same machine or installation. Both running and standing rope parts may need to be inspected on the same trip to some high or inconvenient location on an installation. Each rope then must be given individual attention, and inspected according to its particular criteria. Also, the information on the condition of each rope must be recorded separately.

NOTE: *A proper inspection cannot be made when a wire rope is supporting a load or when it is in motion because a taught line can hide broken wires or damaged cores. This applies to both standing and running lines. For example: To inspect boom pendant lines, the boom must be down with the lines relaxed.*

Throughout an inspection, wire rope should be examined carefully for:

- *Fishhooks, kinks, and worn and corroded spots.* Usually breaks in individual wires will be concentrated in areas where the wire runs continually over the sheaves or bend onto the drum.

- *Deterioration resulting in a suspected loss of original rope strength.* To determine whether continued use of the rope would constitute a safety hazard.
- *Abrasion, wear, fatigue, corrosion, improper reeving, and kinking.* Often more important in determining the stable life of wire rope than are strength factors based on new rope conditions. Abrasion, or reverse and sharp bends cause individual wires to break and bend back. These breaks are known as fishhooks. When wires are slightly worn but have broken off squarely and stick out all over the rope, that condition is usually caused by overloading or rough handling. If the breaks are confined to one or two strands, the strength of the rope may be seriously reduced.

NOTE: *Inspection instructions should be used only as guidelines.*

When an inspection discloses any unsatisfactory conditions in a rope, *destroy* the rope or cut it into small pieces as soon as possible. This precaution prevents the defective rope from being used for hoisting.

IMPORTANT: *Wire rope and rigging inspection is the single most important operational check to be made on hoisting and rigging equipment requiring two critical prerequisites: technical knowledge and experience.*

Condition of Wires

When 4 percent of the total number of wires in the rope is found to have breaks within the length of one lay of the rope, the rope is considered unsafe. Consider the rope unsafe when three broken wires are found in one strand of 6×7 rope, or six broken wires in one strand of 6×19 rope, or nine broken wires in one strand of 6×37 rope. Overloading wire rope will reduce the diameter. Additionally, failure to lubricate wire rope will reduce the diameter. This occurs because the hemp core will eventually dry out and collapse or shrink. The surrounding strands are therefore deprived of support, and the strength and dependability of the rope are equally reduced. Rope that is 75 percent of its original diameter should be removed from service.

When widespread pitting and corrosion of the wires are visible through inspection, the rope should be removed from service. Special care should be taken to examine the valleys and small spaces between the strands for rust and corrosion. Deterioration from corrosion is more dangerous than from wear because corrosion ruins the inside wires—a process hard to detect by inspection. Rust should be removed at regular intervals by wire brushing using petroleum solvent, compressed air, or steam cleaner to remove all foreign material and old lubricant from the valleys between the strands. Wire rope bending around hoist drums and sheaves will wear like any other metal article, so lubrication is just as important to an operating wire rope as it is to any other piece of working machinery.

OSHA *requires* maintaining permanent records, signed by a qualified person, for every wire rope installation.

Removing Wire Rope from Service

Do not wait for a wire rope to fail before inspecting ropes in a rigging spread. Close examination of a rope will indicate not only when it is time to remove the rope from service, but also whether the rope is suited to the job it is doing. *Remove* wire rope from service and replace the rope if *one* of the following conditions exist:

- *Broken wire, but rope showing no wear.* Indicates that the rope has been subject to excessive bending; usually resulting from using sheaves and drums that are too small for the rope diameter, or from using a rope that is too coarse. Replace:
  - *Running ropes.* When there are six or more randomly broken wires in one rope lay, or three or more broken wires in one strand within one lay
  - *Standing rope.* When there are three or more broken wires in one rope lay
  - *Any rigging rope.* When there is one or more broken wires near an attached fitting
- *Crushed or flattened rope or jammed strands.* Indicates that rope has been under too much pressure, which usually is the result of improper winding; also possible that the rope has not been designed to withstand the conditions to which it has been exposed.
- *Breaking of whole strands or entire rope soon after installation.* Either the rope is overloaded, or a stronger rope or larger diameter rope is required.
- *High stranding or unlaying,* as well as bird-caging, kinks, bulges, gaps or excessive clearance between strands, core protrusion, heat damage, torch burns, or electric arc strikes.
- *Reduced rope diameter.* Indicates critical deterioration due to excessive abrasion of outside wires, collapse of core support, or external or internal wire corrosion. Replace any rope whose diameter is reduced by more than:
  - $^1/_{64}$-in. (0.4 mm) for rope up to and including ¾-in. (19.05 mm) diameter $^1/_{16}$-in. (1.59 mm) for rope between $^7/_8$-in. (22.23 mm) and $1^1/_8$-in. (28.58 mm) diameter $^3/_{32}$-in. (2.38 mm) for rope of 1¼-in. (31.75 mm) to 1½-in. (38.1 mm) diameter

- *Excessive wire abrasion.* Wear due to friction on sheaves, rollers, or drums flattens outer wires and reduces tope diameter. The smaller cross section reduces the degree of internal lubrication available to a wire rope, which in turn further accelerates deterioration. Replace any rope whose diameter has been reduced by more than one-third is nominal size.
- *Internal wire damage.* Serious and dangerous condition, which is not virtually apparent. Although lubrication may protect external wires in a rope, lack of penetration may seriously damage internal wires as well as the rope's core.
- *Severe corrosion.* More dangerous than wear, corrosion can exist internally before it can be detected visually. The remaining strength cannot be calculated with safety.
- *Corrosion can be controlled by proper lubrication.* When corrosion is detected, however, the rope must be removed from service.
- *Rope stretch or elongation.* Excessive stretch beyond a wire rope's initial elongation is a deterioration of the rope resulting either from overloading or from loss of strength and is a reason for replacement. Permanent initial stretch, caused by wires and strands tightening into their respective cores, is about 6 in. (152.4 mm) for 100 ft (30.48 m) of six-strand wire rope, about 9 to 10 in. (226.6 to 254 mm) for 100 ft (30.48 m) of eight-strand wire rope.

Safe wire rope operation depends on the rope's strength as it nears the end of its useful life. Where rope is consistently in use, it should be replaced automatically—regardless of condition—after a predetermined amount of time, such as several hundred hours, several weeks, or several months. This will eliminate the risk of rope failure from fatigue. Only inspection, however, will permit determining a rope's condition and whether it should be removed from service sooner.

Inspection Process

Before beginning wire rope inspection,

- Identify all chains, slings, and cables with a tag that shows the load rating, limitations, etc.
- Check the general condition of the wire rope and look for localized damage and wear, especially at wire rope attachments.
  - Surface wear—normal and unusual
  - Broken wires—number and location
  - Kinks, abrasions, crushing, or other obvious damage to the wire rope
  - Lack of lubrication
  - Reduction in diameter
  - Rope stretch (elongation)

Integrity of end attachments
Evidence of abuse or contact with another object
Heat damage
Corrosion, rust damage

- Inspect all parts that come in contact with the wire rope. Poor performance of wire rope can often be traced back to worn or wrong-sized sheaves, drums, rollers, etc.
- Retire the wire rope if any of these conditions exist or if there is any other apparent damage to the wire rope. When in doubt about the extent of the damage, retire the wire rope in question immediately, because without laboratory analysis, it is impossible to determine the strength of damaged or used wire—and therefore whether wire rope with any amount of damage is safe to use. (For specific inspection procedures, check various OSHA and ANSI publications.)
- Destroy, rather than discard, wire rope to be retired by cutting it up into short pieces. Wire rope that is not destroyed might be used again by someone not aware of the hazard associated with its use.
- Observe all wire rope, in continuous service, during normal operations; and visually inspect such rope regularly on a weekly basis (or more if required).
- Check all wire ropes each time, before putting into service and periodically during normal operations.
- Inspect operational wire rope visually each week, and thoroughly examine it at least once a month.
- Inspect thoroughly any wire rope that has been idle for more than a month before putting it back into service. Only a close examination of a rope will indicate not only when it is time to remove the rope from service, but also whether the rope is suited to the job it is doing.
- Allow only the manufacturer to repair chains, slings, and cables; provide certification of proof testing.

14.10 Common Wire Rope Degradation

Wire ropes degrade in service due to hostile operating load conditions. Fatigue of wires in a rope—in combination with other detrimental factors, such as abrasion, tension, gouging, twisting, crushing and corrosion—is frequently responsible for rope deterioration and premature failure during operation.

As a wire rope accumulates bending cycles, its strength first decreases because of wear and then increases due to cold working of the wires. As the amount of cold working increases, more and more of the embrittled wires break and the strength of the rope decreases rapidly (Table 14.6).

Mode	Symptoms	Possible causes
Fatigue	Wire break is transverse—either straight across or Z shape. Broken ends will appear grainy.	Check for rope bend around too small a radius; vibration or whipping; damaged sheaves; rollers too small; reverse bends; bent shafts; tight grooves; corrosion; small drums and sheaves; incorrect rope construction and size (too big); improper installation; poor end terminations. (In the absence of other modes of degradation, all rope will eventually fail in fatigue.)
Tension	Wire break reveals a mixture of cup and cone fracture and shear breaks.	Check for overloads; sticky, grabby clutches; jerky conditions; loose bearing on drum; fast starts, fast stops, broken sheave flange; wrong rope size and grade; poor end terminations. Check for too great a strain on rope after factors of degradation have weakened it.
Abrasion	Wire break mainly displays outer wires worn smooth to knife edge thinness. Wire broken by abrasion in combination with another factor will show a combination break.	Check for change in rope or sheave size; change in load; overburden change; frozen or stuck sheaves; soft rollers; sheaves or drums; excessive fleet angle; misalignment of sheaves; kinks; improperly attached fittings; grit and sand; objects embedded in rope; improper grooving.
Abrasion plus fatigue	Reduced cross section is broken off square, thereby producing a chisel shape.	A long-term condition normal to the operating process. Short term: see Abrasion.
Cut or gouged rough wire	Wire ends cross section is necked down as in a cup and cone configuration. Tensile break produces a chisel shape.	Check on all the above conditions for mechanical abuse, or either abnormal or accidental forces during installation.
Torsion of twisting	Wire ends show evidence of twist and/or cork-screw effect.	Check on all the above conditions for mechanical abuse, or either abnormal or accidental forces during installation.

Table 14.6 Diagnostic Guide to Common Wire Rope Degradation (*Continued*)

Mode	Symptoms	Possible causes
Mashing or crushing	Wires are flattened and spread at broken ends.	Check on all the above conditions for mechanical abuse, or either abnormal or accidental forces during installation. This is a common occurrence on the drum when the lower layer is installed with less tension than the layer going on top.
Corrosion	Wire surfaces are pitted with break showing evidence either of fatigue tension or abrasion.	Indicates improper lubrication or storage, or a corrosive environment.

Source: American Iron and Steel Institute.

Table 14.6 Diagnostic Guide to Common Wire Rope Degradation (*Continued*)

Wire Rope Glossary[1]

BACK-STAY Guy used to support a boom or mast; or that section of a main cable, as on a suspension bridge, cableway, and the like, leading from the tower to the anchorage.

BASKET OF SOCKET The conical portion of a socket into which a broomed rope–end is inserted and then secured.

BECKET An end attachment to facilitate wire rope installation.

BECKET LOOP A loop of small rope or strand fastened to the end of a larger wire rope. Its function is to facilitate wire rope installation.

BIRDCAGE Descriptive of the appearance of a wire rope forced into compression. The outer strands form a cage and, at times, displace the core.

BLOCK A term applied to a wire rope sheave (pulley) enclosed in side plates and fitted with some attachment such as a hook or shackle.

BRIDGE SOCKET A wire rope or strand end termination of forged or cast steel designed with baskets, having adjustable bolts, for securing rope ends. *Closed type* has a U-bolt with or without a bearing block in theU of the bolt; *open type* has two eye-bolts and a pin.

BRIGHT ROPE Wire rope fabricated from wires that are not coated.

CABLE A term loosely applied to wire rope, wire strand, and electrical conductors.

[1]*Source:* Hanes Supply, Inc., and CCISCO Coastal Construction Industrial Supply Co., Inc.

CABLE-LAID WIRE ROPE A type of wire rope consisting of several wire ropes laid into a single wire rope [e.g., 6 × 42 (6 × 6 × 7) tiller rope].

CHOKER ROPE A short wire rope sling that forms a slip noose around an object that is to be moved or lifted.

COME-ALONG Device for making a temporary grip on a wire rope.

COMMON STRAND Galvanized strand made of galvanized iron wire whose grade is common iron.

CREEP The unique movement of a wire rope with respect to a drum surface or sheave surface resulting from the asymmetrical load between one side of the sheave (drum) and the other. It is not dissimilar to the action of a caterpillar moving over a flat surface. It should be distinguished from slip which is yet another type of relative movement between rope and the sheave or drum surface.

DOG-LEG Permanent bend or kink in a wire rope caused by improper use or handling.

END PREPARATION The treatment of the end of a length of wire rope designed primarily as an aid for pulling the rope through a reeving system or tight drum opening. Unlike *end terminations,* these are not designed for use as a method for making a permanent connection.

END TERMINATION The treatment at the end or ends of a length of wire rope, usually made by forming an eye or attaching a fitting and designed to be the permanent end termination on the wire rope that connects it to the load.

EXTRA HIGH STRENGTH STRAND A grade of galvanized or bright strand.

FATIGUE As applied to wire rope, the term usually refers to the process of progressive fracture resulting from the bending of individual wires. These fractures may and usually do occur at bending stresses well below the ultimate strength of the material; it is not an abnormality although it may be accelerated due to conditions in the rope such as rust or lack of lubrication.

FIBER CENTER Cord or rope of vegetable or synthetic fiber used as the axial member of a strand.

FIBER CORE Cord or rope of vegetable or synthetic fiber used as a rope's axial member.

FILLER WIRE Small spacer wires within a strand that help position and support other wires. Also the name for the type of strand pattern utilizing filler wires.

FLAT ROPE Wire rope that is made of a series of parallel, alternating right lay and left lay ropes, sewn together with relatively soft wires.

FLATTENED STRAND ROPE Wire rope that is made of either oval- or triangular-shaped strands in order to form a flattened rope surface.

GUY LINE Strand or rope, usually galvanized, for stabilizing or maintaining a structure in fixed position.

HAWSER Wire rope, usually galvanized, used for towing or mooring marine vessels.

HIGH-STRENGTH STRAND Grade of galvanized or bright strand.

KINK A unique deformation of a wire rope caused by a loop of rope being pulled down tight. It represents irreparable damage to and an indeterminate loss of strength in the rope.

LAGGING External wood covering on a reel to protect the wire rope or strand; or the grooved shell of a drum.

LAY TYPES

Right Lay—Direction of strand or wire helix corresponding to that of a right-hand screw thread

Left Lay—Direction of strand or wire helix corresponding to that of a left-hand screw thread

Cross Lay—Rope or strand in which one or more operations are performed in opposite directions. A multiple operation product is described according to the direction of the outside layer

Regular Lay—Type of rope wherein the lay of the wires in the strand is in the opposite direction to the lay of the strand in the rope. The crowns of the wires appear to be parallel to the axis of the rope

Lang Lay—Type of rope in which the lay of the wires in the strand is in the same direction as the lay of the strand in the rope. The crowns of the wires appear to be at an angle to the axis of the rope

Alternate Lay—Lay of a wire rope in which the strands are alternately regular and lang lay. (*Reverse Lay* is another term for alternate lay.)

Spring Lay—This is not definable as a unique lay; more properly, it refers to a specific wire rope construction

PEENING Permanent distortion resulting from cold plastic metal deformation of the outer wires. Caused by pounding against a sheave or machine member, or by heavy operating pressure between rope and sheave, rope and drum, or rope and adjacent wrap of rope.

PREFORMED STRANDS Strands in which the wires are permanently formed during fabrication into the helical shape they will assume in the strand.

PREFORMED WIRE ROPE Wire rope in which the strands are permanently formed during fabrication into the helical shape they will assume in the wire rope.

REEVE To pass a rope through a hole or around a system of sheaves.

SAFE WORKING LOAD This term is potentially misleading and is, therefore, in disfavor. Essentially, it refers to that portion of the nominal rope strength that can be applied either to move or sustain a load. It is misleading because it is only valid when the rope is new and equipment is in good condition. See Sec. 14.5.

SEALE The name for a type of strand pattern that has two adjacent layers laid in one operation with any number of uniform sized wires in the outer layer, and with the same number of uniform but smaller sized wires in the inner layer.

TAG LINE A small wire rope used to prevent rotation of a load.

WARRINGTON The name for a type of strand pattern that is characterized by having one of its wire layers (usually the outer) made up of an arrangement of alternately large and small wires.

Regulations and Standards

OSHA Regulations

1910.184 and 1910.251—*Slings*

1919 Subpart D—*Gear Certification*

1919.24—*Limitations on Use of Wire Rope*

U.S. Consensus Standards

ANSI/SAME Spec 9A/ISO 10425 *Steel Wire Ropes for the Petroleum and Natural Gas Industries*

ANSI/SAME B30.5-2007 *Mobile and Locomotive Cranes*

ASTM A931-08 *Standard Method for Tension Testing of Wire Ropes and Strand*

Industry Consensus Standards and Guidelines

Wire Rope Inspection Guidelines (*www.alpind.com*)

CHAPTER 15
Wire Rope, Metal Mesh, and Chain Slings

Wire rope slings are the most frequently used type of sling in industry today, offering a strong, dependable, and economical option for most lifting applications. Their popularity is enhanced by the numerous sling configurations available to support a broad range of applications, including single and multipart slings (such as round braid, flat braid, and cable laid), grommets, single-leg slings, multiple-leg bridles, and a wide variety of fittings and attachments.

15.1 Wire Rope Slings

A wire rope sling is a complex machine, with many moving parts, that takes the full brunt of the burden plus additional stresses and forces every time a load is lifted. All wire rope slings should be made of improved plow steel and should have independent wire rope cores to reduce the possibility of the rope being crushed in service.

Wire Rope Sling Types

The Wire Rope Technical Board identifies the three most common types of multipart, wire rope slings as cable laid, round braided, and flat braided.

Cable-Laid Slings

Multipart cable-laid slings are similar in appearance to single-part slings. The body of a standard cable-laid sling, however, consists of six individual wire ropes, each composed of six strands around a steel core. All six ropes are laid around another wire rope of the same construction, making a finished sling body of seven parts. Typical bodies might be described as $7 \times 7 \times 7$ or $7 \times 7 \times 9$.

Variations of the cable-laid type of sling contain multiple strands of ropes, or both, in the same manner, but with varying parts in the sling body. Both hand and mechanical splices form the eyes in sling variations, whereas only mechanical splices are used to form eyes in standard cable-laid slings.

Because there are usually many more component wires in a cable-laid sling body than in a typical single-part wire rope sling with the same rated capacity, by comparison, the cable-laid body will exhibit phenomenal flexibility. For this reason, the D/d ratio can be less for cable-laid slings—they can be in a much tighter arc around a load. A minimum D/d ratio of 10 is recommended by most wire rope and sling manufacturers for sling body in a cable-laid sling.

Round-Braided Slings

This type of sling is perhaps the more common braided construction, offering flexibility far beyond single-part and even some cable-laid slings. In addition, round-braided slings are easy to handle, resist slipping and kinking, and their multiplicity of parts provides early warning signals for inspectors by exhibiting damage among their many wires and parts before critical loss of capacity occurs. Close internal inspection is made easy by the braided design.

Because of their extreme flexibility, eight-part round-braided slings conform quickly to load contour, provide maximum gripping contact with the load, and securely hold slippery, irregular, and unbalanced loads.

Round-braided slings are usually fabricated by hand braiding component ropes. Properly balanced braids provide good control over residual torques in the rope parts. Therefore, such a sling does not tend to rotate or unlay itself under load, as do single-part sling bodies. This reduces the requirement for tag lines on many lifts.

Flat-Braided Slings

Slings with six ropes or parts are the more popular of this type, although other constructions are in use. Flat braids provide a relatively wide bearing area for protection against slippage and marring of the load. These slings are hand fabricated, with special care being taken to balance individual ropes.

As with all multipart slings, flat braids consist of a large number of small wires. Because these component wires are comparatively small in diameter, they are highly susceptible to wear and breakage from abuse, abrasion, and nicking.

In both round- and flat-braided slings, as with cable-laid types, the number of component wires is greatly increased in comparison with single-part slings of the same diameter and capacity. The D/d ratio of braided slings is 20 times the component diameter, allowing lifts of heavy, compact loads that would not be possible with a comparable single-part sling.

15.2 Selection of Wire Rope Slings

When selecting a wire rope sling to give the best service, there are four characteristics to consider: strength, ability to bend without distortion, ability to withstand abrasive wear, and ability to withstand abuse.

Although they have become the workhorses of the sling field, with exceptionally sound safety features, wire rope slings must be chosen carefully for the services in which they are to be placed as they have certain limitations. When selecting a wire rope sling to give the best service, consider its primary characteristics.

Strength

Strength is a function of the rope's size, grade, and construction. It must be sufficient to accommodate the maximum load that will be applied. The maximum load is determined by means of an appropriate multiplier—the number by which the ultimate strength of a wire is divided to determine the working load limit.

As a sling suffers from the rigors of continuous service, however, both the design factor and the sling's ultimate strength are proportionately reduced. If a sling is loaded beyond its ultimate strength, it will fail. As a sling suffers from the rigors of continued service, however, both the design factor and the sling's ultimate strength are proportionately reduced. If a sling is loaded beyond its ultimate strength, it will fail.

For this reason, older slings must be more rigorously inspected to ensure that rope conditions adversely affecting the strength of the sling are considered in determining whether or not a wire rope sling should be allowed to continue in service.

Fatigue

Fatigue of a wire rope refers to its ability to bend without failure. In a wire rope, fatigue results from the development of small cracks resulting from repeated applications of bending loads. It occurs when ropes make small bends. The best means of preventing fatigue failure of wire rope slings is to use blocking or padding to increase the radius of bend.

Abrasive Wear

Abrasive wear is the ability to withstand abrasion, depending on the size, number of wires, and construction of the wire rope. Smaller wires bend more readily and therefore offer greater flexibility, but are less able to withstand abrasive wear. Conversely, the larger wires of less flexible ropes are better able to withstand abrasion than are the smaller wires of the more flexible ropes.

Abuse

All other factors being equal, misuse or abuse of wire rope will cause a wire rope sling to become unsafe long before any other factor. Abusing a wire rope sling can cause serious structural damage to the wire rope, such as kinking or bird caging, which reduces the strength of the wire rope. To prolong the life of the sling and protect the lives of workers;

- *Always* follow manufacturer's recommendations when using wire rope slings.
- *Do not* use wire rope slings in chemically active environments; they are susceptible to damage and strength loss when used.
- *Do not* allow a wire rope sling, fabricated with a hand-tucked splice, to rotate during use—it can unravel and fail.
- *Do not* drag wire rope slings across floors or pull from underneath loads.
- *Avoid* twists, kinks, and knots in wire rope slings before lifting.
- *Store* wire rope slings where they will not be subjected to dirt, moisture, extreme heat, corrosion, or mechanical damage.

Wire Rope Life

Many operating conditions affect wire rope life, including bending stresses, loading conditions, speed of load application (jerking), abrasion, corrosion, sling design, materials handled, environmental conditions, and history of previous usage. Also affecting wire rope sling service life are the weight, size, and shape of the loads to be handled, and flexibility.

Generally, more flexible wire ropes are selected when smaller radius bending is required. Less flexible wire ropes should be used when the rope must move through or over abrasive materials. Frequently or continuously used wire slings give useful service far longer than idle slings.

- A durable identification tag *must* be permanently affixed to wire rope slings, stating manufacturer's name, serial number, size, grade, rated load and angles on which the rating is based, reach, and number of legs.
- Remove slings from service if tags are illegible, and turn in slings for disposal and replacement.
- Wipe slings clean periodically to remove abrasive grit and then lubricate slings.
- *Always* wear leather gloves when working with wire rope to help prevent injury to hands from broken wires.
- Eyebolts *must* only be used for a straight load, *never* for an angled load.
- Swivel eyes *must* be used if there are angled loads.

15.3 Rated Capacity of Wire Rope Slings

Relative to other types of slings, wire rope slings are average in strength/weight ratio, average in abrasion and cut resistance, poorest in elongation and flexibility, and average for high temperatures (IWRC only).

Wire rope sling capacity is based upon the nominal or catalog strength of the wire rope and factors affecting the overall strength. Manufacturers provide tables of rated capacities for their various sling types based on rope diameters, rigging angles, and hook or end fixture sixes.

These factors include termination efficiencies, type of hitch, number of rope parts in the sling body, diameter around which the sling is bent, and diameter of the pin or hook over which the sling eye is placed.

- *Never* force the eye of a sling onto a hook or pin that has a diameter larger than the natural width of the eye.
- Avoid placing a sling eye onto a hook or pin whose diameter is less than the diameter of the sling body.
- Make sure rated capacities of fittings and attachments must be equal to or greater than that of the wire rope sling.
- *Never* "shock load" a sling; the actual force caused by a sudden application of load can easily exceed rated capacities and damage a sling. Abruptly releasing a load can also damage the sling.
- *Always* protect the sling body against sharp edges and corners of loads, protrusions or abrasive surfaces. Sharp bends can distort wire rope and reduce its strength.
- *Never* expose fiber core wire rope slings to temperatures exceeding 200°F (93.3°C). Avoid using IWRC wire rope slings at temperatures above 400°F (204.4°C) or below −60°F (−51.1°C).
- Slings are susceptible to damage and strength loss when used in chemically active environments.
- Slings fabricated with a hand tucked splice can unravel and fail if the sling is allowed to rotate during use.
- *Do not* drag slings across floors or pull from underneath loads.
- Avoid twists, kinks, and knots in wire rope slings before lifting.
- Store wire rope slings where they will not be subjected to dirt, moisture, extreme heat, corrosion, or mechanical damage.

15.4 Wire Rope Sling Splices

Only experienced splicers should make eye spices in wire rope slings, and they should form all splices with the Flemish splice, properly served to conceal the sharp protruding ends of the wire, which can cause serious injury to the riggers handling them.

Securing a splice by swaged or pressed mechanical sleeves will produce an eye as strong as the rope itself. While not absolutely necessary, it is good practice—and strongly recommended—that thimbles be placed in all spliced eyes.

Socketing should be done only by experienced persons. Care should be taken, when socketing a wire sling end, not to allow the twist or lay to come out of the rope as it enters the throat of the socket.

Two types of splices are used to form the eyes for single-part, eye-and-eye slings:

- *Mechanical splice.* The eye is made by forming a loop and pressing or swaging one or more metallic sleeves over the rope junction (Fig. 15.1a).
- *Hand splice.* The eye is made by forming a loop and trucking strands of the dead end into the live end of the rope (Fig. 15.1b).

The efficiency loss in any type of splice is reflected in the rated capacity that the manufacturer applies to a sling.

Mechanical sling splice
(a)

Hand sling splice
(b)

FIGURE 15.1 Wire rope sling splices. (a) Mechanical; (b) hand.

Always rig a sling so that neither the load nor the lifting device can rotate, since the twisting motion may cause eye spices to pull out, or it may result in excessive stress in the splice (Fig. 15.2).

Wire rope slings are most commonly made in diameters from ¼ to 2 in. (6.35 to 50.8 mm)—the greater the diameter, the greater the sling strength—and in lengths from 5 to 20 ft (1.52 to 6.10 m). They are formed of either single-part (made from a single wire) or multipart (made from multiple wire ropes) construction. A multipart sling can have from 3 to 32 parts or ropes.

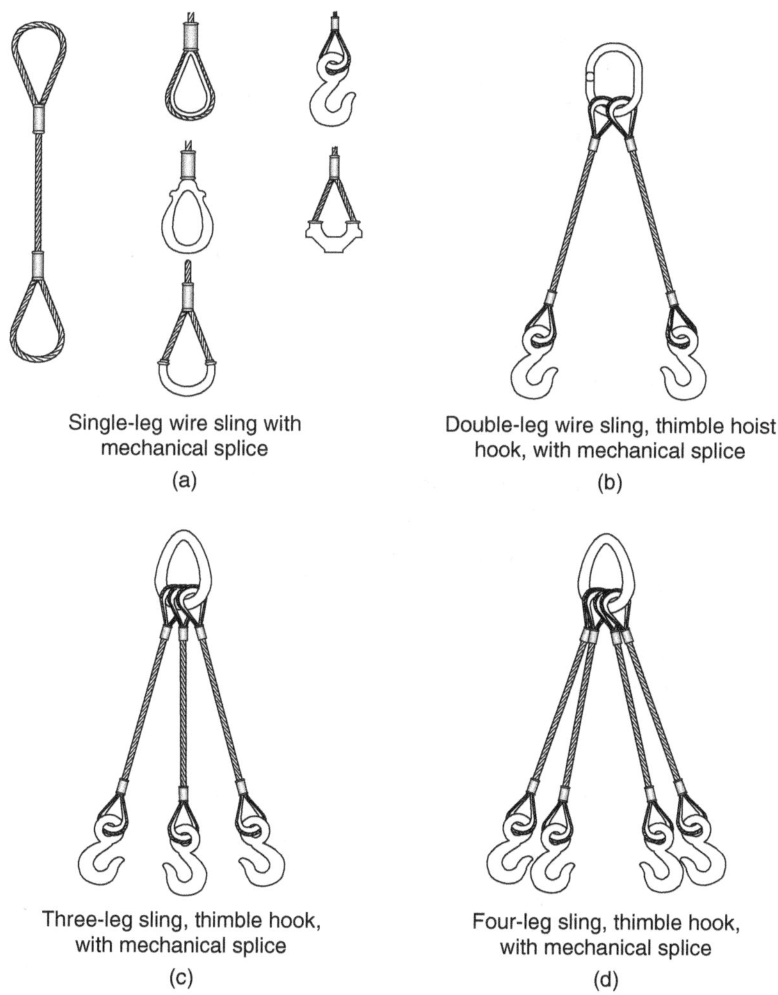

Figure 15.2 Wire rope slings. (a) Single-leg wire sling with mechanical splice; (b) double-leg wire sling, thimble hoist; (c) three-leg sling, thimble hook with mechanical splice; (d) four-leg sling, thimble hook, with mechanical splice. (*Source:* American Riggers Supply, Kansas City, KS.)

15.5 Safe Lifting Practices with Wire Rope Slings

Before making a lift, check to be certain that the sling is properly secured around the load and that the weight and balance of the load have been accurately determined. If the load is on the ground, do not allow the load to drag along the ground. This could damage the sling. If the load is already resting on the sling, ensure that there is no sling damage prior to making the lift.

Next, position the hook directly over the load and seat the sling squarely within the hook bowl. This gives the operator maximum lifting efficiency without bending the hook or overstressing the sling.

Wire rope slings are subject to damage resulting from contact with sharp edges of the loads being lifted. These edges should be blocked or padded to minimize damage to the sling.

After the sling is properly attached to the load:

- Make sure that the rated capacities of fittings and attachments are equal to or greater than that of the wire rope sling.
- *Never* force the eye of a sling onto a hook or pin that has a diameter larger than the natural width of the eye. Also avoid placing a sling eye onto a hook or pin whose diameter is less than the diameter of the sling body.
- Protect the sling body against sharp edges and corners of loads, protrusions or abrasive surfaces. Sharp bends can distort wire rope and reduce its strength.
- *Never* expose fiber core wire rope slings to temperatures exceeding 200°F (93°C).
- Avoid using IWRC wire rope slings at temperatures above 400°F (204.4°C) or below –60°F (–51°C).
- *Do not* use slings fabricated with a hand-tucked splice, which can unravel and fail if the sling is allowed to rotate during use.
- Avoid twists, kinks, and knots before lifting.
- Remember that slings are susceptible to damage and strength loss when used in chemically active environments.
- Make sure that the load is not lagged, clamped, or bolted to the floor.
- Guard against shock loading by taking up the slack in the sling slowly; apply power cautiously so as to prevent jerking at the beginning of the lift, and accelerate or decelerate slowly.
- Check the tension on the sling; raise the load a few inches, stop, and check for proper balance and that all items are clear of the path of travel.

- *Do not* drag slings across floors or pull from underneath loads, accelerate or decelerate.
- *Never* allow anyone to ride on the hood or load.
- Keep all personnel clear while the load is being raised, moved, or lowered. Crane or hoist operators should watch the load at all times when it is in motion.
- *Never* allow more than one person to control a lift or give signals to a crane or hoist operator except to warn of a hazardous situation.
- *Never* raise the load more than necessary. Never leave the load suspended in the air.
- *Never* work under a suspended load or allow anyone else to do so.

Once the lift has been completed, clean the sling, check it for damage, and store it in a clean, dry, airy place. It is best to hang it on a rack or wall. *Remember* that damaged slings cannot lift as much as new or well-cared-for older slings. Safe and proper use and storage of slings will increase their service life.

15.6 Care and Maintenance of Wire Rope Slings

Proper care and use of slings are essential for maximum service and safety. Slings must be protected with cover saddles, burlap padding, or wood blocking as well as from unsafe lifting procedures such as overloading to prevent sharp bends and cutting edges.

Many operating conditions affect wire rope life, including bending stresses, loading conditions, speed of load application (jerking), abrasion, corrosion, sling design, materials handled, environmental conditions, and history of previous usage. Also affecting wire rope sling service life are the weight, size, and shape of the loads to be handled.

Flexibility also is factor. Generally, more flexible wire ropes are selected when smaller radius bending is required. Less flexible wire ropes should be used when the rope must move through or over abrasive materials.

NOTE: *Frequently or continuously used wire slings give useful service far longer than idle slings.*

Lubrication

Although every wire rope sling is lubricated when manufactured, it must also be lubricated "in the field" to increase its useful service life.

Wire rope slings *must* be lubricated according to manufacturer's instructions to prevent or reduce corrosion and wear due to friction and abrasion. Before applying any lubricant, however, the sling user should make certain that the sling is dry. Applying lubricant to a wet or damp sling traps moisture against the metal and hastens corrosion.

Corrosion deteriorates wire rope. It may be indicated by pitting, but it is sometimes hard to detect. Therefore, if a wire rope sling shows any sign of significant deterioration, that sling must be removed until it can be examined by a person who is qualified to determine the extent of the damage.

There is no set rule regarding how often this should be done. It depends on the conditions under which the sling is used. The heavier the loads, the greater the number of bends, or the more adverse the conditions under which the sling operates, the more frequently lubrication will be required.

Storage

Wire rope slings *must* be stored in a well-ventilated, dry building or shed, where they will not be subjected to dirt, moisture, extreme heat, corrosion, or mechanical damage. *Never* store wire rope slings on the ground or allow them to be continuously exposed to the elements, making them vulnerable to corrosion and rust. If it is necessary to store wire rope slings outside, Make sure that they are set off the ground and protected.

NOTE: *Using the sling several times a week, even at a light load, is a good practice. Records show that slings that are used frequently or continuously give useful service far longer than those that are idle.*

Discarding of Slings

Wire rope slings can provide a margin of safety by showing early signs of failure. Factors requiring that a wire sling be discarded include the following:

- Severe corrosion
- Localized wear (shiny worn spots) on the outside
- One-third reduction in outer wire diameter
- Damage or displacement of end fittings—hooks, rings, links, or collars—by overload or misapplication
- Distortion, kinking, bird caging, or other evidence of damage to the wire rope structure
- Excessive broken wires

15.7 Inspection of Wire Rope Slings

No equipment can operate indefinitely at its rated capacity. Wire rope slings *must* be visually inspected before each day's use. The operator *must* check the twists or lay of the sling. If ten randomly distributed wires in one lay are broken, or five wires in one strand of a rope lay are damaged, the sling *must not* be used.

It is not sufficient, however, to check only the condition of the wire rope. End fittings and other components *must* also be inspected regularly for visible damage or distortion, elongation, corrosion, cracking, nicks, or abrasions that may cause failure or reduce the strength or ability of this equipment to perform safely.

To ensure safe sling usage between scheduled inspections, all workers must participate in a safety awareness program. Each operator must keep a close watch on those slings he or she is using. If any accident involving the movement of materials occurs, the operator must immediately shut down the equipment and report the accident to a supervisor. The cause of the accident must be determined and corrected before resuming operations.

Wire rope slings *must* be cleaned prior to each inspection because they are subject to damage hidden by dirt or oil; and they *must* be visually inspected before each use. The operator should check the twists or lay of the sling.

Corrosion deteriorates wire rope. It may be indicated by pitting, but it is sometimes hard to detect. Therefore, if a wire rope sling shows any sign of significant deterioration, that sling must be removed until it can be examined by a person who is qualified to determine the extent of the damage.

The operator of lifting or hoisting equipment *must* visually inspect slings before each use, checking the twists or lay of the sling.

To ensure safe sling usage between scheduled inspections, all workers must participate in a safety awareness program. Each operator must keep a close watch on those slings he or she is using. If any accident involving the movement of materials occurs, the operator must immediately shut down the equipment and report the accident to a supervisor. The cause of the accident must be determined and corrected before resuming operations.

15.8 Wire Mesh Slings[1]

Unlike nylon fiber and wire rope, metal mesh slings resist abrasion and cutting, grip the load firmly without stretching, and easily maintain balanced loads. Because wire mesh slings withstand temperatures up to 500°F (278°C)—they are widely used where loads are abrasive, hot, or tend to cut web slings, such as in metalworking industries (Fig. 15.3).

The capacity of metal mesh slings is determined by wire diameter, mesh width, and type of hitch. A 10-gage steel wire in a 20-in.

[1]*Source:* Safeway Sling USA, Greendale, AI.

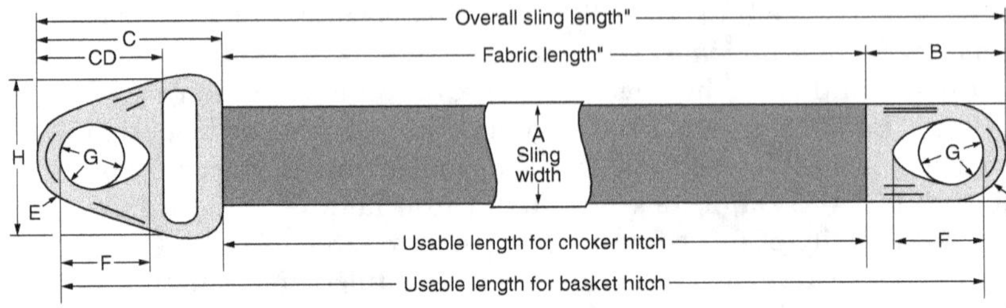

FIGURE 15.3 Wire mesh sling. (*Source:* Safeway Sling USA, Greendale, WI.)

(50.8-mm) wide mesh can lift 12 tons (10.9 metric tons) in a direct connection, and 24 tons (21.8 metric tons) in a basket hitch.

Wire mesh slings combine alloy steel fittings and galvanized high-tensile steel mesh. The mesh is joined to the fitting by means of drilled holes. Fittings are plated for protection and high visibility. Wire or chain mesh slings are available in widths from 2 to 20 in. (50.8 to 508 mm), in lengths from 5 to 20 ft (1.5 to 6.10 m), and in three mesh sizes:

- 10-gage for heavy-duty, general purpose
- 12-gage for medium duty
- 14-gage for light-duty

Heavy-duty 10-gage mesh is recommended for general purpose lifting because it combines strength and abrasion resistance with good flexibility; 12-gage wire mesh is recommended for medium duty application. Because of improved load balance, however, 80 percent of all mesh slings are worked with a choker hitch.

Wire Mesh Sling Properties

The principal characteristics of metal slings are their smooth, flat-bearing surfaces and their ability to conform to irregular shapes; to grip loads tightly; not to kink, tangle, or whip; not to stretch, thereby easily maintaining balanced loads; to resist corrosion, abrasion, and cutting; and to withstand temperatures up to 500°F (278°C).

For handling loads with finishes that could damage the metal mesh, slings can be coated with neoprene or clear PVC plastic that combines the softness of an elastic surface with the strength of steel. Wire mesh slings provide:

- Smooth, flat, wide-bearing surfaces that distribute load to help avoid load damage
- Good flexibility in a grips load's contours
- Firm gripping of loads without stretching to reduce load damage

- Resistance to abrasion, cutting, and corrosion for greater sling life
- Long wear life with plating of alloy steel end fitting
- Resistance to corrosion with zinc-coating of mesh
- Repairability ease of mesh materials

For handling loads with finishes that could damage the metal mesh, slings can be coated with neoprene or clear PVC plastic that combines the softness of an elastic surface with the strength of steel.

Temperature Limits

Metal mesh slings that are not impregnated with elastomers may be used in a temperature range from –20°F (–7°C) to + 500°F (260°C) without decreasing the *working load limit* (WLL). Metal mesh slings impregnated with polyvinyl chloride or neoprene may be used only in a temperature range from 0°F (–18°C) to 200°F (93°C). For operations outside these temperature ranges or for metal mesh slings impregnated with other materials, the sling manufacturer's recommendations shall be followed.

Coatings that diminish the rated capacity of a sling *must* not be applied.

Sling Handles

Handles *must* have a rated capacity at least equal to the metal fabric and exhibit no deformation after proof testing. Attachments of handles to fabric *must* be joined so that the rated capacity of the sling is not reduced.

Make sure the mesh handles meet minimum requirements of the sling.

Sling Markings

Each metal mesh sling *must* have permanently affixed durable identification providing:

- Manufacturer's name or trademark
- Rated capacity in vertical basket hitch and choker hitch

15.9 Rated Capacities of Wire Mesh Slings

Wire mesh slings capacity is determined by wire diameter, mesh width, and type of hitch. For example: a 10-gage steel wire in a 20-in. (508-mm) wide mesh can lift 12 tons (10.9 metric tons) in a direct connection; and 24 tons (21.8 metric tons) in a basket hitch. Because of improved load balance, however, 90 percent of all mesh slings are worked with a choker hitch.

To handle loads with finishes that could damage the metal mesh, slings can be coated with neoprene or clear PVC plastic that combines the softness of an elastic surface with the strength of steel.

These slings combine alloy steel fittings and galvanized high tensile steel mesh with mesh joined to the fitting by means of drilled holes. Fittings are plated for protection and high visibility. Heavy duty 10-gage mesh is recommended for general purpose lifting, combining strength and abrasion resistance with good flexibility; and 12-gage mesh, for medium duty applications.

15.10 Care and Maintenance of Wire Mesh Slings

Remove the sling from service if any of the following is visible:

- Broken weld or brazed joint along the sling edge
- Broken wire in any part of the mesh
- Reduction in wire diameter of 25 percent due to abrasion or 15 percent due to corrosion
- Lack of flexibility due to distortion of the mesh
- Visible distortion or wear of either end fitting
- Cracked end fitting

Wire mesh slings with normal usage eventually will need repair and parts replaced.

15.11 Alloy Steel Chain Slings

Chains are commonly used for slings because of their strength, ruggedness, abrasion resistance, high-temperature resistance, and ability to adapt to the shape of the load. Chain slings are the best choice for lifting very hot materials, because they can be heated up to 1000°F (538°C); however, when alloy-steel chain slings are exposed to service temperatures in excess of 600°F (316°C), operators *must* reduce the WLLs in accordance with the manufacturer's recommendations.

Alloy Steel Chain Characteristics

Chains have limited application as slings. They are used primarily in foundries where they are exposed to high temperatures and where they must pick up rough castings that would quickly destroy other types of slings (Fig. 15.4). As with Manila rope, a chain should preferably be padded where bearing on sharp edges or metal parts. Otherwise, some of the links may be subjected to severe bending stresses. The chain may also bruise or otherwise damage the edges of the load being lifted.

Only alloy steel chain is suitable for slings used in overhead lifting, and is identified by a letter (usually A) marked on each link. This type of sling is made of wire or bar steel links from 1/4 to 1½ in. (6.35 to 38.1 mm) diameter, in lengths of 5 to 20 ft (1.52 to 6.10 m).

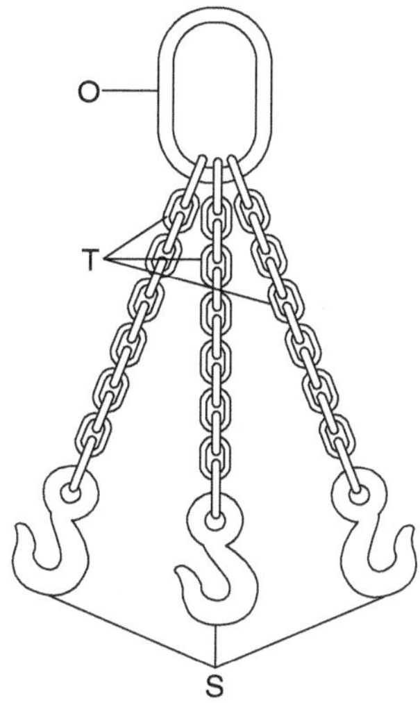

FIGURE 15.4 Basic type of chain sling construction. O, oblong master link, recommended for all assemblies; T, triple-leg sling; S, sling hook.

Alloy steel chain slings are flexible, durable, and long lasting, however, considerable care *must* be taken when using chain slings because sudden shocks will damage them. They will adhere securely to the contours of the load, but misuse of a chain sling could damage the sling, resulting in sling failure and possible injury to an employee.

Although chain will show little or no stretch at rated capacity, it does have the ability to elongate up to 20 percent prior to failure. This elongation serves as a visual warning sign prompting users to remove the sling from service before injury or damage is sustained from a broken chain sling.

Perhaps the most advantageous feature of a chain sling is its ability to be repaired. Nearly any damaged components discovered during an inspection can be repaired and restored to useful condition. During the repair process, slings are refurbished, proof tested, and recertified, offering an economical alternative to new chain sling replacements.

Do not use any chain sling for overhead lifting before verifying that it is constructed of alloy steel. Different "types" of chain slings can be fabricated and are usually designated by a three-character symbol, based upon the number of legs and types of components used in the assembly:

Rated Capacities of Chain Slings

The most commonly used chain slings for overhead lifting are rated at Grade 80, although stronger Grade 100 and Grade 120 are widely used; WLL and lifting capacity increase as the grade increases. Whichever grade of alloy steel chain is used, however, it must meet the OSHA capacity requirements of *29 CFR 1926.251*. Riggers often prefer Grade 80 because it meets or exceeds *1926.251* requirements for the various sizes, angles, and configurations, although NACM, ANSI, and OSHA recommend for overhead lifting only Alloy Chain Grade 80 or 100 (see Table 15.1).

Chain dia. (in.)	Single branch sling 90° loading	Double sling			Triple and quadrupal sling[3]		
		Vertical angle[1]			Vertical angle[1]		
		30°	45°	60°	30°	45°	60°
		Horizontal angle[2]			Horizontal angle[2]		
		60°	45°	30°	60°	45°	30°
1/4″	3,250	5,650	4,550	3,250	8,400	6,800	4,900
3/8″	6,600	11,400	9,300	6,600	17,000	14,000	9,900
1/2″	11,250	19,500	15,900	11,250	29,000	24,000	17,000
5/8″	16,500	28,500	23,300	16,500	43,000	35,000	24,500
3/4″	23,000	39,800	32,500	23,000	59,500	48,500	34,500
7/8″	28,750	49,800	40,600	28,750	74,500	61,000	43,000
1″	38,750	67,100	54,800	38,750	101,000	82,000	58,000
11/8″	44,500	77,000	63,000	44,500	115,500	94,500	66,500
11/4″	57,500	99,500	81,000	57,500	149,000	121,500	86,000
13/8″	67,000	116,000	94,000	67,000	174,000	141,000	100,500
11/2″	80,000	138,000	112,500	80,000	207,000	169,000	119,500
13/4″	100,000	172,000	140,000	100,000	258,000	210,000	150,000

[1] Rating of multileg slings adjusted for angle of loading measured as the included angle between the inclined leg and the vertical.
[2] Rating of multileg slings adjusted for angle of loading between the inclined leg and the horizontal plane of the load.
[3] Quadruple sling rating is same as triple sling because normal lifting practice may not distribute load uniformly to all four legs.
Source: Peerless Chain Co.,Winona, MI.

TABLE 15.1 Rated Capacity (Working Load Limit) in Pounds, for Alloy Steel Chain Slings

Despite its lower WLL, Type 347 Stainless Steel Alloy is preferred for its superior service longevity where some corrosive or extreme heat environments can affect Grade 80 or Grade 100 chain.

Before using any chain sling, check to make sure that it is the correct grade. Chain slings *must* have *ID* tags that contain grade of chain, nominal chain size, number of chain legs, and the working load limit for the sling assembly along with serial number and manufacturer's name or symbol. Always check for proper and clear identification, especially the WLL, and ensure that the grade on the tag matches the grade markings on the chain. Do not use a chain if there is no tag on it.

Most alloy steel chains are fabricated with a 4 to 1 safety factor, with chain manufactured into sling forms having safety factors between 3.5 to 1 and 5 to 1. *Do not use* chains that are not made of alloy steel—such as those used to tie down equipment and material on flatbed trucks—that have no good ductile properties and can fail without warning. Nonalloy chains are *prohibited* for overhead hoisting.

NOTE: *Any chain sling used for lifting must have suitable characteristics for the type of load hitch and environment used, and must be in accordance with* ASME B30.9, Secs. 9-1.5 *and* 9-1.8.

While a $1^1/_4$ in.-diameter single chain sling can lift 40 tons, a four-leg chain sling can easily handle more than 100 tons (Table 15.2). But, when a chain sling is hooked back on itself (into the chain, rather than to the master link or master coupling link), the capacity of the sling should be reduced by one-fourth. (Thus, choker-hitch ratings in manufacturers' safe working load tables are less than ratings for vertical hitches.)

Chain size (in.)	Minimum allowable chain size (in.)	Chain size (in.)	Minimum allowable chain size (in.)
1/4″	13/64″	1″	13/16″
3/8″	19/64″	$1\frac{1}{8}$″	29/32″
1/2″	25/64″	$1\frac{1}{4}$″	1″
5/8″	31/64″	$1\frac{3}{8}$″	$1\frac{3}{32}$″
3/4″	19/32″	$1\frac{1}{2}$″	$1\frac{3}{16}$″
7/8″	45/64″	$1\frac{3}{4}$″	$1\frac{13}{32}$″

Source: Peerless Chain Co., Winona, MI.

TABLE 15.2 Minimum Allowable Chain Size at Any Point of Link

15.12 Selecting Chain Slings

Only alloy steel chain is suitable for slings used in overhead lifting—identified by a letter marked on each link. This type of sling is made of wire or bar steel links from ¼ in. (6.35 mm) to 1½ in. (38.1 mm) diameter; in lengths of 5 ft (1.52 m) to 20 ft (6.10 m).

WARNING: *DO NOT use hooks, rings, oblong links, or makeshift fasteners formed from bolts, rods, and so forth, or other such job-fabricated attachments.*

All chain slings *must* have permanent affixed durable identification, stating starting size, grade, rated capacity, and sling manufacturer.

- Determine the weight and configuration of the load to be lifted.
- Determine the type of chain sling required, based on the load weight and configuration.
- Use *only* WLLs to determine the size of body chain required.
- Determine the matched components and accessories needed to complete the sling.
- Determine the length of chain needed—variable affected by chain size and attachments used.
- *Never* apply the WLL (maximum load in pounds) to a chain, even when chain is new, and when load is uniformly applied in direct tension to a straight length of chain.

Chain Sling Assembly

Make sure chain slings *always* have a metal ID tag attached showing name of manufacturer, serial number, size, reach, rated capacity, and angle of lift. Each sling manufacturer must provide a complete certificate of test.

When assembling a chain sling measure the reach of lift from the master link's load-bearing surface to the hook or lower terminal load-bearing surface (Fig. 15.5).

- *Single-leg sling.* If the required measurement falls in the middle of a link, cut the next link.
- *Double-leg sling* (when assembling a clevis system sling):
  Cut chain to length and count the links.
  There *must* be an *even* number of links so hooks hang in the correct plane.
  Hooks should always point out.
- *Triple- or quadruple-branch sling:*
  Cut chain to length and count the links.
  There *must* be an *odd* number of links so hooks hang in the correct plane.
  Hooks should always point out.

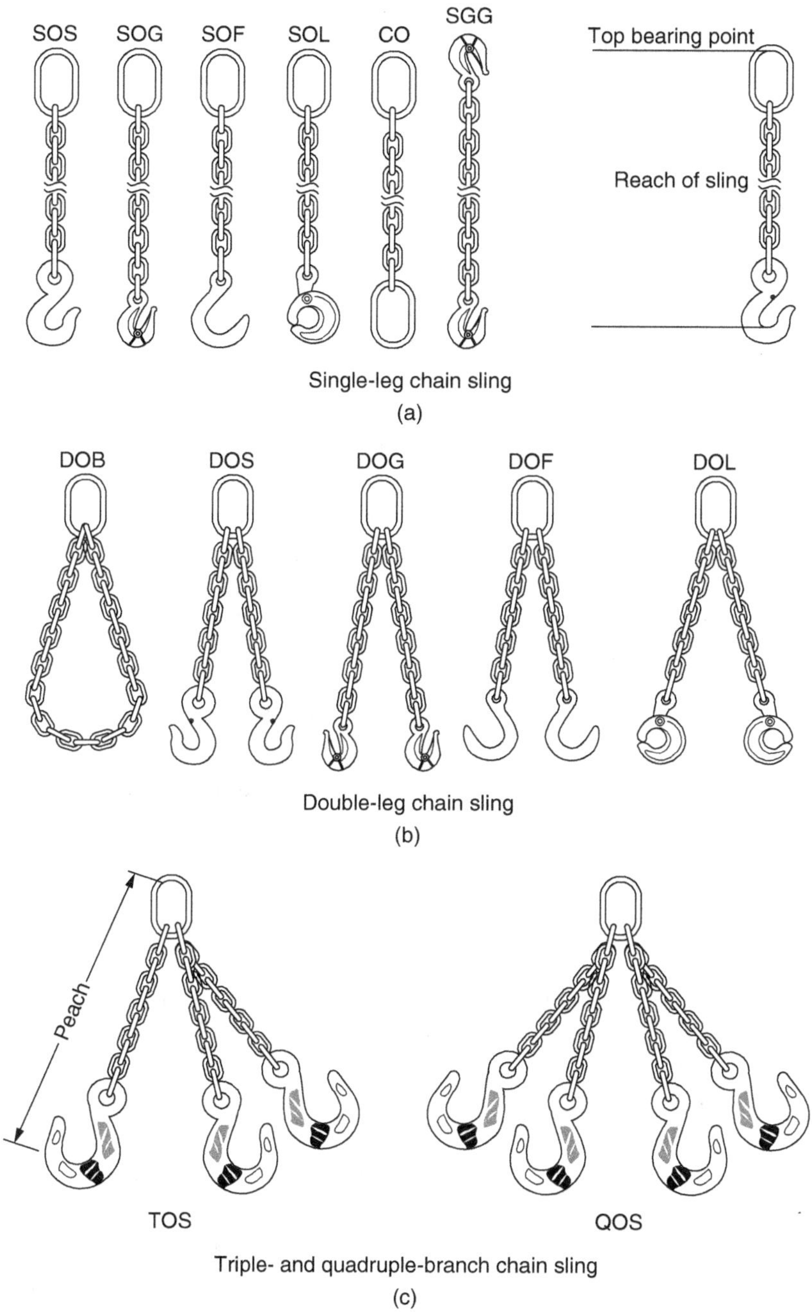

Figure 15.5 Typical chain sling configurations. (a) Single-leg; (b) double-leg; (c) triple- and quadruple-branch. (*Source:* American Sling Company, Hurst, TX.)

15.13 Safe Chain Sling Work Practices[1]

Be aware of the intended use and rated capacities of ropes, chains, cables, fittings, and attachments when installing overhead rigging, including the compounding effect of triangulation.

Failure of any component or link in the chain or chain sling can result in catastrophic failure, serious personal injury, and equipment and property damage. The safety of anyone working with or near an overhead lifting device depends on implementing safe work practices, including the following:

- Make sure that all chain rigged for overhead lifting is approved and rated for that function, such as a Grade 8 or Schedule 80 alloy chain.
- Always check that the grade on the tag matches the grade markings on the chain.
- *Do not* use a chain sling if there is no tag on the chain.
- Be aware of the intended use and rated capacities of ropes, chains, cables, fittings, and attachments when installing overhead rigging, including the compounding effect of triangulation.
- Make sure that workstation locations limit workers' exposure to injury should any part of the rigging fail.
- Replace overhead booms used to guide lines, whenever possible, with outboard davits that would fall overboard should a failure occur.
- Make sure only adequately trained persons use chain slings as specified by *OSHA 1926.20(a)(4).*
- Operate a chain sling in strict accordance with *OSHA 1910.184* and *ASME B30.9.*
- *Do not* allow any portion of a person's body to be placed between the sling and the load, or the sling and the crane or hoisting hook.
- Under *no* circumstance should a person be allowed to stand or pass under a suspended load.
- Shorten or lengthen slings only by methods approved by the manufacturer or a qualified person.
- Use padding to protect the sling from contact with sharp edges.
- *Do not* allow slings to be constricted, bunched, or pinched by the load, hook, or any fitting.

---

[1]*Source:* National Association of Chain Manufacturers, Tucson, AZ.

15.14 Care and Maintenance of Chain Slings

Always inspect chains before use—look for wear, abrasions, collapse, and any damage (visible or not) no matter how insignificant. Remove such chains from service.

Make sure that hooks, rings, links, or any coupling devices have the same or higher rating as the chain to which they are attached. *Never* use makeshift links or coupling devices. Observe these precautions when using chain slings to protect both operator and materials or equipment (or both).

- Inspect chain and attachments before use.
- Do not exceed WLL.
- Be aware of factors that can reduce the load the chain will hold, including:
  - Acceleration in rate of load application which can produce dangerous overloading
  - Variation in the angle of the load to the sling—as the angle decreases, the working load of the sling will increase
  - Twisting, knotting, or kinking subjects chain links to unusual loading, decreasing the working load of the sling.
- Center load in hook (s); hook latches must not support the load.
- Avoid sudden jerks when lifting and towering.
- Balance all loads, avoid tipping of loads.
- Use pads around sharp corners.
- *Do not* drop the load on chains.
- Match the size and WLL of attachments such as hooks or rings to the size and WLL of the chain.
- For overhead lifting, use only alloy chain and attachments (at least Grade 80).
- Pad chain slings where bearing on sharp edges or metal parts, otherwise, some of the links may be subjected to severe bending stresses for which they are not designed. The chain may also bruise or otherwise damage the edges of the load being lifted.
- *Never* apply the WLL (maximum load in pounds) to a chain, even when chain is new, and when load is uniformly applied in direct tension to a straight length of chain.

Although chains are commonly used because of their strength and ability to adapt to the shape of the load, care *must* be taken, when using alloy steel chain slings because sudden shocks will damage them. Misuse of chain slings could damage the sling, resulting in sling failure and possible injury to an employee.

Chain slings are the best choice for lifting very hot materials. They can be heated to temperatures of up to 1000°F (538°C). However, when alloy chain slings are consistently exposed to service temperatures in excess of 600°F (316°C), operators must reduce the WLLs in accordance with the manufacturer's recommendations. (For use of alloy chains in extreme temperatures, consult the manufacturer for load limitations.)

All sling types *must* be visually inspected prior to use; when inspecting alloy steel chain slings, however, pay special attention to any stretching, wear in excess of the allowances made by the manufacturer, and nicks and gouges. These signs indicate that the sling may be unsafe and must be removed from service. Whenever wear at any point of any chain link exceeds that specified in Table 15.2, the assembly shall be removed from service.

All chain slings shall have permanent affixed durable identification, providing starting size, grade, rated capacity, and sling manufacturer.

WARNING: Do not *use hooks, rings, oblong links, or makeshift fasteners formed from bolts, rods, and so forth, or other such job-fabricated attachments.*

Storage and Maintenance

Chain and chain slings require careful storage and regular maintenance:

- Store chains on an A-frame in a clean, dry place.
- Oil chains before prolonged storage to avoid corrosion.
- *Do not* heat alloy chain as it will alter its thermal treatment.
- *Do not* plate or change the surface of chain.

Repair of Chain Slings

- Lay the chain sling without any twists and verify that the links all face the same direction, multiple-legged chain slings as follows: two-legged, the hooks should oppose each other; three-legged, two hooks in one direction, and one hook opposing; four-legged, two hooks facing and two hooks opposing.
- Measure the chain reach and count the links, per leg; reject the chain assembly if the chain is stretched, especially in a few links or fittings (or both). Reject if the chain assembly is shorter than the tags, in links or length.
- Accept a chain sling if there are no visible defects such as cracks or twists; discoloration due to heat exposure; weld splatter; visible chemical or acid burns; and the diameter of

the chain complies with OSHA tables and gages. Properly identify the chain as Grade 80 or 100.

- Reject any chain that has stretch, twists, visible heat damage, weld splatter, chemical or corrosion cracks, is worn or abraded, or has gouge areas that fall below the OSHA tables and gages. Also, the chain must be rejected if it is not easily identifiable as Grade 80 or 100.
- End fittings must be rejected if they are not rated for lifting; the hooks or end fittings are twisted and distorted, have excessive wear, cracks, heat or corrosion effects; makeshift fasteners are used; or the chain was repaired without proper authorization. All connectors must comply with the proper WLL, and cannot be frozen from rust or debris.

All chain slings *must* comply for certification to be acceptable: serial number must completely match the tag and the certificate; the reach must be exact along with the link count; all repairs must be listed on the service card along with who repaired the assembly.

15.15 Inspection of Chain Slings

Every chain or chain sling is only as strong as its weakest link, thus it is very important that every component of the sling meets the manufacturer's WLL. When a sling becomes deformed to the extent that its individual links cannot move within each other freely, the condition is known as *binding*. It is an indication that the sling is unsafe.

Chain sling wear generally occurs on the load-bearing inside ends of the links. Pushing links together so that the inside surface becomes clearly visible is the best way to check for this type of wear. Wear may also occur, however, on the outside of links when the chain is dragged along abrasive surfaces or pulled out from under heavy loads. Either type of wear weakens slings and makes accidents more likely.

Periodically, the operator *must* inspect the total length of the sling, looking for stretching, binding, wear, or nicks and gouges. Heavy nicks or gouges (or both) must be filed smooth, measured with calipers, then compared with the manufacturer's minimum allowable safe dimensions. When in doubt, or in borderline situations, do not use the sling. *Never* attempt to repair the welded components on a sling. If the sling needs repair of this nature, the supervisor must be notified.

All sling types *must* be visually inspected prior to use; however, when inspecting alloy steel chain slings, however, pay special attention to any stretching, wear in excess of the allowances made by the manufacturer, and nicks and gouges. These signs indicate that the sling may be unsafe and *must* be removed from services.

Designate a qualified person to inspect slings and all fastenings and attachments each day before use for damage or defects—and additional periodic inspections where service conditions warrant, as determined on the basis of:

- Frequency of sling use
- Severity of service conditions
- Nature of the lifts being made
- Experience gained during the service life of slings used in similar circumstances

Make periodic inspections of alloy steel chains slings at intervals no greater than 12 months: yearly for normal service use; monthly to quarterly for severe service use; and as recommended by a qualified person for special and infrequent service use. Thoroughly inspect chain slings and attachments in accordance with *ASME B30.9, Section 9-1.2 – Inspection and Repair*.

- *Always* clean chain slings before each inspection, as dirt or oil may hide damage.
- Use nonacid/noncaustic solvent to clean chain slings so that marks, nicks, wear, and other defects are visible.
- *Always* inspect chain slings daily before and after every use.
- Keep an accurate record of all inspections. State entities and OSHA require a written record of inspections at least once a year. Additional inspections must be performed as conditions of usage change.
- Make sure the chain is alloy steel and the correct grade—either 80 or 100.
- Make sure the tag displaying the WLL is on the chain and specifies the required safety factor.
- Inspect individual links for signs of damage check for:
  Stretch
  Twists or bends
  Deformity—cracks, nicks, knots or gouges, stretch, bends, and other signs of abuse
  Whether or not the chain hangs perfectly straight
  Discoloration due to excessive heat
  Excessive pitting or corrosion
  Defective welds
  Excessive wear, especially where on link bears on another
  Distorted or damaged master links, coupling links, or attachments—especially spread in throat openings of hooks
  Missing or illegible sling identifications
  Other conditions that cause doubt as to continued safe use of the sling

- Mark plainly with paint each link or attachment showing any of the conditions listed here to indicate rejection/removal from service until properly repaired.
- Make sure hooks have positive locking latches, openings within tolerance, and be straight, not twisted
- Make sure master links are in good condition.

Where any defect or deterioration is present, remove the sling or attachment from service immediately. Whenever wear at any point of any chain link exceeds that specified in the manufacturer's table, the assembly *must* be removed from service. If a sling has stretched so that it is more than 3 percent longer than it was when new, it is unsafe and *must* be discarded.

NOTE: *Do not use worn or damaged alloy steel chain slings or attachments; discard them.*

Damaged slings may be used *only* after they have been repaired, reconditioned, and proof tested by the sling manufacturer or a qualified person using the following criteria:

- Make sure that slings and attachments conform to the original strength requirements.
- Mark those slings or attachments to identify who made the repairs.
- Replace rather than repair cracked, broken, or bent links.
- *Do not* use mechanical coupling links or carbon steel repair links to repair broken lengths of alloy chain.

RECORDKEEPING: *State entities and OSHA require a written record of inspections at least once a year; additional inspections must be performed as conditions of usage change.*

Regulations and Standards

OSHA Regulations

OSHA Guidance on Safe Sling Use

Natural and Synthetic Fiber Rope Slings
Synthetic Web Slings
Synthetic Round Slings
Wire Rope Slings
Metal Mesh Web Slings
Alloy Steel Chain Slings

1910.184 Subpart N—*Slings*

Includes tables and figures for sling configurations, sling construction, sling diameters, and maximum load capacities

1926.251 Subpart H—*Materials Handling, Storage, Use, and Disposal*

1926.251 *Sling/Wire Rope Inspection*

1926.251(b) *Alloy Steel Chains*

1926.251(b)(6) *Inspection*

1926.251)(c) *Wire Strand Rope and Metal Mesh Slings*

1926.251(c)(1) Tables H-3 through H-14 shall be used to determine the safe working loads of various sizes and classifications of improved plow steel wire rope and wire rope slings with various types of terminals. For sizes, classifications, and grades not included in these tables, the safe working load recommended by the manufacturer for specific, identifiable products shall be followed, provided that a safety factor of not less than 5 is maintained.

1926.251(c)(13) *Minimum Sling Heights*
1926.251(d) *Natural Rope and Synthetic Fiber Slings*
1926.251(e) *Synthetic Webbing*

1926.550 Subpart N—*Materials Handling and Storage Cranes, Derricks, Hoists, Elevators, and Conveyors*

U.S. Consensus Standards

ANSI/ASTM

A391/A391M-01 *Standard Specification for Grade 80 Alloy Steel Chain*

A906/A906M-02 *Standard Specification for Grade 80 and Grade 100 Alloy Steel Chain Slings for Overhead Lifting*

A952/A952M-02 *Standard Specification for Forged Allloy Steel Grade 80 and Grade 100 Lifting Components and Welded Attachment Links*

A973/A973M *Standard Specification for Grade 100 Alloy Steel Chain*

ANSI/ASME

B30.9 – 2006 *Slings (SEC.4.5 - 10/8/2007)*
B30.23 – 2005 *Personnel Lifting Systems*

Industry Consensus Standards and Guidelines

American Chain Association

American Chain Manufacturers Association

American Wire Producers Association

Associated Wire Rope Fabricators (*www.awrf.org*)

Committee of Domestic Steel Wire Rope and Specialty Cable Manufacturers (*domesticwirerope.org*)

Cordage Institute

International Organization for the Study of the Endurance of Ropes (*www.oipeec.org*)

National Association of Chain Manufacturers (*www.nacm.info*)

Web Sling and Tie Down Association (*www.wstda.com*)

Wire Rope Technical Board (*www.domesticwirerope.org*)

SECTION 5

Rigging Devices

CHAPTER 16

Connections, Fittings, and End Attachments

A rigger should know not only what types of fittings and end attachments to use for a particular lifting, hoisting, or rigging application, but also what their safe working load (SWL) limits are in relationship to the rope or chain being used.

Although there are a wide variety of connections, fittings, and end attachments, only forged alloy steel, load-rated types should be used for overhead lifting and hoisting to assure the highest degree of safety. Such attachments have their SWL stamped directly on them.

Lifting and rigging attachments for chains or lifting slings can be configured to handle bulky or heavy loads, such as with one or more individual chains fitted with hooks or consist of looped chains that are configured to lift barrels or drums. Special hoist rings are designed to handle different types of lifting configurations, such as a side load and spreader beams.

16.1 Seizing Wire Rope

It is most important that all rope ends be tightly seized to prevent wires and strands from becoming slack with use. Seizing is the process of securing one rope to another, two or more parts of the same rope to itself, or fittings of any kind to a rope by binding with small stuff or with annealed iron wire. It is most important that all rope ends be tightly seized to prevent wires and strands from becoming slack with use.

In the manufacture of wire rope, great care is taken to lay each wire in the strand and each strand in the rope under uniform tension. If the ends of the rope are not properly secured, the original balance of tension will be disturbed and maximum service will not be obtained, because some strands will carry a greater portion of the load than others.

Rope diameter		Annealed iron seizing wire diameter		Seizing length	
in.	mm	in.	mm	in.	mm
<1/2	12.7	0.035	0.889	1/2	12.7
9/16–7/8	14.3–22.2	0.063	1.600	3/4	19.05
$1–1^1/_2$	25.4–38.1	0.092	2.337	$1^1/_4$	31.75
$1^5/_8–2^1/_8$	41.27–54	0.120	3.048	$2^1/_2$	63.5
$2^1/_4$ >	57.15	0.135	3.429	$2^1/_2$	63.5

TABLE 16.1 Seizing for Wire Rope (Diameters and Lengths)

Before cutting wire rope proper seizing *must* be applied on both sides of the place where the rope is to be cut, with the length of seizings *never* less than the diameter of the wire rope being seized:

- For preformed wire rope, one seizing on each side normally is enough.
- For wire ropes that are not preformed, a minimum of two seizings is required, placed six rope diameters apart.

Length of wire rope seize should never be less than the rope diameter (Table 16.1).

Always use soft annealed wire for seizing. With galvanized rope, it should be zinc-coated steel wire. Only two methods of seizing are acceptable for wire ropes:

- *Greater than 1 in. (25.4 mm) diameter*. Place one end of the seizing wire in the valley between two strands. Turn its long end at right angle to the rope and closely and tightly wind the rope back over itself and the rope until the proper length of seizing has been applied. Twist the two ends of wire together and, by alternately pulling and twisting, draw the seizing ends tight (Fig. 16.1).
- *Less than 1 in. (25.4 mm) diameter.* Wind the seizing wire in a tight spiral around the rope end to the desired length of seizing. Twist the two ends of the seizing wire together at the approximate center of the seizing, alternately twisting and pulling until the proper tightness is achieved.

Always apply seizing in the opposite direction from the lay of the rope to prevent loosening when the wire rope shrinks as a result of loading. The size of wire rope determines how the seizing is made. Make sure to use the proper size and grade of wire for seizing.

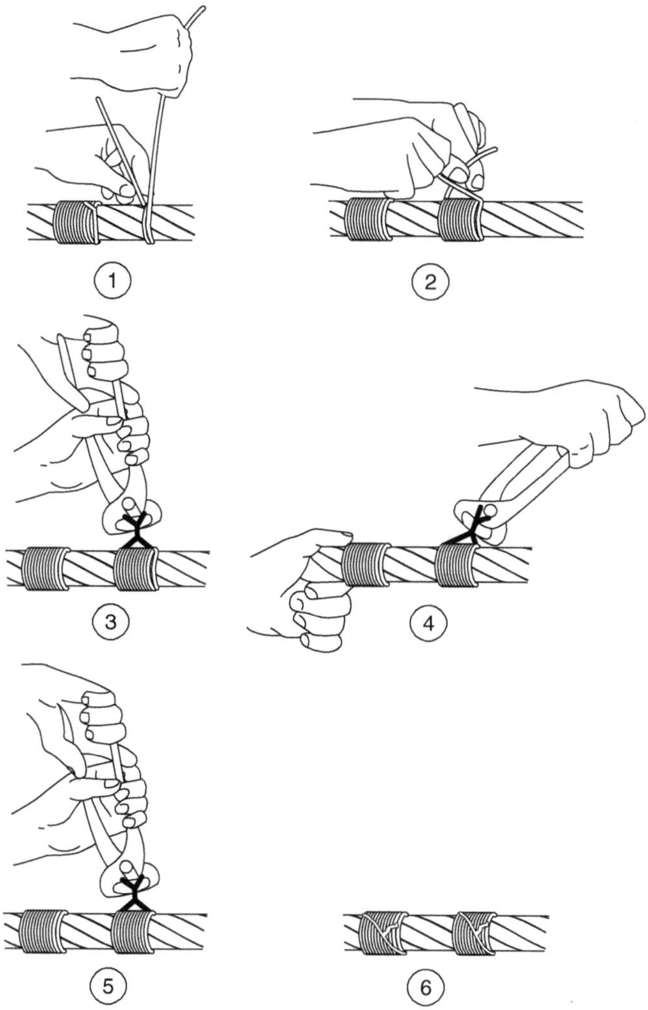

FIGURE 16.1 Putting seizing on wire rope. (*Source:* U.S.Army FM 55-501, Army Department, Washington, DC.)

Do not try to tighten the seizing by twisting. Instead, draw up on the seizing and repeat if needed, Cut the ends and pound them down on the rope. If the seizing is to be permanent, or the rope is $1^5/_8$ in. (41.28 mm) or more in diameter, use a serving bar, or iron, to increase tension on the seizing wire when putting on the turns.

16.2 Connections

The end of a wire rope tends to fray readily, and cannot be easily connected to plant and equipment. A number of different mechanisms exist to secure the ends of wire ropes to make them more useful.

Spliced Eyes

The most common and useful type of end fitting for a wire rope is when the end is turned back to form a loop. The loose end is then fixed by any number of methods spliced back to the wire rope.

Wire rope splicing, however, is a skill that requires training and proper tools to be done safely and effectively; thus, *only* experienced splicers should do eye splicing. Always refer to the wire rope manufacturer's handbooks and manuals for the proper splicing types, methods, and procedures to use.

An eye splice may be used to terminate the loose end of a wire rope. The eye *must* always be combined with a thimble to maintain wire rope strength and reduce wear and distortion in the eye end of the rope. If a thimble is not used on a spliced eye, the efficiency of the connection can be reduced by as much as 10 percent because the rope flattens under loading.

The efficiency of a well-made eye splice with a heavy-duty thimble varies from about 95 percent for 5/16-in. (7.94-mm) wire rope to 88 percent for ¾-in. (19.05-mm) wire rope; 76 percent for 1¼-in. (31.75 mm) wire rope; and 70 percent for 1½-in. (38.1 mm) wire rope. All splices should have at least four tucks, and the completed splice should be carefully wrapped with a wire serving to cover the protruding wire ends to eliminate the danger of lacerating the hands of riggers.

When forming a loop, the strands of the end of a wire rope are unwound a certain distance and plaited back into the wire rope, forming the loop, or an eye, called an eye splice. While there are great differences in the efficiencies of eye splices, there is little difference in appearance. Every wire rope manufacturer has different names for its particular type of eye splices, but they are usually variations of the three basic types.

Flemish Eye (Rolled Eye)

These are the best and most secure of the various eye splices. The Flemish eye with pressed metal sleeve is the best of all and the one recommended for all rigging and hoisting use; when properly made it develops almost 100 percent of the catalogue breaking strength (Fig. 16.2).

The strand ends of the spliced eye are secured against the live portion of the rope by means of a steel or aluminum sleeve set in place under pressure. The splice's strength is not wholly dependent on the sleeve since it has the hand splicing to fall back on should the sleeve fail—with the tapered end of the sleeve allowing the splice to pass over obstructions without snapping.

Tucked Eye

These splices are less secure than the Flemish eye type, with the exception of the tucked eye with pressed metal sleeve, which is almost

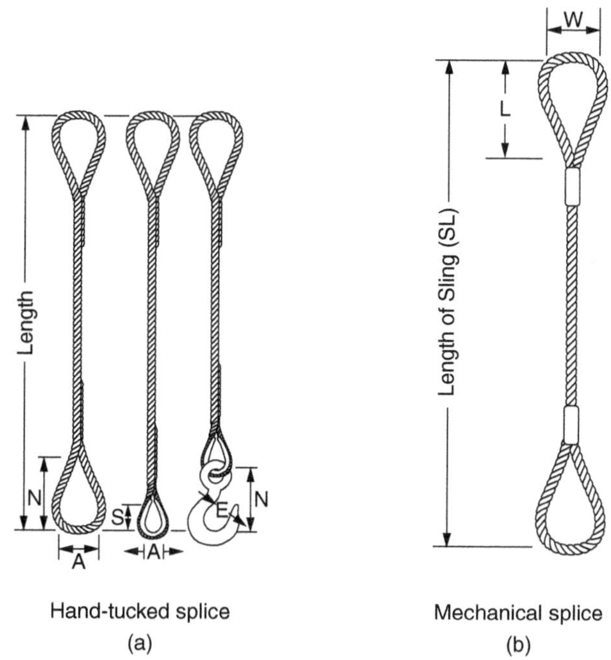

Figure 16.2 Single part sling. (a) Hand-tucked splice; (b) mechanical splice.

100 percent efficient. Tucked eye splices develop only 70 percent of the strength of the wire rope and tend to come free as the wire rope unwinds. The tucks under the eye then begin to pop free and, on short lengths of cable or sling, one or two turns will seriously weaken the eye; therefore, they should not be used for hoisting loads.

Tucked eye splices should have at least five tucks and the completed splice should be carefully and tightly wrapped with a wire serving to cover the whole of the splice. Adding wire seizings protects the rigger's hands, but adds no strength.

Fold Back Eye

These splices, similar to the fold back eye splices with pressed metal sleeve, are made by bending the wire rope to the eye dimension required and securing the free or dead end of the rope against the live portion of the rope by means of a steel or aluminum sleeve set in place under pressure (Fig. 16.3).

Never use these splices for overhead hoisting because improper swaging or split sleeves will result in complete failure without warning. Only the Flemish eye with pressed metal sleeve is recommended for all rigging and hoisting use. When properly made, this eye develops almost 100 percent of the catalog rope breaking strength.

FIGURE 16.3 Fold back eye with pressed metal sleeve.

Ferrules

Hand splicing wire rope requires skilled labor and long hours, and splices may cause injury to operators and damage to materials. On the other hand, mechanical spliced slings are made in minutes using a splicing machine and aluminum alloy or steel ferrules (Fig. 16.4).

In addition, with mechanical ferrule splicing:

- Wires cannot unravel and fray while making slings.
- Slings can easily be identified with SWLs or rated capacity, or any other information stamped directly into the ferrule.
- Moisture cannot enter the joint, thus eliminating the risk of internal corrosion.
- No servings are required, making a very neat attachment.
- There are no jagged ends of wires to snag or injure the hands.
- Attachment is very strong, when properly made—stronger than the wire rope.

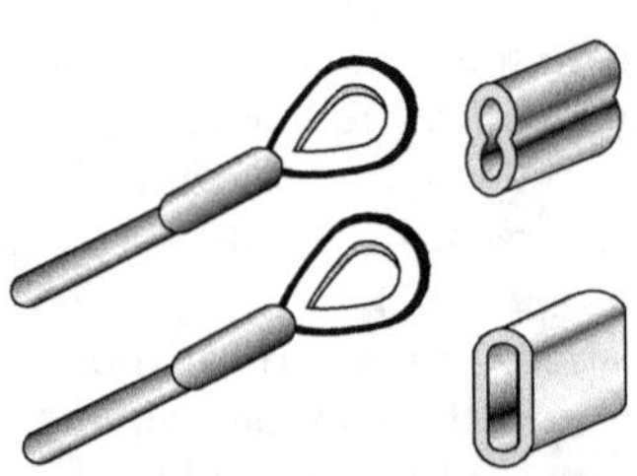

FIGURE 16.4 Mechanical splice ferrules. (*Source:* Talurit AB, Vastra Frolunda, Sweden.)

Forged regular swivel (a) Forged chain swivel (b) Forged jaw end swivel (c)

FIGURE 16.5 Types of swivels. (a) Regular; (b) chain; (c) jaw end. (*Source:* The Crosby Group, Tulsa, OK.)

Mechanical Splice

To make a mechanical splice, pass an aluminum alloy or steel ferrule of the correct size (straight or conical) over the end of the wire rope. Then, reinsert the end of the rope in the ferrule, forming an eye of the requisite size, or over a thimble. Finally, place the ferrule in a swaged press and apply pressure to cause the metal to flow around both parts of the rope.

Swivels

These coupling devices permit either half of a connection to rotate independently, thus preventing chain or rope (or both) from twisting; and allowing untwisting. Variable swivel designs include eye and eye, eye and clevis (shackle), hook and eye, hook and clevis, and chain types (Fig. 16.5).

When selecting a swivel, *make sure* that the swivel capacity is sufficient for loading applications.

16.3 Fittings

Wire rope becomes an assembly when its ends are altered by some form of splice or addition of fittings.

Swaged Terminations

Swaging is a method of wire rope termination that refers to the installation technique. The purpose of swaging wire rope fittings is to connect two wire rope ends together, or to otherwise terminate one end of wire rope to something else. A mechanical or hydraulic swager is used to compress and deform the fitting, creating a permanent connection.

Collets

These hollow clamping devices—a subtype of chucks—form a collar around the wire rope to be held exerting a strong clamping force on the rope when it is tightened via a tapered outer collar, thus securing the rope's strands and wires. The connections are similar to the fold back eye except instead of the sleeve being swaged on, it is formed of a split collet, sleeve, and nut arrangement.

To make a collet connection place the nut on the wire rope then make a loop of the eye size required and place the collets on the wire rope. Extend the rope's dead end at least one-half the rope diameter beyond the collets.

Hold the collets together (by hand), and turn on the nut as far as possible (by hand). Finally, close the loop around the thimble by holding the live end of the rope and pulling on the clamp. Keep the thimble one-half to one rope diameter from the clamp, then place assembly in a vise and torque the nut.

Sockets

A socket is useful when a termination fitting needs to be replaced frequently. For example, if the end of a wire rope is in a high-wear region, the rope may be periodically trimmed, requiring the termination hardware to be removed and reapplied.

An example of this is on the ends of the drag ropes on a dragline. The end loop of the wire rope enters a tapered opening in the socket, wrapped around a separate component called the *wedge*. The arrangement is knocked in place, and load gradually eased onto the rope. As the load increases on the wire rope, the wedge becomes more secure, gripping the rope tighter.

The strongest wire rope end attachment is the zinc-poured socket that, if properly attached, will develop 100 percent of the wire rope's strength. However, *never* think that the splice is stronger than the rope. Only trained and properly qualified personnel should be permitted to make these connections, because it is a skill that requires good facilities, and a thorough understanding of the manufacturer's instructions.

Remember that the rope is always stronger than indicated in the manufacturer's table. Therefore, it is possible for the splice to be as strong as the strength of the rope, as indicated in the catalog, and still fail before the body of the rope fails.

Zinc (Spelter) Sockets

These are the most reliable of all terminal fittings and, when properly attached, these standard drop-forged sockets are 100 percent efficient. They are efficient and permanent terminal attachments, recommended for all standing ropes and whenever service conditions are severe.

These sockets are suitable, however, only for standing ropes subject to little or no movement, as rope movements tend to crack the ropes at entry to sockets. Rapid hoisting, road traveling, and impact loads, acceleration, or braking loads always set up severe vibration in the rope directly above the point of the load attachment. There is no way to prevent rope fatigue and eventual wire breaks from occurring at the point where the rope enters the socket.

To make a socket fitting, put a seizing on the rope at a distance from the end equal to the length of the socket basket, and two additional seizings spaced one and one-half rope diameters apart immediately in back of the first seizing. Unlay the strands of the rope and cut off the hemp center near the first seizing, then carefully unlay and broom out all the wires in the several strands. It is not necessary to straighten the wires.

After separating the wires, clean them with a solvent (benzine or gasoline). Dry the wires with a clean cloth, then dip them for about three-quarters of their length into a 50 percent solution of commercial muriatic acid for about half a minute, or until each wire is thoroughly cleaned. *Do not* allow the acid to reach the hemp center of the rope. Rinse by immersing in boiling water that contains a small quantity of neutralizer (bicarbonate of soda). Use care not to allow acid to come into contact with any other part of the rope.

Bind the wires together and insert the end of the rope into the socket so that the ends of the wires are even with the top of the basket; then remove this temporary binding wire. Spread out the wires so as to equally occupy the entire space within the basket.

Clay should be applied to the annular space between the neck of the socket and the periphery of the rope to prevent the molten metal from running out. Hold the socket and several feet of rope in a vertical position, and pour in molten zinc until the basket of the socket is full. The zinc must be at the correct temperature. Before pouring, dip the end of a soft pine stick into the ladle for a few seconds and remove. If the metal adheres to the stick, the metal is too cold. If it chars the stick, the metal is too hot. Allow the socket to cool, then remove the seizings, except the one closest to the socket.

Never use babbitt (*white metal*) or lead for securing sockets on hoisting ropes. Both metals are far too soft for the exposure required, and efficiency can be as low as 25 percent of rope strength.

Resocket all hoisting ropes on elevators and mine hoists at frequent intervals, varying from monthly to semiannually, according to the severity of service. Before resocketing, however, the socket should be annealed by heating it to a cherry-red color and then allowing it to cool. After the sixth annealing, the socket should be discarded.

Sometimes wedge-type sockets are used. In these, the end of the wire rope is fed through the socket, bent around the wedge, then reinserted through the socket again. Under tension, the wedge pulls tight against the bend in the wire rope, holding it solidly in place.

Efficiency of such sockets, however, is low—only 70 percent of the strength of the rope. In using the wedge socket, care must be exercised to install the rope so that the pulling part is directly in line with the clevis pin. Otherwise, a sharp bend will be produced in the rope as it is.

Swaged Sockets

Swaged sockets make efficient and permanent terminal attachments for wire rope. They are made by compressing a steel sleeve over the rope with a hydraulic press. Properly made, they provide an efficiency of 100 percent.

Careful inspection of the wires leading into these socketed terminals is most important because the strength of this section (from 5 to 10 ft above the load attachment) is bound to be greatly reduced because of fatigue within the wires. On visual inspection, one broken wire is sufficient to cause it to be removed from service and discarded.

It is therefore advisable to cut off this damaged section periodically even though it may look sound. About one-third of the estimated life of the rope is a good interval. When inspecting any existing socketed connection, make sure to examine the rope closely for corrosion at the socket base. This corroded wire is very susceptible to fatigue.

Cappel Sockets

Properly installed and frequently inspected, these sockets give 100 percent efficiency (depending on the wedges being kept tight); among the simplest devices for anchoring a wire rope for any purpose. They are intended for "on the job" attachment and for quick rope replacement. Principal advantages are simplicity, and ease and speed of applying and detaching.

They are also used where conditions are such that spliced eyes could not be reeved, and would have to be made after the rope was in place.

Wedge Sockets

These are among the simplest devices for anchoring a wire rope for any purpose. They are used to temporarily terminate the end of a wire rope, specifically designed to allow the alteration of the rope's length (Fig. 16.6).

Wedge sockets are intended for "on the job" attachment and for quick rope replacement. Principal advantages are their simplicity, and ease and speed of applying and detaching. They are also used where conditions are such that spliced eyes could not be reeved, and would have to be made after the rope was in place.

Efficiency of a wedge socket, however, is low—having only 70 percent of the strength of the rope. Care must be taken so that moving loads do not force the wedges out and accidental slackening

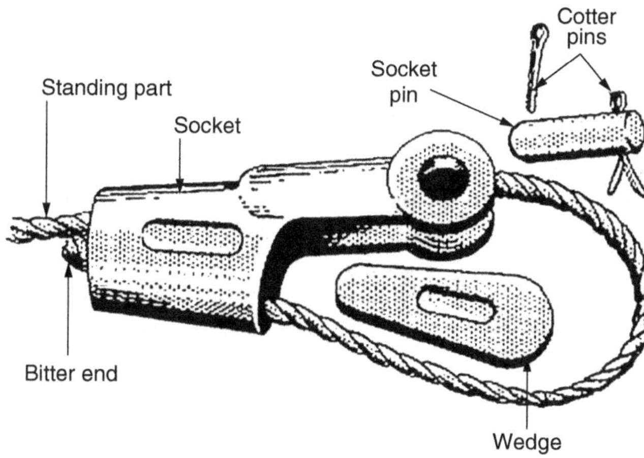

Figure 16.6 Parts of a wedge socket.

of the rope does not release the socket. To prevent this from happening, the wedge sockets that are used should be the positive locking types. Also, the dead or short end of the rope should either have a clip attached to it, or be looped back and secured to itself by a clip. The loop thus formed must not be allowed to enter the wedge. *Do not* attach the dead end of the rope to the live side of the rope with the clip as this will seriously weaken the attachment. The clip will ultimately take the load and could deform and break the wire rope (Fig. 16.7). These sockets, however, *must not* be used on nonrotating

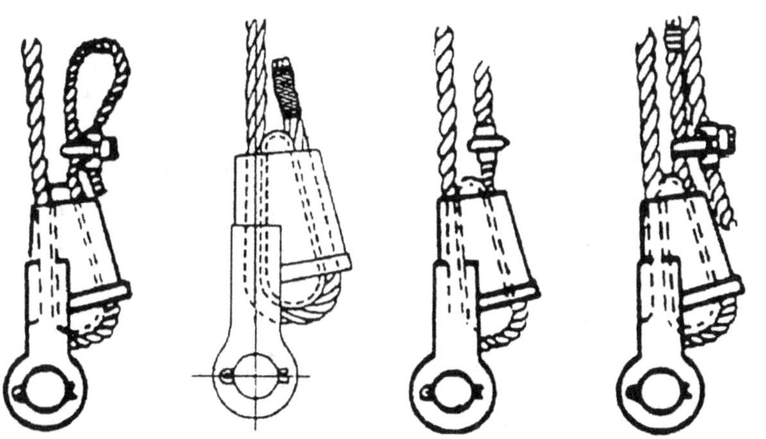

Figure 16.7 Wedge socket clip methods. (a) Looped back; (b) no clip; (c) special clip; (d) extra piece of same size rope clipped to main rope.

wire ropes, because of the small radius of wedge sockets and the possibility of severe core slippage, unless extreme care is used in their installation, and the rope end is very tightly seized.

When using wedge sockets it is good practice to start out with a longer rope than is required so that the socket can be renewed periodically without having to splice the rope. The wedge can be punched out of the socket, with the bad wire cut off and refitted in the socket. The rope end should always protrude at least 6 to 9 in. beyond the socket.

It is very important to ensure that the loaded part of the rope is not kinked where it leaves the wedge. The sockets must be installed so that the load line is in a straight line pull with the eye of the socket (Fig. 16.8). If wedge-type sockets are to be used, feed the end of the wire rope through the socket, bent around the wedge, then reinsert the end through the socket again. Under tension, the wedge pulls tight against the bend in the wire rope, holding it solidly in place.

When using a wedge socket, care must be exercised to install the rope so that the pulling part is directly in line with the clevis pin. Otherwise, a sharp bend will be produced in the rope as it enters the socket.

Care *must* be taken that moving loads *do not* force out the wedges; and accidental slackening on the cable *does not* release the socket. Positive locking wedges will eliminate the problem; wedge sockets should be installed so that the load line is in a straight line pull with the eye of the socket.

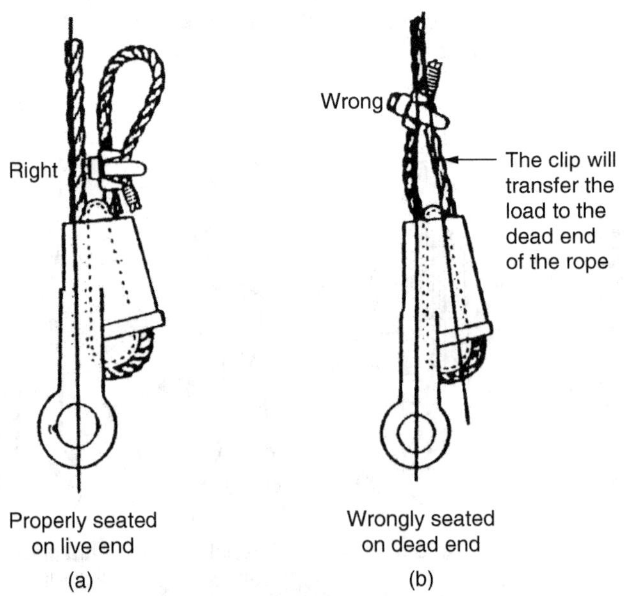

FIGURE 16.8 Wedge socket fitted on wire rope. (a) Properly seated on live end; (b) wrongly seated on dead end. (*Source:* Mantis Cranes, Span Deck Inc., Franklin, TN.)

Thimbles

When the wire rope is terminated with a loop, there is a risk that the wire rope can bend too tightly, especially when the loop is connected to a device that spreads the load over a relatively small area. A thimble can be installed inside the loop to preserve the natural shape of the loop and protect the cable from pinching and abrasion on the inside of the loop. The use of thimbles in loops is industry' best practice. The thimble prevents the load from coming into direct contact with the wires.

Clips (Clamps)

The most common method of making an eye or attaching a wire rope to a piece of equipment is with cable or with clips (Fig. 16.9). When properly made, and according to the manufacturer's Clip Installation

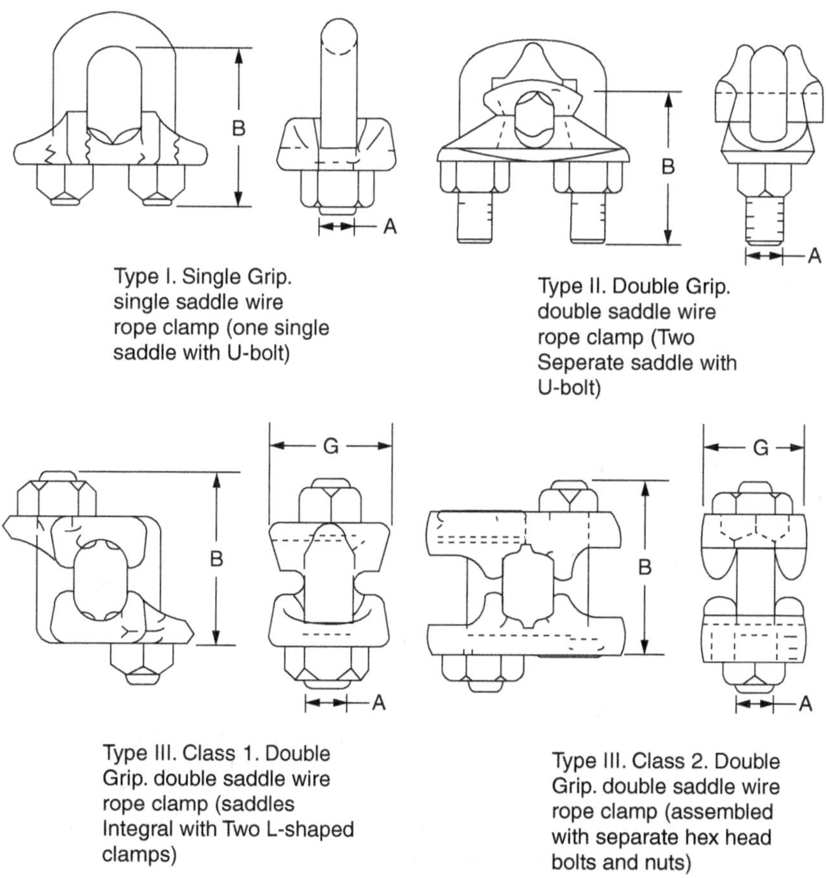

FIGURE 16.9 Wire rope clips. (a) Type I; (b) Type II; (c) Type III; (d) Type IV.

Rope diameter		Number of clips for eye attachment	Spacing between clips	
in.	mm		in.	mm
1/4	6.35	2	$1^1/_2$	38.1
5/16	7.93	2	2	50.8
3/8	9.53	2	$2^1/_4$	57.2
7/16	11.11	2	$2^1/_2$	63.5
1/2	12.7	3	3	76.2
5/8	15.88	3	4	101.6
3/4	19.05	4	$4^1/_2$	114.3
7/8	22.23	4	$5^1/_4$	133.4
1	25.4	4	6	152.4
$1^1/_8$	28.58	5	7	177.8
$1^1/_4$	31.75	5	8	203.2

Source: The Crosby Group, Tulsa, OK.

TABLE 16.2 Installation of Wire Rope Clips (Numbers and Spacing)

Tables, clipped eye terminations will develop approximately 80 percent of the wire rope strength.

Always use a heavy-duty thimble to form an eye in the end of a wire rope. (Table 16.2).

U-Bolt Clips. These clips are used to fix the loose end of the loop back to the wire rope. A U-bolt clip consists of a U-shaped bolt, a forged saddle, and two nuts. The two layers of wire rope are placed in the U-bolt. The saddle is then fitted over the ropes on to the bolt (the saddle includes two holes to fit to the U-bolt). The nuts secure the arrangement in place. Three or more clamps are usually used to terminate a wire rope (Fig. 16.10).

The U-bolt section *must* be set on the dead or short end of the wire rope and the saddle on the live or long end of the rope. Placing the U-bolt on the live instead of the dead end, or even one clip, can reduce the efficiency of the connection up to 40 percent.

Double Saddle Clips (Flat Grip Clips). These clips are preferable to the U-bolt clips because it is impossible to install them incorrectly, and because they cause less damage to the rope. In addition, less turn back is required with these connections, and fewer saddle clips than U-bolt clips are required in some rope sizes (Fig. 16.11).

- Make sure when using clips that the corrugation in the saddles matches that of the rope lay being used—left-lay or right-lay clips for left-lay or right-lay wire ropes, respectively. Otherwise, the ridges between the corrugations in the forging

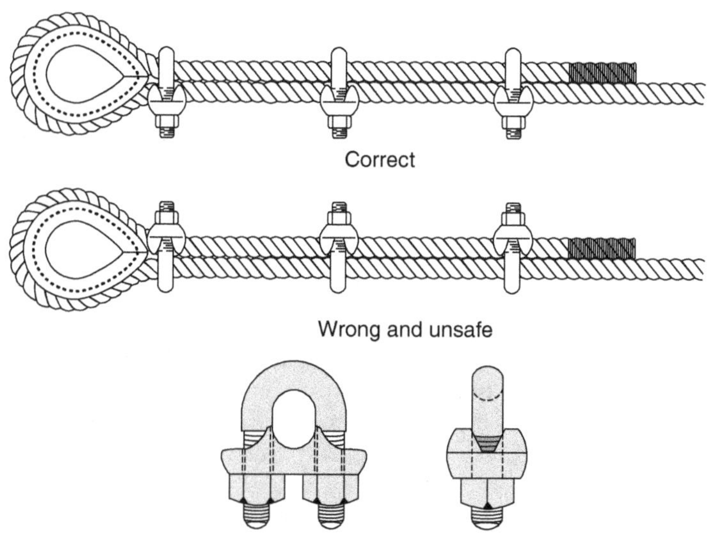

Figure 16.10 U-bolt clips on dead end of wire rope. (*Source:* Innovative Engineering, Heswall, Wirral, UK.)

Figure 16.11 Twin base wire clip.

will run crossways, rather than parallel to and between the strands of the wire rope. This will result in cutting wire stands when the clip nuts are properly tightened.

- *Do not* allow any slack in either part of the wire rope between clips.
- *Never* use any kind of clip to directly connect two straight lengths of rope. If such a connection is necessary, use clips to form an eye (with thimbles) in each length and connect the eyes together.

- *Never* connect two straight lengths of rope using clips.
- *Always* use a thimble to form an eye to prevent the rope from wearing in the eye and to provide a safer connection.
- *Always* use new clips for connections, as reused clips will not develop the proper efficiency.
- *Never* use fewer than the number of clips recommended by manufacturer tables.
- *Always* turn back the correct amount of rope for dead ending to permit proper spacing of the clips.

These terminations have the advantage of allowing thorough examination and ease of field installation. When applied with proper care, thimbles and clipped eye terminations will develop approximately 80 percent of the rope strength.

Clamp and Thimble

This connection combines both clamp and thimble in one unit (Fig. 16.12) capable of developing approximately 80 percent of the wire rope strength.

Shackles (Clevises)

The main types of shackle commonly used in rigging operations include:

- *D and Bow.* Available with screw pins, round pins, or bolt type closures (Fig. 16.13).
- *Plate.* Mainly used for joining lengths of luffing rope (pendants) when extra lengths of boom are added to any wire rope luffing boom crane (Fig. 16.14).

(a) Cable clamp

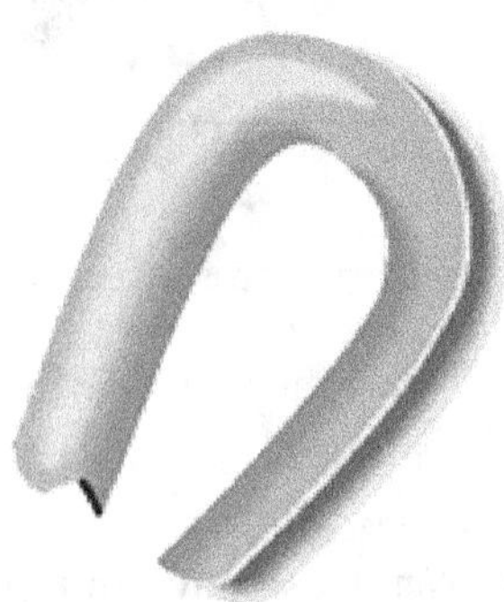

(b) Thimble

FIGURE 16.12 Clamp and thimble connections.

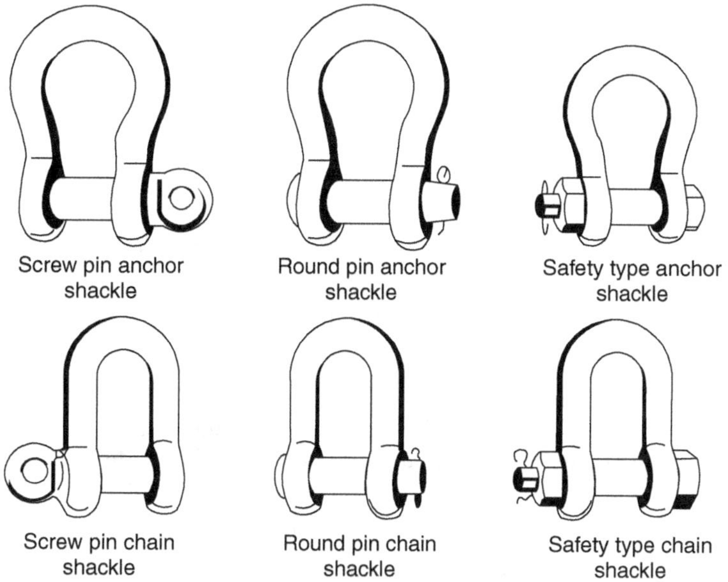

FIGURE 16.13 Typical rigging shackles. (a) Screw pin; (b) round pin; (c) safety type.

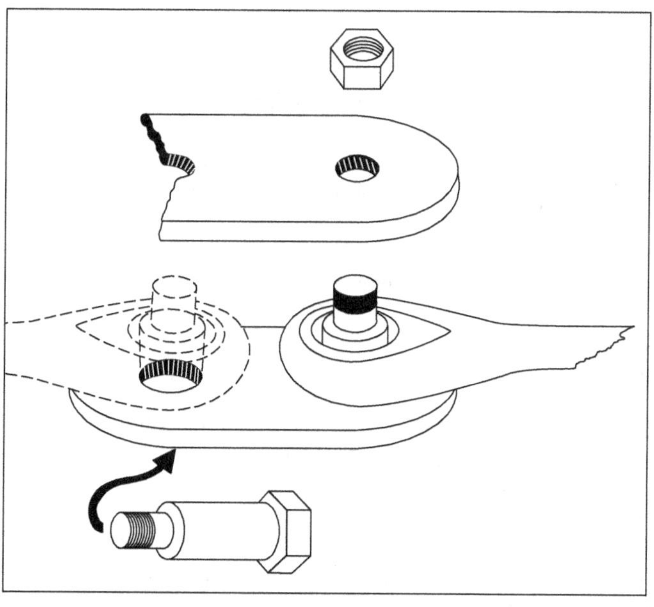

FIGURE 16.14 Small plate shackle.

- *Ratchet release shackles.* These devices allow workers to erect steel columns without having to climb the columns to release the lifting shackle used by the crane to lift the column into place. They use a ratchet system that can be operated from below to release the shackle.

All shackles used for lifting *must* be stamped with the SWL.

- *Do not* use a shackle that does not have the SWL marked.
- Make sure that the SWL of the shackle is at least as great as the chain, links, and rings in the sling you are using.
- *Never* use a bolt and nut in place of the proper shackle pin. A bolt that does not fit tightly is likely to bend and break.
- *Never* replace a shackle pin with anything but a properly fitting cotter pin that is capable of carrying the loading normally applied to it.
- *Always* use cotter pins with any round pin shackle.
- *Do not* use a screw-type pin shackle if there is a chance that the pin might roll when loaded and unscrew—move the pin of the screw shackle or use a shackle fitted with a nut and bolt.
- *Never* use a shackle that is bent, deformed, or damaged. Deformed shackles probably have microscopic cracks that can lead to complete failure during lifting.
- Destroy all shackles that show crown or pin wear greater than10 percent of the fitting's original diameter.

To prevent jamming, tighten shackle pins finger tight and then release a quarter turn.

Use washers or ferrules to center thimbles and hooks on the shackle pin to prevent unnecessary side strain. Where several sling eyes are to be connected to a lifting hook, use a large bow shackle so that all the sling eyes can be safely accommodated. The pin must rest on the hook and the sling eyes in the bow section.

16.4 Attachments

Lifting and rigging attachments are used on a crane or hoist between the hook and the item to be lifted. Lifting and rigging attachments must be properly configured for the weight of the load, the type of crane or hoist, and the type of chain, rope, or hook being used to lift the load.

There are many different attachment types, such as hooks, sling release systems, spreader beams, and equalizer beams. Important specifications to consider include maximum load capacity, maximum lifting volume, and application or use.

Hooks

There are a wide variety of hooks in use for chain slings, mostly 80 grade alloy steel and stamped with the SWL (Fig. 16.15). Every hoisting hook should be fitted with a safety catch, particularly where there is a chance of the slings being displaced (Fig. 16.16).

Make sure that all hoisting hooks, except grab and sorting types, are equipped with safety catches. Every hoisting hook *must* be equipped with swivel and headache ball; and the ball *must* be securely attached to either hook or rope, preventing it from sliding on the load line. When selecting a hook for a chain sling (Fig. 16.17):

- *Always* refer to the manufacturer's load ratings for specific values of specific hooks.
- Make sure that the hook has at least the same SWL as the chain.

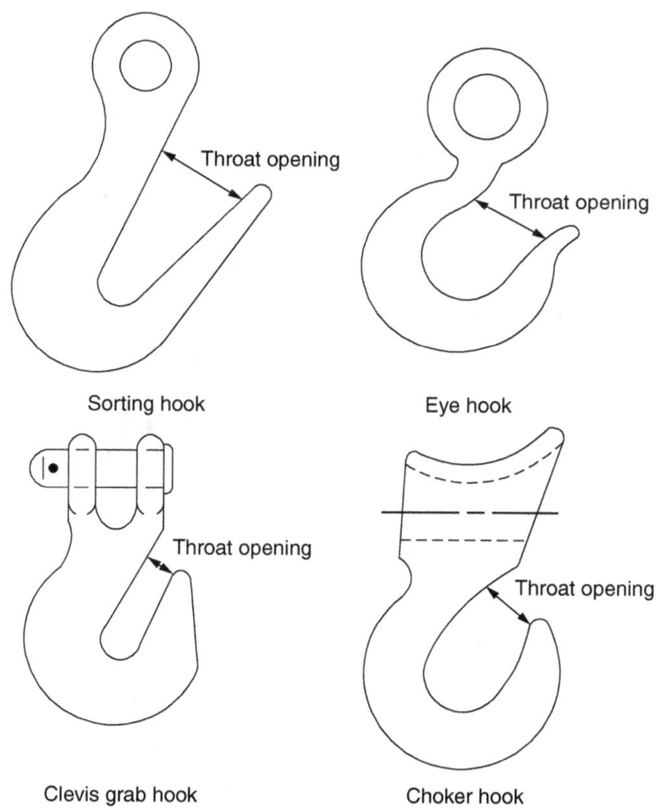

FIGURE 16.15 Miscellaneous rigging hook. (a) Sorting hook; (b) eye hook; (c) clevis grab hook; (d) choker hook.

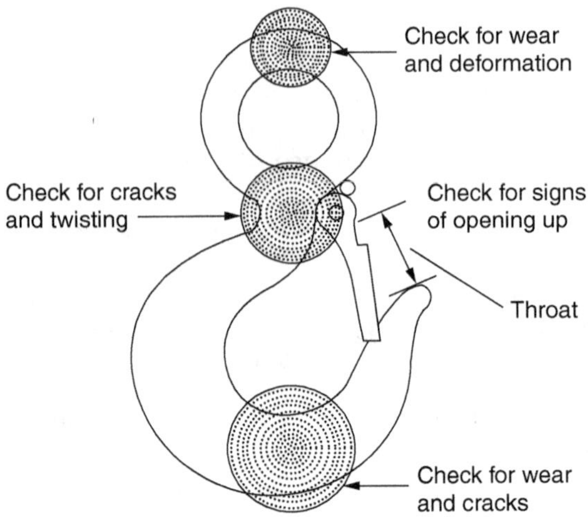

FIGURE 16.16 Typical safety hook with safety latch.

FIGURE 16.17 Typical swivel and headache ball, attached to hook. (*Source:* Gunnebo Johnson Corp., Tulsa, OK.)

- Remember that a hook's SWL applies only when the load is applied directly on the saddle of the hook. Eccentrically loading the hook, or applying a load anywhere between the saddle and the tip of the hook, will reduce the hook's rated SWL considerably.
- Make sure that the hook:
  - Opening is wide enough to accept the largest rope, ring, link or shackle that has to be placed over the hook. If a chain hook opening is stretched more than 5 percent it must be withdrawn from service.
  - Inside, or "bight", is rounded so that it does not cut into or damage slings and fittings.
  - Freely rotates at all times. If the load exceeds 2 tons (1.82 metric tons), there must be a ball or roller thrust bearing between the trunnion and the nut.
- *Always* discard bent or distorted hooks.
- *Do not* attempt to weld or repair them.
- *Do not* weld any fittings to the hooks.

It is extremely important to inspect hooks frequently, checking their body for cracks (especially on the neck and saddle), corrosion, and twisting the saddle for wear, and the throat for possible opening up beyond new hook dimensions. Lubricate hook bearings regularly:

- *Antifriction.* Every 14 days for frequent swiveling; every 45 days for infrequent swiveling.
- *Bronze thrust bushing or no bearing.* Every 16 hours for frequent swiveling; every 21 days for infrequent swiveling.

Rings and Links

Most chain slings *must* be fitted with some type of end fitting, usually consisting of a master ring or link on the end that fits over a crane hook; and a hook or link at the other end that attaches to the load. These master rings and links have large inside dimensions that permit the fitting to slip easily onto the thicker crane hook section. Types of rings and links include:

- *Master rings.* These end fittings require a section diameter about 15 percent larger and an inside with about 33 percent greater than a comparable oblong link to withstand the same loading without being deformed.
- *Pear-shaped master links.* These end fittings, once commonly used for chain end connection, are not as versatile as oblong links, and thus they are seldom used. A significant danger

with this type of link occurs when the pear shape accidentally reverses in use, thus subjecting the narrow end to severe bending when the fitting is jammed down on the crane hook's thick saddle.

- *Oblong master link.* These end fittings, either welded or mechanical, permit easy attachment of the sling chain to the hook or end links. They are the preferred links in rigging practices because they are less bulky and far more versatile than other comparable end shapes.
- *Connecting links.* These end fittings, either welded or mechanical, permit attachment of the sling chain to the hook or end links.
- *Rings.* These *must* have at least the same SWL as the chain, hook, and other parts of a sling.

Always discard any ring that has been stretched by more than 5 percent.

Do not place a ring (or shackle or eye bolt) over a crane hook unless it hangs freely. Welded connecting links *must* be attached by the manufacturer to assure the connection's compatibility with a chain sling's other components. Mechanical couplers, however, enable a rigger to assemble (in the field) reliable customized slings from component parts.

Eye and Ring Bolts

All eye and ring bolts *must* be equipped with shoulder or collars for lifting use. *Do not* use uncollared eyebolts for any lifts other than vertical lifts because they can break off when side strain is applied. The shoulderless types are designed for vertical loading only, and will either bend or break when a load pulls on them at an angle.

Even eye and ring bolts with shoulders, however, have a reduced SWL when subjected to angled loading. To assure the maximum SWL of the bolt, its shoulder must be at right angle to the axis of the hole, and the shoulder must be completely in contact with the bearing surface. To assure firm contact, washers may have to be used with the bolt (Fig. 16.18).

- Make sure that eyebolts are securely screwed into the ferrule or nut before use.
- Bolts secured by nuts must be properly torqued; and screw-type bolts should be set to a depth at least one and one-half ties the diameter of the bolt.
- *Do not* lift the load if the ferrule is loose.
- *Do not* hammer an eyebolt to tighten; use a podger bar.

FIGURE 16.18 Hot galvanized drop-forged eyebolts and ring bolts. (*Source:* Chicago Hardware and Fixture Co., Franklin Park, IL.)

- Make sure that the crown of a collared eyebolt is in line with the strain when using multilegged slings.
- *Do not* put a sling through two or more eyebolts. Use two slings attached to the eyebolts with shackles.
- *Do not* attach slings to eyebolts with hooks because the hook is usually too small.
- *Always* apply loads in the plane of bolt eyes to minimize the possibility of bending, especially when using bridle slings, which always develop an angular pull in the eyebolts.
- *Always* use a shackle when connecting a sling leg to an eyebolt.
- *Never* reeve slings through eyebolts for a lift.
- *Always* turn eyebolts in line with each other, using washers under bolt collars if necessary, to permit alignment when the bolt is tightened.

A typical use for an eyebolt is for lifting precast concrete panels which have ferrules cast into them.

Turnbuckles and Bottlescrews

A turnbuckle or bottlescrew is a device for tensioning or fine length adjustment of flexible steel wire rope (FSWR). They also are called rigging, union, or stretching screws. They have a limited application in rigging, however, and should be used with great care, for there are many inherent dangers associated with their use.

A variety of end fittings are used as turnbuckles: eye and eye, hook and hook, jaw and jaw, or stub and stub, as well as any combination of these types (Fig. 16.19).

The most common use of turnbuckles (bottlescrews) is in leveling loads. Some loads, such as when placing a rotor into a stator in a power station, must be placed into position perfectly level. This is done by having one sling on one side of an equalizing beam and a

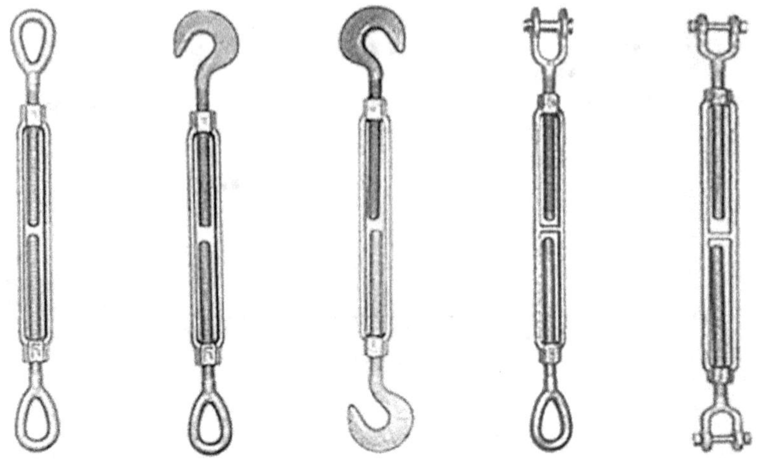

Figure 16.19 Rigging screws and turnbuckles (bottlescrew). (*Source:* Roughrider Rigging Ltd., Regina, Sk, Canada.)

shorter sling attached to a bottlescrew on the other side both attached to the load. The load can be made completely level by adjusting the screw thread on the bottle.

The rated load capacity of a turnbuckle depends on the type of end fitting as well as the outside diameter of the fitting's threaded section. All turnbuckles of comparable size have equal, rated load capacities except hook and eye types which, because of their configurations, have lower ratings.

- Make sure all hook end fittings are fitted with safety catches.
- *Always* secure a turnbuckle frame to its end fitting if the turnbuckle is subject to vibrations, thus preventing the fitting from turning loose.
- *Always* use wire to lock the fitting in place, instead of heavier and less effective lock nuts or jam nuts.
- Inspect all turnbuckles frequently.
- Check end fittings, rods, and bodies for cracks, bending, and deformation; check threaded portions for thread damage.

When using turnbuckles and bottlescrews:

- *Do not* use turnbuckles with a hook to support loads.
- *Do not* weld reinforcing or steel across the hook to stop it from opening.

- *Do not* subject turnbuckles with a clevis to a side load.
- Make sure bottlescrews have inspection holes to give a view of the amount of thread left in the bottle.
- *Do not* use bottlescrews if inspection holes are absent or the thread is not fully engaged.
- *Do not* use turnbuckles or bottlescrews if the SWL is not branded on them.
- Inspect threads to ensure that they are in good condition and will take the SWL branded on the side.
- Lock bottlescrews to prevent them from unscrewing where there is likely to be vibration. *Do not* try to lock them using lock nuts which do not prevent unscrewing and puts extra strain on the thread.
- Use eye or clevis types only for supporting loads.
- *Always* secure a turnbuckle's frame to its end fitting if the turnbuckle is subject to vibrations, thus preventing the fitting from turning loose.
- *Always* use wire to lock the fitting in place, instead of heavier and less effective lock nuts or jam nuts.
- Inspect all turnbuckles frequently.
- Check end fittings, rods, and bodies for cracks, bending, and deformation; check threaded portions for thread damage.

Spreader and Equalizing Beams

Among the number of lifting and rigging attachments used on a crane or hoist, between the hook and the item to be lifted are the spreader beam (lifting beam) used to lift hard-to-handle loads and the equalizing beam (load leveler) commonly used to stabilize the load.

Spreader beams support long loads during lifts and prevent them from tipping, sliding, or bending. The beam also helps eliminate low sling angles and prevents a load from being crushed. Equalizing beams distribute a load equally between two sling legs or two hoist lines when making a lift in tandem (Fig. 16.20).

Overhauling Weights

Provide extra weight to the crane hoist wire and hook when there is no load attached. To minimize the spinning of the overhauling weight and assembly, it is preferable to have a ball or roller bearing swivel between the rope and overhauling weight.

- Make sure that where a hoist limit is fitted, the crane is set so that the swaged fitting at the top of the overhauling weight assembly does not come into contact with the boom head sheave.

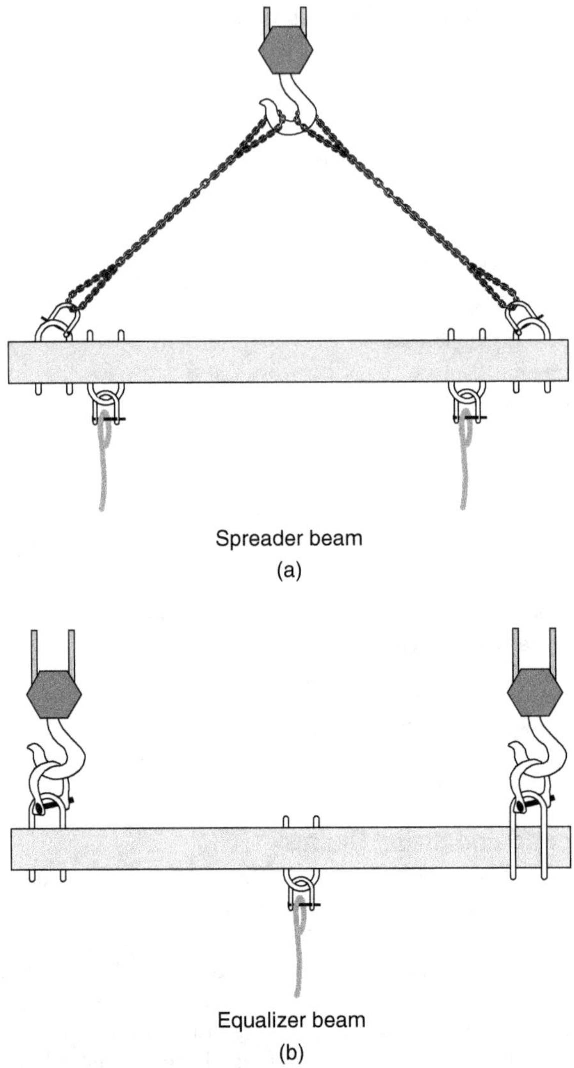

FIGURE 16.20 Typical application of (a) spreader beam, and (b) equalizer beam. (*Source:* Electronic Library of Construction, Occupational Safety & Health, *www.cdc.gov.*)

- Check for signs of corrosion and fatigue where the overhauling weight is attached to the rope, that is, around the edges of a wedge or socket or other attachment. The overhauling weight assembly should never be set up to rest on a swaged rope fitting.

- Make sure all attachments are properly configured for the weight of the load, the type of crane or hoist, and the type chain, rope, or hook being used to lift the load.

16.5 Care and Maintenance of Parts and Assemblies

Any connection, fitting, or attachment will fail if it is abused, misapplied, worn out, or improperly maintained. Failure can cause loads to swing or fall out of control, and may result in personal injury or death, as well as property damage. Proper training is important. Riggers should be properly trained in the design and use of rigging assemblies. Conduct all lifting operations in such a manner that if an equipment failure should occur, no person will be injured.

- Keep out from under a load and keep out of a load's line of force.
- *Never* permit anyone under a lifted load.
- Avoid irregular loading. *No* rigging fitting should be loaded principally in bending, torsion, or shear. Avoid these conditions by loading only in the normal, straight-line pull for which the fitting is designed. Avoid side loading.
- Inspect fittings and attachments regularly—*no* part will last indefinitely. Corrosion, wear, deformation, gouges, misuse, abuse, alteration, and other conditions will occur and will reduce the working load limit (WLL) of a part.
- *Do not* exceed the WLL stated by the rope manufacturer. Extreme environmental conditions of heat, cold, corrosion or any combination of the three, or high-risk applications may necessitate a reduction of the WLL.
- *Never* exceed the WLL for the connection, fitting, or attachment.
- Avoid shock loads. Avoid jerking, impacting, swinging, or handling loads in such a manner that the WLL of the rigging can be exceeded.
- *Do not* weld or modify fittings. Wear and normal use will reduce the WWL of any assembly. If you have any doubts about a new or used assembly, appropriate inspection and testing should be conducted to ensure adequate performance.
- Inspect all parts periodically to determine when to replace a part and avoid a rigging hazard. The frequency of inspection depends on users' specific applications and local conditions; when in doubt, inspect prior to use. If a part is judged defective, destroy the part.
- *Never* simply discard a defective connection, fitting, or attachment—destroy it.

Regulations and Standards

OSHA Regulations

1926.251(c)(15) *End Attachments*

1926.251(f) *Shackles and Hooks*

U.S. Consensus Standards

ASME/ANSI B30.10 – 1988 *Hooks*

B30.11 – 1988 *Monorails and Underhung Cranes*

B30.16 – 1987 *Overhead Hoists (Underhung)*

B30.17 – 1985 *Overhead and Gantry Cranes (Top Running Bridge, Single Girder, Underhung Hoist)*

B30.20 – 2006 *Below-the-Hook Lifting Devices*

ANSI/ASME HST-1M – 1982 *Performance Standard for Electric Chain Hoist*

HST-2M – 1993 *Performance Standard for Hand Chain Manually Operated Chain Hoists*

HST-3M – 1985 *Performance Standard for Manually Lever Operated Chain Hoists*

HST-4M – 1985 *Performance Standard for Overhead Electric Wire Rope Hoists*

HST-5M – 1985 *Performance Standard for Air Chain Hoists*

HST-6M – 1986 *Performance Standard for Air Wire Rope Hoists*

ASME *Crane Standards*

B30.2 – 1990 *Overhead and Gantry Cranes (Hand Powered and Power Driven)*

B30.21 – 1982 *Manual Lever Operated Hoists (Ratchet, Pawl, and Friction Brake)*

ASTM/ANSI B30.9 *Ratchet Tiedown*

Industry Consensus Standards and Guidelines

American Chain Manufacturers Association

Associated Wire Rope Fabricators

Cordage Institute

Web Sling and Tiedown Association

Wire Rope Technical Board

CHAPTER 17

Sheaves

Sheaves are used in rigging operations to change the direction of travel of fiber or wire rope. When assembled in multiples, such as in blocks, sheaves can provide almost any required mechanical advantage.

Although sheaves should be mounted in exact alignment with each other, this rarely occurs in rigging operations. To compensate for the misalignment, sheave grooves are shaped to provide some tolerance.

Sheaves rotate about their mounting shafts on bushings or bearings. In a system of blocks, the sheaves rotate at different rates or speed, and have different loads. When raising and lowering, however, the line tension is not equal throughout the system. Calculations can be made to find the maximum allowable line speed for a given total sheave load. If the required speed is greater than the maximum allowable line speed calculated, then increase the shaft size or the hub width (or both) and recalculate. Continue the process until the maximum allowable line speed is equal to or exceeds the required line speed.

Formulas for calculating bearing pressure, bearing velocity, and line speed appear below:[1]

Bearing pressure $$BP = \frac{Line\ pull \times Angle\ factor}{Shaft\ size \times Hub\ width}$$

Bearing velocity $$BV = \frac{PV}{BP}$$

Line Speed $$LS = \frac{BV(Tread\ diameter) - Rope\ diameter}{Shaft\ diameter}$$

where BP = bearing pressure
BV = bearing velocity
LS = line speed
PV = pressure velocity

[1]*Source:* The Crosby Group Inc., Tulsa, OK.

Angle	Factor	Angle	Factor
0°	2.00	100°	1.29
10°	1.99	110°	1.15
20°	1.97	120°	1.00
30°	1.93	130°	0.84
45°	1.87	135°	0.76
40°	1.84	140°	0.68
50°	1.81	150°	0.52
60°	1.73	160°	0.35
70°	1.64	170°	0.17
80°	1.53	180°	0.00
90°	1.41	—	—

Source: Kennedy Wire Rope and Sling Co., Corpus Christi, TX.

TABLE 17.1 Angle Factor Multipliers

NOTE: *Due to material property relations the maximum BP times the maximum BV is NOT equal to the maximum PV.*

The WLL is the maximum load that should be exerted on the sheave block and its connecting fitting. This total load value may be different from the weight being lifted or pulled by a hoisting or hauling system.

To properly determine the rated capacity block to be used, it is necessary to determine the total load being imposed on each block in the system. A single sheave block used to change load line direction can be subjected to total loads greatly different from the weight being lifted or pulled. Total load values vary with the angle between the incoming and departing lines to the sheave block. To obtain the total load on the sheave block, use the correct angle factor multiplier (Table 17.1).

17.1 Sheave Selection

Selecting the proper bearing is critical to the performance of any application using sheaves. Sheave load, line speed, frequency of use, and the operating environment are important considerations when making bearing selection.

Types of Sheaves

- *Plain bore.* High bearing friction; cast iron; provides its own bearing. For very slow line speeds and very infrequent use; used on fiber ripe blocks. For light intermittent use only; must be oiled frequently.

- *Bronze-bushed.* Self lubricating. For slow line speeds and infrequent use, and more frequent use at greater loads; moderate bearing friction.
- *Roller-bushed.* Unground rollers, without races; for light infrequent use only. Must be lubricated periodically with a heavy grease.
- *Bronze bearings.* Pressure lubricated; for use where service requirements are extra heavy and continuous. Must be lubricated frequently.
- *Roller bearings.* Ground rollers and full races; recommended for medium-duty use where operation is high speed.
- *Tapered bearings.* Taper angles allow the bearing to handle a combination of radial and thrust loads—the steeper the cup angle, the greater the ability of the bearing to handle thrust loads.
- *Antifriction bearings.* Precision bearings that combine long life with minimum maintenance; minimum bearing friction. For use in operations with continuous high speeds and heavy loads.

Sheave Diameter

Although no minimum sheave diameter would prevent a hoisting mechanism from operating, the diameter nevertheless has a direct bearing on rope life. When selecting the proper sheave for a standard wire rope application, three principal factors must be considered: sheave diameter, sheave material, and groove size and shape.

The tread diameter of a sheave determines the bending stress in the wire rope and the contact pressure between it and the sheave surface (Table 17.2). Properly sized sheaves for the size and construction of a particular wire rope can prolong the lives of both sheave and rope.

	Ratio of cable diameter to pulley diameter					
Cable dia. (in.)	**6 × 7 or 7 × 7**	**6 × 19 or 7 × 18**	**6 × 31 7 × 31 6 × 37**	**6 × 7 or 7 × 7**	**6 × 19 or 7 × 18**	**6 × 31 7 × 31 6 × 37**
1/4	10-1/2	6	3	7	4-1/2	2-1/2
5/16	13-1/8	7-1/2	3-3/4	8-3/4	5-5/8	3-1/8
3/8	15-3/8	9	4-1/2	10-1/2	6-3/4	3-3/4
7/16	18-3/8	10-1/2	5-1/4	12-1/4	7-7/8	4-3/8
/2	21	12	8	14	9	5

Source: Tie Down Engineering, Atlanta, GA.

Table 17.2 Minimum Tread Diameter for Sheaves (or Pulleys)

If a wire rope is used with inadequately sized sheaves, the severe bending stresses imposed on the wire rope will cause wires to break from fatigue, even though actual wear may be slight. One of the fastest ways to destroy a wire rope is to run it over small sheaves. Excessive and repeated bending and straightening of the wires will lead to premature failure from fatigue.

Small sheaves will accelerate wear of both wire rope and sheave grooves. Because the pressure per unit area of wire rope on a sheave groove for given load is inversely proportional to the size of the sheave, the smaller the sheave diameter, the greater the rope pressure per unit area on the groove.

To determine the unit radial pressure between a wire rope and a sheave, use the following formula:

$$P = \frac{2L}{Dd}$$

where P = unit radial pressure
L = wire rope length
D = tread diameter of sheaves
d = nominal diameter of wire rope

If the unit radial pressure exceeds the maximum values, the material from which the groove is manufactured is too soft for the operating conditions, and rapid wear of the grooves will result (Table 17.3).

The sheave diameter can also influence the wire rope strength. When a wire rope is bent around a sheave, there is a loss of effective strength due to the inability of the individual strands and wires to adjust themselves entirely to their changed position. The rope strength efficiency decreases to a marked degree as the sheave diameter is reduced with respect to the diameter of the rope.

Wire rope manufacturers have established standards for sheave sizes that should be used with various wire rope constructions (Table 17.4).

Sheave Material

Hard sheave surfaces offer the best bearing surfaces for wire rope, thus prolonging sheave and rope life. If a sheave is forged or cast from a material softer than the wire rope, the sheave life and the wire rope life will be shortened. The sheave will have a tendency to take on the impression of the rope, causing scoring and corrugation of the line groove.

Cast manganese steel sheaves offer the ultimate in sheave material. This surface actually hardens to the use of wire rope and provides greatly extended sheave life and increased service time of the wire rope.

Material		Regular lay rope				Lang-lay rope			Lang-lay rope
		6 × 7	6 × 19	6 × 37	8 × 19	6 × 7	6 × 19	6 × 37	Flattene strandd
Wood	(psi)	150	250	300	350	165	275	330	400
	(kPa)	1034	1723	2068	2413	1137	1896	2275	2758
Cast iron 125 Brinell	(psi)	300	480	585	680	350	550	660	800
	(kPa)	2068	3309	4033	4688	2413	3792	4550	5515
Carbon steel casting 160 Brinell	(psi)	550	900	1075	1260	600	1000	1180	1450
	(kPa)	3792	6205	7411	8673	4136	6694	8135	9997
Chilled cast iron	(psi)	650	1100	1325	1550	715	1210	1450	1780
	(kPa)	4481	7584	9136	10.687		4930	8343	9997
Manganese steel, induction hardening, or flame-hardened ground grooves	(psi)	1470	2400	3000	3500	1650	2750	3300	4000
	(kPa)	18,135	18547	20,684	24,132	11,378	18,961	22,753	27,579

TABLE 17.3 Allowable Wire Rope Bearing Pressures for Various Sheave Materials

Construction	Suggested D/d* ratio	Minimum D/d* ratio
6 × 7	72	42
19 × 7 or 18 × 7	51	34
Rotation resistant		
6 × 9 Seale	51	34
6 × 27 H	45	3
Flattened strand		
6 × 31 V	45	30
Flattened strand		
6 × 21	45	30
Filler wire		
6 × 25	39	26
Filler wire		
6 × 31	39	26
Warrington seale		
6 × 36	35	23
Warrington seale		
8 × 19	41	27
Seale		
8 × 25	32	21
Filler wire		
6 × 41	32	21
Warrington seale		
6 × 42	21	14
Filler wire		

*D = tread diameter or sheave; d = nominal diameter or rope.
Source: Hanes Supply Inc., Buffalo, NY.

TABLE 17.4 Proper Sheave and Drum Sizes

Groove Size and Shape

Sheave grooves must be smooth and slightly larger than the nominal wire rope diameter to provide maximum rope support. Check groove conditions with a calibrated groove gage.

NOTE: *Check if gage contour matches the groove bottom. Proper fitting sheave groove should support the rope over 125° to 150° of rope circumference (Fig. 17.1).*

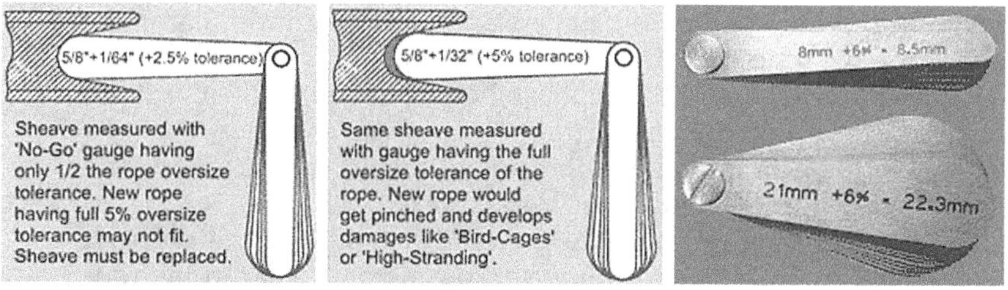

FIGURE 17.1 Calibrated groove gage. (*Source*: Python Wire Rope, Clifton NJ.)

Newer gages, made to include the oversize tolerances for minimum groove conditions, are marked accordingly. If the groove diameter is too large, the rope will not be properly supported. It will tend to flatten and become distorted, thus accelerating the bending fatigue in individual wires and causing premature failure.

Conversely, if the sheave groove is too narrow for the wire rope, the loading tension will pull the rope deeply into the groove, causing it to be pinched. Both wire rope and sheave will be subjected to severe abrasive wear.

Proper sheave design requires that the groove diameter have a maximum tolerance varying from 1/64 to 5/32 in. in excess of the nominal wire rope diameter, depending on the rope size. The depth of the sheave groove should be at least one and one-half times the wire rope's nominal diameter.

The wire rope in service will wear and decrease in diameter; with the resulting distortion causing poor distribution of the stress throughout the cross-section of the rope, with a corresponding overloading of some portions.

As a rope is bent around a sheave, the strands and wires bind against each other and do not move freely. This increases internal and external abrasion and friction, as well as hindering the rope from readjusting to its load. As a result, some of the wire and strands are forced to carry more load than they should, thus shortening wire rope's service life.

In addition, an undersized rope tends to wear an undersized groove in the sheave. Then, when it is necessary to install a new rope, the sheave groove will not provide proper working clearances for the replacement rope. The resulting pinching or wedging of the rope causes binding that prevents proper load distribution and produces excessive internal and external friction.

To assure long and efficient wire rope life, sheave grooves should be smoothly contoured, be free of surface defects, and have rounded edges.

Alignment

Sheaves should be mounted so that they are aligned exactly with each other. Because in actual use this is usually impossible, sheaves are manufactured with grooves that provide tolerance for misalignment and guide the rope into place.

A lead angle of only about 2° can be accommodated without difficulty. Any appreciable misalignment, however, will cause the wire rope to rub against the sides of the groove, resulting in both rope and sheave wear thus shortening the useful life of both.

After every wire rope change, check that sheaves are properly aligned and running true; otherwise, considerable wear in the rope and sheaves will result. Always equip sheaves with cable keepers where an unloaded wire rope might possibly leave the sheave groove.

Bearings

Sheave bearings should be either permanently lubricated or equipped with a means for proper lubrication. Inadequate lubrication of a sheave that is too heavy for a load will cause the wire rope to slip in the sheave whenever the rope velocity changes. The momentum of the heavy sheave will cause it to continue turning after the wire rope has stopped, producing a grinding wheel action that can cause severe wire rope abrasion and wear flat spots in the sheave, further damaging the rope.

Friction losses in sheaves are a function of the style of wire rope, the ratio of sheave to wire rope diameter, and the type of bearing. When computing friction loses at sheave bushings, it is reasonable to assume about a 4 ½ percent loss; for sheave bearings, the loss is about 1 to 2 percent, depending on the quality of the bearings. These losses are based on using a wire rope that makes a bend of 180° over the sheave. For smaller turning angles, these values can be reduced.

17.2 Care and Maintenance of Sheaves

Sheaves must be regularly inspected, lubricated, and maintained for peak efficiency and extended usefulness. The frequency of inspection is dependent on the frequency and periods of use, environmental conditions, and the user's good judgment.

Maintenance Checklist

Sheaves

- *Wear* on pins or axles, rope grooves, side plates, bushing or bearings, and fittings; excessive wear may be a cause to replace parts or remove sheave from service.
- *Deformation* of side plates, pins and axles, fitting attachment points, trunnions, etc.; deformation can be caused by abusive

service or overload (or both) and may be a cause to remove sheave from service.

- *Misalignment* or wobbles in sheaves.
- *Security* of nuts bolts, and other locking methods, especially after reassembly following a tear down inspection; original securing method should be used, such as staking, set screw, cotter pin, and cap screw.

Components

- Pins retained by snap rings for missing or loose rings.
- Pin nuts for proper positioning:
  - *Tapered roller bearing*—should be tightened to remove all play during sheave rotation
  - *Bronze bushing and straight rollers*—should have a running clearance of 0.031 in. per sheave of end play; and should be adjusted accordingly.
- Pins of bronze bushings and straight roller bearings for a running clearance of 0.031 in. (0.787 mm) per sheave of end play; adjust pins accordingly.
- Tapered roller bearings to remove all end play during sheave rotation.
- Hook or shackle case clearance for clearances set at 0.031 to 0.052 in. (0.787 to 1.321 mm) at the factory; increased clearance can result from component wear; clearance exceeding 0.12 to 0.18 in. (3.048 to 4.572 mm) requires disassembly and further inspection.
- Deformation or corrosion of hook and nut threads.
- Surface condition and deformation of hook.
- Welded side plates for weld corrosion or weld cracking.
- Hook latch for deformation, proper fit, and operation.

Remove from service any bushing with cracks on inside diameter or bushing end; bushings that are cracked or extended beyond sheave hub (or both) are indications of bushing overload.

Lubrication

The frequency of lubrication depends on frequency and period of sheave use as well as environmental conditions, which are contingent on the user's good judgment.

Assuming normal sheave use, the following schedule is suggested when using lithium-base grease of a medium consistency:[1]

---

[1]*Source:* The Crosby Group, Tulsa, OK.

Sheave Lubrication

- *Bronze bushings (not self lubricated).* Every 8 hours of continuous operation or every 14 days of intermittent operation.
- *Roller bearings.* Every 24 hours of continuous operation or every 14 days of intermittent operation.
- *Self-lubricating bronze bushing.* Used for slow line speeds and infrequent use (moderate bearing friction). Frequent inspection is required to determine the condition of bushing.
- *Tapered roller bearings.* Every 40 hours of continuous operation or every 30 days of intermittent operation.

Hook Lubrication

- *Antifriction.* Every 14 days for frequent swiveling; every 45 days for infrequent swiveling.
- *Bronze thrust bushing (or no bearing).* Every 16 hours for frequent swiveling; every 21 days for infrequent swiveling.

17.3 Inspection of Sheaves

Almost every rope installation has one or more sheaves ranging from traveling blocks with complicated reeving patterns to equalizing sheaves where only minimal rope movement is noticeable.

Each sheave should receive an individual examination at periodic inspections. Sheaves rotate on either bearings or bushings that *must* be inspected for discoloration (due to excessive heat), metallic particles, chips or displaced metal, broken or distorted bearing retainer or seals, adequate lubrication, and tight bearing caps. The sheave inspection is for wear and damage, wear in the wire rope sheave groove, loose or damaged sheave guards, and worn bearings and pins.

The frequency of sheave inspection depends on the frequency and periods of use, environmental conditions, and the user's good judgment (Fig. 17.2).

Inspect sheaves carefully for:

- *Uneven wear of groove contours.* Sheaves that have worn out of round or have developed flat spots will set up vibrations in the wire rope that will result in premature wire fatigue. Remachine the groove or replace the sheave.
- *Any sign of cracks in sheave flanges.* Broken flanges can permit a wire rope to jump the sheave or become badly cut as it rubs against the rough broken edge. Replace the sheave.
- *Depth and flare of the groove. Make sure* that the wire rope does not rub against the sheave flange. Such rubbing may cause

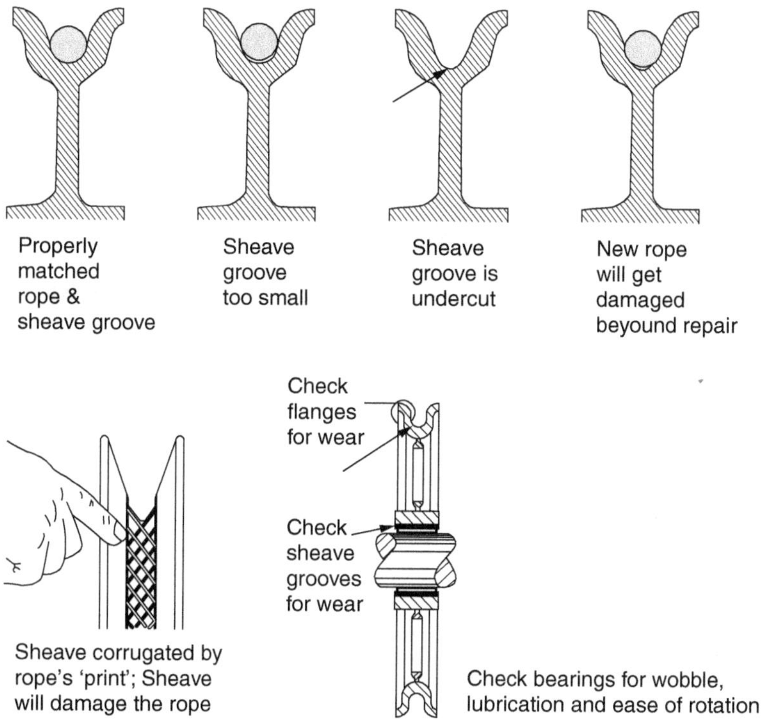

FIGURE 17.2 Common sheave troubles. (*Source*: Python Wire Rope, Clifton NJ.)

the wire rope to fail prematurely. Remachine the groove or replace the sheave.

- *Wear on one side of the groove.* Often due to excessive fleet angle or poor alignment. To correct, move the sheave.
- Broken or chipped flanges; cracks in hubs, spokes, etc.
- *Wear on bearings.* Worn bearings will cause a sheave to wobble, putting vibration into the wire rope and increasing wire fatigue. Repair bearings or replace the sheave.
- Signs of rope contact with guards
- Out-of-round condition
- Alignment with other sheaves.

NOTE: Never *use wire rope on sheaves designed for fiber rope.* Never *use fiber rope on sheaves that have been used for wire rope service.*

Assessing the general physical condition of a sheave, such as groove smoothness, freedom from cracks and knicks, existence of wear on guards, etc., is a matter of careful and knowledgeable observation.

A proper fitting sheave groove should support the rope over 135–150 degrees of rope circumference. Observe the groove so that it may be clearly seen whether the contour of the gauge matches the contour of the bottom of the groove.

In a field inspection, when the gauge for worn grooves fits perfectly, the groove is at the minimum permissible contour. Anything narrower is unsuitable for use.

Under normal operating conditions, as a groove wears it tends to become deeper and narrower. Excessive wear in an over-width manner frequently indicates some operating abnormality such as alignment.

Regulations and Standards

Industry Consensus Standards and Guidelines

American Wire Cloth Institute
Associated Wire Rope Fabricators
Cordage Institute
Web Sling and Tiedown Association
Wire Rope Technical Board
Woven Wire Products Association

CHAPTER 18
Reeving Blocks and Tackles

Reeving refers to the configuration of passing fiber or wire rope through a system of blocks—pulleys or sheaves—to provide a mechanical advantage that permits lifting heavy loads with a reasonable pull of the haul rope. The mechanical advantage of a multipart reeved system equals the number of parts of line actually supporting the running block and the load.

18.1 Blocks and Tackles

A block and tackle is a system of two or more pulleys with a rope or cable threaded between them, usually used to lift or pull heavy loads.

The most common arrangement is to have a set of fixed pulleys or "sheaves" all mounted on a single axle, and another set left to move. Each set is called a *block* (the whole assembly, with the rope, is the *tackle*).

The mechanical advantage of a block and tackle is equal to the number of lines running between the two blocks. For example, a tackle consisting of three fixed and three moving pulleys has six lines going between the pulleys, giving it a mechanical advantage of 6. This enables the lifting of weights six times heavier than would be possible otherwise.

Blocks

A block consists of one or more sheaves fitted in a wood or metal frame supported by a shackle inserted in the strap of the block. Blocks are constructed for use with fiber rope or wire strand rope (Fig. 18.1).

- *Wire rope blocks.* Heavily constructed, having large sheaves with deep grooves. A large sheave is needed with wire rope to prevent sharp bending.
- *Fiber line blocks.* Generally not as heavily constructed, having smaller sheaves with shallow, wide grooves. Because fiber

rope is more flexible and pliable, it does not require a sheave as large as that required by wire rope.

- *Crane blocks*. Identified by the number of sheaves (pulleys) per block—single, double, or triple—fitted in a wood or metal frame and supported by a shackle inserted in the strap of the block. They are fitted with a number of attachments, such as hooks, shackles, eyes, and rings (Fig. 18.2).
- *Snatch blocks*. Designed so that they can be readily opened for the insertion of the bight (as distinct from the end) of a rope (Fig. 18.3).

Each block has one or more steel straps that strengthen the block and support the sheave pin. A load is applied to the block by means of a hook or shackle inserted in the top of the strap. The strap may continue through the block and form a projection, called the *becket*, to attach another line. The becket usually has a thimble to prevent chafing of the line.

Tackle

This is an assembly of blocks and lines used to gain a mechanical advantage in lifting and pulling. In a tackle assembly, the line is reeved over the sheave(s) of blocks. The mechanical advantage of a machine is the amount the machine can multiply the force used to lift or move a load. To move any load heavier than the force a worker can exert

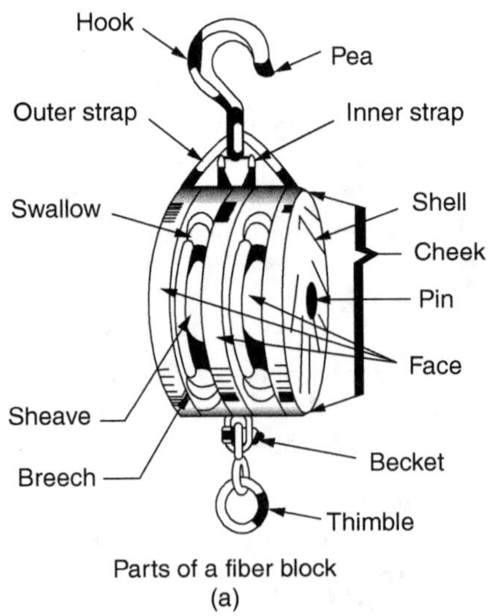

FIGURE 18.1 Block construction. (a) Parts of a fiber rope block; (b) components of metal blocks. (*Source*: *U.S. Army Field Manual 55-17*, Washington, DC.)

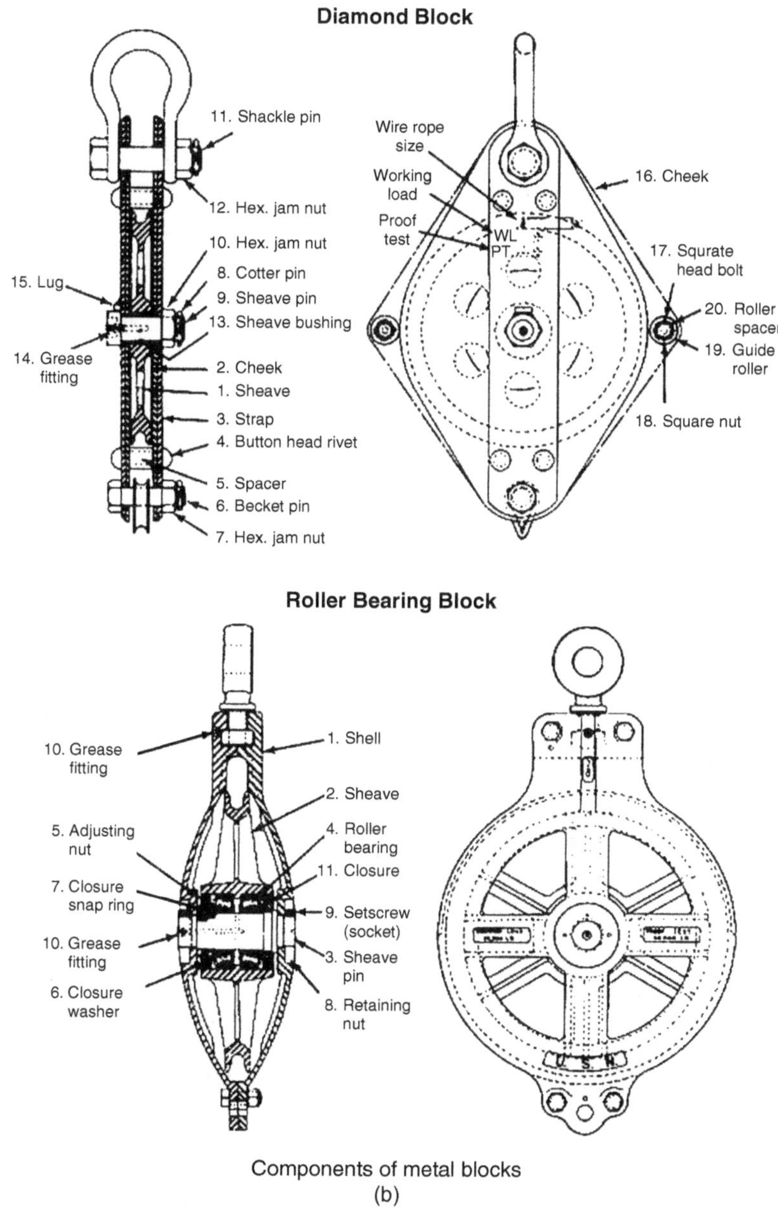

FIGURE 18.1 Block construction. (a) Parts of a fiber rope block; (b) components of metal blocks. (*Source*: *U.S. Army Field Manual 55-17*, Washington, DC.) (*Continued*)

requires the use of a machine that can provide a mechanical advantage to multiply the applied force. For example, if the downward pull on a block-and-tackle assembly requires 10 lb (4.54 kg) of force to raise100 lb (45.4 kg), the assembly has a mechanical advantage of 10.

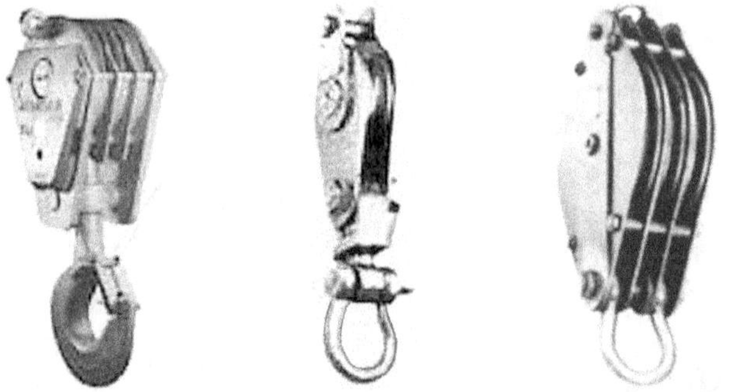

Crane block (a) | Snatch block (b) | Construction block (c)

FIGURE 18.2 Typical blocks. (a) Crane and hook; (b) snatch; (c) construction. (*Source*: Eastern Rigging Supply Co. Inc., Cos Cob, CT.)

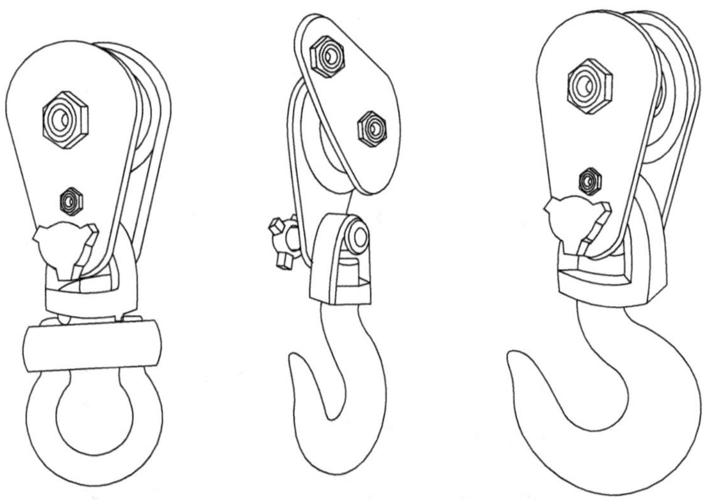

With swivel shackle (closed position) | Rear view-showing (open position) | With swivel hook (closed position)

FIGURE 18.3 Dead end snatch blocks.

There are two types of tackle systems: simple and compound. A simple tackle system is an assembly of blocks in which a single line is used. A compound tackle system is an assembly of blocks in which more than one line is used (Fig. 18.4).

A tackle may be rigged to advantage—where the pull on the rope is in the same direction as the load to be moved. Or, it may be rigged to disadvantage—where the pull on the rope is in the opposite direction of the load is to be moved.

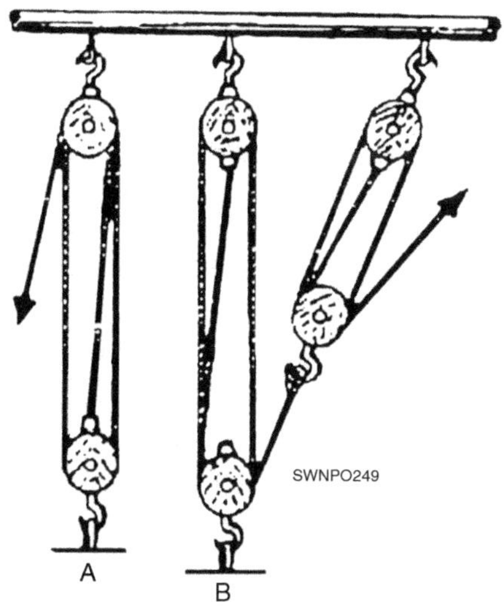

Figure 18.4 Types of tackle. (a) Simple; (b) compound.

While it may seem logical to rig to advantage in every case, the decision of which to use depends on pragmatic consideration for the total ergonomics of working with a particular situation.

In a tackle assembly, the line is reeved over the sheaves of blocks (Fig. 18.5).

- *Block(s).* In a tackle assembly change(s) the direction of pull, provides mechanical advantage, or both.
- *Fall.* Either a wire rope or fiber line reeved through a pair of blocks to form a tackle.

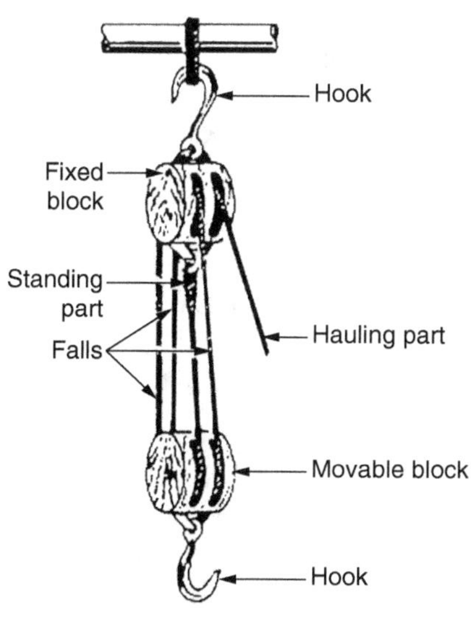

Figure 18.5 Parts of a tackle. (*Source*: *U.S. Army Field Manual 55-17*, Washington, DC.)

- *Hauling part of the fall.* Leads from the block on which the power is exerted.
- *Fixed (or standing) block.* The end that is attached to a becket.
- *Movable (or running) block of a tackle.* The block attached to a fixed objector support; when a tackle is being used, the movable block moves and the fixed block remains stationary.
- *Frame (or shell).* Made of wood or metal, houses the sheaves.
- *Sheave.* Grooved wheel over which a rope or wire runs; usually the blocks have one, two, three, or four sheaves—with some blocks having up to eleven sheaves.
- *Cheeks.* Solid sides of frame or shell.
- *Pin.* Metal axle on which the sheave turns; running from cheek to cheek through the middle of the sheave.
- *Becket.* Metal loop formed at one or both ends of a block; the standing part of the line is fastened to the becket.
- *Straps (inner and outer).* Hold the block together and support the pin on which the sheaves rotate.
- *Swallow.* Opening in the block through which the line passes.
- *Breech.* Part of the block opposite the swallow.
- *To overhaul.* To lengthen a tackle by pulling the two blocks apart.
- *To round.* To bring the blocks of a tackle toward each other, usually without a load on the tackle (opposite of overhaul).
- *Two-blocked.* Both blocks of a tackle are as close together as they can go (*chock-a-block*).

18.2 Selecting the Block

It is most important that the correct block is selected. Most blocks are marked on the shell with the SWL and the size of rope for which they are designed. If it is not marked, a suitable block can be selected by remembering that:

- Sheave diameter should be not less than 9 times the diameter for cordage and 18 times the diameter for SWR.
- Approximately one-third of the circumference of the rope should be in contact with the sheave.

Types of Blocks

- *Standing block.* Connected to a fixed object.
- *Traveling block.* Connected to the load that is being lifted; it also moves with the load.

- *Snatch block.* Single sheave block fabricated so the shell opens on one side at the base of the hook to allow a rope to slip over the sheave without threading the end through the block; used when it is necessary to change the direction of pull on the line.

When selecting a block or sheave for the system in a specific application, consider the following points as well as the features of the blocks and sheaves:

- Are there regulations that could affect the choice of blocks or sheaves, such as federal or state OSHA, elevator safety, insurance, etc.?
- What is the weight of the load, including any dynamics of impacts that add to a load value?
- How many parts of line are required?
- What is the size of line to be used?
- What is the speed of the line?
- What type of fitting is required for the particular application?
- How will the block be reeved, and does it require a dead end becket?
- If the block is to be a traveling block, what weight is required to overhaul the line?
- What is the fleet angle of the wire line?
- How will the block or sheave be maintained?
- Do conditions in the particular application require special maintenance considerations?

Most importantly, *make sure* block fittings are made of forged alloy steel. Because shackles and eyes are inherently stronger than hooks, blocks equipped with them are rated at higher working loads. They are especially recommended for block connections where the block is to be mounted permanently; where standing blocks must support not only the load, but also the hoisting strain of the load line; and where it would be dangerous to have a hook connection become accidentally disengaged.

Mechanical Advantage

When two or more blocks are used with a rope or chain for hoisting or hauling, the whole system is known as a tackle. The purchase (mechanical advantage) of a tackle equals the number of parts pulling on the load, with a part identified as either a line leading into or out of the sheave of the moving block; or a line attached to the moving block's becket. A line attached to the load is not included.

In practice, the purchase is never quite that much because of friction in the sheave, which has traditionally been calculated as one-tenth of the load for every sheave over which the line passes. In blocks with roller or needle bearings, this figure is probably excessive, but it provides a safety factor in the calculations.

A fixed block does not add power—it merely changes the direction of pull. But when one tackle is applied to the hauling part of another, the total purchase gained is the power of the first multiplied by the power of the second.

The increased force produced by a tackle is offset by both the increased length of rope needed and the friction in the system. To raise a block and tackle with a mechanical advantage of 6 a distance of 1 ft, it is necessary to pull 6 ft of rope through the blocks.

Frictional losses also mean there is a practical point at which the benefit of adding a further sheave is offset by the incremental increase in friction, which would require additional force to be applied to lift the load. Too much friction may result in the tackle not allowing the load to be released easily or the reduction in force needed to move the load being judged insufficient because undue friction has to be overcome as well. Tackles can be made up in various combinations to provide a mechanical advantage.

Tackle may be rigged to advantage—where the pull on the rope is in the same direction as that in which the load is to be moved; or it many be rigged to disadvantage—where the pull on the rope is in the opposite direction of that in which the load is to be moved (Fig. 18.6).

While it may seem logical to rig to advantage in every case, the decision of which to use depends on pragmatic consideration for the total ergonomics of working with a particular situation.

REMEMBER: *The anchorage point (becket) for the tackle or the blocks must be able to carry the total weight of the load plus the weight of the blocks, as well as the pull exerted on the lead line.*

The mechanical advantage of a tackle can be calculated using the formula:

$$P = WG\,(1 + fn)$$

where P = pull required on the running end of the fall
W = weight to be moved (in the same units as P)
G = theoretical gain
f = coefficient of friction
n = total number of sheaves including any leading blocks

(Manufacturers publish tables and charts for quick referencing of safe loads for common tackle configurations, using both cordage and steel wire rope.)

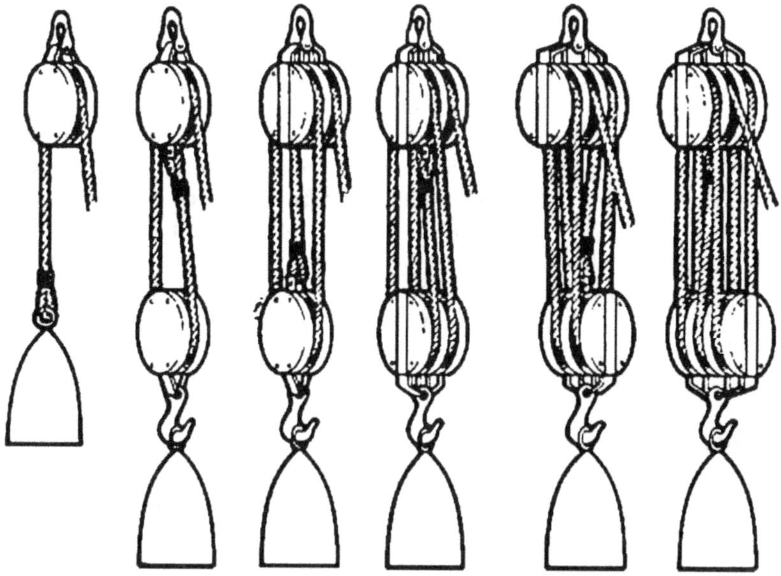

FIGURE 18.6 Various configurations of blocks and tackles. (*Source*: *U.S. Army Field Manual 55-17*, Department of Army, Washington, DC.)

Inverting Tackle

Inverting a tackle gains a mechanical advantage of 1, because the number of parts at the moveable block is increased. When a tackle is inverted, however, the direction of pull is difficult.

This can be overcome easily by adding a snatch block, which changes the direction of the pull but does not increase the mechanical advantage. A *single-luff* tackle consists of a double and a single block; a *double-luff* tackle has one triple and one double block.

Theoretical Gain

Theoretical gain is the factor by which the load weight is reduced to the force required, and is equal to the number of returns in the tackle configuration.

For example, in a 2/2 configuration, the theoretical gain is four—the load could be lifted using a force equal to one-quarter of the load weight. However, friction in the system means that the practical gain will always be slightly less than the theoretical gain, and slightly more force will be required.

Coefficient of Friction

The increased force produced by a tackle is offset by both the increased length of rope needed and the friction in the system. To raise a block and tackle with a mechanical advantage of 6 a distance of 1 ft, it is necessary to pull 6 ft of rope through the blocks.

The coefficient of friction is the proportion of overall effort that is lost due to friction in a particular tackle system. With blocks in good condition, the following coefficients of friction are good estimates: 10 percent for cordage, 15 percent for flexible steel wire rope (FSWR).

Frictional losses also mean there is a practical point at which the benefit of adding a further sheave is offset by the incremental increase in friction, which would require additional force to be applied to lift the load. Too much friction may result in the tackle not allowing the load to be released easily, or the reduction in force needed to move the load being judged insufficient, because undue friction has to be overcome as well.

Loads on Blocks

The manufacturer's working load limit (WLL) indicates the maximum load that should be exerted on the block and its connecting fitting. This total value may be different from the weight being lifted or pulled by a hoisting or hauling system.

It is necessary to determine the rated capacity of block to be used because a single sheave block used to change load line direction can be subjected to total loads greatly different from the weight being lifted or pulled. The total load value varies with the angle between the incoming and departing lines to the block.

To obtain the total load on the block, multiply the line pull by the proper angle factor (Table 18.1).

Angle (deg.)	Factor	Angle (deg.)	Factor
0	2.00	100	1.29
10	1.99	110	1.15
20	1.97	120	1.00
30	1.93	130	0.84
40	1.87	135	0.76
45	1.84	140	0.68
50	1.81	150	0.52
60	1.73	160	0.35
70	1.64	170	0.17
80	1.53	180	0.00
90	1.41	—	—

Source: The Crosby Group, Tulsa, OK.

Table 18.1 Angle Factor Multipliers

Wire rope dia. (in.)	Factor A wire rope wt (lb)	Number of line parts	Factor B overhaul factors	
			Roller bearing sheaves	Bronze bushed sheaves
1/2	.46	1	1.03	1.05
5/8	.72	2	2.07	2.15
3/4	1.04	3	3.15	3.28
1	1.85	4	4.25	4.48
$1^1/_4$	2.89	5	5.38	5.72
		6	6.64	7.00
		7	7.73	8.39
		8	8.94	9.80
		9	10.20	11.30
		10	11.50	12.80

Source: Crosby Group, Tulsa, OK.
Required block weight = [(Boom Length × Factor A) + Drum Friction] × Factor B

TABLE 18.2 Determining Overhaul Weight

Overhaul Weights

Determining the weight of the block or overhaul ball that needed to freely fall, the block requires knowing the size of wire rope, number of line parts, type of sheave bearing, length of crane boom, and drum friction (use 50 lb, unless other information is available (Table 18.2).

18.3 Reeving Tackle Blocks

The formulas for computing BS and SWL* are as follows:

$$TW = W \times (1 + F)$$

$$SHP = TW - MA$$

$$BS = SHP \times SF$$

Where W = Weight to be lifted
TW = Total weight
SHP = Strain on hauling part
BS = Breaking stress
SWL = Safe working load
MA = Mechanical advantage
SF = Safety factor
F = Friction

Size In inches	Manila		Sisal	
	SWL	BS	SWL	BS
1	200	1,000	160	800
$1^1/_2$	450	2,250	360	1,800
2	800	4,000	640	3,200
$2^1/_2$	1,250	6,250	1,100	5,500
3	1,800	9,000	1,440	7,200
$3^1/_2$	2,450	12,250	1,960	9,800

Source: U.S. Army FM 55-17, Washington, DC

TABLE 18.3 Line Strength (new line safety factor of 5)

[Table is computed in pounds for a new line. For a line that has been used, these figures will decrease. Old line may have only 60 percent of the strength, shown in pounds, for a given size of line. (Table 18.3)]

To determine the SWL for a line of known size to be rove into a tackle, use one of the following formulas as appropriate, where "C" denotes circumference and "D" denotes diameter. The formulas for manila and nylon will give the SWL in pounds. The formulas for wire rope will be in tons. (Table 18.4)

If unsure which type of wire rope is used, *always* use the formula for mild steel when figuring the SWL. This will ensure ultimate safety since the different strengths of wire rope cannot be identified visually.

There are many methods of reeving tackle blocks (Fig. 18.7). To reeve blocks in simple tackle, first lay the blocks a few feet apart, placed down with the sheaves at right angles to each other and the becket bends pointing toward each other. Start reeving by leading the

Manila line:	SWL = C2 × 200
Three strand nylon line:	SWL = C2 × 500
2-in-1 braided nylon line:	SWL = C2 × 600
Extra improved plow steel wire rope:	SWL = D2 × 10
Improved plow steel wire rope:	SWL = D2 × 8
Plow steel wire rope:	SWL = D2 × 7
Mild steel wire rope:	SWL = D2 × 6

TABLE 18.4 Formulas to Determine the SWL for a Line of known Size

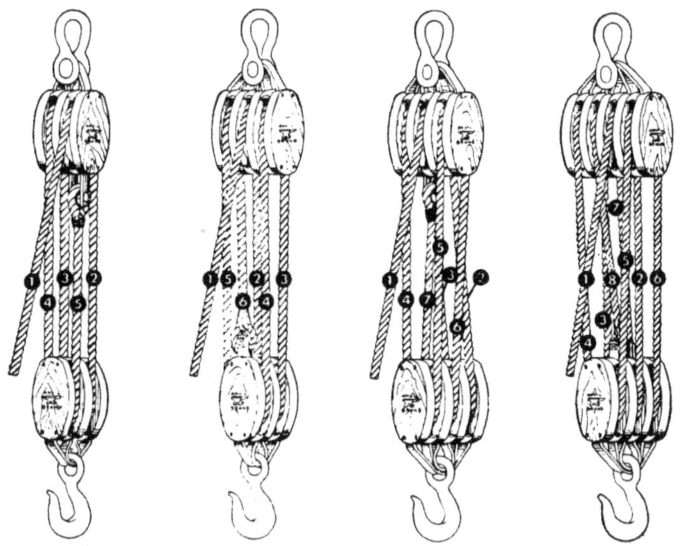

FIGURE 18.7 Various methods of reeving tackle blocks. (*Source*: The Crosby Group, Tulsa, OK.)

standing part of the falls through one sheave of the block that has the greatest number of sheaves.

When reeving a pair of tackle blocks, one of which has more than two sheaves, *make sure* the hoisting rope leads from one of the center sheaves of the upper block to prevent toppling and avoid injury to the rope. In this way, the hoisting strain is put on the center of the blocks, thus preventing them from twisting and toppling, which could damage the rope as it cuts across the edges of the block shell.

The two blocks should be placed so that the sheaves in the upper block are at right angles to those in the lower one. The upper block should be a shackle block; the lower one, a hook block. The shackle will prevent the block from becoming detached accidentally should the block or its attaching point be jarred. The lower block's hook makes it easier to attach or detach the block from the load.

Use new as well as very long tackles and reeves systems cautiously because they tend to twist—usually because of the lay of rope, rather than the method of reeving. This tends to increase the power required to hoist, and causes damage to sheave and rope. Only braided fiber rope or nonrotating wire rope will help eliminate such problems.

Do not lace the blocks in reeving systems having five or more parts of line. Lacing will cause the traveling block to tilt, with resultant excessive wear and damage to both sheaves and rope.

Balancing Reeving

Blocks having more sheaves than needed should be reeved symmetrically, with the rope distributed equally across the block so that loading on the system is balanced.

- When blocks in a system have an equal number of sheaves, *make sure* the dead end of the rope is securely fastened to the becket of the standing (upper) block.
- When the system has an unequal number of sheaves, always fasten the rope to the becket of the block having the least number of sheaves.

Begin reeving at the block fitted with the becket. Use a shackle block as the upper one of a pair and a hook block as the lower one. Next, pass the standing part around the sheaves from one block to the other, making sure that no lines are crossed until all sheaves have a line passing over them. Now secure the standing part of the falls at the becket of the block having the fewest number of sheaves, using a becket hitch for temporary securing or an eye splice for permanent securing.

When blocks have two or more sheaves, the standing part of the fall should be led through the sheave closest to the center of the block. This places the strain on the center of the block and prevents the block from toppling and the lines from being chafed and cut through by rubbing against the edges of the block.

Falls are normally reeved through 8-in. (203.2-mm) or 10-in. (254-mm) wood or metal blocks, in such a way as to have the lower block at right angles to the upper. Two 3-sheave blocks are the traditional arrangement. The hauling part must go through the middle sheave of the upper block; otherwise, the block will tilt to the side and the falls will jam under load.

Sheaves in a set of blocks revolve at different rates of speed. Those nearest the lead line revolve at the highest rate of speed and wear out more rapidly.

Effect of Friction

Sheaves in a system of blocks rotate at different rates of speed and have different loads. Thus, when raising and lowering a load, the line tension is not equal throughout the system.

When a load is at rest, the tension on the rope is equal to the load divided by the number of the parts of rope supporting the lower block (running block). When the haul rope is hoisting, however, the strain becomes somewhat greater due to sheave friction.

Each time a rope passes over a sheave friction is produced, which reduces a block's efficiency (usually expressed in terms of a percentage). On well-maintained sheaves, friction losses are assumed to be about 10 percent per sheave having plain bores, used with Manila-rope blocks; 5 percent per sheave having bronze bushings; and 3 percent per sheave having roller bearings.

Simply multiplying the number of sheaves by the appropriate percentage of friction and then adding this result to the lead-line static load produces a calculated load that is significantly lower than the actual lead-line stress. Thus, when calculating the effect of friction, it is imperative to calculate loads progressively—line by line.

It is not necessary for the rigger, however, to repeat these calculations each time the lead-line pull must be determined. Instead, using manufacturer's tables of multiplication factors, or ratios, for specific types of sheaves can help determine the lead-line pull with one simple calculation.

Number of Line Parts

To determine the number of parts of line to be used for a given load, or the line pull required for a given load, the Crosby Group has developed a set of factors and ratios applicable for blocks and independent sheave systems (Tables 18.5, 18.6, and 18.7).

1. To determine required block weight:

$$\text{weight} = [(\text{boom length} \times \text{factor A}) + \text{drum friction}] \times \text{factor B}$$

2. To determine number of parts of line:

   Calculate ratio A or B

$$\text{ratio A or B} = \frac{\text{total load to be lifted}}{\text{single line pull (lb)}}$$

Wire rope Size (in.)	Wire rope weight (lb/ft – 6 × 19 IWRC)
3/8	26
7/16	35
1/2	46
9/16	59
5/8	72
3/4	1.04
7/8	1.42
1	1.85
1-1/8	2.34
1-1/4	2.89

Source: The Crosby Group, Tulsa, OK.

TABLE 18.5 Wire Rope Weight—Factor A

Number of line parts	Roller bearing sheaves	Bronze bushed sheaves
1	1.03	1.05
2	2.07	2.15
3	3.15	3.28
4	4.25	4.48
5	5.38	5.72
6	6.54	7.03
7	7.73	8.39
8	8.94	9.80
9	10.20	11.30
10	11.50	12.80

Source: The Crosby Group, Tulsa, OK

Table 18.6 Overhaul Factors—Factor B

Select ratio that line is reeved through, to determine number of parts of line.

Or:

3. To determine the number of parts of line to be used for a given load when the weight of load and single line pull are known:

$$\text{Number of parts of line} = \frac{\text{load to be lifted}}{\text{lead-line pull}} \times F$$

4. To determine the lead-line pull when the load to be lifted and the number of parts of line are known:

$$\text{Lead-line pull} = \frac{\text{load to be lifted}}{\text{parts of line}} \times F$$

or

$$\text{Lead-line pull} = \frac{\text{load to be lifted}}{R}$$

where F = multiplication factor
R = ratio of line parts divided by F

5. To determine the maximum load that can be lifted with a particular reeving arrangement:

$$\text{Total to be lifted} = \text{lead-line pull} \times R$$

Ratio A bronze bushed sheaves	Ratio B antifriction bearing sheaves	Number of line parts
0.96	0.96	1
1.87	1.98	2
2.75	2.88	3
3.59	3.81	4
4.39	4.71	5
5.16	5.60	6
5.90	6.47	7
6.60	7.32	8
7.27	8.16	9
7.91	8.98	10
8.52	9.79	11
9.11	10.60	12
9.68	11.40	13
10.20	12.10	14
10.70	12.90	15
11.20	13.60	16
11.70	14.30	17
12.20	15.00	18
12.60	15.70	19
13.00	16.40	20

Source: The Crosby Group, Tulsa, OK.

Table 18.7 Ratios R to Determine Number of Line Parts or Line Pull for a Given Load

Note: *Increasing the number of sheaves and line parts increases the mechanical advantage of reeved system. Because friction losses will also be greater, the efficiency of the system will be lower. On the other hand, by reducing the number of sheaves and parts of line, friction losses can be reduced—and the system's efficiency is increased. The mechanical advantage, however, will be lower.*

BEWARE—sometimes increasing the number of parts on blocks does not necessarily mean an improved lifting capacity, because there is a limit beyond which the effort required to overcome friction becomes greater than that necessary to lift the load.

For blocks having plain-bore sheaves, reeving beyond 6 or 7 parts of line provides little advantage; and reeving beyond 10 or 11 parts actually decreases the mechanical advantage due to the cumulative effect of sheave friction.

Likewise, with blocks having bronze bushings, reeving beyond 9 or 10 parts has no particular advantage; and sheaves with roller bearings reach their practical advantage with 15 or 16 parts of line.

In addition, increased load-carrying ability means a slower travel speed. Thus, to maintain a mechanical advantage of 5, a system's lead-line must be able to move five times faster than the load.

IMPORTANT: *The size and capacity of blocks and reeved systems must be sufficient to carry the loads to which they will be subjected.*

A block hook will usually start bending when the block is loaded to 70 percent of its maximum strength; thus, any such hooks showing signs of opening up are probably being overloaded. To avoid overloading, it is necessary to know exactly how a reeved system works and to understand how friction affects a system's capacity.

18.4 Safety Precautions

For maximum safety and efficiency, tackle block systems must be properly designed, used, and maintained. Only a qualified person should rig a tackle block system—as defined by ANSI/ASME B.30—incorporating good rigging practices; and only a qualified person should recommend changes from full sheave reeving arrangement.

The following safety precautions apply to the operation of both lifting and running tackles:

- Read, understand, and follow manufacturer's instructions to select, use, and maintain tackle block systems; but *remember*—ratings shown in manufacturers' literature are applicable only to new or "in as new" components.
- Instruct workers to keep hands and body away from blocks, sheaves, swivels, and pinch points where wire rope touches block parts or loads.
- Instruct workers to be alert and to wear proper safety gear in areas where loads are moved or supported with tackle block systems.
- Take into account shock loading and extraordinary conditions when selecting components for use in tackle block systems.
- Tackle blocks and sheaves are intended for tension or pull. *Do not* side load assemblies because it exerts additional force or loading, which the system is not designed to accommodate.
- Always make sure the hook supports the load; the latch must NEVER support the load.

- Welding of load supporting parts or systems can be hazardous; only a qualified person will have the knowledge of materials, heat treatment, and welding procedures that are absolutely necessary for proper welding. (Consult the manufacturer for information.)
- When overhauling a tackle, a turn should be taken with the running end round a belaying picket to prevent the weight on the tackle taking charge. Similarly, when rounding in, one person should be detailed to take in the spare rope and to keep a turn round the belaying picket.
- When the fall is taut, it should not be jarred by being struck or being trodden on.
- The returns should not be handled when moving unless absolutely necessary.
- Suspended weights should be lowered uniformly and not with a jerking motion.
- When working near the safe limit of a tackle, and before leaving it with the strain on, the tackle should be eased off slightly to distribute stress equally between the returns.

Working load limit ratings indicate the greatest force or load a component can carry under usual environmental conditions and are based on all sheaves of the tackle block system being used. If all sheaves are not used, balance must be maintained and the WLL must be reduced proportionally to prevent overloading sheave components.

Safe loads for common tackle configurations, using both cordage and SWR rope, are tabulated in reference tables and charts.

18.5 Care and Maintenance of Tackles

The best wire rope cannot give the service expected of it unless the equipment on which it is used is properly designed and maintained. This is especially true of blocks and sheaves. For maximum safety and efficiency, tackle block and sheave systems must be properly designed, used, and maintained.

If tackles are to work efficiently, all friction losses must be reduced to a minimum. This can be achieved by observing the following rules.

- Blocks must be well lubricated and free from grit and dirt. If blocks complain, that is, make a noise, they are not in good working order.
- Fall must be free from kinks and twists, and must run freely over the sheaves.
- Tackles must be carried, not dragged along the ground.

- Tackles must be prevented from twisting. The power required is increased by over 40 percent if a tackle is allowed to twist one complete turn.
- The correct size blocks for the rope must be used.
- Keep all sheaves well lubricated when in operation to reduce friction and wear.

18.6 Inspection of Blocks

When inspecting blocks, *make sure* to check:

- Beckets, end connections, shave bearings, and center pins of blocks for excessive wear. Replace any defective parts.
- For elongated links, eyes, or shackles; bent shackles, links or center pins; and enlarged hook throats. Such defects are a sure indication of overloading. Replace the block.
- Sheaves to be sure that they are lubricated and that they rotate properly and freely.
- Sheave grooves for smoothness; excessive wire rope wear will occur if a wire sheave shows the imprint of the rope.
- Clearance between the sheaves and cheek and partition plates to ensure that the wire rope will not slip between them while in use.

18.7 Reeving Glossary

Fixed block The block that is fastened to a nonmovable object; it remains stationary while the tackle is in use.

Moving block The block to which the weight is attached; it moves when the tackle is in use.

Leading block A block used to change the direction of the running end of a tackle in order that the power may be more conveniently applied. Snatch blocks are normally used.

Return Length of rope running between the blocks. Collectively, all returns are known as the fall.

Standing end The end of the fall that is anchored to one of the blocks in the tackle system.

Running end The loose end of the fall that is pulled in order to lift the load.

Reeving The action of threading or passing the fall around the sheaves of the blocks when making up the tackle.

Rounding in The action of bringing together the two blocks in a tackle system. When the two blocks in a tackle system are rounded in to their maximum, they are termed chock-a-block.

CHAPTER 19

Winches and Drums

Winches are designed as pulling devices, not for lifting loads—hoists are lifting devices. Winches, however, are often used for lifting, usually because the cost per pound of winches tends to be a lot less than an equivalent hoist.

A hoist or winch is a winding drum mounted on a frame with a power plant and the necessary controls. A disconnect clutch between the engine and gear train, with a friction clutch mounted on the drum, controls the drum. For additional safety, a ratchet-and-pawl system at the drum provides a means of positively locking the drum against inadvertent spooling out of rope.

19.1 Winches

Winches are rated in line pull (usually measured in pounds or kilograms). To determine the proper size winch, consider the length of wire rope required, as the rated line pull is quoted with the first layer of wire rope on the winch drum. As wire is winched in, the winch becomes less powerful with each layer of wire rope on the winch drum. And, of course, winch cables never spool onto the winch drum evenly, resulting in cable bunching, which further compounds the problem of layer buildup and, in turn, a loss of pulling power.

Remember: *Deduct 10 percent of the rated line pull for each layer of wire rope on the drum. A winch with a line pull of 6000 lb will have a line pull of 5400 lb on the second layer, 4860 lb on the third layer, and 4374 lb on the fourth layer, etc. Thus, to pull 6000 lb for any distance will require a winch with a greater initial rated line pull.*

The line-pull capacity of a winch, however, can be effectively doubled by using a snatch block/pulley block. But, while the line pull of the winch doubles, the winching speed will be halved, due to the gearing ratio. Usually this is not important and is often safer if the winch is pulling an abnormal load. For every pulley used, however, there is a frictional loss of approximately 10 percent per pulley.

Calculating the size of winch required to pull a load requires knowing:

- *Weight of load.* Rolling or dead weight.
- *Support.* On wheels or not on wheels.
- *Surface conditions.* Sand, hard-core, etc.
- *Travel gradient and distance.*
- *Required length of wire rope.* Rated line pull is quoted with the first layer of wire rope on the winch drum. *Remember*—the winch becomes less powerful with each layer of wire rope on the winch drum.
- *Loss of winch power.* Cable bunching compounds the problem of layer buildup and in turn a loss of pulling power. As a guide for each layer of wire rope on the drum, deduct 10 percent off the rated line pull.

Depending on the lifting configuration, it is a good practice, and ***recommended,*** that when lifting use two pulley blocks, one at the top of the lifting tower and one at the base. The winch wire rope should be routed from the winch through the pulley at the base of the tower, up to the upper pulley and then down to the load. *Remember* that each pulley/snatch block will reduce the available power by approximately 10 percent per pulley due to friction losses, these figures must be taken into account when calculating the size/line pull of winch used.

NOTE: *If a roller fairlead is fitted to the winch and the wire rope is routed directly from the winch to a single pulley block mounted at the top of the tower and then down to the load, the wire rope will be bent around the roller fairlead. This small radius will eventually cause damage to the wire rope and is* **NOT** *recommended.*

Winch Reevings

Work-rated winches can be reeved in various ways to gain desired mechanical advantages. Winch reeving affects capacity, lifting distance, lifting speed, and cable size. Winches are available in single and double line arrangements (Fig. 19.1).

Work-rated winches are designed with a standard range or ranges of lead off angles. The lead off angle is a critical specification in winch applications.

Single and Double Line Pull

A single line pull system uses a pulley block with pulling power being the winch pulling rating.

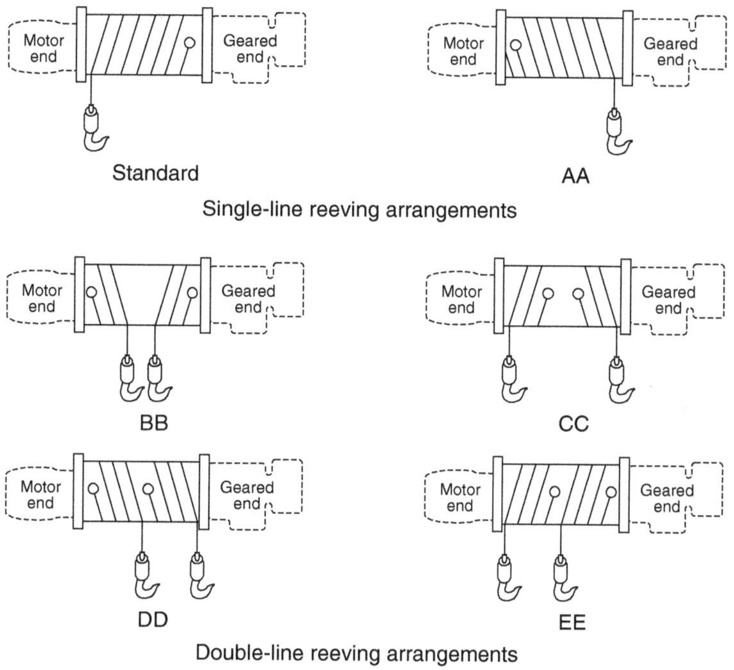

FIGURE 19.1 Winch drum reeving arrangements. (*Source*: Canco Cranes and Equipment, North Vancouver, BC, Canada.)

With a double line pull, a snatch block is attached to the load, with the wire rope from the winch drum wound out and around the snatch block pulley, then back and attached to a place adjacent to the winch. Where there are obstacles in the path of travel, indirect pulling will be required, with the pulley block attached to the load and the wire rope attached to an anchor point offset from the direction in which the load is to be moved (Fig. 19.2).

NOTE: *This procedure is* not recommended *unless absolutely necessary, because the winch pulling power and the line speed decrease as the angle between the wire rope increases.*

Lifting with a Winch

To calculate the size of winch required to pull a load requires knowing the load's weight (rolling or dead weight) and type of surface over which the load is being winched.

Four criteria must be considered before using a winch for lifting:

- *X safety ratio.* Usually a minimum of 5:1 must be applied to ensure that the winch will lift its rated load safely (rated on

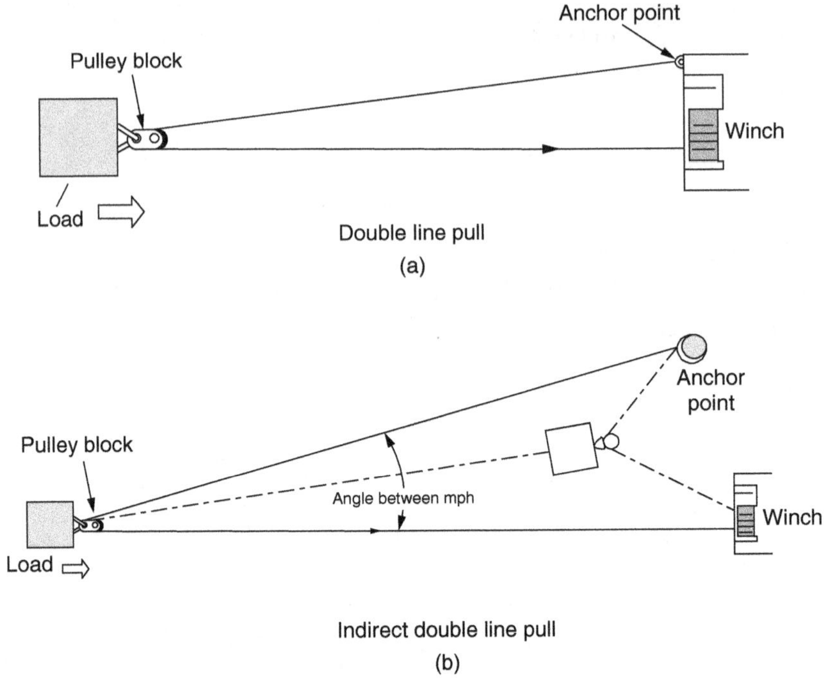

FIGURE 19.2 Winch line pull. (a) Double line; (b) indirect double line. (*Source*: U.S. Army FM 55-501, Department of Army, Washington, DC.)

the first layer of wire rope on the drum). Check and confirm the amount of safety ratio according to regulations governing the use being adopted.

- *No freespool clutch mechanism can be on the winch.* Usually as standard that must be totally disabled or removed.
- *Secondary brake must be fitted.* Most winches rely on the gearing alone to act as the brake; a secondary brake—drum brake—supporting the gear brake effectively provides a twin braking system.
- *Safety holding chain or device should be fitted if the load is to be suspended.* Once a load has been raised and then left suspended, a suitably-rated chain fitted to secure the load will act as a safety device. *Never* allow the load to hang from the winch in free air.

Before using a winch for lifting purposes, *make sure* it complies with all construction safety methods and regulations as well as your insurer's requirements.

Never—under any circumstances—use a winch for lifting or transporting personnel.

19.2 Drums

When wire rope is run off one reel onto another or onto a winch or drum, it should be run from *top to top*, or from *bottom to bottom*, as shown (Fig. 19.3). Drums should be designed to conform to the rules for sheaves; however, a drum can be made slightly smaller than the minimum recommended size for sheaves because a rope is flexed only once by the drum, whereas it is flexed twice (bending and straightening) each time it passes over a sheave.

A drum should be designed, if possible, to handle all the rope in one smooth, even layer. Two and sometimes three layers are permitted, but more than three layers may cause crushing of the rope on the bottom layer as well as the end of any layer where pinching occurs.

The proper way to wind a rope on a drum depends on the lay of the rope. Rope should wind in a helix, that is opposite to its lay (Fig. 19.4).

- *Top winding*. Since in most installations the rope winds into the top of the drum, extend a clenched fist downward—right fist for right lay, left fist for left lay. Extending the index finger will point in the direction of the lead line; the thumb, to the drum anchorage point.

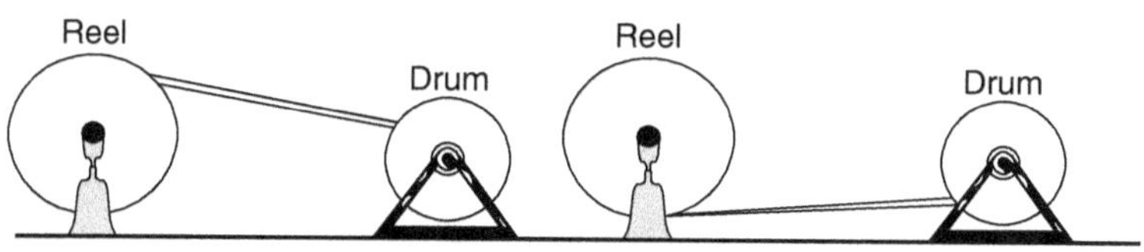

FIGURE 19.3 Correctly transferring wire rope from reel to drum. (*Source*: *U.S. Army Field Manual 55-17*, Army Department, Washington, DC.)

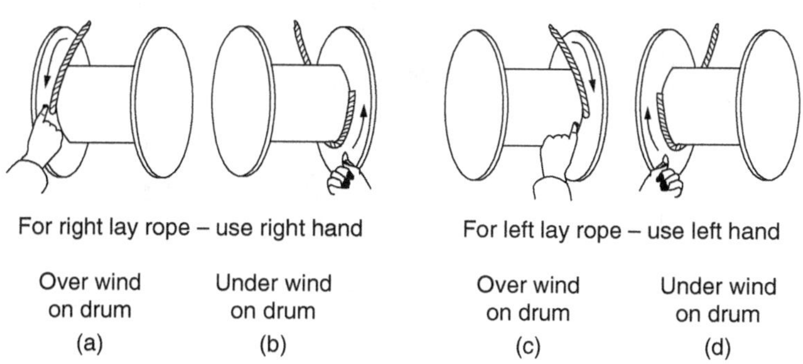

FIGURE 19.4 Hand-rule for determining proper method of winding wire rope on a drum—top winding and bottom winding. (*Source*: *U.S. Army Field Manual 55-17*, Washington, DC.)

- *Bottom winding*. Where the rope winds onto the bottom of the drum, simply turn the clenched fist over palm upward and extend the index finger and thumb to determine the direction of the lead line and location of the anchorage point, respectively.

The drum anchorage should be positioned to favor right-lay rope, since left-lay rope is not always available from stock. The manner in which a rope spools on the drum also depends on:

- Drum size, ratio of drum diameter to rope diameter, and rope speed
- Amount of rope to be spooled, type of hoisting equipment, and load on rope

Make sure that the wire rope spools evenly on the drum, either by using a spooling device or by keeping the fleet angle to within the correct limits, maintaining tension on the wire rope at all times—approximately 10 percent of the working load (Fig 19.5).

The drum capacity factor *F* applies to nominal wire rope size and level winding. Since new wire ropes are usually oversized by 1/32 in. per inch of rope diameter, the results obtained by the formula must be decreased to account for oversized new wire rope, or random or uneven winding, or both, accordingly (Fig. 19.6).

To determine the approximate capacity of a given drum or reel for a particular wire rope diameter:

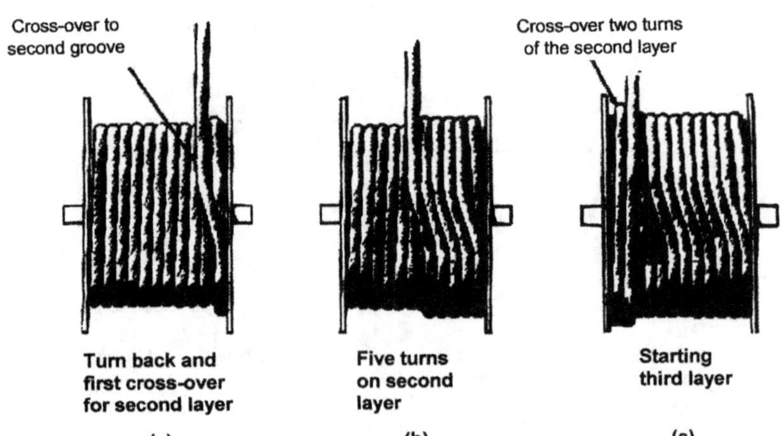

FIGURE 19.5 Winding rope layers on a drum. (*Source*: *U.S. Army Field Manual 5-125*, Department of Army, Washington, DC.)

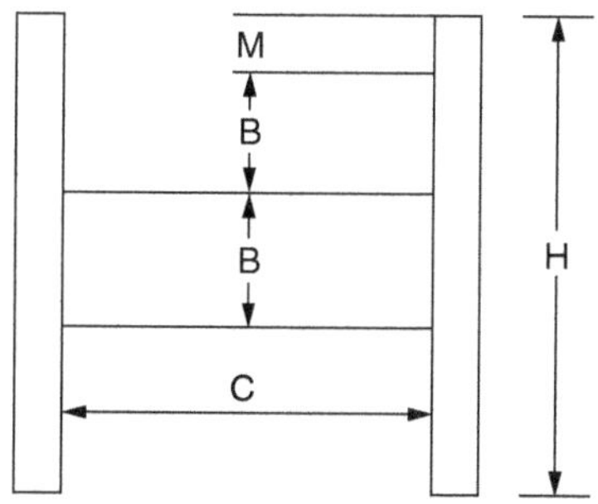

FIGURE 19.6 Computing rope capacity (L) in feet for any size drum (*Source*: Wire Rope Industries, Point Claire, Quebec, Canada.)

$$L = (A + B) \times A \times C \times F$$

where L = drum capacity (rope length in ft)
A = depth of rope layer (in.) = ½ $(H - B) - M$
B = diameter of drum (in.)
C = width between reel flanges (in.)
M = desired clearance = min. 1 dia. rope (D)
D = nominal rope diameter (in.)
$K = \dfrac{0.2618}{D_2}$
H = height of reel flange

Depth of rope layer:

$$A = ½\,(H - B) - M$$

where H = diameter of the flange (in.)
M = desired clearance (in.)

For *oversized wire ropes*, decrease calculated length from 0 percent to 6 percent; for *random wound wire ropes*, decrease calculated length from 0 percent to 8 percent.

Fleet Angle

Where a wire rope passes over a sheave and onto a drum, the rope will remain in alignment with the sheave, but will deviate to either side of the drum, depending on the drum's width and distance from the first fixed sheave. This angle between the centerline through the sheave and the centerline of the wire rope leading to the drum is called the *fleet angle* (Fig. 19.7).

Because drums are often offset in some installations, the fleet angle on one side of the drum could be larger than that on the other side. Always consider the larger side of the two angles in making calculations.

To avoid excessive wear on the sheave and to prevent excessive chaffing of the oncoming wire rope against previous wraps on the drum, it is desirable to keep the fleet angle as small as possible. If the fleet

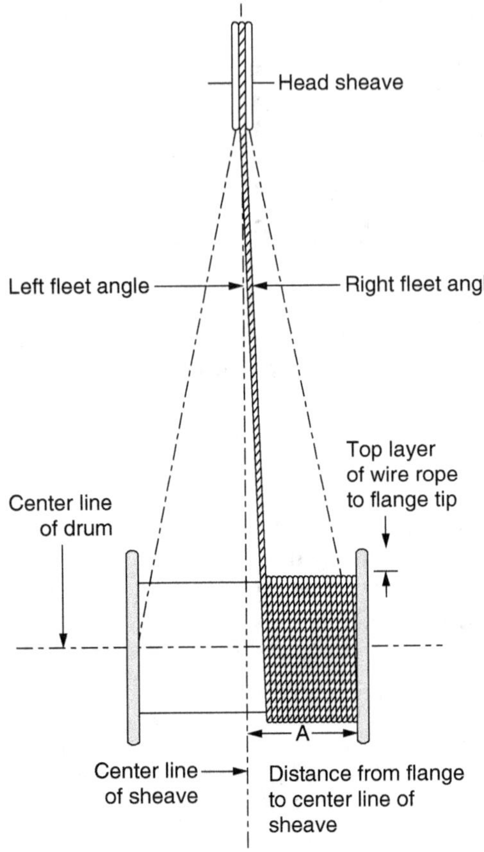

FIGURE 19.7 Fleet angle related to drum's sides. (*Source*: *U.S. Army Field Manual 5-125*, Department of Army, Washington, DC.)

angle is too small, it will cause the wire rope to pile up against the flange head, damaging both rope and equipment.

Too small an angle will also produce considerable vibrations with subsequent deterioration of the wire rope. Intermediate idler sheaves can help eliminate this particular problem.

For average conditions, good practice would be to keep the fleet angle within 2° for a grooved drum, and within 1½° for a drum having a smooth face. This is equivalent to a distance of 29 ft (9.84 m) and 38 ft (11.58 m), respectively, for each foot of drum width on either side of the centerline of the sheave.

Flat-Faced Drum

When using a flat-faced drum, it is important that the wire rope be wound on the drum in a straight helix at its proper angle. Use a steel starting piece, cut in a tapered shape that fills the space between the first turn and the flange, to guide the rope into its correct position.

Make sure that all turns of the first layer of new wire rope on a drum are wound taut and correctly. If the siding is loose or nonuniform, serious damage will result to multiple windings. Tap each turn of the wire rope against the preceding one with a wooden mallet. Use care—*do not* jam the strands together.

Each succeeding layer of wire rope should lie in the groove formed by the preceding layer. *Never* permit wire rope to cross wind on a drum.

Always fit the correct length of wire rope to a drum, for if the rope is too short, it will pay out and place the full load on the rope anchorage. Conversely, if the rope is too long, it might exceed the drum's spooling capacity, spilling over the flanges, and be damaged, crushed, or fail.

There should always be at least two or three full turns of wire rope left on a drum when the hook is at its highest point. Flanges on grooved drums should project either twice the wire rope diameter or 2 in. (50.8 mm) beyond the last layer of rope, whichever is greater. Flanges on ungrooved drums should project either twice the wire rope diameter or 2½ in. (62.5 mm) beyond the last layer of rope, whichever is greater.

Kinking

Kinking of wire rope can be avoided if ropes are properly handled and installed. Kinking is caused by the rope taking a spiral shape as the result of unnatural twist in the rope. One of the most common causes for this twist is improper unreeling and uncoiling. Even though the kink may be straightened so that the damage appears slight, the relative adjustment between the strands has been disturbed so that the rope cannot give maximum service.

Great care *must* be taken when removing wire rope from reels or coils. Looping the rope over the flange of the reel or pulling the rope off a coil while it is lying on the ground will create loops in the line; if these loops are pulled tight, kinks will result, permanently damaging the wire rope.

Unreeling and Uncoiling Wire

To unreel wire rope, the reel should be revolved and the rope taken off the same way it is put on the reel. One method is to put a shaft through the center of the reel and jack it up so that the reel will revolve freely. Pull the wire rope straight ahead, keeping it taut to prevent it from loosening up on the reel.

To uncoil wire rope, remove the ties and roll the coil along the ground so that the rope lies straight behind. With this uncoiling method, there will be no twist or kink in the wire rope. The most common and easiest method of uncoiling wire rope is to hold one end of the rope while the coil is rolled along the ground (Fig. 19.8).

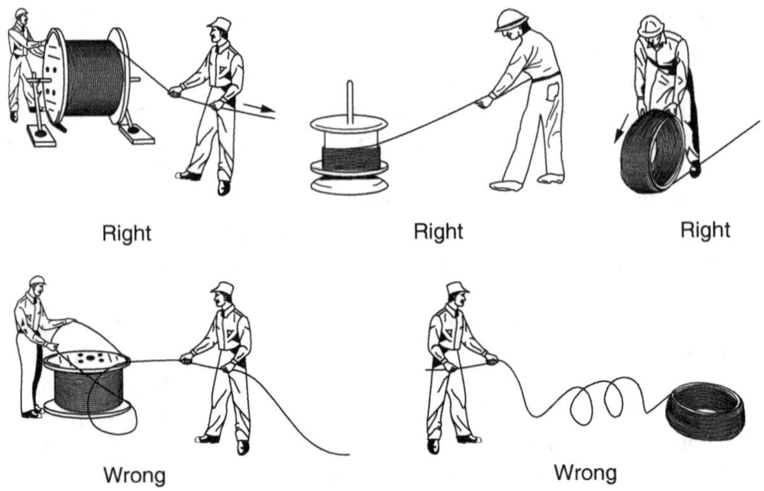

FIGURE 19.8 Right and wrong ways of unreeling and uncoiling wire rope. (*Source*: Crown Wire & Cable Co., Millville, NJ.)

NOTE: *Crown Wire and Cable engineers remind users to exercise great care when removing wire rope from reels or coils.* Do not *drop reels or coils, as this can damage the wire rope and collapse the reel, making the removal of the wire rope extremely difficult, they say—if not impossible.*

Regulations and Standards

OSHA Regulations

1910.184 and 1926.231 *Slings: Steel Chain, Wire Rope, Metal Mesh*

1926 Subpart N—*Cranes, Hoists, Elevators, and Conveyors*

1926.553 *Base-Mounted Drum Hoists*

U.S. Consensus Standards

ANSI/ASME

B30.7 – 2006 *Base-Mounted Drum Hoists*

Industry Consensus Standards and Guidelsines

Cordage Institute
Web Sling and Tiedown Association
American Wire Cloth Institute
Associated Wire Rope Fabricators
Wire Rope Technical Board
Woven Wire Products Association

SECTION 6

Rigging Tools

CHAPTER 20

Hand Tools and Portable Power Tools

Virtually every type of tool can be a hand tool, although many have also been adapted as power tools—powered by an electric motor, a compressed air motor, or a gasoline engine—and are classified as either stationary or portable (hand-held).

Portable electric tools may be either corded or battery powered. Compressed air is the customary power source for nailers and paint sprayers. Powder-actuated tools are powered by explosive cartridges. Gasoline-powered tools include chain saws made for outdoor use.

Although hand tools and power tools help to easily perform tasks that otherwise would be difficult or impossible, these simple tools can be hazardous and have the potential for causing severe injuries when used or maintained improperly. Special attention toward hand tool and power tool safety, therefore, is necessary in order to reduce or eliminate these hazards.

20.1 Safe Use of Hand-Held Tools

Among many unsafe practices are the striking of hardened faces of hand tools together (such as using a carpenter's hammer to strike another hammer, hatchet, or metal chisel), using a file for a pry, a wrench for a hammer, or using pliers instead of the proper wrench. Workers using hand tools in this manner are likely to be exposed to the hazards of flying debris, harmful dusts, or noise.

- *Always* use the correct tool for the specific job.
- Keep bystanders at a safe distance from workers using hand or power tools.

Personal Protective Equipment

Always provide workers with personal protective equipment and *make sure* that workers wear appropriate protective equipment—such as glasses, goggles, dust masks, face shields, hearing protection, etc. —to protect themselves from hazards.

Prohibit workers from wearing clothing with strings or loose ends to catch on things.

Keep Tools in Good Condition

Tools in poor condition include wrenches with cracked work jaws, screw drivers with broken points or broken handles, hammers with loose heads, dull saws, and extension cords or electric tools with broken plugs, improper or removed grounding lugs, or split insulation. Tighten handles, correct any defects, sharpen cutting tools, and eliminate *mushroom heads* on wedges and punches. Take out of service immediately any tools that have deteriorated and are irreparable.

Use Tools the Correct Way

Make sure that the force applied to a hand tool is directed away from the body, and always follow the manufacturer's instructions. Don't apply force to a screwdriver, for example, when the object is held in hand; and cut away from the body when using a knife or other sharp tool.

Use and maintain power tools according to the manufacturer's instructions. *Make sure* that power tools are properly grounded, or are double insulated. *Never* cut the three-pronged plug off, or use a two prong adapter.

Keep all guards and shields in place. Switch off and unplug power tools before changing blades or servicing and repairing. Unplug and put tools away after use. Consider locking out power tools to prevent other from using them without permission—especially young children.

Keep Tools in a Safe Place

Do not carry sharp tools in pockets or leave sharp tools in toolboxes with the cutting edges exposed.

20.2 Accident Control

- *Make sure* that employees are trained to select the right tools for each job.
- *Establish* daily tool inspection procedures and provide good repair facilities to ensure that tools will be maintained in a safe condition.
- *Establish* a procedure for control of tools such as a checkout system at tool cribs.

- *Provide* proper storage facilities in the tool room and on the job.
- *Make complete checks* of operations to determine the need for special tools that will do the work more safely than ordinary tools.
- *Make sure* all dull or damaged tools are *not* returned to stock.
- *Never* carry tools that in any way might interfere with using both hands freely on a ladder or while climbing on a structure.
- *Always* use a strong bag, bucket, or similar container to hoist tools from the ground to the job.
- *Always* return tools in the same manner.
- *Do not* bring tools down by hand, carried in pockets, or dropped to the ground.
- *Do not* leave tools where employees are moving or walking; mislaid and loose tools cause a substantial portion of hand tool injuries.
- *Never* carry chisels, screwdrivers, and pointed tools in a pocket.
- *Always* carry tools in a tool box or cart, in a carrying belt (sharp or pointed end down) like those used by electricians and steel workers, in a pocket tool pouch, or in the hand with points and cutting edges pointing away from the body.
- *Pay close attention* when carrying tools on shoulders, especially to clearances when turning around, and handle the tools so that they will not strike other employees.

20.3 Care and Maintenance of Tools

Proper maintenance and repair of hand and power tools is the responsibility of the tool crib attendant in charge of storeroom for portable tools and related supplies. Hand and portable tools *must* be inspected before each use, oiled prior to each use (or daily), and cleaned after each use. Any tool that is not operating correctly or shows signs of defects *must* be removed from service immediately, tagged, and turned in to the tool crib for repair or replacement.

Line supervisors *must* ensure a complete tool inspection is completed at least every six months for tools within their purview. Tool inspections include (but are not limited to):

- Housekeeping in tool supply room
- Tool maintenance and repair
- Removal from service of any tool(s) requiring maintenance until maintenance is complete; or broken or malfunctioning tools are repaired

Hand-Held Tools

To avoid accidents at the workplace resulting from the incorrect use of hand tools, it is important that workers understand the proper use of these items.

The employer *must* discuss with workers concerns about the dangers involved with using hand tools. During the session, each worker should demonstrate the correct methods for operating hand tools.

Prepare exercises in advance that will allow each employee to use the hand tools; and demonstrate the proper care and storage of the tools. Major hand-held tools common to most construction worksites include:

- *Wrenches*. When placing an adjustable wrench on a nut, make sure the adjustable jaw faces the operator; then pull the wrench toward the operator. Use socket wrenches for hard-to-reach places.

  Never use a pipe wrench on nuts because the corners of the nuts or bolts are likely to break the teeth of the wrench jaws, making it unsafe for future use. Manufacturers make wrenches of different sizes; so the amount of leverage obtained with the wrench handle is the maximum application.

  Never add more leverage to any tool by use of an extension; it is unsafe to add more leverage with a length of pipe.

- *Hammers*. When replacing hammer handles, make sure they fit the hammer head. Wedge the handle securely in the head and make sure that it is free of splinters and cracks.

  Never strike hardened steel surfaces with a steel hammer. Use a soft metal hammer or one with a plastic, wood, or rawhide head when striking steel surfaces.

  Always wear safety glasses to protect your eyes from flying objects. Inspect sledge hammers carefully before each use. Use the right type of hammer for the specific job.

  Hand-held portable circular cutoff saws are widely used for cutting concrete, masonry products, sheet metal products (both steel and aluminum), and light steel sections such as angles and channels.

- *Quick-cut saws. Must* be serviced and maintained in accordance with the manufacturer's instructions. Guards and air intakes should be cleaned regularly and often. Excessive blade vibration should be corrected before trying to make a cut.

- *Pliers. Never* substitute pliers for another tool, such as a wrench, to complete the task. It may cause the bolt heads to become chewed. Pliers cannot grip nuts and bolts securely and will slip.

Use *hand insulated grips* when working with electricity. *Make sure* the protective coverings are free from cracks or holes.

Use a *vise* when cutting wire with the pliers. Hold the open end of the wire with your free hand to prevent the cutoff piece from flying through the air. If a vise is not available, use your foot to secure the wire and always use safety glasses.

- *Files. Must* be cleaned with a file card when finished using. *Do not* strike the file against another piece of metal.
- *Scrappers. Must* be kept in good, sharp condition for best results.
- *Crowbar.* Should be used only for jobs that require prying.
- *Hand hooks. Must* be kept sharp to prevent slipping when in use; and they *must* be stored with the point in cork to reduce accidents.

REMEMBER:

Always *use the correct tool for the intended job.*
Keep tools in good condition.
Keep cutting tools sharp.

Check for unsafe tools, including wrenches with cracked or worn jaws; screwdrivers with broken or worn tips, split or broken handles; hammers with chipped, mushroomed, or loose heads, broken or split handles; mushroomed heads on chisels, wedges, and punches; and dull saws—among other flaws.

Portable Power Tools (Electric and Pneumatic)

The misuse of common hand tools is a major source of injury to construction and maintenance workers because it is assumed that *everyone knows* how to use common hand tools.

Observation and the record of injuries, however, show that this is not the case. Before handling portable power tools, employees *must always*

- Be familiar with the requirements outlined in *OSHA 1910. 241-244*
- Use, store, and handle portable power tools consistent with those requirements
- Read, know, understand, and comply with all manufacturer safety practices prior to using equipment.

- Wear all required personal protective equipment at all times when using equipment.
- Make sure power tools are properly grounded or are double insulated, and never cut the three-prong plug off or use a two-prong adapter.
- Switch off and unplug power tools before changing blades or servicing and repairing.

Only persons qualified by training and experience should be in charge of the tool room and pass judgment on the condition of tools for further use; otherwise tools should be sent out for repairs.

20.4 Power Tool Safety

Before each use and during power-tool inspections:

- Examine for damage.
- Check plugs for damage and missing ground terminals.
- Replace damaged or inadequate plugs.
- Check switches with the tool unplugged; switches should be free of excessive drag, binding, loose mountings, and obvious defects.
- Check plugs, collets, etc. for proper operating condition and ensure that the correct keys and wrenches are available.
- Visually check housing for defects and damage.
- Guards must be used, installed properly, and function smoothly.
- Replace missing hardware and tighten any loose bolts, etc.
- Adjust according to the manufacturer's recommendations.
- Resharpen blades, bits, etc. and reinstall according to manufacturer's recommendations.

20.5 Inspection of Tools

All tools should be inspected prior to each use by the operator. Inspections should include but are not limited to the following items.

Hand Tools and Equipment

- Take out of service all hand tools—such as chisel, punches, etc.—that develop "mushroomed" heads and recondition them before using.

- Replace prior to use:
  Handles on hammers, axes, and similar equipment that are cracked or fractured; *make sure* that heads are properly and securely attached.
  Wrenches whose handles are bent or whose gripping surfaces are worn
  Screwdrivers that are bent or whose ends are chipped
- Store tools in a dry, secure location where they won't be tampered with, and in such a way that sharp edges do not present a danger when reaching into tool cribs and storage areas.
- Maintain tool cutting edges sharp so that the tool will move smoothly and will not bind.
- Make sure all handles are free of burs and splinters and are firmly attached to the working head of the tool.

Portable Power Operated Tools and Equipment

- Make sure all grinders, saws, and similar equipment are fitted with appropriate machine guarding as specified by the manufacturer.
- Properly guard all rotating or moving parts of equipment to prevent physical contact by the operator.
- Make sure all cord-connected and electrically operated tools and equipment are effectively grounded by either a grounding prong or an approved double-insulated case.
- Check all prongs to assure that they are not bent or otherwise damaged.
- Check all cases to assure that they are not cracked or damaged.
- Make sure all electric cords are in good condition free of frays or other physical defects.
- Make sure pneumatic hoses are free of damage or deterioration.

Abrasive Wheel Equipment

Make sure that:

- The work rest is not more than 1/8 in. from the surface of the wheel.

- The adjustable tongue on the top side of the grinder is within 1/4 in. of the wheel.
- The grinder is mounted in such a way that it is secure and will not shift or tip.
- The On-Off control switches are clearly marked in red and readily accessible to the operator to easily deactivate equipment in case of emergency.
- The maximum RPM rating of the grinder is clearly posted and the maximum rating of the wheel does not exceed grinder rating.
- Grinding wheels are not cracked or otherwise damaged.
- Grinders that use a coolant are equipped with splash guards to prevent coolant from coming into contact with operator.

Powder-Actuated Tools

Make sure that:

- Powder-actuated tools are stored in their own locked container when not being used.
- All powder-actuated tools are left unloaded until they are actually used.
- Only trained and authorized employees will use powder-actuated tools.

Machine Guarding

- Keep machine guards clean, secured, and arranged so that they do not offer a hazard in their use.
- Properly guard all moving chains, gears, pulleys, etc.
- Make sure all emergency "STOP" buttons are colored red and easily accessible to the operator in an emergency.
- Properly ground all noncurrent-carrying metal parts of electric equipment will be properly grounded.
- Maintain sufficient clearance around equipment to assure for safe operation, maintenance, and waste removal.

20.6 Portable Power Tools Safety Inspection Checklist

Hand and portable power tools				
1.	Are all portable hand or power tools maintained in a safe condition?	Y	N	N/A
2.	If compressed air is used for cleaning purposes, is it used at pressures less than 30 psi and only with effective chip guarding and personal protective equipment?	Y	N	N/A
3.	Are power tools equipped and used with guards whenever possible?	Y	N	N/A
4.	Are all belts, gears, shafts, pulleys, sprockets, spindles, drums, flywheels, chains, or other reciprocating, rotating or moving parts of equipment guarded if operator is exposed to contact or if they otherwise create a hazard?	Y	N	N/A
5.	Is all necessary personal protective equipment provided whenever the use of hand and power tools could create falling, flying or splashing debris, or harmful dusts, fumes, mists, vapors, or gases?	Y	N	N/A
6.	If tools and equipment are brought in from home, are they subject to the same safety requirements as supplied tools and equipment?	Y	N	N/A
7.	Are all chain saws, percussion tools, and hand-held powered circular saws (with blade diameter greater than 2 in.) equipped with a constant pressure switch that shuts off power when released?	Y	N	N/A
8.	Are all hand-held powered drills; tappers; fastener drivers; horizontal, vertical, and angle grinders (with wheel diameter greater than 2 in.); disc sanders (with disc diameter greater than 2 in.); belt sanders; reciprocating saws; saber, scroll, jig saws (with blade shanks greater than a nominal 1/4 in.); and other similarly power tools equipped with a constant pressure switch or control?	Y	N	N/A
Note: They may be equipped with a lock-on control provided the turnoff can be accomplished by a single motion (by the same finger or fingers) that turns it on. The Construction standard requires a "momentary contact on-off control" instead of a constant pressure switch or control. This means that if the switch is pressed, the tool turns on and if the switched is pressed again, the tool turns off.				

TABLE 20.1 Portable Power Tool Safety Checklist (*Continued*)

9.	Are all hand-held powered platen sanders, grinders (with 2 in. wheels diameter or less), routers, planers, laminate trimmers, nibblers, shears, scroll saws, and jig saws (with blade shanks 1/4 in. wide or less) equipped with a positive "on-off" control?	Y	N	N/A
Note: A positive "on-off" control means a switch that you must push to turn the tool on and then push again to turn it off. Control switches as described in questions seven and eight may also be used.				
10.	On hand-held power tools, is the operating control located so as to minimize the possibility of accidental operation?	Y	N	N/A
Note: This requirement does not apply to concrete vibrators, concrete breakers, powered tampers, jackhammers, rock drills, garden appliances, household and kitchen appliances, personal care appliances, medical or dental equipment, or to fixed machinery.				
11.	Are all portable power-driven circular saws (with blade diameter greater than 2 in.) equipped with guards above and below the base plate or shoe?	Y	N	N/A
Note: This requirement does not apply to meat cutting saws.				
12.	Does the upper guard on a circular saw cover the saw to the depth of the teeth, except for the minimum arc required to permit the base to be tilted for bevel cuts?	Y	N	N/A
13.	Does the lower guard on a circular saw cover the saw to the depth of the teeth, except for the minimum arc required to allow proper retraction and contact with the work?	Y	N	N/A
14.	When a circular saw is removed from the material being sawed, does the lower guard automatically and instantly return to the covering position?	Y	N	N/A
15.	Are belt sanding machines provided with guards at each nip point, where the sanding belt runs onto a pulley?	Y	N	N/A
16.	If a saw cracks, is it immediately removed from service?	Y	N	N/A
17.	Are all portable, electrically powered tools properly grounded or double insulated?	Y	N	N/A
18.	Are impact tools, such as drift pins, wedges, and chisels, kept free of mushroomed heads?	Y	N	N/A
19.	Are the wooden handles of tools kept free of splinters or cracks, and are they fixed tightly in the tool?	Y	N	N/A

Table 20.1 Portable Power Tool Safety Checklist (*Continued*)

20.	Is it prohibited to lower or hoist a tool by the cord?	Y	N	N/A
21.	Do woodworking tools meet the American National Standards Institute (ANSI) safety codes?	Y	N	N/A
Pneumatic power tools and hose				
22.	Are pneumatic power tools secured to the hose or whip by some positive means, so as to prevent the tool from being accidentally disconnected?	Y	N	N/A
23.	Are safety clips or retainers used on pneumatic impact (percussion) tools to prevent attachments from being accidentally expelled?	Y	N	N/A
24.	Are all pneumatically driven nailers, staplers, and other similar equipment with automatic fastener feeds that operate at more than 100 psi pressure at the tool equipped with a safety device on the nozzle to prevent the tool from ejecting fasteners, unless the muzzle is in contact with the work surface?	Y	N	N/A
25.	Are all compressed air hoses and hose connections designed for the pressure and service to which they are subjected?	Y	N	N/A
26.	Is it prohibited to lower or hoist tools by the hose?	Y	N	N/A
27.	Do all hoses (exceeding 1/2 in. inside diameter) have safety devices at the source of the supply or branch line, to reduce pressure in case of hose failure?	Y	N	N/A
28.	Are airless spray guns—of the type that atomize paints and fluids at high pressure (1000 pounds or more per square inch)—equipped with automatic or visible manual safety devices which prevent accidental release of paint or fluid?	Y	N	N/A
Note: In lieu of the above, a diffuser nut which will prevent high pressure, high velocity release while the nozzle tip is removed, plus a nozzle tip guard which will prevent the tip from coming in contact with the operator (or other equivalent protection) shall be provided.				
29.	Are all fuel-powered tools stopped while being refueled, serviced, or maintained?	Y	N	N/A
30.	Is all fuel transported, handled, and stored in accordance with applicable regulations?	Y	N	N/A
31.	When fuel-powered tools are used in enclosed spaces, are measures taken to prevent the buildup of toxic gases?	Y	N	N/A

Source: National Institute for Occupational Safety & Health (*www.cdc. gov/niosh*)

TABLE 20.1 Portable Power Tool Safety Checklist (*Continued*)

Regulations and Standards

OSHA Regulations

1926 Subpart I—*Tools – Hand and Power*

1926.301	*Hand Tools*
1926.302	*Power-Operated Hand Tools*
1926.303	*Abrasive Wheels and Tools*
1926.305	*Jacks-Lever and Ratchet, Screw, and Hydraulic*
1926.306	*Air Receivers*
1926.307	*Mechanical Power-Transmission Apparatus*

1910 Subpart P—*Hand and Portable Powered Tools*

1910.242	*Hand and Portable Powered Tools*
1910.243	*Guarding of Portable Powered Tools*
1910.244	*Other Portable Tools and Equipment*

U.S. Consensus Standards

ANSI B7.1—*Safety Requirements for Use, Care, and Protection of Abrasive Wheels*

Industry Standards Resources

National Safety Council, Data Sheet 230 – Powder-Actuated Hand Tools

CHAPTER 21

Welding and Cutting

While there are many methods for joining metals, welding is one of the most convenient and rapid methods available. Welding is the process of joining metals by heating them to their melting temperature and causing the molten metal to flow together.

One of the most popular welding processes is *shielded metal arc welding* (SMAW). It dominates other welding processes in the maintenance and repair industry because of its versatility and the simplicity of its equipment and operation. Although the flux core process has grown in popularity, SMAW continues to be used extensively in the construction of steel structures.

21.1 Types of Welding Processes

The primary differences between the various welding processes are the methods by which heat is generated to melt the metal. Following are some of the most common types of welding used by steelworkers.

Oxyfuel Gas Welding (OFW)

This process uses a gas flame as a source of heat, which is produced by burning a combustible gas, such as methylacetylenepropadiene (MAPP) or acetylene, mixed with oxygen. Gas welding is widely used in maintenance and repair work because of the ease in transporting oxygen and fuel cylinders. The oxyfuel process is adaptable to brazing, cutting, and heat treating all types of metals (Fig. 21.1).

Arc Welding

This process uses an electric arc to join the metals being welded. A distinct advantage of arc welding over gas welding is the concentration of heat. In gas welding the flame spreads over a large area, sometimes causing heat distortion. Arc welding's characteristic concentration of heat is an advantage because less heat spread reduces buckling and warping (Fig. 21.2).

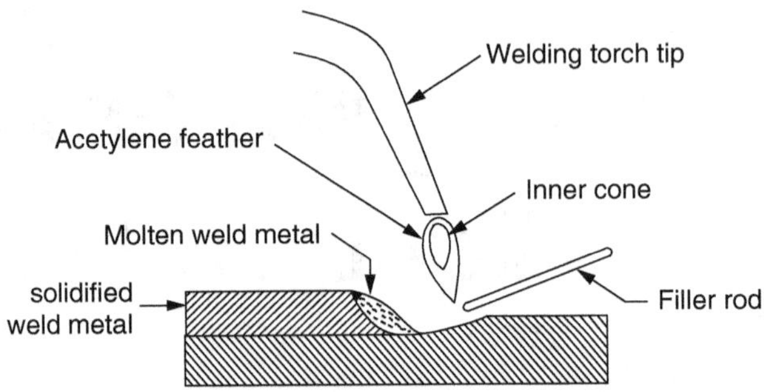

FIGURE 21.1 Oxyfuel gas welding (OGW). (*Source:* Naval Construction Force Capability Program, Department of Navy, Washington, DC.)

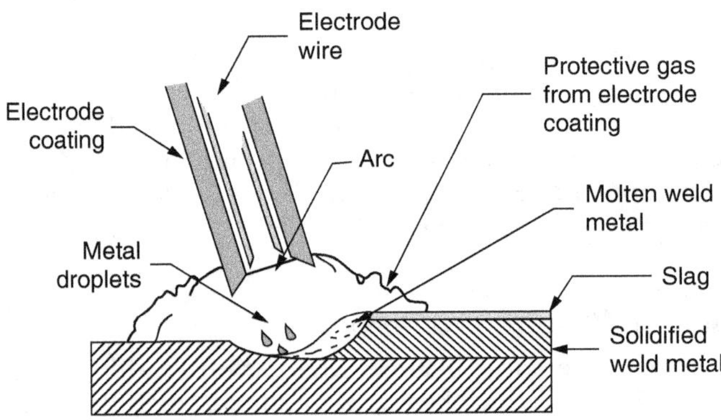

FIGURE 21.2 Shielded metal arc welding (SMAW). (*Source:* Naval Construction Force Capability Program, Department of Navy, Washington, DC.)

This heat concentration also increases the depth of penetration and speeds up the welding operation; therefore, you will find that arc welding is often more practical and economical than gas welding.

All arc-welding processes have three things in common: a heat source, filler metal, and shielding. The source of heat in arc welding is produced by the arcing of an electrical current between two contacts. The power source is called a *welding machine* or, simply, a *welder*. This should not be confused with the same term that is also used to describe the person who is performing the welding operation. The welder (welding machine) is either electric- or motor-powered.

Shielded Metal Arc Welding (SMAW)

This process is performed by striking an arc between a coated metal electrode and the base metal. Once the arc has been established, the molten metal from the tip of the electrode flows together with the molten metal from the edges of the base metal to form a sound joint (a process known as *fusion*) (Fig. 21.2).

The coating from the electrode forms a covering over the weld deposit, shielding it from contamination. The main advantages of the shielded metal arc welding process are that high-quality welds are made rapidly at a low cost.

Gas-Shielded Arc Welding (GSAW)

The primary difference between shielded metal arc welding and gas shielded arc welding is the type of shielding used. Both the arc and the molten puddle are covered by a shield of inert gas that prevents atmospheric contamination, thereby producing a better weld. The primary gases used for this process are helium, argon, or carbon dioxide. In some instances, a mixture of these gases is used.

The processes used in gas shielded arc welding are known as *gas tungsten arc welding* (*GTAW* or *TIG*) and *gas metal arc welding* (*GMAW* or *MIG*). They can be used to weld all types of ferrous and nonferrous metals of all thicknesses (Fig. 21.3).

Filler Metals

When welding two pieces of metal together, a space is often left between the joint that is filled during the welding process with filler metal, the most commonly used being welding rods and welding electrodes.

- *Welding rod* refers to a form of filler metal that does not conduct an electric current during the welding process; its only purpose is to supply filler metal to the joint. This type of filler metal is often used for gas welding.
- *Electrode*, in electric-arc welding, refers to the component that conducts the current from the electrode holder to the metal being welded.

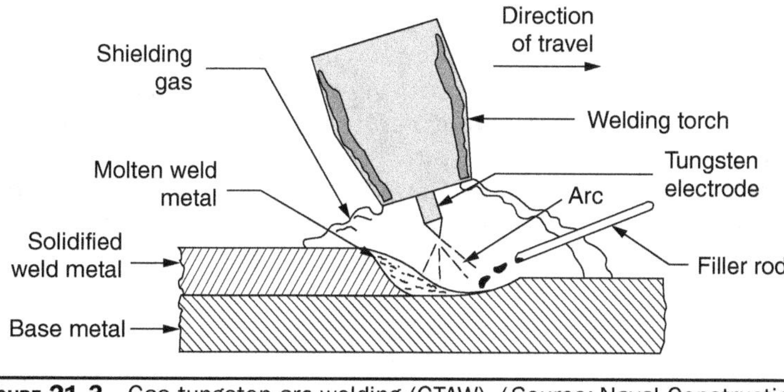

FIGURE 21.3 Gas tungsten arc welding (GTAW). (*Source:* Naval Construction Force Capabillity Program, Department of Navy, Washington, DC.)

21.2 Welding Hazards

If welding is being done in an area where a fire hazard exists, a welding permit (hot work permit) *must* be used in accordance with established procedures. These precautions are based on regulatory requirements.

In addition, a trained fire watch, *must* be posted to look for fires during and after the welding job. Combustible and flammable materials must be cleared from the welding area, because a spark or a piece of hot slag could easily ignite these materials and cause a tragic fire.

The extent of welding hazards—including harmful metal fumes, toxic gases, ultraviolet and infrared radiation, loud noises, intense heat, and glaring light—depends on the type of welding being performed and the precautions that are taken. The content of the welding rods, coatings, filler metals, and base materials also greatly impacts the welder's health.

Welding Gases

In gas welding, leaking oxygen can enrich the atmosphere so that a naked flame, cigarette, spark or electrical fault can be dangerous.

Fumes and Explosions

Welding, cutting, or heating empty drums *must not* be permitted due to undetectable fuel residues that may vaporize and explode. Always determine what has been inside and, if necessary, clean the drum thoroughly before cutting, welding or heating. Welding heat can also generate toxic fumes from chemical residues.

Coated Metal Surfaces

Avoid welding on coated metal surfaces, such as galvanized iron.

Fumes

Some components of welding fumes can be fatal. *Avoid* welding in confined spaces where it's not possible to ensure good ventilation. Normally, good local or general ventilation is required to control exposures to the metal fumes and gases of welding operations, for example, local exhaust ventilation in which an exhaust hood is placed near the welding arc or flame, and the contaminants are drawn away from the welder's breathing zone. Exhausted air should be discharged to the outdoors, if possible. Welding in confined spaces such as tanks, cabs of mobile equipment, and large shovels may be especially hazardous and requires additional ventilation.

Burns

Welding temperature can reach 6000°C, with welding operations involving the hazard of burns from flame, arc, molten metal, hot metal surfaces, and metal splatter (fragments and sparks), as well as the arc itself reflecting off surrounding materials and burning co-workers working nearby. In general, welding operations require face, neck, and eye protection for the welder to protect against sparks and splatter of molten metal—normally meaning that a welder *must* wear a welding hood or helmet—though in some cases adequate goggles, gloves, and other protective clothing of neck and arms suffices.

Radiation and Intense Light

These can be harmful to both the welder and anyone else nearby. Damage to uncovered skin can be similar to severe sunburn. Unprotected eyes can become extremely red and sore and, in extreme cases, suffer permanent damage. Intense light associated with welding can cause damage to the retina of the eye; ultraviolet radiation, from the arc, can cause "arc eye" or "welder's flash" after less than a minute of exposure; and infrared rays may damage the cornea and result in the formation of cataracts, and can cause burns to the skin. Welding hoods and special welding goggles with UV filter lenses and side shields are designed to protect eyes and face from UV exposure, while appropriate gloves and aprons *must* be used to protect exposed skin. Welding curtains may be used for the same purpose to protect others in the vicinity of the welding area. This equipment *must* be used faithfully for every welding job in order to prevent UV burns. Flash burns to the eyes are extremely painful and can cause permanent damage, including blindness.

Noise

Prolonged exposure to loud noise can permanently damage a welder's hearing. Follow company policies for using personal protective equipment (PPE) to prevent hearing loss and UV burns, and follow them consistently. Correct any situations that pose a fire or electrical shock hazard. If a safety concern about welding hazards arises, don't let it become an accident waiting to happen—report it to the work supervisor or the company's safety office.

Respiratory

When personal respiratory protection is required, this may be provided by a supplied-air welding hood; or when the components and concentration of the fume are known, by a filter-type respirator having filter for protection against fumes. It is preferable, of course, to provide adequate ventilation so as to make the use of respirators unnecessary.

Electric Shock

Although welding generally uses low voltage, the risk of electric shock in welding is high as a result of wet or cramped spaces. Falls and other accidents can result from even a small shock; brain damage and death can result from a large shock. Any electrical hazards should be identified and immediately addressed.

Always wear dry gloves to protect against shock; wear rubber-soled shoes; and use an insulating layer such as a board or a rubber mat for protection on surfaces that can conduct electricity.

- Ground the piece being welded and the frame of all electrically power machines.
- Keep dry and in good condition insulation on electrode holders and electrical cables.
- *Do not* change electrodes with bare hands, with wet gloves, or when standing on wet floors or grounded surfaces.

21.3 Operation and Maintenance of Welding Equipment

Field welders *must* have current, local jurisdiction certification for the thickness and position to be used in the work.

- *Welding process.* Field welding is routinely limited to manual SMAW
  - Flux cored arc welding (FCAW) is *not* recommended for use in the field.
  - Submerged arc welding (SAW) is used in the field for certain applications.
- *Preheat requirements.* Based on grade of steel and thickness of base metal for shop welding.
  - Minimum preheat for field welding is 120°C for all but ASTM A7M steel.
  - Minimum preheat of 200°C *must* be used for ASTM A7M steel, unless carbon equivalent tests are performed.

Operation

Workers assigned to operate or maintain arc welding equipment *must* be acquainted with the OSHA regulations covering welding machines and their operation. And workers in charge of the oxygen or fuel-gas supply equipment, including generators, and oxygen or fuel-gas distribution piping systems *must* be instructed and judged competent by their employers for this work before being left in charge.

- *Make sure* that rules and instructions covering the operation and maintenance of oxygen or fuel-gas supply equipment are readily available, including generators and oxygen or fuel-gas distribution piping systems.

- Perform hot work, if possible, in a properly designated area equipped with all necessary controls and adequate ventilation.

Before starting welding operations:

- Move combustible materials at least 35 ft away from the work site. If this is not possible, protect combustible materials with metal guards or by flameproof curtains or covers (other than ordinary tarpaulins).
- Cover floor and wall openings within 35 ft of the work site to prevent hot sparks from entering walls or falling beneath floors or to a lower level.
- Use fire resistant curtains or tinted shields (or both) to prevent fire, employee burns, and ultraviolet light exposure.
- Check all connections to the machine to make certain they are properly made.
  - Work lead *must* be firmly attached to the work
  - Magnetic work clamps *must* be freed from adherent metal particles of spatter on contact surfaces.
  - Coiled welding cable *must* be spread out before use to avoid serious overheating and damage to insulation.
- Check grounding of the welding machine frame, giving *special attention* to safety ground connections of portable machines.
- *Make sure* there are no leaks of cooling water, shielding gas, or engine fuel.
- *Make sure* that proper switching equipment for shutting down the machine is provided.
- Manufacturer's instructions, covering operation of its equipment, *must* be strictly followed.
- Electrode holders, when not in use, *must* be placed so that they cannot make electrical contact with persons, conducting objects, fuel, or compressed gas tanks.
- Cables with splices within 10 ft (3 m) of the holder *must not* be used.
- Welder *must not* coil or loop welding electrode cable around parts of his body.

Maintenance

The operator should report any equipment defect or safety hazard to his supervisor, and use of the equipment *must* be discontinued until its safety has been assured.

- Repairs *must* be made only by qualified personnel.
- Machines that have become wet *must* be thoroughly dried and tested before use.
- Cables with damaged insulation or exposed bare conductors *must* be replaced.
- Joining lengths of work and electrode cables *must* be done by the use of connecting means specifically intended for the purpose.
- Connecting means *must* have insulation adequate for the service conditions.

21.4 Safe Welding Practices

Helmets or hand shields *must* be used during all arc welding or arc cutting operations, excluding submerged arc welding; helpers or attendants *must* be provided with proper eye protection.

Arrange helmets and hand shields to protect the face, neck, and ears from direct radiant energy from the arc. Goggles *must* be ventilated to prevent fogging of the lenses as much as practicable.

- *Make sure* that all operators and attendants of resistance welding or resistance brazing equipment use transparent face shields or goggles, depending on the particular job, to protect their faces or eyes, as required.
- *Make sure* helmets and hand shields are made of a material that is an insulator for heat and electricity. Helmets, shields, and goggles
    - *Must not* be readily flammable
    - *Must* be capable of withstanding sterilization
    - *Must* have filter plates and cover plates designed for easy removal.
- *Always* use goggles or other suitable eye protection during all gas welding or oxygen cutting operations. Spectacles without side shields having suitable filter lenses are permitted for use during gas welding operations on light work, for torch brazing, or for inspection.
- *Make sure* that all parts are constructed of a material that will not readily corrode or discolor the skin.
- Use *only* approved apparatus such as torches, regulators or pressure-reducing valves, acetylene generators, and manifolds.
- *Guard* against mixing of fuel gases and air or oxygen; mixtures may be explosive.

- *Do not* allow the use of any device or attachment that facilitates or permits mixtures of air or oxygen with flammable gases prior to consumption, except at the burner or in a standard torch—unless approved for the purpose.
- *Prohibit* the use of liquid acetylene.

Welding Standards

The voluntary character of ***welding standards*** becomes mandatory when so established by procurement documents, or when specified by governmental authorities having jurisdiction, for public safety, upon that type of regulated subject.

Welding standards include codes, specifications, recommended practices, classifications, methods, and guides. The various documents, although similar in content, are not interchangeable.

- Codes, usually applicable to processes, explicitly indicate mandatory action.
- Specifications provide requirements, generally for products.

Regulations and Standards

OSHA Regulations

1926 Subpart J—*Welding and Cutting*

1926.350 *Gas Welding and Cutting*

1926.351 *Arc Welding and Cutting*

1926.352 *Fire Prevention*

1926.353 *Ventilation and Protection in Welding, Cutting, and Heating*

1926.354 *Welding, Cutting, and Heating in Way of Preservative Coatings*

1910 Subpart Q—*Welding, Cutting, and Brazing*

1910.253 *Oxygen-Fuel Gas Welding and Cutting*

1910.254(a)(2) *Installation of Welding Equipment*

1910.254(a)(3) *Welder Instruction and Qualification*

National Institute for Occupational Safety and Health (*www.cdc.gov/niosh*)

U.S. Consensus Standards

AWS

D1.1/D1.1M–2004 *Structural Welding Code - Steel*

CC-RM – D1.1 *Code Clinic Reference Manual*

D1.1/D1.1M Bundle A *Structural Welding Code Steel Bundled Set A*

D1.1/D1.1M Bundle B *Structural Welding Code Steel Bundled Set B*

D1.1/D1.1M Bundle C *Structural Welding Code Steel Bundled Set C*

D1.5M/D1.5 Bundle D *Bridge Welding Code Bundled Set D*

D1.2/D1.2M Bundle E *Structural Welding Code Aluminum Bundled Set E*

AWS/ANSI Z 49.1–2005 *Safety in Welding and Cutting*
ANSI/AWS D1.2/D1.2M-03 *Structural Welding Code - Aluminum*

D1.3-98 *Structural Welding Code - Sheet Steel*

D1.4-98 *Structural Welding Code - Reinforced Steel*

D1.5M/D1.5-2002 *Bridge Welding Code*

D1.6-99 *Structural Welding Code - Stainless Steel*

ANSI/API 1104 *Welding of Pipelines and Related Facilities*
ANSI/AWS API-M – API 1104 *Code Clinic Reference Manual*
ASNT SNT-TC-1A-2001 *Recommended Practice No. SNT-TC-1A - Non-Destructive Testing*
ASME – 2004 *Boiler and Pressure Vessel Code – Section IX: Welding and Brazing Qualifications*
API 620 *Design and Construction of Large, Welded, Low-Pressure Storage Tanks*

Industry Standards Resources

American Welding Society (*www.welding-advisers.com*)

Naval Safety Center (*www.safetycenter.navy.mil*)

CHAPTER 22

Jacks, Rollers, and Air Skids

These portable, compact devices can be used for raising or lowering heavy loads over short distances; to lift a load a short distance for placing cribbing, skids, and rollers; to lift concrete slabs or metal plates into place; for precise placement of heavy loads, such as beams; or to exert outward pressure against opposing restraints, such as trench shoring.

Because they are small and easily transported, jacks can be used in very restricted areas; used individually or joined to operate collectively; and may become a permanent part of a structure or facility.

Jack manufacturers provide each device with a lever or handle for rigging or construction jobs.

22.1 Basic Jack Categories

Although there are a number of different styles of jacks available, only heavy-duty hydraulic jacks or screw jacks should be used for rigging operations. Small capacity jacks are normally operated through a rack bar or screw, while large capacity jacks are usually operated hydraulically and generally are available in capacities from 5 to 100 tons or greater.

Portable jacks for heavy-duty use in rigging operations or in construction fall into three general categories.

Hand- or Power-Operated Hydraulic Jacks

This type of jack lifts heavy loads with little effort. The unit pressure exerted on a small area of incompressible fluid is transmitted undiminished to a larger area in the same closed circuit, thus multiplying the exerted force. It is not as fast as a ratchet-type jack, but it is excellent for easy lifting of heavy loads.

A variation of this type of pump is the hydraulic ram with separate hydraulic pump operating on the same principle as the integral pump-type jack. Rams enable users to operate the pump at a safe distance from the load being lifted.

Mechanical Ratchet Jacks

This is the simplest jack construction, employing the basic lever and fulcrum principle. A downward stroke of the level raises the rack bar one notch at a time and the pawls spring into position, automatically holding the load and releasing the lever for the next lifting stroke.

This type of jack should be used on lighter loads, usually up to 20 tons, because of the physical effort required to operate it. These devices are less expensive than other types of jacks and operate quickly.

Hand- or Power-Operated Mechanical Screw Jacks

For heavier loads, the screw and nut principle is employed in two general types: regular and inverted. For lighter loads, a simple lever bar will apply sufficient power to turn the screw. As loads increase, gear reductions and ratchet devices serve to multiply the operator force.

In the heaviest of jacks, the screw is operated by an air motor for faster lifting or lowering with minimum effort.

Overall individual lifting capacity of specific jacks may well range over 400 tons (363 metric tons). Stroke movement, usually a function of the length (or height) of the jack frame, may be anywhere from 1/4 in. (6 mm) up to over 3 ft (1 m). Regardless of the type of jack used, however, it is very important *not* to overload it.

Never, under any circumstances, should an automobile-type jack be used for any rigging or construction job.

With long-term use under continued pressure, the hydraulic jack may leak slightly, allowing a lowering of the load. Although screw-type or rack-and-level mechanical devices better serve the long-range requirements, hydraulic jacks are usually simpler to operate than the others and require less effort.

22.2 Types of Jacks

Portable jacks for construction jobs, requiring lifting, pushing, bending, and straightening, include mechanical and hydraulic jacks (Fig. 22.1).

Hydraulic Bottle Jacks

These multipurpose hand jacks are used for lifting, bracing, shoving, pushing, and spreading for any construction and industrial application include extension screws to accommodate work height, ribbed steel lifting saddles to prevent slippage damage, wide base for stability, and a built-in bypass device to prevent jack damage from overpumping (Fig. 22.2).

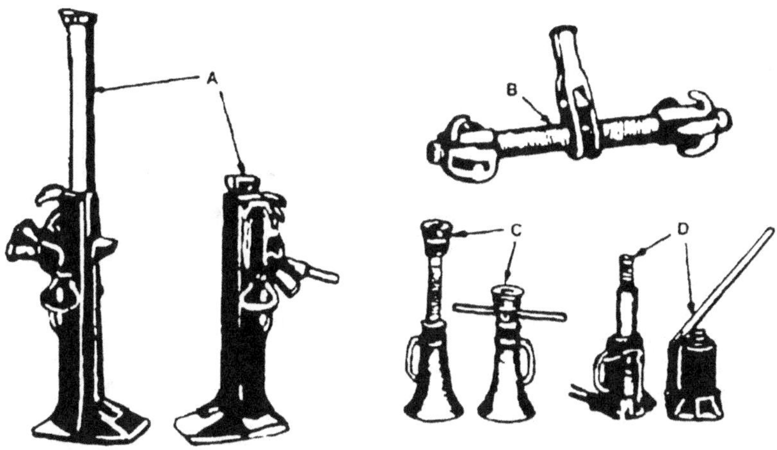

FIGURE 22.1 Mechanical and hydraulic jacks. (a) Ratchet lever jack with foot lift; (b) steamboat ratchet; (c) screw jack; (d) hydraulic jack.

FIGURE 22.2 Hydraulic bottle jack. (*Source:* Torin Jacks Inc., Yakima, WA.)

Hydraulic bottle jacks are available from 2- to 30-ton (1.8- to 26.8-metric tons) capacity, with lift ranges from 7 to 19 in. (17.78- to 48.3-cm).

Hand Hydraulic Jacks

These jacks are used for construction and industrial applications for lifting and moving heavy loads—and horizontally, for pushing, spreading, bending, and straightening—available from 1½- to 20-ton

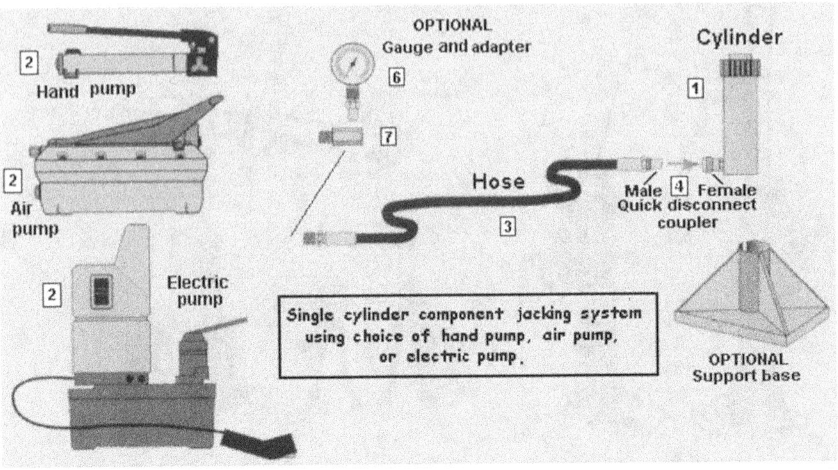

FIGURE 22.3 Simple single cylinder, hydraulic jack system. (*Source:* Otto Service Co., Burbank, CA.)

(1.35- to 17.86-metric tons) capacity with 7- to 22-in. (17.8- to 56-cm) lift range (Fig. 22.3).

Air-Hydraulic Service Jacks

Jacks for rigging operations, having capacities of 20 to 22 tons (17.9 to 19.6 metric tons), are completely automatic, remote control, and monitoring systems that enable:

- Mechanically fail-safe gripping mechanisms
- Pre-set overload protection
- Automatic synchronization of lifting speeds irrespective of loads
- Ample operating speeds
- Load to be held mechanically not hydraulically any time movement is stopped
- Movement to be stopped and held indefinitely at any point in jack stroke
- Extremely precise adjustment of the lift for setting of weld gaps or the fitting of steelwork connecting bolts
- Automatic lowering without any jack loads
- Minimal relubrication of gripping mechanisms
- Gripping mechanisms fully accessible at all times
- Full efficiency in all attitudes from vertical through to horizontal
- Complete self-containment; very rugged

Steelworker Jacks

Ratchet-Lever Jacks

These are rack-bar type jacks that employ the lever and fulcrum principle to lift and lower the load having a rated capacity of 15 tons. A foot lift enables engaging loads close to the base of the jack (Fig. 22.4).

NOTE: *Track jacks look similar to ratchet jacks, but are trip-lowering and are not intended for use as general-purpose jacks.*

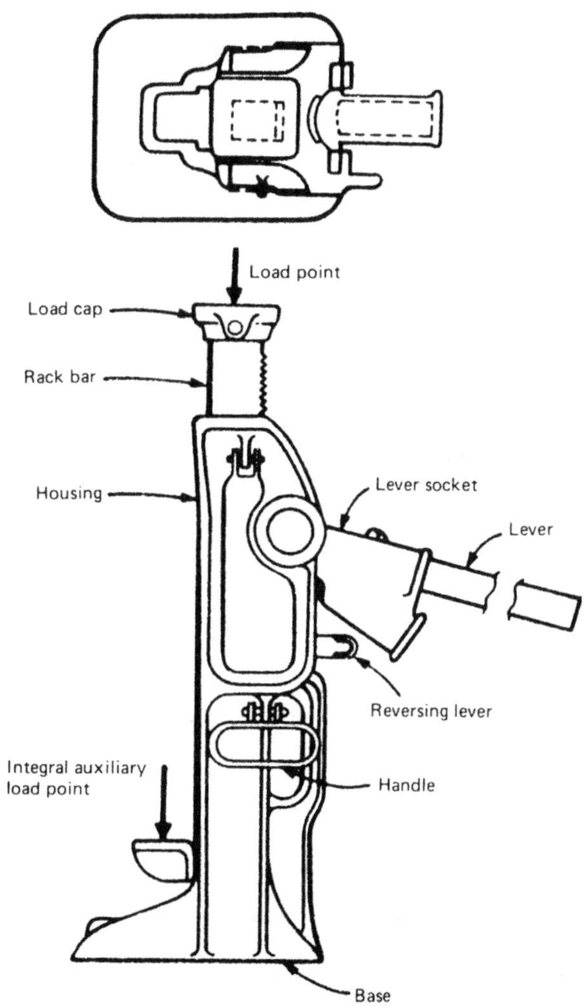

FIGURE 22.4 Typical ratchet jack. (*Source:* American National Standards Institute, New York, NY.)

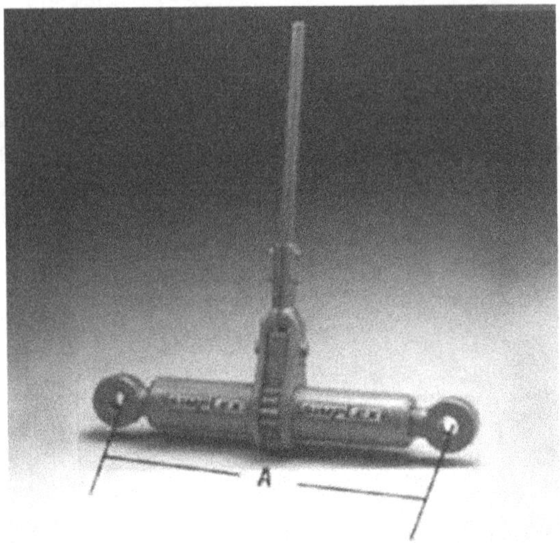

FIGURE 22.5 Steamboat ratchet (push-pull) jack. (*Source:* Simplex, Broadview, IL.)

Steamboat Ratchets

Often referred to as pushing and pulling jacks, steamboat ratchets are ratchet screw-type jacks of 10-ton (9-metric-ton) rated capacity having end fittings that permit pulling parts together or pulling them apart. They are primarily used for tightening lines or lashings and for spreading or bracing large parts, such as in bridge construction (Fig. 22.5).

Screw Jacks

Screw jacks are fully enclosed units packed in grease and sealed to prevent leakage, having a rated capacity of 12 tons (10.7 metric tons). They are approximately 13-in. (33-cm) high when closed and have a safe rise of 7 in. (17.78 cm); used for general purposes, including steel erection (Figs. 22.6 and 22.7).

Hydraulic Jacks

These jacks have outweighed conventional screw jacks that were predominantly in use. They consist of two cylinders joined together, working on the principle that when there is an increase in pressure at any point in a container of fluid, there is an equal increase in pressure at every other point in the container (Fig. 22.8).

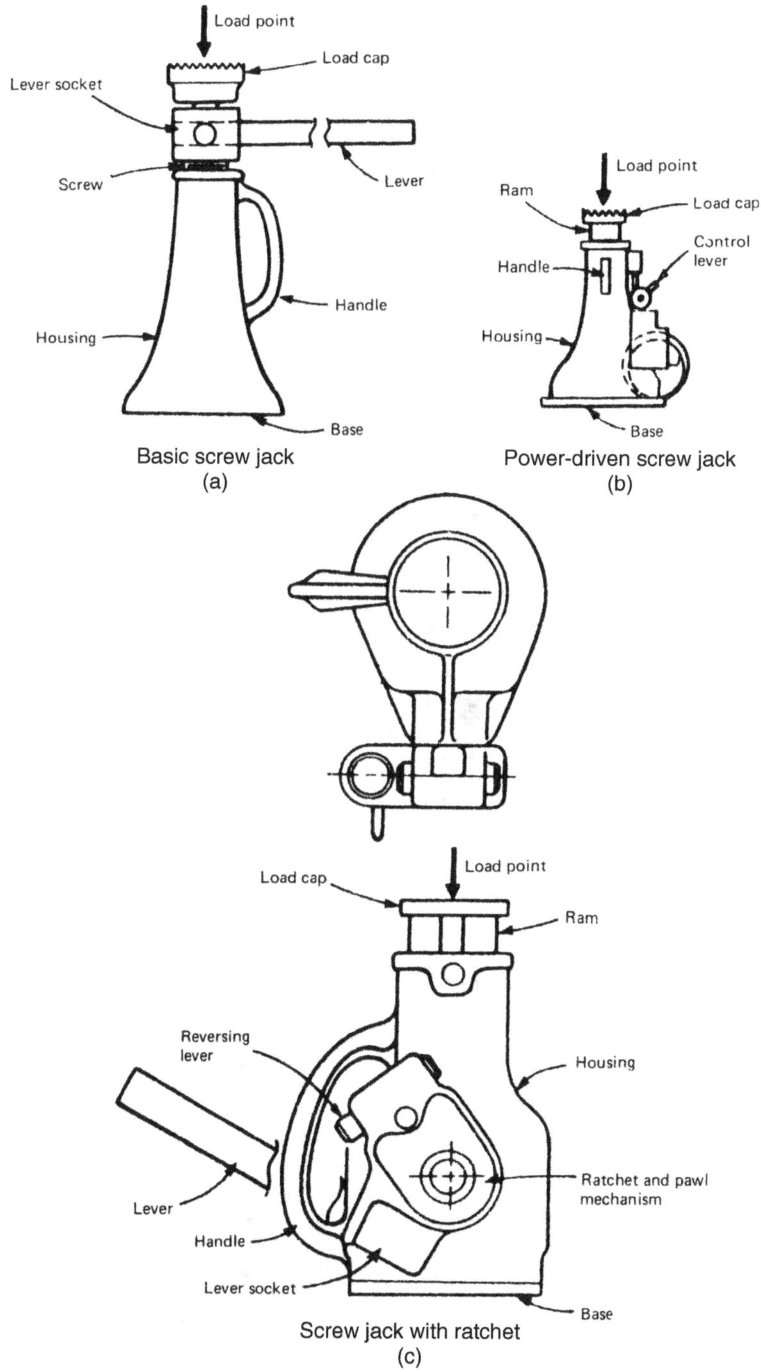

FIGURE 22.6 Screw jacks. (a) Basic; (b) power-driven; (c) with ratchet. (*Source:* American National Standards Institute, Washington, DC.)

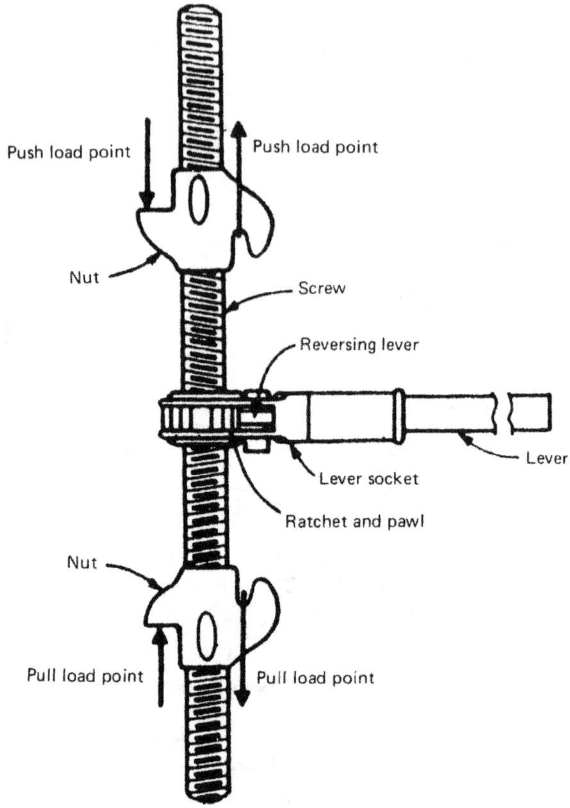

FIGURE 22.7 Push-pull-type jack. (*Source:* American National Standards Institute, Washington, DC.)

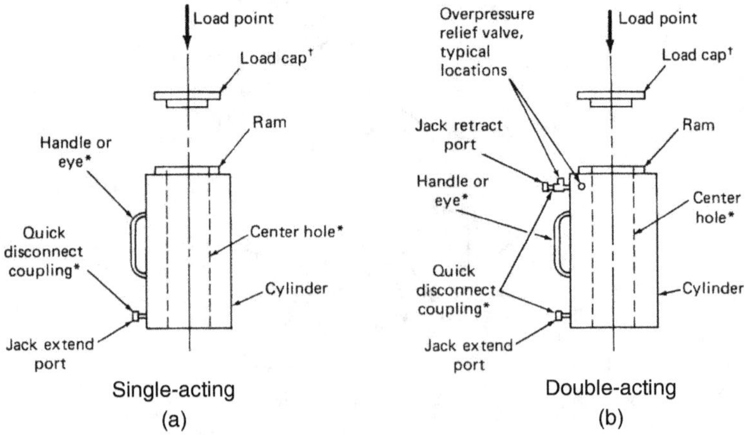

FIGURE 22.8 Hydraulic jack. (a) Single-acting; (b) double-acting. (*Source:* American National Standards Institute, Washington, DC.)

Hydraulically actuated jacks are available in many different capacities, and are used for general purposes. The number of jacks used is determined by the weight of the load and the rated capacity of the jacks. Before lifting a load, *make sure* that the jacks have a solid footing and are not susceptible to slipping.

22.3 Operation of Jacks

When using jacks, it is more important that the operator be familiar with the particular equipment and the manufacturer's instructions for testing, operation, and maintenance. All jacks should be visually inspected before each shift or each use—whichever is less frequent—as well as during operation, for any deficiencies that might constitute a hazard.

Any jack should be removed from service immediately and not used until any of the following deficiencies are corrected:

- Improper engagement of extreme wear of pawl and rack
- Cracked or broken rack teeth
- Cracked or damaged housing
- Excessive wear, bending, or other damage to threads
- Leaking hydraulic fluid
- Scored or damaged plunger
- Improperly functioning swivel heads and caps
- Loose bolts or rivets
- Damaged or improperly assembled accessory equipment
- Other items that may affect the jacking operation, as specified in the manufacturer's instructions.

Before operating a jack it is important to determine the load to be lifted and to assure that it is within the jack's load rating. All hydraulic jacks have the safe load indicated on the nameplate. Screw jacks occasionally are not labeled. Mechanical jacks sometimes have two ratings—sustaining and lifting—that should be indicated by the manufacturer.

The hydraulic pressure, or lever arm length and force to be applied, should also be clearly marked on the jack.

A rough estimate of the safe load on a screw jack can be made using the formula:

$$W = 31{,}400dt$$

or

$$W = 14{,}000d^2$$

where W = safe load on jack (lb)
d = diameter of screw at root of threads (in.)
t = thickness of nut or length of thread engaged (in.)

Choose the lower value of W computed from the formula.

The load that a screw jack should be able to lift is approximately:

$$W = \frac{24PL}{r^2}$$

where W = load (lb)
L = length of lever (in.)
P = pitch of screw threads (in.)
r = average radius of screw thread (in.)

The force on the lever can be assumed to be 91.2 kg (120 lb) for most cases.

When preparing for a lift, make sure that the jack is firmly supported at the base in such a manner that it cannot shift under the load. Always make certain that the jack is in a true vertical position (when lifting and resting on good footing). *Never* place a jack directly on the ground, even when the soil appears to be firm.

If the load is to be raised in its entirety by several jacks, it should be secured laterally by struts to prevent all the jacks from upsetting in unison. When using a jack in a horizontal position to move an object, it should be lashed or blocked up to prevent it from falling should the object unexpectedly move faster than the jack.

After the load has been raised to the required height, shoring or cribbing blocks should be placed under the load and wedged to take the load off the jacks, or to be ready to take the load should a jack fail. When required, use shims or constraints to prevent the base or load from slipping.

Never raise a load on a jack so high that the ram, screw, or ratchet runs out of the base, thereby dropping the load. Jacks designed in accordance with the ANSI safety requirements for jacks have positive stops or other methods to prevent overtravel. These will not alter the operating characteristics of the jack.

Hydraulic jacks are filled with a special hydraulic fluid that will not freeze.

WARNING: *Never use water in a jack—under NO circumstances.*

A pressure-relief valve should be used and set at, or below, the rated pressure at the rated load where remotely operated hydraulic jacks are used. Double-acting hydraulic jacks should be marked to indicate the need for a relief valve as indicated.

Operating Checklist

- Make sure all operators are instructed in the proper use of jacks, including signals and other procedures for multiple jacks or other special lifts.
- Make sure the recommended operating handle is used and properly seated in its socket.
- Make sure operators do not straddle the operating lever of a mechanical jack.
- Make sure there is sufficient swing area for the operating lever or levers.
- Remove operating levers when jack is not in use to avoid accidental dislocation of the jacks, thereby reducing the tripping hazard.
- Make sure all personnel are clear of the load before lowering.
- Follow every load with cribbing or blocking where practical.
- *Do not* allow personnel to work or pass under a jacked load until the load is secured by cribbing, blocking, or some other means against accidental lowering.
- Avoid off-center loading of jacks.
- *Do not* use extenders, unless authorized by a qualified person.

22.4 Safety precautions using jacks

Read, understand, and follow the operating instructions and the *ANSI B30.1* and *ASME/ANSI PALD* safety standards for jacks.

- Wear eye protection when using jacks.
- Inspect the jack before each use.
- Do not use the jack if it is damaged, altered, or in poor condition.
- Prevent tipping by setting jack up on a hard, level surface.
- Lift only dead weight.
- Do not exceed the rated lifting capacity of the jack.
- Center the load on the jack saddle, because off-center loads can damage the seal and cause hydraulic failure.
- Use the jack for lifting purposes only, *not* for supporting loads.
- Support a lifted load immediately with jack stands.
- Use only approved hydraulic fluid, such as Chevron AW hydraulic oil MV or equivalent.

22.5 Maintenance of Jacks

- *Lubrication.* Regularly lubricate all moving parts of a jack requiring such service. Check lubricating systems for delivery of lubricant. Follow the manufacturer's recommendations as to the points and frequency of lubricant, maintenance, or lubrication levels, and types of lubricants to be used.
- *Hydraulic fluid.* Use only hydraulic jack fluid that is fully compatible with the jack manufacturer's specifications.
- *Cleaning.* Keep the threads of a screw jack free from grit and dirt. Lubricate the screw frequently. Clean all jacks exposed to rain, sand, or grit-laden air prior to use. Make sure the jack-operating level and load-bearing surfaces are free of slippery material or fluid.
- *Storage.* Jacks should be stored where they are protected from the elements, abrasive dust, and damage; always store hydraulic jacks in a vertical position.
- *Repairs.* Purchase repair parts only from the original manufacturer or authorized service center. If jack parts are other than those obtained from the original manufacturer, they must conform to ANSI specifications.

Testing

After a jack has been repaired and before it is returned to service it should be tested by:

- Lifting a load equal to the rated capacity of the jack to within 1 in. of the full travel under the prescribed operating conditions for the particular type of jack being tested.
- Adding to this original load an additional load equal to one-quarter of the rated load already applied to the jack.
- Checking for any evidence that any part of the jack is stressed beyond the yield point of the material forming any part; such evidence automatically disqualifies the jack for future service.
- Placing blocks under the load as it is lifted to hold it in the event the jack should fail.

WARNING: *Keep hands and fingers out from under the load during testing.*

22.6 Inspection of Jacks Checklist

• Do jacks in use have a rating sufficient to lift and sustain the loads?	Y	N	N/A
• Is the rated load legibly and permanently marked in a prominent location on the jack by casting, stamping, or other suitable means?	Y	N	N/A
• In the absence of a firm foundation, is the base of the jack blocked or cribbed?	Y	N	N/A
• If the cap could slip, is a block placed between the cap and the load?	Y	N	N/A
• Do all jacks have a positive stop to prevent overtravel?	Y	N	N/A
• Are operators instructed to watch the stop indicator (which must be kept clean) to determine the limit of travel?	Y	N	N/A
• After the load has been raised, is it required to be cribbed, blocked, or otherwise secured at once?	Y	N	N/A
• Are hydraulic jacks that are exposed to freezing temperatures supplied with adequate antifreeze liquid?	Y	N	N/A
• Are all jacks properly lubricated at regular intervals?	Y	N	N/A
• Is each jack thoroughly inspected?	Y	N	N/A
• Are jacks that are used constantly or intermittently at one locality thoroughly inspected at least every 6 months?	Y	N	N/A
• Are jacks that are sent out thoroughly inspected when they are returned?	Y	N	N/A
• Are jacks that are subjected to abnormal load or shock thoroughly inspected immediately before and immediately after use?	Y	N	N/A
• Are repair or replacement parts examined for defects?	Y	N	N/A
• Are repairs made on disabled jacks before they are used again?	Y	N	N/A

Source: Construction Standards 1926.305 and *1926.554; .General Industry Standards 1910.244* and *1910.308*. Occupational Safety & Health Administration (*www.osha.gov*).

22.7 Hydraulic Jack Troubleshooting

Hydraulic jack repair procedures must be performed in a dirt-free environment by qualified personnel who are familiar with the particular equipment. (Table 22.1 Hydraulic Jack Troubleshooting Checklist.)

Trouble	Cause	Solution
Erratic Action	Air in system	With jack sitting on its base and ram retracted, bleed air by opening release valve. Pump for 10 seconds. Close release valve, and pump ram to full height. Completely open release valve, and let ram retract quickly.
	Viscosity of oil too high	Change to a lower viscosity oil
	Ram sticking or binding	Look for dirt, gummy deposits, leaks, misalignment, worn parts, or defective packing
	Internal leakage in ram	Take jack to an authorized service center for repair
Ram does not advance	Release valve is open	Close release valve
	Low/no oil in reservoir	Fill with oil to fill line and bleed system
	Air locked system	Bleed the system
	Load is above capacity	Use correct equipment of system
Ram only extends partially	Low oil level in reservoir	Fill with oil to fill line and bleed system
	Piston rod is binding	Look for dirt, gummy deposits, leaks, misalignment, worn parts, or defective packing
Ram advances slowly	Pump not working correctly	Take jack to authorized service center for pump repair
	Leaking seals	Take jack to authorized service center for seal replacement
Ram advances but doesn't hold pressure	Pump check valve not working	Take jack to authorized service center for seal replacement

TABLE 22.1 Hydraulic Jack Troubleshooting Checklist

Trouble	Cause	Solution
	Ram seals are leaking	Take jack to authorized service center for seal replacement
Jack leaks oil	Worn or damaged seals	Take jack to authorized service center for seal replacement
Ram will not retract, or retracts slowly	Release valve is closed	Open release valve
	Reservoir too full	Drain oil to correct level
	Ram damaged internally	Take jack to authorized service center for repair

Source: SPX Corp., OTC Tools, Owatonna, MN.

TABLE 22.1 Hydraulic Jack Troubleshooting Checklist (*Continued*)

22.8 Plank Skids and Rollers

Timber skids, for all practical purposes, are simple wood beams, commonly used under heavy machinery or other equipment that is being moved on rollers.

Plank Skids

Oak planks—15 ft (4.57 m) long and 2- to 3-in. (5.1- to 7.6- m) thick—are appropriate for most operations involving plank skids. They distribute the weight of a load and also provide a smooth runway surface on which to skid the load along or on which to use rollers to ease the effort required to move the load.

Timber skids (planks) are placed longitudinally under heavy loads to distribute the weight over a greater area (Fig. 22.9). The angle of the skids must be kept low to prevent the load from drifting or getting out of control. Skids can be greased, but only when horizontal movement is involved. *Extreme care* must be exercised because, in most circumstances, greasing is inherently dangerous—it can cause the load to drift sideways suddenly, causing injuries to personnel and damage to equipment.

Rollers

For moving heavy loads across a floor or ground, hardwood (usually maple) rollers 7½ in. (19.05 cm) in diameter by 10 ft (3.05 m) long are commonly used. Pipe rollers may also be used. *Never*, under any condition, use an oxygen cylinder or any other high-pressure gas

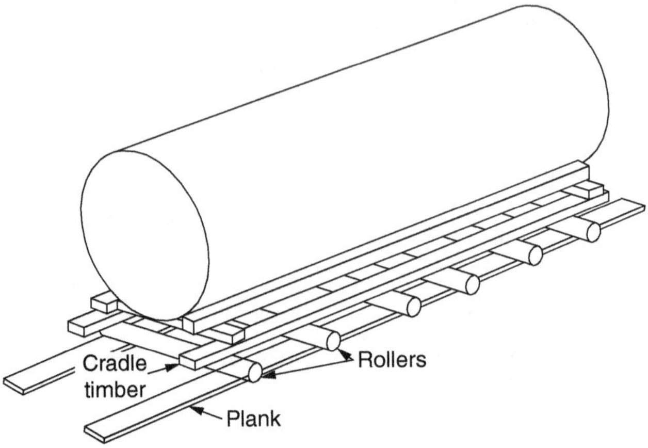

FIGURE 22.9 Planks and rollers.

cylinder in place of a roller. When full, such cylinders may have a pressure in excess of 2000 psi (146 kg/cm^2).

Rollers offer advantages over other transporting systems because of their low profile and a reduced requirement for motivation. While they may be used individually, they generally are used collectively. In some cases, they may satisfy an end in themselves, and they can be readily added to an existing system to increase load-carrying capacity.

Because of a load's relatively unaltered center of gravity when raised on rollers, control is easily maintained and the tipping potential is reduced. The roller's low physical profile minimizes the height to which the load must be raised for insertion of the carrying system. And because of the load-spreading aspects of roller, which can come in almost any length, they are able to traverse varying terrain configurations with minimal site preparation.

Rollers come in two broad classifications: those with axles and those without axles. The axle may be fixed, around which the roller rotates (on bearings), or the axle itself may rotate with the roller (secured at the ends in a race-type bearing). When wide rollers are used to carry a rigid load, mobility is somewhat limited. It is difficult to turn or redirect the load. If the rollers are employed with a dolly configuration for added flexibility, the height-saving advantage is reduced.

Without an axle, the roller might be of the endless-chain configuration. This track-laying principle eliminates or reduces the torque encountered with use of the axle. Such an endless-chain system drastically lessens rolling resistance and maintenance of the bearings or other wearing parts common to the axle roller.

Roller systems may be used in the temporary needs of construction (such as movement into position of steel members or large generators) or incorporated into the permanent structure itself (facilitating movement of the stands at the Superdome sports complex).

Hardwood or pipe rollers can be used in conjunction with plank skids for moving heavy loads into position. Planks are placed under the rollers to provide a smooth continuous surface to enable them to roll easily. The rollers must be smooth and round to aid in the ease of movement and long enough to pass completely under the load.The load should be supported by longitudinal wooden members to provide a smooth upper surface for the rollers to roll on. The skids placed underneath must form continuous support. Normal practice is to place four to six rollers under the load to be moved.

Several rollers are placed in front of the load, and the load is then slowly rolled onto these rollers. As the load passes the rollers that are left clear of the load, they are picked up and moved in front of the load, creating a continuous path of rollers.

Turns can be made using rollers; but first, the front rollers must be inclined slightly in the direction of the turn and the rear of the rollers in the opposite direction. This inclination of the rollers can be made by striking them sharply with a sledge hammer.

Rollers can be fabricated and set on axles in side beams as a semipermanent conveyor for lighter loads. Permanent metal roller conveyors are normally fabricated in sections that can be joined together.

22.9 Air Skids and Air Jacks

The basic principle behind this type of skid is the elevation of the skid or skids with the load so that the overall weight rests on a thin film of air. Once elevated, the load is moveable in any direction with minimum effort, against negligible friction facilitating precise positioning. For use of these air casters, the load is elevated by a system of leveling screws, jacks, or lifting eyes, or it may be prepositioned to allow insertion beneath it of the casters.

The skids are then inflated using sufficient pressure depending on the load to be lifted and the number of casters used. When the air pressure exceeds the weight of the load, the air will seep beneath the air bag to elevate the system. The load is then moved to position where some type of lifting device is used to suspend it, permitting removal of the casters for further use.

The requirements for such a system might include an overload compressed air drop, or other access to compressed air, and a nonporous floor. A floor covering that assumes even air escape is satisfactory. Where heavy loads are to be carried, several casters may be combined and formed to support a pallet that provides a firm load-spreading base for the object to be transported (Fig. 22.10).

Heavy-load moving devices using a cushion of air to separate the load from the floor or ground are designed to lift and move loads up to 300 tons and more and place the load with pinpoint precision. The

Steel air skid
(a)

Lift glide air skid
(b)

Figure 22.10 Air skids. (a) Steel air skid; (b) lift glide air skid. (*Source:* AirFloat Systems, Decatur, IL.)

devices are powered by a series of air cushions (bearings and casters) that reduce friction to allow one to move a 1-ton load with approximately 1 kg of force.

Floating a load on an "air cushion" dramatically reduces resistance to movement—based on the fact that the force required to move a load over a surface is the product of the weight of the load times the coefficient of friction. And, because air buoyancy is omnidirectional, there are no directional restraints (as in the case of trolleys and conveyors). Loads can be easily rotated or moved (or both) in any direction with virtually frictionless effort and very accurate positioning.

Among the benefits to using air bearings and casters are as follows:

- Environmentally friendly—powered by compressed air
- Easily lift and safely move loads up to 100+ tons
- Omnidirectional—turn all types of loads on a dime
- Move odd shapes, sizes, uneven weights
- Spaces forklifts and cranes cannot access
- Affordable and cost effective

Air cushion devices, however, do have some limitations. The surface over which the equipment is to be operated *must* be reasonably smooth, flat, and continuous. Rough surfaces will require more air volume (to maintain air pressure).

A floor surface should be nonporous to minimize loss of air. Floor expansion joints, cracks, air gaps, etc. should be filled with epoxy or mastic; and abrupt changes in surface slopes should be eliminated. Thin steel plates can provide a smooth surface for travel over rough ground.

Pneumatically powered air-jack lifting systems have capacities to 140 tons (127 metric tons) and, when used in series, provide virtually limitless capacities. They are omnidirectional in operation, requiring no additional force to change direction.

These low profile devices require only 1 in. (25.4 mm) clearance for placement—inflating to full height in seconds. A control console provides individual pressure regulation to compensate for unequal load distribution.

22.10 Blocking and Cribbing[1]

Timbers are commonly used to provide a foundation for heavy loads or jacks. Cribbing *must* be used when a heavy weight must be supported at a height greater than blocking can provide. Cribbing is made up by aligning timber in tiers that run in alternate directions (Fig. 22.11).

Blocking and cribbing is often necessary as a safety measure to keep an object stationary to prevent accidents and injury to personnel working near these heavy objects. When selecting blocking as a

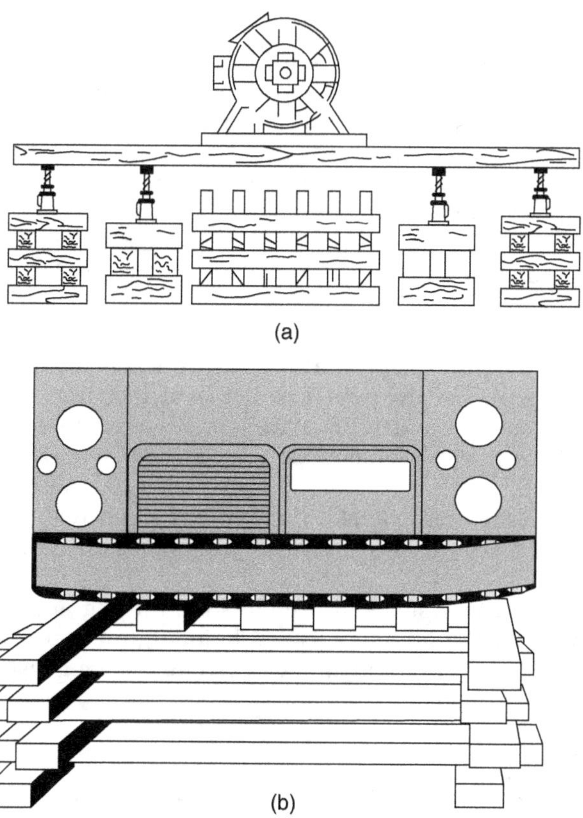

Figure 22.11 Timber cribbing.

[1]*Source:* Integrated Publishing—Construction (*www.tpub.com*).

foundation for jacks, *make sure* it is sound and large enough to support the load safely. In addition, it must be free from grease and thoroughly dry; and it must be placed firmly on the ground with the load distributed evenly. Equally critical is placing the bottom timbers so they rest evenly and firmly on the ground.

Lifting loads by jacking stages requires placing blocking under the jacks, lifting the load to the maximum height the jacks can safely accommodate, placing the cribbing under the load in alternating tiers, with no personnel under the load, and then lowering the load onto the cribbing.

When cribbing is not high enough or at the correct height, build up the blocking under the jacks until the jacks can bear against the load while in their lowered position. Raise the jacks again to their maximum safe height and lower onto the added cribbing. This procedure can be repeated as many times as necessary to build up the cribbing to the desired height.

22.12 Specialized Hydraulic Jacking Systems

Today's heavy construction activity often involves considerable lifting and lowering of heavy loads. More and more extremely large and heavy components often are either prefabricated on site or delivered on special vehicles. They are subsequently installed as units, lifted, jacked horizontally, or lowered into their final position.

As soon as the weight of a load or its size exceeds the capacity of available cranes or lifting equipment, specialized hydraulic jacking systems must be used. These systems, developed by several manufacturers, combine hydraulic jacks with a regripping mechanism that allows the movement of almost unlimited loads quickly, over any distance, and in any direction. Loads of 1000 tons or more can be lifted evenly, jacked horizontally, or lowered to their desired locations in situations where space is limited or ordinary cranes cannot reach.

This equipment is available on a for-hire basis, with installation, servicing, and maintenance handled exclusively by the manufacturer or its licensees.

The two basic systems in use are strand lifting and rod lifting. Both systems make use of similar techniques, but each has its own characteristics. For a certain application, one system is usually more suited than the other.

Strand Lifting

Strand jacks range from 15- to 560-ton (13.6- to 508-metric ton) capacity with a minimum breaking load of 38 tons (34.5 metric tons) per strand. These jacks may be used singly, in pairs, or in groups to give any lifting capacity required.

The equipment consists of a moving unit with strand cables as tensile members. The moving unit is basically a hydraulic center-hole jack with two stranded anchorages, an upper anchorage attached to the piston and a lower anchorage fixed to the support.

During the upward stroke of the piston, the strands are gripped by means of wedges seated in conical holes in the upper anchorage, which moves with the piston. During piston retraction, however, the strands are held at the lower anchorage affecting a stage-by-stage advance of the cable, and its attached load is achieved.

A special gripping mechanism assures an automatic and equal seating of the wedges. During lifting, the wedges of the lower anchorage remain in rubbing contact with the strand and will automatically grip, even if the oil pressure suddenly drops. This safety feature always ensures the load is secure.

High-tensile seven-wire prestressing strand with a nominal diameter of 0.6 in. (15.24 mm) is used to form the tensile member. The guaranteed breaking load per strand varies between 56.93 tons (23.1 metric tons) and 58.4 tons (26.5 metric tons).

On the structure to be moved, the strands are anchored by means of compression fittings bearing against an anchor head. Both wedge and compression fitting anchorage develop the full breaking load of the strand.

The number of strands per moving unit can vary depending on the magnitude of the load. The same strand can be reused several times.

Each moving unit normally has its own hydraulic pump. The pumps can be either controlled directly or connected to a central console. Synchronized demand-regulated pumps ensure equal travel, irrespective of the load on individual moving units. Thus large, but relatively fragile loads as well as hyperstatically supported structures can be moved without introducing undue erection stresses.

Cogged Jack-Rod System

This system consists of a moving unit and a cogged lifting rod. During the upward stroke of the pistons, pawls on the upper yoke engage the lifting rod and advance it upward. As the lifting stroke ends, the pawls on the lower yoke engage the cogged lifting rod, supporting the load while the pawls on the upper yoke disengage and the jacks retract. This process is then repeated in a continuous automatic sequence, eliminating the need for resetting. Lowering a load is accomplished by reversing the process.

Because of the inherent rigidity of the cogged lifting rods, the jacking system may be operated in either direction pushing rather than pulling the load to its desired position. The lifting rods come in easily transportable lengths and may be coupled to give any desired length.

A large number of moving units may be connected to a central console, as with the strand-type lifting system, ensuring equal travel, irrespective of the load on the individual moving unit. Selective operation of the individual units allows corrections and adjustments to be made.

Spacer disks between the jack piston and the yoke provide mechanical security against possible oil leakage in the system.

The control console also contains automatic monitoring devices that survey such mechanical and hydraulic operations as the correct engagement of pawls and the right hydraulic pressure distribution. In the event of any deviation, automatic shutoffs stop the lifting procedure. Minor deviations between the jacks cannot accumulate because such disparities are automatically eliminated after each incremental advance.

Precise cog spacing on the lifting rods assures parallel movement, even over long distances, and provides at any time an accurate measurement of the distance covered.

Steel-Bar Jack-Rod System

This system uses standard square steel bars and a grip-jaw arrangement that is essentially similar to that of a slip-form jack, except that the jaws are housed in thrust blocks above and below two parallel hydraulic cylinders. This allows the jack rod to be inserted or removed laterally from the lift-climbing device, or vice versa.

The lifting/lowering stroke is gage-set within the climber unit and is 4 in. (100 mm) for 12-, 16-, and 25-ton (10.89-, 14.51-, and 22.68-metric tons) climbers; 8 in. (200 mm) for 42- and 100-ton (36.1- and 90.72-metric tons) units.

The most commonly used units in this system are hydraulically operated double-acting lift climbers. These climbers are provided with an up/down device. The changeover from lifting to lowering involves a simple manual operation with the up-down device at the point of directional change. Thereafter, the lower operation is carried out fully automatically through the pump only.

A further device can be provided where the lowering of rigid loads is undertaken, thus avoiding differences in synchronization between particular lift climbers. The device is an equalizer that involves all the climbers being interconnected to a supplementary oil-cushion system. Equalizer devices are only used with lift climbers having the up/down arrangement fitted to allow for a considerable lowering distance.

A jacking system with the climbers can be arranged by:

- Mounting climbers on a static structure with the jack rods attached to the particular load; thus the rod will pass through the climber.
- Assembling climbers under a load with the rods suspended from a supporting structure or fixed to a supporting trestle or mast structure.

For systems with lower lift capacity, the grip-jaw arrangement can be modified to allow using selected steel wire ropes with certain types of lifting operations.

A wide range of single- and double-acting hydraulic pumps are available with varying capacities to meet the range of requirements that may result from a particular lifting/lowering operation. In general, heavy lifting jobs require the use of high-pressure rubber hoses that will accommodate the normal working pressure of these pumps operating at between 1422 psi (100 kg/cm^2) and 2148.2 psi (170 kg/cm^2).

Regulations and Standards

OSHA Regulations

1926.305 Subpart I—*Jacks – Lever and Ratchet, Screw, and Hydraulic*

Industry Consensus Standards and Guidelines

ANSI B30.1 *Jacks*

ASME/ANSI *PALD-1*

ASTM D1194-04 *Plate Bearing Test*

AASHTO–245 *Plate Bearing Test*

SECTION 7

Lifting and Hoisting Machinery

CHAPTER 23

Derricks and Cranes

The crane, an invention of ancient origin (both the Greeks and the Egyptians used various forms of cranes to erect their monumental temples and tombs), is used to move heavy weights—as much as several hundreds of tons depending on their design capacity—in both the vertical and horizontal directions; to load and unload heavy objects; and to construct tall buildings.

The first true cranes, founded on the principles of levers and counterweights, used a pulley system fixed to a single mast or boom. Lifting power was provided by humans or draft animals operating a treadmill or large wheel. Eventually, a second mast and guy wires were added to increase the strength and stability of this early form of crane.

One of the most significant developments in crane design, which probably occurred during medieval times with the advent of Gothic architecture, was the jib crane (also known as the derrick crane), which features a pivoting horizontal arm (a jib) that projects outward from the top of the boom. The addition of hinged movement to the outermost section of the jib allows for even further versatility and movement.

NOTE: *Derrick is the term originally applied to the gallows structures, named after the well-known Elizabethan-era English executioner Godfrey Derrick.*

Today, the derrick is a large hoisting machine similar in most respects to the crane, except for its typically stationary foundation.

23.1 Types of Derricks

A derrick is a device consisting of one mast, or equivalent member, hinged freely at the bottom and held at the head by guys or braces—with or without a boom. It is controlled by (usually four) lines powered by such means as manual-hauling or motors, so that the pole can move in all four directions. A line runs up the mast and over its top with a hook on the end, like with a crane. The basic derricks used in construction and maintenance operations are described in the following sections.

Basic Derrick

Typically, a two-drum winch powers a derrick's hoisting and topping motions. When the boom is used only for lifting and swinging, the topping motion is not required, and a fixed rope guy line is used instead. The derrick's winch can be located at any level, but more often it is located at the boom foot's floor level. Heavy winches are usually located on the ground (Fig. 23.1).

Chicago Boom

Used primarily during new construction of high-rise projects, the Chicago boom allows economical and efficient movement of materials, which may be too large or heavy to be lifted using conventional hoist cages.

The device is usually attached to an outside upright member of a structure serving as the mast, with the boom being stepped in a fixed socket clamped to the upright, and structural beam connections, or bracing, serving as the stiff legs. The derrick is complete with load, boom, and boom point swing line falls. Common hoisting blocks are used with this type of derrick—one strapped at the boom tip, the other at the pivot fitting (Fig. 23.2).

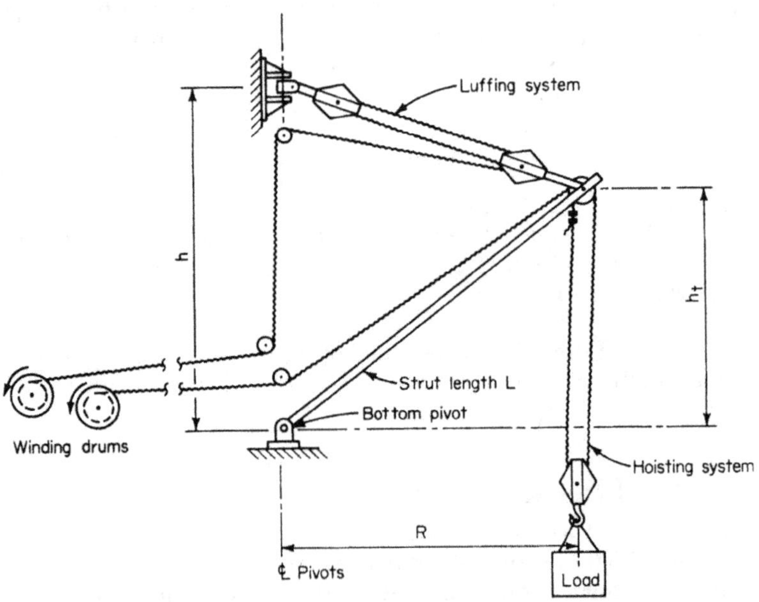

FIGURE 23.1 Basic derrick arrangement. (*From* H.I. Shapiro, *Cranes and Derricks*, McGraw-Hill, 1980.)

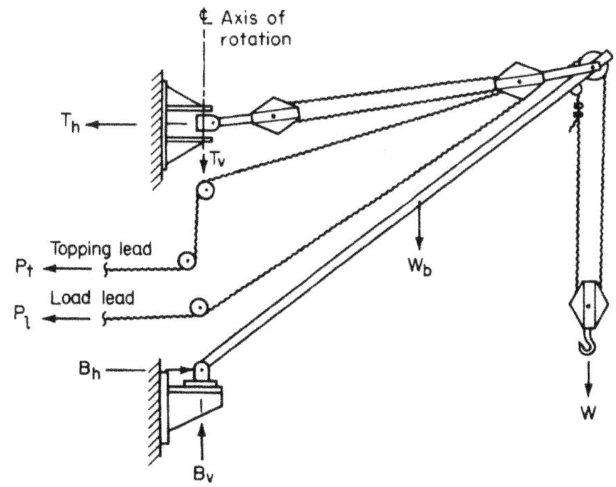

FIGURE 23.2 Chicago boom derrick. (From H.I. Shapiro, *Cranes and Derricks*, McGraw-Hill, 1980.)

Chicago boom lengths range from 10 to 125 ft, with lifting capacities of ¼ ton to about 35 tons. Most consist of trussed or lattices angle iron or steel tubing, or a combination of both.

If heavy loads are to be lifted, steel guy ropes should be installed to carry the stress from the anchorage of the topping lift, to a beam at the approximate elevation of the boom hinge or socket. Swing guys, running from each side of the boom tip to a point of anchorage, enable the boom to swing.

Basket Derrick

In the Chicago boom family, this derrick is without a boom; its base is supported by four ropes attached to corner posts or other parts of the structure. Its boom foot is located below the support anchorages. Additional ropes provide stability (Fig. 23.3).

The location of the base of a basket derrick can be changed by varying the length of the rope supports. The top of the pole is secured with multiple reeved guys to position the top of the pole to the desired location by varying the length of the upper guy lines. The load is raised and lowered by ropes through a sheave or block secured to the top of the pole.

Guy Derrick

Basically, the guy derrick is a Chicago boom with its own integral mast and is held vertically by guy ropes. Standard practice is to use six guy wires spaced evenly around the derrick (Fig. 23.4).

A guy derrick's mast and boom are usually constructed of latticed steel members. Common lengths of mast and boom are 25 and 100 ft, respectively, although these can be larger or smaller, depending on their particular application. Lifting capacities range up to 200 tons.

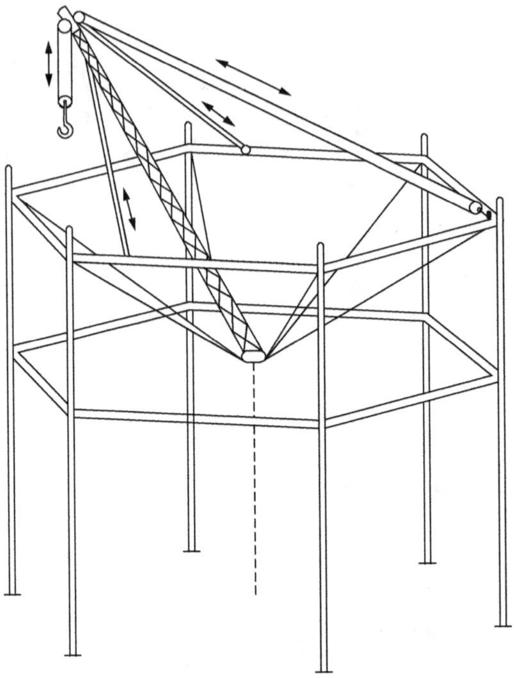

FIGURE 23.3 Basket derrick. (*Source: U.S. Army Field Manual 5-125*, Department of Army, Washington, DC.)

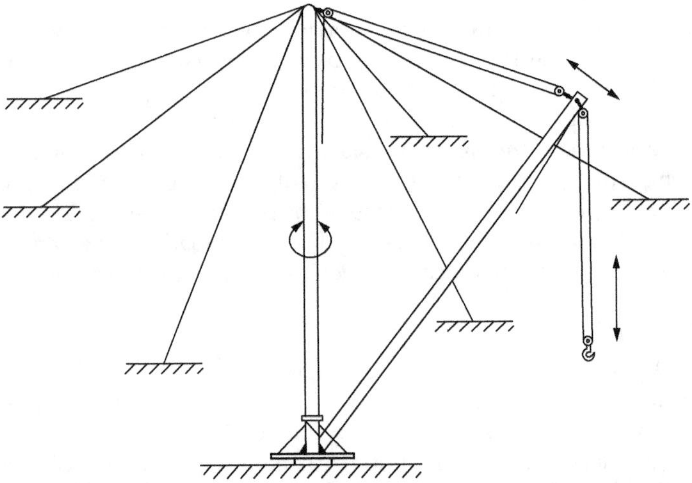

FIGURE 23.4 Guy derrick. (*Source: U.S. Army Field Manual 5-125*, Department of Army, Washington, DC.)

Mast and boom are mounted on a ball and socket joint that enables the mast to lean in the direction of loading. Because the beam is shorter than the mast, this derrick is able to swing through 360°. The mast top is fitted with a pivot pin (gudgeon) to which a ring fitting (spider) is attached from which the guy wires radiate. A large horizontal wheel (bull wheel) fitted to the bottom of the mast permits the system to be swung or slewed. The bull wheel is swung by wrapping rope from a winch around the wheel. A bull pole (or hand wheel with gearing) permits manual operation.

Topping and load lead lines are threaded through the mast, entering at a point above the boom foot and emerging from the mast top. Sheaves, over which lead lines run to the winch, are mounted in the derrick base.

When installing a guy derrick, the mast must be plumbed as closely as possible. The mast base is usually mounted on a steel grillage to transfer the loads to the building. Six to eight guys radiating in a horizontal circle from the mast are used to plumb the member (Fig. 23.5).

Guys are anchored to the building frame and hung in predetermined catenary controlling the amount of mast lean when the system is under load. Because the mast is allowed to lean, design of the guying system is not as critical as that used for tower crane masts.

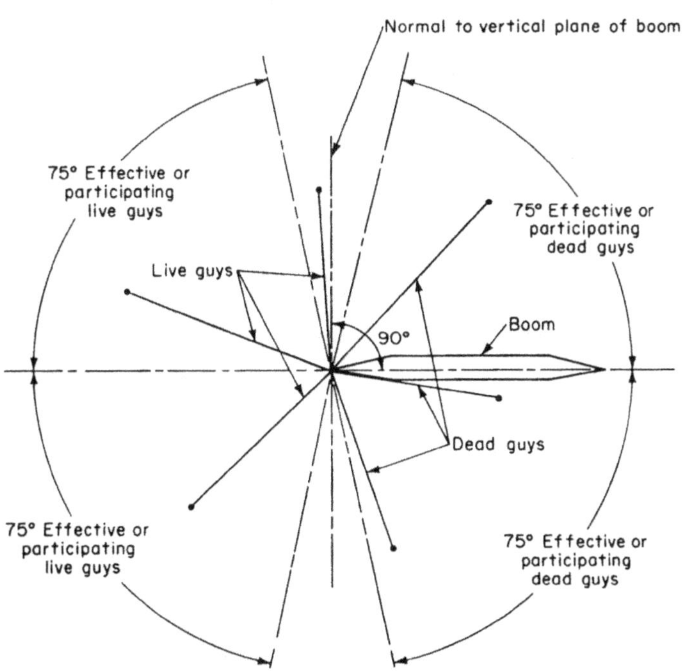

FIGURE 23.5 Guying systems for a guy derrick. (From H.I. Shapiro, *Cranes and Derricks*, McGraw-Hill, 1980.)

Because guys within 90° of either side of the boom do not support loading, they are considered dead guys. Those within 70° of either side of the boom, however, impose a dead-load moment on the system. The remaining guys in the system are considered live guys, with those lying within a 150° area opposite the boom actually supporting the mast.

One of the most practical aspects of guy derricks is the ability of the derrick to jump itself to succeeding floors as the structure's frame erection proceeds. With the winch operated from the base floor, the boom is first lifted to the next floor level, where it is then used to lift the mast to a new grillage base. Initially a mobile crane must be used to erect the guy derrick. While installation costs are considerable, the two-person derrick crew provides an operational economy that more than offsets these preliminary costs.

Gin-Pole Derrick

A gin-pole derrick resembles a guy derrick without a boom and is held in position with guys fastened on top to its spider so as to permit leaning the mast in any direction. The load is raised and lowered by ropes reeved through sheaves or blocks at the top of the mast. Although only two guys, about 60° to 90° apart (in plain view) attached to the top of the gin pole, are necessary to take the strain due to the load, normally four to six guys are installed to prevent the pole from falling over backward in the event of an unexpected jerk or a sudden release of the load (Fig. 23.6).

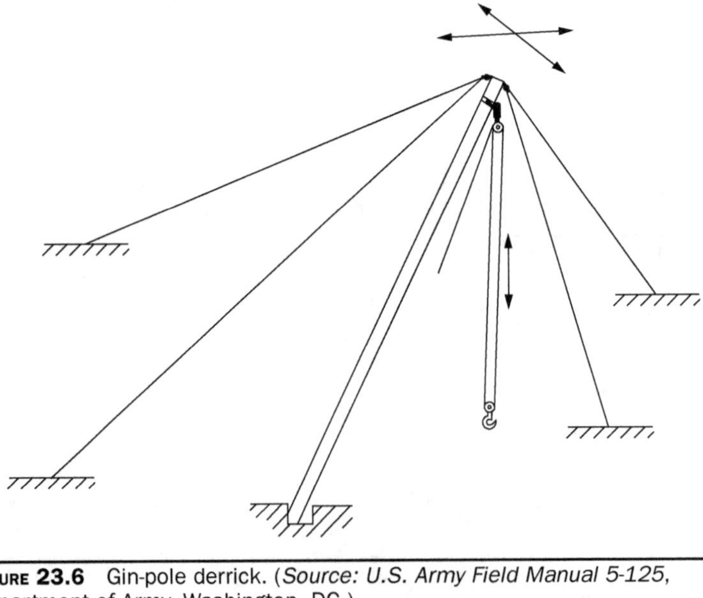

FIGURE 23.6 Gin-pole derrick. (*Source: U.S. Army Field Manual 5-125,* Department of Army, Washington, DC.)

Gin poles do not swing under load, but are used to lift loads, such as installed machine or structural components, vertically at a fixed radius. Heavy gin poles lean only 5° or 10° from the vertical; lighter ones can lean more when fitted with a topping lift. And gin-pole lengths can be 250 ft or more, with lifting capacities up to about 300 tons.

When installed, the gin pole should be plumbed nearly vertical (±10°). Because this type of derrick lifts without radius change, the pole height must be considerably greater than the height of the lift, and its guys should be heavily preloaded to hold to a minimum the radius change, resulting from loading. Care must be taken not to preload the guys to the point of overloading the pole in compression.

For erecting loads taller than a single gin-pole derrick, or where working space is confined, a pair of poles can be used. The poles are erected in tandem, leaning slightly away from each other. An open space is provided between guys at the top end of the load to provide clearance for the load.

Stiff-Leg Derrick

This derrick is similar to a guy derrick except that the mast is supported or held in place by two or more stiff members, called stiff legs, capable of resisting either tensile or compressive forces. Sills are generally provided to connect the lower ends of the stiff legs to the foot of the mast (Fig. 23.7).

Other Derrick Forms

A-frame Derrick

Similar to the Chicago boom, with the A-frame taking the place of a column. The boom is hinged from a cross member between the bottom ends of two upright members spread apart at the lower ends and joined at the top. The boom point is secured to the junction of the side members, and the side members are braced or guyed from this junction point (Fig. 23.8).

The A-frame derrick can swing through a 180° arc and is commonly mounted on barges for marine work.

Gallows Frame

Consists of two gin poles assembled with a horizontal beam across the pole heads. The beam supporting the upper load blocks is arranged to ensure concentric loading and to provide the gallows frame a greater capacity (about 50 percent more) than would the use of double gin poles. Gallows frames have lift heights of 200 ft with capacities up to 600 tons. Shorter frames can lift 1200 tons.

The boom is hinged from a cross member between the bottom ends of two upright members spread apart at the lower ends and

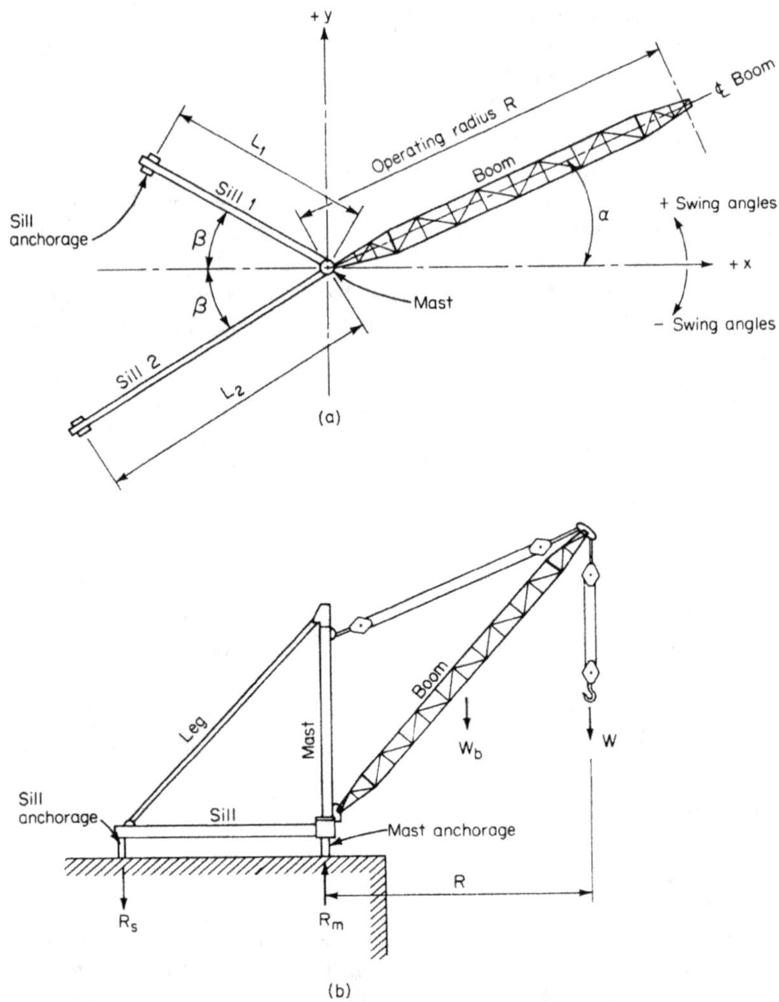

FIGURE 23.7 Installation of stiff-leg derrick. (From H.I. Shapiro, *Cranes and Derricks*, McGraw-Hill, 1980.)

joined at the top. The boom point is secured to the junction of the side members, and the side members are braced or guyed from this junction point.

Column Derrick

Modified guy derrick, consisting of a structural frame (instead of guys), installed at the mast top spider and attached to building columns or other substantial bracing means. Without guys in the way, the derrick can be installed at locations where space is limited or there is insufficient clearance for a full set of guys, and eliminates having to maneuver to boom-in and duck under the guys.

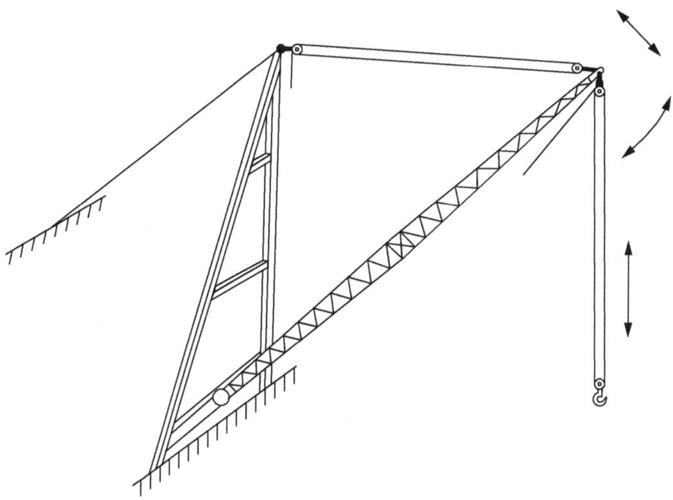

Figure 23.8 A-Frame derrick. (*Source: U.S. Army Field Manual 5-125*, Department of Army, Washington, DC.)

23.2 Types of Cranes

A crane is a mechanical lifting device equipped with a winder, wire ropes, and sheaves that can be used both to lift and lower materials and to move them horizontally. It uses one or more simple machines to create mechanical advantage and thus move loads beyond the normal capability of a human.

Cranes exist in an enormous variety of forms, each tailored to a specific use. Sizes range from the smallest jib cranes, used inside workshops, to the tallest tower cranes, used for constructing high buildings, and the largest floating cranes, used to build oil rigs and salvage sunken ships.

Mobile Cranes

Most basic types of mobile cranes consist of a steel truss or telescopic boom mounted on a mobile platform, which may be rail, wheeled (including "truck" carriers), or caterpillar tracks. The boom is hinged at the bottom, and can be raised and lowered by cables or by hydraulic cylinders. A hook is suspended from the top of the boom by wire rope and sheaves. The wire ropes are operated through a variety of transmissions, including electric motors and internal combustion engines; however, electronic control of hydrostatic drives is today's standard.

Crawler-Mounted Cranes

These cranes are a single engine in the superstructure, which both propels the machine and lifts the loads. A system of shafts, sprockets, and drive chains working off the engine propel this type of crane (Fig. 23.9).

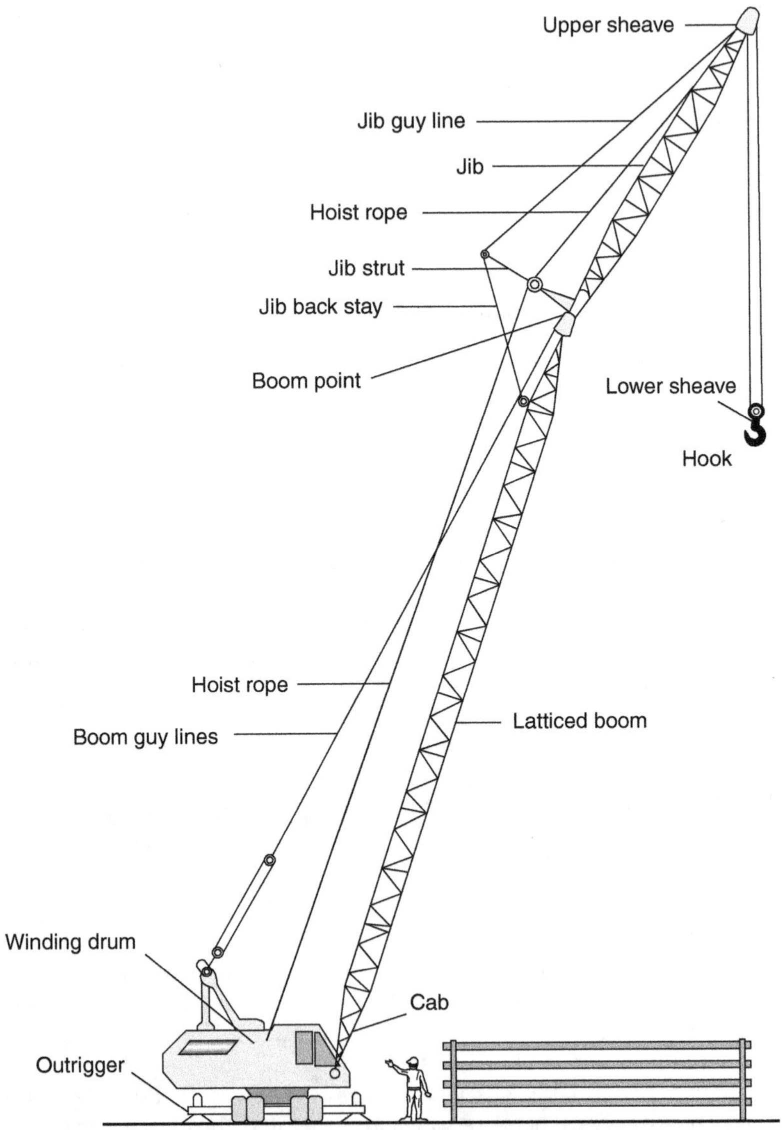

FIGURE 23.9 Crawler-mounted crane with outriggers and lattice boom fitted with a jib. (*Source:* Wikimedia Foundation Inc., San Francisco, CA.)

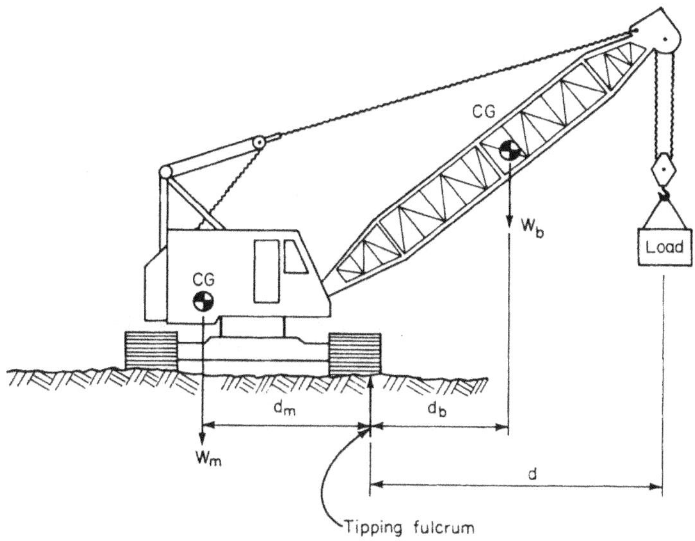

FIGURE 23.10 Crawler-mounted, mobile crane annotated for consideration of stability. (From H.I. Shapiro, *Cranes and Derricks*, McGraw-Hill, 1980.)

Crawler cranes are capable of rotating through 360° and have travel speeds ranging from about 0.5 to 1.5 mph, depending on the machine's size. Their lifting capacities range up to 300 tons, with various boom configurations and base modifications able to increase this considerably. To ensure stability under loading, however, crawler cranes require a wide-track spread. The wider the spread, the greater the crane's lifting capacity (Fig. 23.10).

However, because track spread is limited by over-the-road restrictions, manufacturers usually provide their machines with hydraulically extendible crawlers, which can increase track spread by as much as 48 in., providing the stability required for lifting heavier loads.

Although crawler cranes have no leveling mechanism, stability very much depends on their being operated on a level plane. This often requires laying timber planking to provide a level crane path.

A crawler crane's stability further depends on the position from which the load is lifted. When making a lift from the side of a crane, the side fulcrum line is located on the crane rollers that ride on the tracks. But because the tracks are loosely pinned and resting on the ground, the set opposite the tripping fulcrum cannot resist the crane's tendency to overturn during a side lift (Fig. 23.11). Thus counterweights are necessary to gain the full capacity of a crawler crane. These are removed for travel, or if the crane must be operated in a narrow configuration.

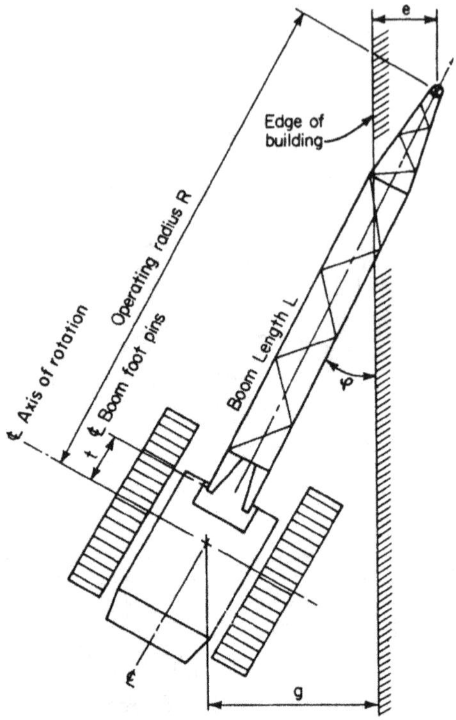

FIGURE 23.11 Plan view of a crawler crane showing the position parameters used for clearance calculations when the boom is at an angle to the wall. (From H.I. Shapiro, *Cranes and Derricks*, McGraw-Hill, 1980.)

Lifting a load over the end of a crane is more stable because the tipping fulcrum is located vertically below the centerline of the drive or idler tumbler shaft (Figs. 23.12 and 23.13).

Wheel-Mounted Cranes

These cranes are mounted on a specially designed chassis that may have as many as nine sets of wheels. To accomplish the dual strain of operating around a job site's rough terrain—as well as traveling on a highway—truck crane transmissions can have as many as 33 forward gears in addition to special high-ratio creeping gears (Fig. 23.14). Depending on the crane model, truck crane travel speeds can range from 35 to 50 mph over the road. They can easily negotiate job site slopes of 20 to 40 percent.

This type of crane is properly stabilized only if all weight is removed from the tires throughout the lifting operation. Manufacturers supply these cranes with hydraulically extendable outriggers that have knuckle-mounted floats on their ends.

The crane is raised on its outriggers by hydraulic rams that either extend down from the outrigger end or push the outrigger beam downward from the chassis in a scissor-like manner.

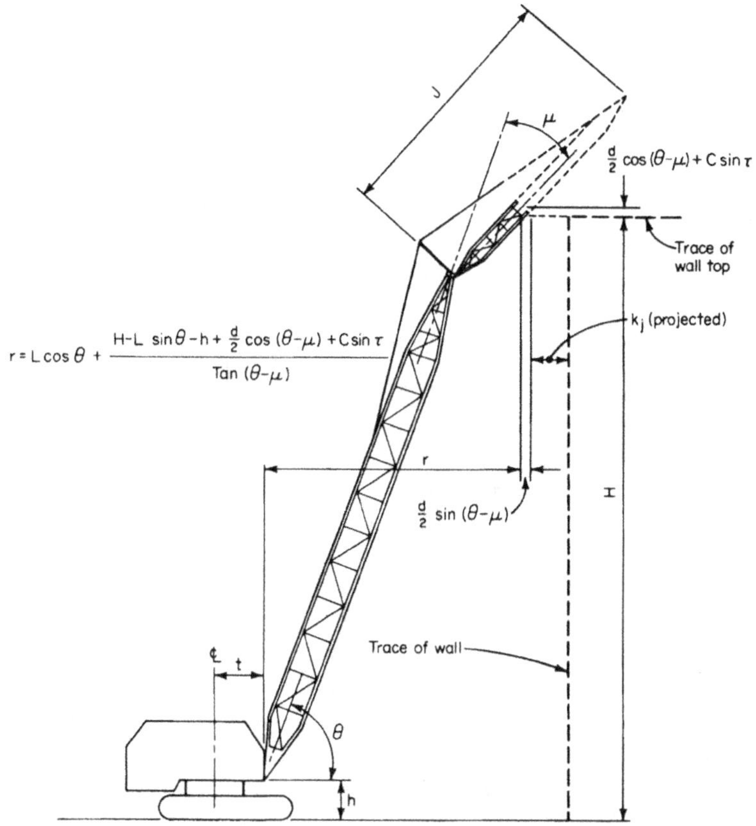

FIGURE 23.12 Elevation of a crane with jib viewed normal to the boom. (From H.I. Shapiro, *Cranes and Derricks,* McGraw-Hill, 1980)

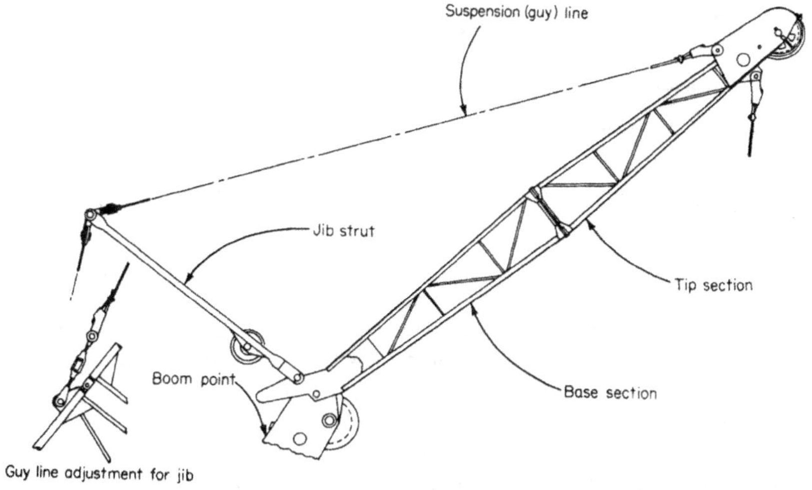

FIGURE 23.13 Typical jib mounting, showing the jib in its shortest configuration; offset 20° from the boom. (From H.I. Shapiro, *Cranes and Derricks,* McGraw-Hill, 1980)

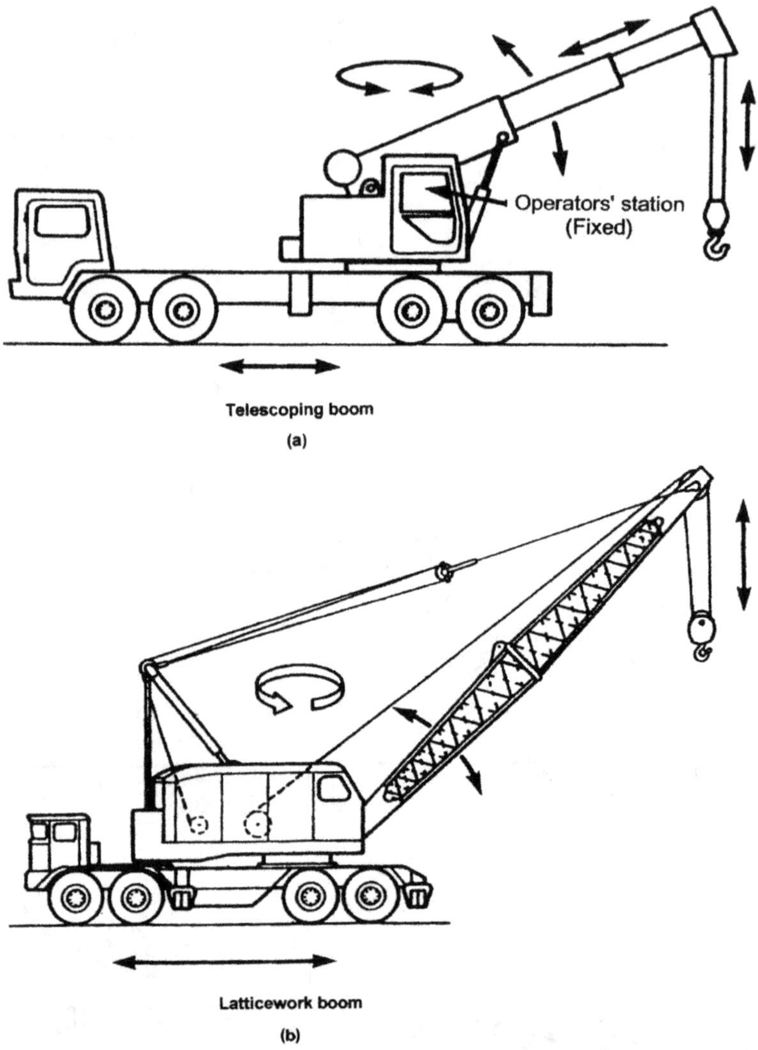

FIGURE 23.14 Wheel-mounted cranes. (a) telescoping boom; (b) latticework boom. (*Source:* Labtrain.noaa.gov—Washington, DC.)

To determine the location of the tipping fulcrum on a crane's outrigger, the machine should be raised so that the outriggers carry the crane's full weight. If the tires remain in contact with the ground, they reduce the outrigger beams' resisting moment and increase the crane's overturning moment.

With tire-mounted cranes having spring-mounted axles, the spring position determines the location of the fulcrum lines. If, however, the two axles are mounted on (walking) beams that parallel the crane's centerline and pivot to the crane's frame (bogie), the frame pivot controls the fulcrum location.

For cranes mounted on axles that are solidly fixed to the frame against rubber shock pads, the center of the tire or the center of a pair of dual tires denotes the location of the tipping line. This mounting arrangement provides the greatest operating stability and permits the highest rated load of the three configurations (Fig. 23.15).

Mechanical blocking can be used to increase the stability of the rubber-tired cranes. Blocking locks a crane's axles to the frame so that the tires define the fulcrum line. This is done automatically on some newer cranes and manually on older machines.

Cranes are transported on carriers having widths that range up to 13 ft, and overall weights to 350,000 lb. Heavier machines are shipped by rail because of load restrictions on highways.

All-Terrain Carriers

These cranes feature the capabilities of both truck cranes and rough terrain cranes. Large-capacity, multi-axle models have high road speed, off-road maneuverability, pick-and-carry ratings, and drive positions in both the chassis-mounted cab and in the operators' cab mounted on the rotating superstructure.

To achieve maneuverability, these cranes typically have all-axle drive and steering as well as crab steering. Many machines are furnished with sophisticated suspension systems that maintain equalized axle loading on uneven terrain, while the crane is in motion or is static. Outriggers are provided for increasing capacity (Fig. 23.16).

Current capacities are available up to 300 tons (272 metric tons), with more models becoming available.

Rough-Terrain Crane

This type of crane is mounted on a two-axle carrier with the operator cab either fixed or rotating—one that swings with the boom. All rough-terrain cranes are fitted with hydraulically telescoping booms consisting of a number of tubes fitted one inside the other. A hydraulic or other powered mechanism extends or retracts the tubes to increase or decrease the total length of the boom (Fig. 23.17).

These types of booms are often used for short-term construction projects, the relative compactness of telescopic booms making them adaptable for many mobile applications.

Oversized tires make these cranes extremely maneuverable over a job site's irregular terrain. Over-the-road travel is limited to about 30 mph. Because of the machine's high center of gravity and mounting system, long-haul over-the-road travel is very uncomfortable. As a result, these machines are often transported by low-bed trailers.

Latticed-Boom Cranes

Mounted on either truck or crawler bases, these cranes are highly versatile, capable of high productivity and a wide range of applications. They are used to load sand and gravel at concrete batching plants, erect high electric transmission towers, set vessels at refineries,

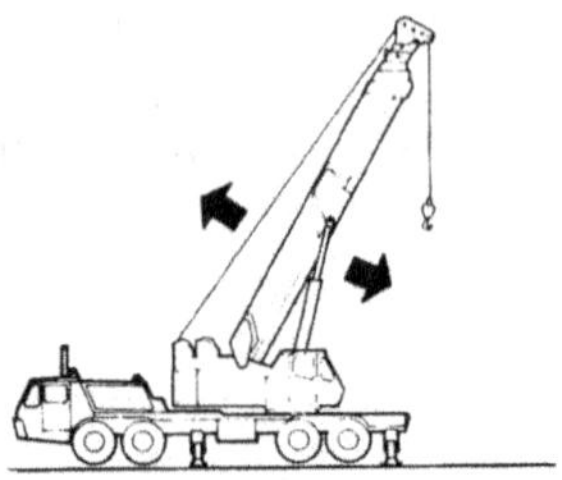

SETTING BOOM ANGLE. Position the boom at the proper angle for the load and working conditions. Refer to the rating plate for proper angle. Pull the boom hoist lever back, or depress the heel of the boom hoist pedal, to raise the boom. Push the boom hoist lever forward, or depress the toe of the boom hoist pedal, to lower the boom. Be sure to pay out line from the main and/or auxiliary winch, to prevent the hook blocks from coming in contact with the boom point.

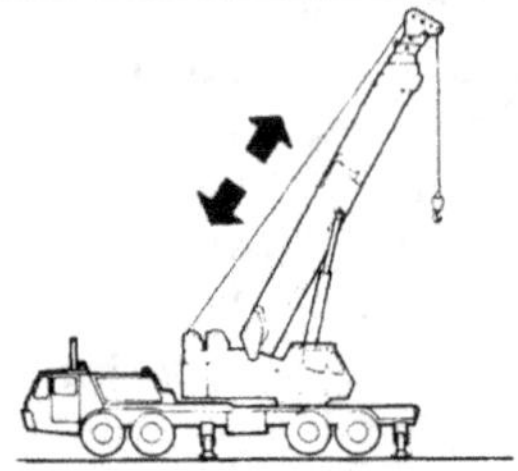

SETTING THE BOOM LENGTH. Push or pull the telescope lever to position the boom at the required length. Check to see that all boom sections are equally telescoped. If section lengths are uneven, depress and hold the individual button for the section to be telescoped. Move the telescope lever forward or back to control the movement in or out for the individual section. Be sure to pay out line from the main and/or auxiliary winch to prevent the hook block(s) from coming into contact with the boom point. Always refer to the rating plate for the proper boom length for the load being lifted.

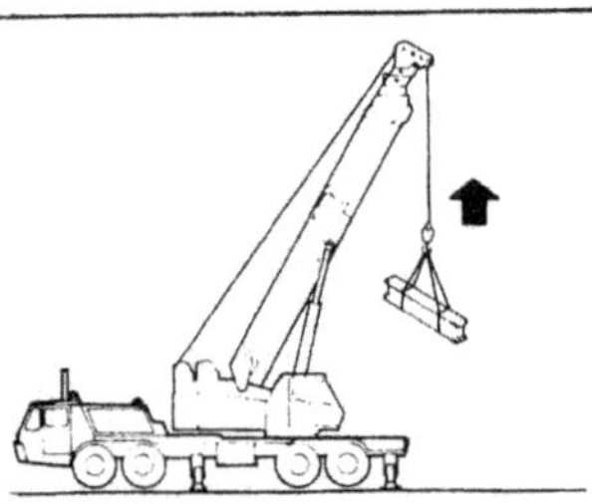

LIFTING THE LOAD. Push the main or auxiliary winch control lever forward to lower the hook block. Attach the load to the hook block. Then pull the main or auxiliary winch lever back to raise the load. Depress and hold the button on the side of the winch lever to operate the winch in high speed. The winch will operate in low speed when the button is not depressed.

SWINGING. Push the swing lever forward to swing the upper to the left (toward the boom). Pull the swing lever back to swing the upper to the right (away from the boom). Plug the swing lever to bring the upper to a stop, and then depress the swing brake pedal to hold the upper. Place the swing brake lock, with the swing pedal depressed, in the ON position to hold the upper stationary without the use of your foot.

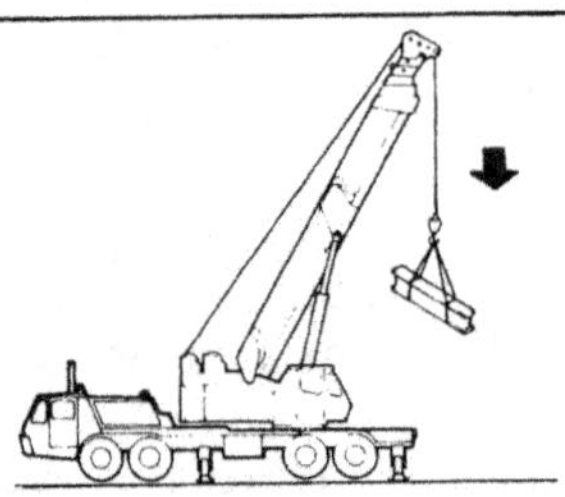

LOWERING THE LOAD. Push the main or auxiliary winch lever forward, to detent, to power down the boom. Push the winch lever past the detent to lower the load under controlled free-fall conditions.

CAUTION

Return the winch lever to the power down position slowly to avoid shock loading the winch and winch line.

Depress the button along side the winch lever to lower the load at high speed. The winch will lower at low speed when the button is not depressed.

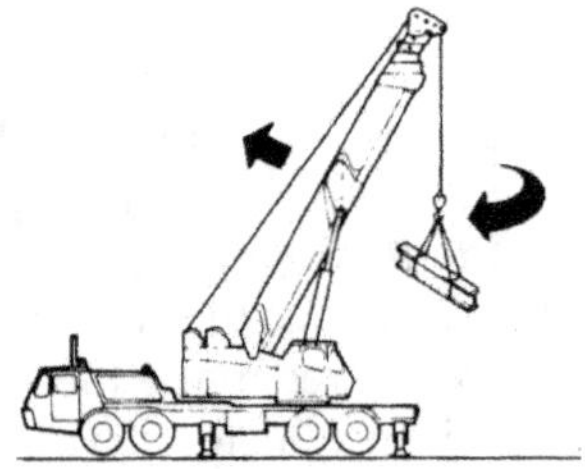

SPOTTING THE LOAD. Spotting the load requires accurate control of the boom and swing movements. It takes practice to locate the load at the exact spot without hunting or overshooting. Adjust the boom as required to accurately locate the load. Never extend the boom out so far that rating is exceeded. See the rating plate.

Figure 23.15 Telescopic crane operating cycles. (From H.I. Shapiro, *Cranes and Derricks*, McGraw-Hill, 1980.)

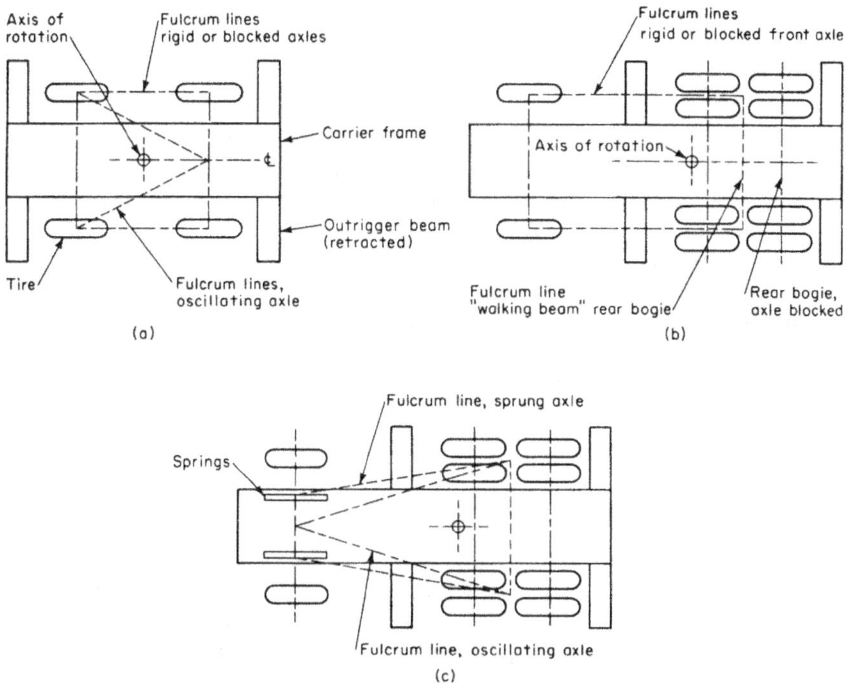

Figure 23.16 Tipping lines for cranes on tires with and without blocking. (a) Rough-terrain crane; (b) truck crane with rigid front axle; (c) truck crane with oscillating or sprung front axle. (From H.I. Shapiro, *Cranes and Derricks,* McGraw-Hill, 1980.)

Figure 23.17 Rough-terrain, wheel-mounted crane, telescoping boom. (*Source:* GROVE, Manitowoc Cranes, Manitowoc, WI)

load and unload ships, demolish building, erect steel, place concrete, and perform almost unlimited lifting tasks. Current machines are available with capacities greater than 1000 tons (900 metric tons).

A rotating structure (superstructure or upper) is supported on the carrier or base and includes the power plant and drives, winches, counterweights, operator's station, and boom-mounting accommodations.

When booms are assembled to long lengths, very lightweight tips with only one or two sheaves are used. Any boom length can be assembled—as specified by the manufacturer—normally in increments of 10 ft (3 m). With few exceptions, mobile cranes are capable of raising the maximum boom lengths assigned to them without assistance from other equipment.

Smaller cranes mount basic booms of 30 ft (9 m) or so, but for large cranes they can be 70 to 100 ft (20 to 30 m) in length. Maximum boom lengths vary with the crane model; the largest currently in use are about 350 ft (107 m).

Support

Ratings for mobile cranes are determined with the machine standing on a firm surface, capable of uniformly supporting the weight of the crane plus its load. Thus chart ratings should be used only if the outriggers, tracks, or ties are supported throughout the lifting operation to ensure maintaining the crane level to within ±1 percent.

For most job sites, once the crane is put in place, the initial settlement is the only settlement that need be considered. This is easily corrected by the first leveling of the crane. If the crane, however, is to be installed in the same location for any length of time (weeks or months), releveling may be required periodically.

It is important to remember that the crane must be maintained level throughout all lifting operations. Crane support should be an essential part of planning all lifts to prevent tipping accidents.

Additional consideration must be given to outrigger floats that bear pressures on soil, because weight stowage restrictions limit the size and area of such floats. Standard floats often produce high bearing pressures that can trigger sudden soil shear failure with subsequent tipping of the crane. Cribbed floats often must be used to prevent such failure.

A rough-terrain crane should never be operated on soil with flat steel plate floats without first providing cribbing to support the floats. The sharp edges of these floats will cut into the soil and thus increase the possibility of crane tipping. Floats having turned up edges do not normally involve the same risk, because they tend to readjust themselves to a level position as the plate digs into the soil.

Stability

Mobile crane stability is extremely critical because it determines the load rating. The rating itself is established by testing the

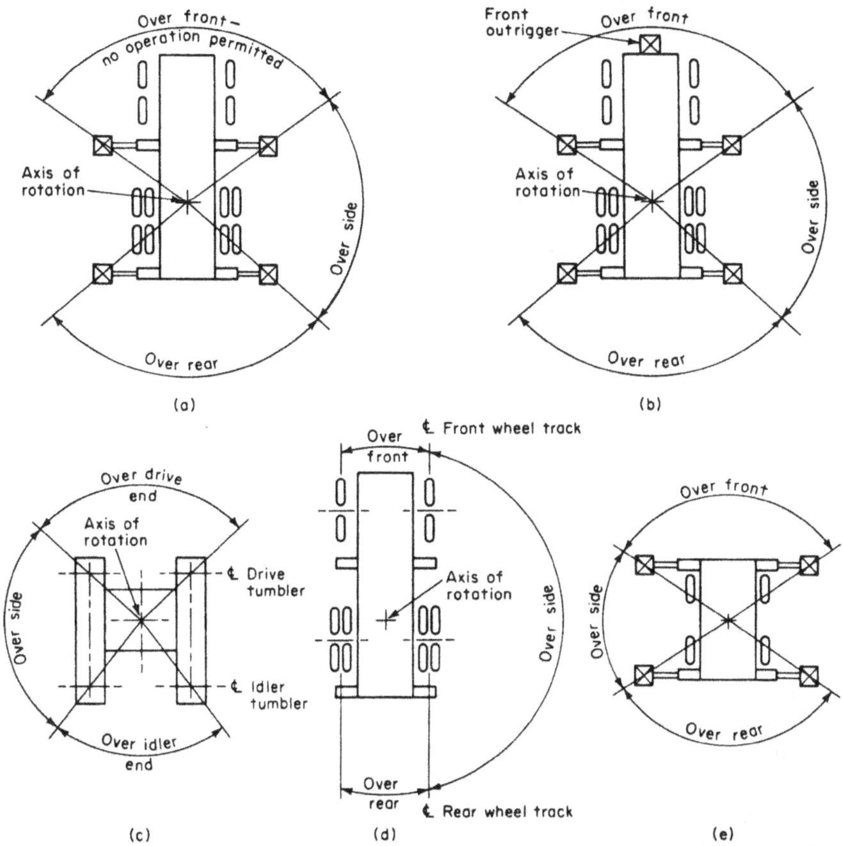

Figure 23.18 Mobile-crane operating sectors for various crane configurations. (a) truck crane on outriggers, normal configuration; (b) truck crane on outriggers with front outrigger; (c) crawler-mounted crane; (d) truck crane on tires; (e) rough-terrain crane on outriggers. (From H.I. Shapiro, *Cranes and Derricks*, McGraw-Hill, 1980.)

machine in calm air with the crane in a level position on firm supports (Fig. 23.18). A change in any one of these test conditions will change the crane's rated load capacity. Most tipping accidents can be attributed directly to the operator's failure to consider the effects of the machine or to operating conditions that vary from these test conditions.

Even though a crane is properly leveled, operating in calm air, and adequately supported with its unloaded hook set precisely to radius, it will exceed its rated capacity the instant it lifts the rated load for that radius. This can happen due to one of the following causes:

- Machine deflecting under loading causes the frame to twist and the lifting radius to increase.
- Boom pendant lines and hoist ropes, elongating under load, further increase the lifting radius, to allow for deflections and rope stretch, it is imperative that a lift at full rated load be made at a shorter radius. The load then can be boomed out to the required radius.
- Wind can be a critical factor in maintaining crane stability
  Blowing from behind the crane operator, wind is extremely critical since it applies force to the boom as well as to the load, adding to the overturning moment.
  Blowing into the operator's face, wind tends to reduce the overturning moment.
  Blowing from the side of the boom, wind becomes critical to boom strength.

Booms and Jibs

Crawler cranes are always fitted with lattice booms, while rubber-tired cranes may have booms that are either lattice or hydraulically telescoping.

The most common boom attachment consists of a base section plus inserts of varying lengths (normally 10-ft increments) and a tip section. Any boom length can be assembled using manufacturer-specified combinations of inserts. When only the base and the tip section are used, the assembly is capable of lifting its maximum capacity.

Hammerhead tips are often used to increase the lifting height. In this configuration, the upper block sheaves and the hook are located away from the boom centerline. Lightweight tips, having one or two sheaves, are used to provide even long booms.

The basic boom for smaller cranes is about 30 ft long; for larger cranes, it runs from 70 to100 ft. Maximum boom lengths will vary with the crane model—to as much as 350 ft with cross sections up to 8 ft^2, or more.

A jib is a relatively low-capacity structure, not very strong with respect to side loading, that is used to extend the vertical reach of a crane. This lightweight boom-like structure is mounted at the boom tip to increase the height of the crane lift.

When a jib is used, the backstay attachment is located on the boom, and fixed stay ropes are used to hold the jib in place. Jib mountings can be parallel to the boom centerline or offset up to 45° from the centerline.

A mobile crane will sometimes be fitted with a tower attachment, especially on sites where the crane must be located close to a structure yet be able to lift and place loads over the structure. The tower is pinned to the crane's upper structure where the boom is normally mounted. A luffing boom is attached to the tower top, and a jib is sometimes fitted at the end of the boom.

Tower attachments usually are assembled and erected without the help of an auxiliary crane, although a small hydraulic unit often will speed unloading and placing boom and tower sections. Varying-length inserts permit the assembly of a number of tower height-and-boom combinations.

Tower Cranes

Even with an extremely long boom, there is a practical limit to the height of buildings that can be erected using a mobile crane. Thus, manufacturers developed a variety of tower cranes, for erecting tall buildings (Fig. 23.19).

All functions on these cranes operate using electric motors that include high- and low-speed ranges, with stepped increments in each range. Newer models have solid-state electronic, stepless controls with friction or eddy-current brakes (or both) that engage automatically as power is withdrawn.

Larger tower crane models have automatic acceleration and deceleration rate control devices for the various motions. Some industrial-type tower cranes even operate under remote control.

These cranes are fixed to the ground (or "jacked up" and supported by the structure as the structure is being built). Tower cranes often give the best combination of height and lifting capacity, and are

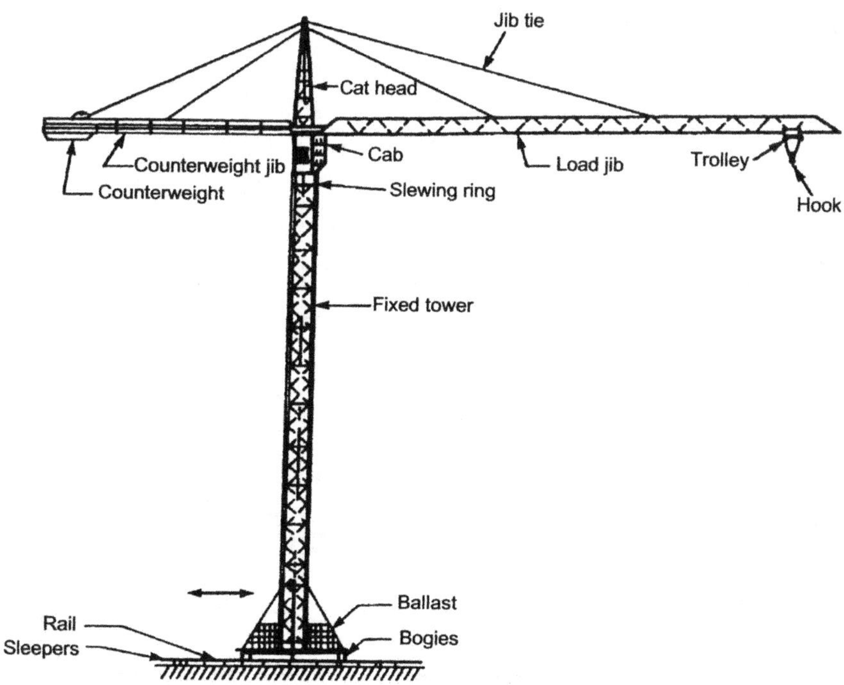

FIGURE 23.19 Hammerhead tower crane. (*Source:* Labtrain.noaa.gov, Washington, DC.)

used in the construction of tall buildings. To save space and to provide stability, the vertical part of the crane is often braced onto the completed structure which is normally the concrete lift shaft in the center of the building (Fig. 23.20). Tower cranes have various base mountings, depending on crane model, size, or application.

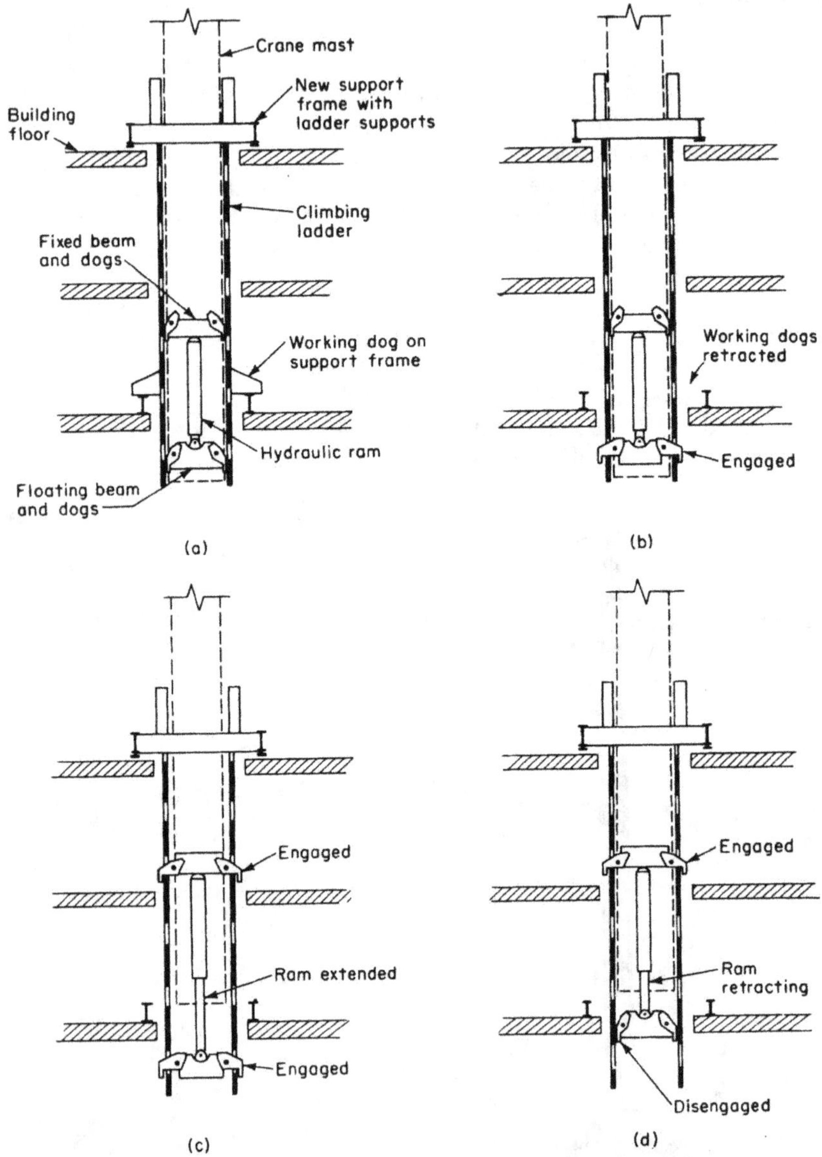

FIGURE 23.20 Climbing sequence of tower crane. (a) Initial condition; (b) load transferred to climbing ladders; (c) crane pushed up by hydraulic ram; (d) ram retracted to start another cycle. (From H.I. Shapiro, *Cranes and Derricks*, McGraw-Hill, 1980.)

Traveling Base

Mounted on this type of base, a tower cranes operate on wide-gage railroad tracks that run adjacent to a building's wall. Such cranes are used almost exclusively on construction of buildings having limited height but considerable length.

The crane tower is rigidly fixed to a platform fitted with four sets of electrically driven wheels (bogies) that ride the steel rails. When installing a traveling-base crane, it is absolutely imperative that the rails be set on a level grade to maintain the tower plumbness throughout its operation, and that they be supported throughout to prevent possible bending or settling under loading.

With this type of crane, ballast weights are stacked on the tower platform.

The crane itself consists of a triangular boom that is raised to the vertical position and acts as the mast. At its top is an auxiliary boom, or jib, that can be raised or lowered to change the crane's working radius.

The operator's cab, in the form of an elevator car, is located within the telescoping mast and can be raised or lowered to change the crane's working radius.

Static Base

This mounting it used with either saddle or articulated jib cranes. The tower is set into a concrete foundation block, or bolted to it, throughout the life of the job.

Acting as a freestanding cantilevered structure, the tower, must be able to resist both vertical and lateral forces, as well as overturning moments. The crane's total weight plus the load it lifts comprise the vertical forces on the block.

While wind load imposes only a minor lateral loading, a combination of wind and the load being lifted will add a significant overturning moment to the crane, which must be resisted.

Climbing Base

This is perhaps the most versatile type of tower crane mounting. It is usually installed at the center of a building under construction, preferably in an elevator shaft. When this is not possible, however, a floor opening to be closed later is provided during construction.

The crane usually is mounted on a foundation of I-beams bolted together to form a cross. A rotary crane is placed on top of the steel tower.

Initially, the base beams are secured to a foundation block, or loaded on the outer ends counterweights. After the crane has erected several stories of the structure, its tower base is raised and bolted to a newly completed floor. When anchored to the floor frame, the tower base transmits the crane's vertical loads to the structure.

NOTE: *Tower cranes are also classified as either fixed type—having their swing circle near the tower top, or slewing type—having their swing circle mounted near the base.*

Tower Crane Jibs

Unlike booms on mobile cranes, tower crane booms are called *jibs*. When mounted in a horizontal or slightly upward sloping position from the tower, they are called *saddle* (or hammerhead) jibs. Those mounted with a pivot and derrick-like system are called *luffing* jibs (Fig. 23.21).

The jib (horizontal boom) is balanced asymmetrically across the top of the tower. Its short arm carries a counterweight of concrete blocks, and its long arm carries the lifting gear. The crane operator either sits in a cabin at the top of the tower or controls the crane by radio remote control from the ground, usually standing near the load.

In the first case the operator's cabin is located at the top of the tower just below the horizontal boom. The boom is mounted on a slewing bearing and is rotated by means of a slewing motor. The lifting hook is operated by a system of sheaves.

Installation

A tower crane is usually assembled by a telescopic crane of smaller lifting capacity but greater height. In the case of tower cranes that have risen while constructing very tall skyscrapers, a smaller crane (or derrick) will sometimes be lifted to the roof of the completed tower to dismantle the tower crane afterwards. A self-assembling tower crane lifts itself off the ground using jacks, allowing the next section of the tower to be inserted at the ground level (Fig. 23.22).

Saddle or hammerhead jib
(a)

Luffing jib
(b)

Figure 23.21 Jib types. (a) Saddle or hammerhead jib (*Source:* Terex Corp., Wetport, CT.); (b) luffing jib. (*Source:* Liebherr Cranes, Newport News, VA.)

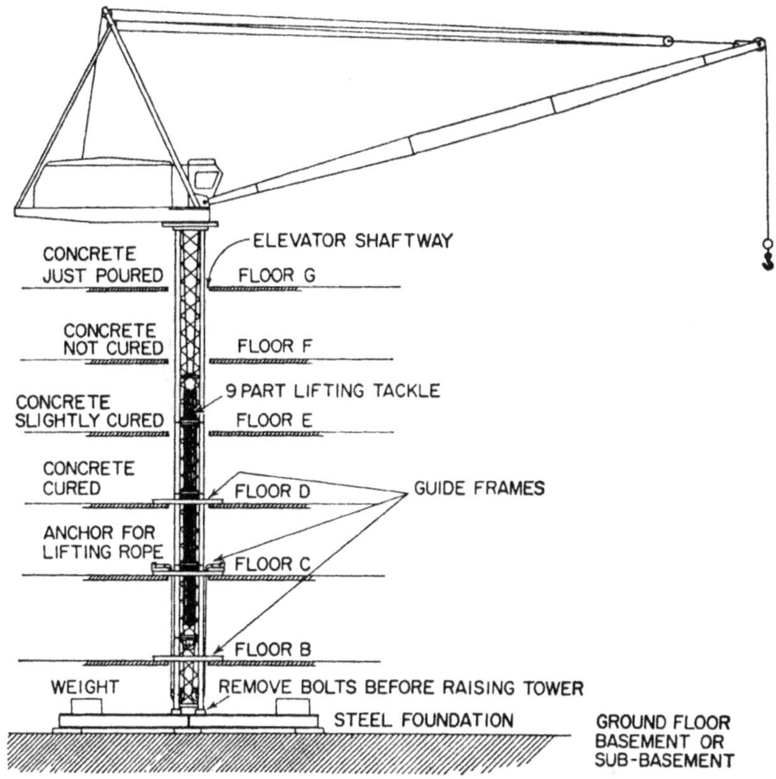

FIGURE 23.22 Raising a climbing tower derrick periodically. (From H.I. Shapiro, *Cranes and Derricks*, McGraw-Hill, 1980.)

Giant Cantilever Crane Also known as a hammerhead crane, the giant cantilever crane is a fixed jib crane consisting of a steel braced tower, on which revolves a large horizontal double cantilever. The forward part of this cantilever or jib carries the lifting crab, and the jib is extended backwards in order to form a support for the machinery and counterbalance.

Besides the motions of lifting and revolving, there is a so-called "racking" motion, by which the lifting crab, with the load suspended, can be moved in and out along the jib without altering the level of the load. Hammerhead cranes are generally constructed in large sizes, up to 350 tons.

Other Types of Cranes

Gantry Crane

This type of crane has a hoist in a trolley that runs horizontally along gantry rails, usually fitted underneath a beam spanning between

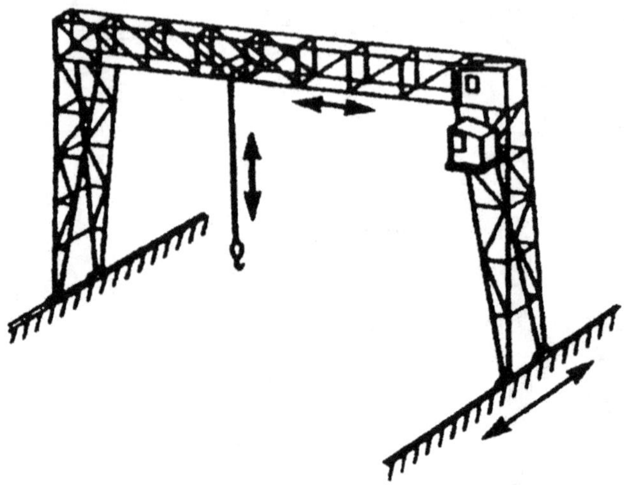

FIGURE 23.23 Gantry crane.

uprights, which themselves have wheels so that the whole crane can move at right angles to the direction of the gantry rails. These cranes come in all sizes, and some can move very heavy loads, particularly the extremely large examples used in shipyards or industrial installations (Fig. 23.23).

Overhead Crane

Also known as a "suspended crane," this type of crane works in the same way as a gantry crane but without uprights. The hoist is on a trolley that moves in one direction along one or two beams which move at right angles to that direction along elevated tracks, often mounted along the side walls of an assembly area in a factory, some capable of lifting very heavy loads (Fig. 23.24).

Floating Crane

Floating cranes are used mainly in bridge building and port construction, but they are also used for occasional loading and unloading of especially heavy or awkward loads on and off ships. Some floating cranes are mounted on a pontoon, others are specialized crane barges with a lifting capacity exceeding 10,000 tons.

Aerial Crane

Aerial cranes usually extend from helicopters, which are able to travel to and lift in areas that are more difficult to reach by a conventional crane.

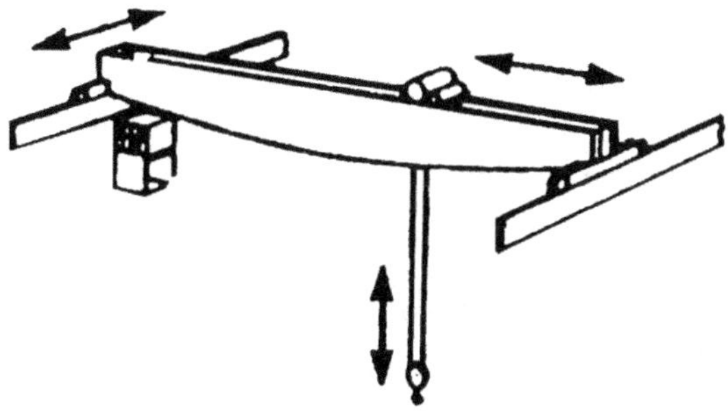

FIGURE 23.24 Overhead traveling crane.

Jib Crane

The horizontal member (jib or boom) of the jib crane, supporting a moveable hoist, is fixed to a wall- or floor-mounted pillar. The jib may swing through an arc to give additional lateral movement, or be fixed.

23.3 Operational Considerations

When working at boom lengths or radii between the figures shown on the load capacity chart, the next lower capacity rating should be used. It is dangerous to guess the capacity for boom lengths or radii between those listed on the rating plate.

It is very dangerous to lift a load without knowing whether it is within the rated capacity while expecting the crane to start to tip to warn of an overload. Cranes may suddenly tip over or the boom may collapse if the load is too heavy.

Always stay within the rated capacity. Operators must reduce the load capacity under adverse field conditions until it is determined that the machine can safely handle the lift.

Manufacturer's operating notes supplied with the machine contain important information concerning proper set-up, operation, and additional points that need to be considered when calculating load handling capacities of cranes. Mistakes in calculating capacity can cause accidents. There are several factors to be considered when calculating a crane's load capacity:

- Loads *must not* be allowed to exceed rated load capacity and working radius.
- Load radius: the horizontal distance between the center of the crane rotation to center of the load.
- Boom length: including the jib, swing away extension or any other attachments that may increase length of the boom.

- Parts of line.
- Quadrant of operation: the area of operation in which the lift is being made; note that different quadrants usually have lower lifting capacities.
- Boom angle: the angle formed between the horizontal plane of rotation and center line of the boom.
- Weight of any attachments: jib, lattice extension, or auxiliary boom point.
- Weight of handling devices: ball, block, or any necessary rigging.
- *Do not* use counterweights heavier than the manufacturer's recommended weight.
- Even a light wind can blow the load out of control, collapse booms, or tip machines.
- Winds aloft can be much stronger than at ground level. Proper precautions *must* be taken when the velocity of wind exceeds 20 mph.
- *Do not* lift loads when winds create an unsafe or hazardous condition. Booms should be lowered, if possible, under high wind conditions.
- *Make sure* the machine is set level; capacity can be adversely affected when it is not level.
- *Make sure* foot pedal brake locks are furnished on some cranes to allow the operator to rest his legs when suspending the load for short periods of time. Operators should keep their feet on the pedals while foot pedal brake locks are in use. Brakes may cool, allowing the load to fall.
- *Never* allow anyone—except the oiler, instructor, or designated person—on a crane with the operator when the crane is in operation.
- Equip each crane with a load chart, visible to the operator, posting rated load capacities, operating speed, and instructions.
- *Make sure* the operator understands and uses the load chart, and can determine the angle and length of the crane boom at all times.
- Post illustrations of hand signals to crane and derrick operators on the job site.
- *Make sure* the signal person uses correct signals for the crane operator to follow.
- Inspect crane machinery and other rigging equipment daily and prior to use to make sure that it is in good condition.

- *Make sure* accessible areas within the crane's swing radius are barricaded.
- *Always* use tag lines to prevent dangerous swing or spin of materials when raised or lowered by a crane or derrick.
- *Make sure* crane outriggers are extended when required.
- *Make sure* crane platforms and walkways have antiskid surfaces.
- Remove broken, worn, or damaged wire rope from service.
- Provide guardrails, handholds, and steps for safe and easy access to and from all areas of the crane.
- Properly bolt and torque tower crane mast bolts to the manufacturer's specifications.
- *Make sure* overload limits are tested and correctly set.
- Post the maximum acceptable load and the last test results on the crane.
- *Make sure* initial and annual inspections are performed and reports are maintained.
- Keep load testing reports/certifications readily available.

Safety Guidelines

Significant and serious injuries may occur if cranes are not inspected before use and not used properly. Often these injuries occur when a worker is struck by an overhead load or caught within the crane's swing radius. Many crane fatalities occur when the boom of a crane or its load line contact an overhead power line.

- Check all crane controls to ensure proper operation before use.
- Inspect wire rope, chains, and hook for any damage.
- Know the weight of the load the crane is to lift.
- Ensure that the load does not exceed the crane's rated capacity.
- Raise the load a few inches to verify balance and the effectiveness of the brake system.
- Check all rigging prior to use; do not wrap hoist ropes or chains around the load.
- Fully extend outriggers.
- *Do not* move a load over workers.
- Barricade accessible areas within the crane's swing radius.
- Watch for overhead electrical distribution and transmission lines, and maintain a safe working clearance of at least 10 ft from energized electrical lines.

23.4 Crane and Derrick Safety

- Allow only properly trained and qualified operators to work with hoisting and rigging equipment.
- *Make sure* that cranes and derricks are restricted from operating within 10 ft of any electrical power line.
- *Make sure* the upper rotating structure supporting the boom and materials being handled is provided with an electrical ground while working near energized transmitter towers.
- Post and make visible to the operator the rated load capacities, operating speed, and instructions.
- *Make sure* cranes are equipped with a load chart, and that operators understand and use the load chart.
- *Make sure* the operator can determine the angle and length of the crane boom at all times.
- *Make sure* crane machinery and other rigging equipment is inspected daily prior to use to ensure that it is in good condition.
- Barricade accessible areas within the crane's swing radius.
- Use tag lines to prevent dangerous swing or spin of materials when raised or lowered by a crane or derrick.
- Post illustrations of hand signals to crane and derrick operators on the job site.
- *Make sure* the signal person uses correct signals for the crane operator to follow.
- *Make sure* crane outriggers are extended when required.
- *Make sure* crane platforms and walkways have antiskid surfaces.
- Remove broken, worn, or damaged wire rope from service.
- Provide guardrails, handholds, and steps for safe and easy access to and from all areas of the crane.
- *Make sure* load testing reports/certifications are available.
- *Make sure* tower crane mast bolts are properly torqued to the manufacturer's specifications.
- *Make sure* overload limits are tested and correctly set.
- Post the maximum acceptable load and the last test results on the crane.
- *Make sure* annual inspections of all hoisting and rigging equipment are performed; and the reports are initialed and maintained.

NOTE: *Chapter 23 was developed from* Crane and Derricks, *H.I. Shapiro, McGraw-Hill, 1980, with permission from the publisher.*

Regulations and Standards

OSHA Regulations

1926 Subpart N—*Cranes, Derricks, Hoists, Elevators, and Conveyors*

1926.550 *Cranes and Derricks*

1926.555 *Conveyors*

U.S. Consensus Standards

ASME B30.2 1990 *Overhead and Gantry Cranes*
ANSI/ASME

B30.9 – 1984 *Slings*

B30.10 – 1987 *Hooks*

B30.11 – 1988 *Monorails and Underhung Cranes*

B30.16 – 1987 *Overhead Hoists (Underhung)*

B30.17 – 1985 *Overhead and Gantry Cranes (Top Running Bridge, Single Girder, Underhung Hoist)*

B30.20 – 1985 *Below-the-Hook Lifting Devices*

HST-4M – 1985 *Performance Standard for Overhead Electric Wire Rope Hoists*

HST-5M – 1985 *Performance Standard for Air Chain Hoists*

HST-6M – 1986 *Performance Standard for Air Wire Rope Hoists*

Industry Consensus Standards and Guidelines

Crane Certification Association of America

Crane Manufacturers Association of America
(Section of Material Handling Industry of America)

CHAPTER 24

Forklifts

Rough-terrain forklifts are convenient tools for moving things from one place to another. As with any other tool, however, it is important to recognize the dangers, practice safety precautions, and use common sense when operating one.

A rough-terrain forklift can carry several tons of material at a time. It is carefully calibrated by the manufacturer to provide a counterbalance at the back of the lift that keeps the load from making the lift tip forward. Some models of rough-terrain forklifts have four-wheel drive as well as four-wheel steering. Most rough-terrain forklifts have a hydraulic system that lifts and lower the forks (Fig. 24.1).

The tires on a rough-terrain forklift are probably the machine's most important component—enabling it to ride over uneven ground because of the traction the tires provide. Some rough-terrain forklifts are equipped with tires that enable driving over rocks, wood, sand, mud, snow, and ice—making them virtually unstoppable. Although most of these tires are filled with air, some are made with solid rubber.

The power of forklifts makes them indispensable machines around construction sites, able to lift and transport just about any load; however, they are also very dangerous. Forklift injuries and accidents most often result from improper use of and lack of training about the forklift. Turnover accounts for a significant number of these fatalities; others include being crushed between the forklift and a surface, being struck by the forklift, being struck by material falling off the forks, and running the forklift off an elevated surface.

Therefore, employees *must* be trained to work safely with forklifts—not just when they're hired but periodically thereafter. Refresher training for the experienced operator is just as important as first-time training for new employees. It is also a good idea for employees to receive basic first aid training in case of injuries. Better training and certification of operators, to ensure that they operate forklifts safely, can reduce operator errors.

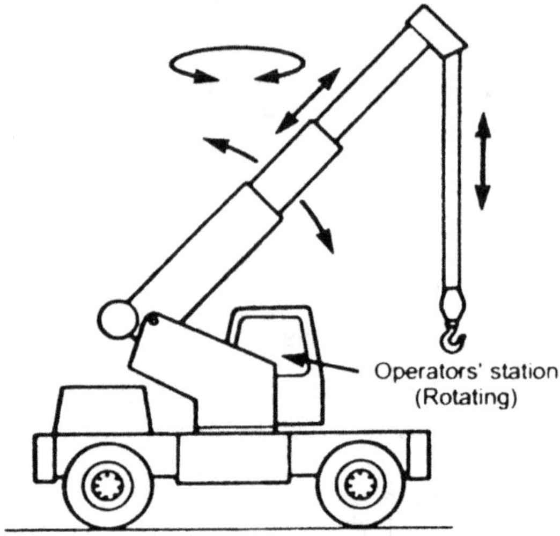

FIGURE 24.1 Rough-terrain forklift.

24.1 Forklift Operational Practices

Safe and efficient use of forklifts depends on both proper operation and proper maintenance. The employer is responsible for making sure that all forklifts used in the workplace comply with current regulations (OSHA, state, and local) to safely do the job required. The forklift operator, however, should check to see that the forklift is in good working condition before starting each shift. Establish a simple daily routine using this checklist:

- Essential for safe forklift operation are the use of personal protective equipment (PPE) and taking precautions to ensure fire safety and electrical safety.
- It is especially important for forklift operators to be alert and prepared for unexpected hazards. Installation of barriers and guards, along with DANGER, WARNING, or CAUTION signs are essential to ensure safety against existing hazards.
- Every forklift must be equipped with an operator-controlled warning device such as horn, whistle, gong, or other sound-producing device(s). The employer *must* determine if operating conditions require the forklift to be equipped with additional sound-producing or visual warning devices (lights or blinkers), and is responsible for providing and maintaining such devices.

Loading

It is important to know the recommended load limit of the forklift (shown on the data plate) and the capacity of the forklift—and never to exceed these limits.

- *Position load* according to the recommended load center (*do not* add extra weight to counterbalance an overload). Keep the load close to the front wheels to maintain the machine's stability. When inserting the fork, keep the mast of the forklift in an upright position before inserting the fork into a pallet. *Make sure* to level the fork before inserting it.
- *Check load* before attempting to lift it. *Make sure:*

  Load is stacked correctly; if not, organize restack.

  Pallet is in good condition; remove damaged pallets from service.

  Load is within the load limit of the forklift; check load plate.

  Special load requirement precautions are considered, such as lifting from the other side or tying it to the carriage or backrest.

  Point of balance is in the middle of the forks when the load is lifted, especially if the load is made up of pieces of different lengths.
- *Never raise or lower* the fork unless the forklift is stopped and braked.

  Check for adequate overhead clearance before raising a load, and maintain a safe working distance from overhead power lines. Lift the load straight up, then tilt the mast back slightly.

  Avoid lifting a load that extends above the load back rest if there is any risk of the load, or part of it, sliding back toward the operator.

  Make sure that the load doesn't catch on adjacent loads or obstructions. *Do not* back up until the forks are free.
- *Loading or unloading* capacity of the forklift is the weight it can safely lift at a specific load center, which is the distance from the heel of the tire to the center of gravity of the test load that the manufacturer uses to establish the forklift's safe working load. Increasing the load center means the forklift become less stable. The manufacturer's load charts show the guaranteed load that can safely be lifted. Carefully follow the forklift manufacturer's capacity guidelines.

  Do not overload the forklift to prevent possible loss of steering control.

  Make sure personnel and equipment on the platform *do not* exceed the available space.

Make sure the combined weight of the platform, load, and personnel does not exceed one-half of the capacity as indicated on the nameplate of the forklift on which the platform is used.

Do not tilt forward when a load is elevated—this can cause the forklift to tip over.

Do not lift a load that extends above the load backrest, unless the load is secured and cannot fall back on the driver.

Traveling

- *Moving the load*. Weight, as well shape and size of load, will affect how load should be moved. Height or width of the load may require taking a different route. When traveling, loads should always be placed against the heel of the tires and fork arms should be below the axle level.

  Keep the load low enough when traveling so as not to block the forward view. At the same time, however, the forks must not be so low that they will scrape the ground, floor, or obstacles near ground level.

  Keep hands, arms, head, feet, and legs inside the forklift.

  Obey posted traffic signs.

  Decrease speed at all corners, sound horn, and watch the swing of both the rear of the forklift and the load.

  Avoid sudden stops.

  Always look in the direction of travel; if load blocks forward vision, travel slowly in reverse.

  Watch for oil and wet spots, loose objects, holes, rough surfaces, people, and vehicles on the floor or roadway.

- *Steering*. Support the load by the front wheels and turn with the rear wheels.

  Do not turn the steering wheel sharply when traveling fast.

  If the forklift is overloaded, steering will be difficult.

  Do not exceed load limits.

  Do not add a counterweight as an attempt to improve steering.

- *Tilt*. Consider the amount of forward and rearward tilt to be used, which is governed by the application; use of maximum rearward tilt is allowable under certain conditions, such as traveling with the load lowered.

- *Stopping distances*. When descending a grade, stopping distance will be greater than on-level operation. Operators must provide methods to allow for these conditions, including reducing speed, limiting loads, and allowing adequate clear space at the bottom of the grade.

- *Parking*. Lower the forks; turn off the engine; engage the parking brake when the forklift is not in use. *Never* park a forklift where it will block entrances, exits, stairways, or fire lanes.

Raising or Lowering Workers

Only operator-up high forklift trucks are designed to lift personnel. A forklift, designed and intended for handling materials, however, may be used to raise and lower personnel—but *only* if the following requirements are met:

- Forklift *must* be approved for elevating personnel.
- Provide a platform that complies with the design requirements in *ANSI B56.1*, para. 7.37.3.
- *Make sure* the work platform is properly designed, constructed, and used to safely support the maximum load that the platform is expected to support. (*Recommendation:* Design should support four times the maximum safe working load.) Combined weight of the platform, load, and personnel *must not* exceed one-half of the capacity as indicated on the nameplate of the truck on which the platform is used.
- *Make sure* a label is prominently attached to the work platform indicating its maximum safe working load.
- *Make sure* the work platform attachment means are applied and the platform is securely attached to the forklift's lifting carriage or forks to prevent accidental lateral or vertical movement. The carriage or forks (or both) *must* be secured to prevent them from pivoting over.
- Equip work platform with guardrails and toeboards that meet OSHA requirements.
- *Make sure* work platform is equipped with a screen.
- Provide protection for personnel in their normal working position on the platform from moving parts of the forklift that represent a hazard, especially a screen to prevent workers from contacting the mast drive mechanism.
- *Make sure* that required restraining means such as railings, chains, cable, body belts with lanyard(s), or deceleration devices are in place and properly used.
- Provide overhead protection as indicated to be necessary by the operating conditions.

Stability

The shape affects a load's center of gravity and the load's stability. Operators *must* evaluate any operation for potential impact to the stability of forklift. Its stability depends on a number of factors, including:

- Ground and floor condition
- Grade
- Speed
- Loading (forklifts equipped with attachments behave as partially loaded machines, even when operated without a load on the attachment)
- Battery weight, on electric forklift trucks [use *only* a battery, or batteries, having a total service weight within the minimum/maximum range as specified on the forklift nameplate (*ANSI B36.1*, para. 7.5.8)]
- Dynamic and static forces
- Operator judgment

To provide forklift stability and avoid tipping over, take into account:

- Weight and shape of load
- Height the load is to be lifted
- Forward or backward tilt of the load
- Area over which the forklift must travel

Visibility

If the size of load limits forward vision, *do not* drive the forklift forward if visibility is not clear around the workplace. Raising the load to see under it is *not* a safe practice because the forklift becomes unstable. Instead, drive the forklift in reverse, but *never* in reverse up an incline. Someone *must* guide the operator while traveling in reverse.

Be aware of "blind spots" created by the mast, the lift cylinder, or other parts of the forklift.

REMEMBER: *Something or someone may have moved since passing through an area.*

24.2 Environmental Considerations

Protective Gear

Operators *must* use the forklift seatbelts provided. They should also wear a helmet and safety glasses for extra protection, even if they are not required at the place of employment.

Noxious Gases and Fumes

Ventilation *must* be provided in enclosed areas where internal combustion-powered equipment is used to maintain an atmosphere that will not exceed the contamination levels specified by the American Conference of Governmental Industrial Hygienists: *Threshold Limit Values of Airborne Contaminants* (*OSHA 1920.1000, Table Z-1*). This includes the atmosphere within the truck cab, when a cab is provided.

Contaminants of carbon monoxide levels in the workplace are dependent on ventilation and proper maintenance of carbon monoxide producers, including internal combustion-powered equipment. Carbon monoxide can become concentrated in area such as trailers, containers, coolers, freezers, and poorly ventilated rooms or buildings. Therefore, limit the use of internal combustion engines in those areas.

Sound

Powered industrial trucks can contribute to the ambient sound in the work area. Consideration should be given to the sound exposure of personnel in the work area.

Work Platform

Whenever a forklift is equipped with a work platform (*does not* include operator-up high lift forklifts), take the precautions previously specified, as well as the following precautions—as required by the operating conditions. Always *make sure:*

- A trained operator is in position to control the forklift or is available to operate the controls.
- A platform is provided that complies with *ANSI B56.1* (para. 7.37.3) design requirements.
- Platform attachment means are applied and the platform is securely attached to the lifting carriage or forks.
- Lifting carriage or forks (or both) are secured to prevent them from pivoting over when the lifting carriage or forks (or both) are supporting the platform used to elevate personnel.
- Mast is vertical–*do not* operate on a side slope.
- Platform is horizontal and centered, *not* tilted forward or rearward when elevated.
- Forklift has a firm and level footing
- Travel controls are in neutral and parking brake is set.

In addition:

- Mark areas with cones or other devices to warn of work by elevated personnel before elevating personnel.
- Lift and lower personnel smoothly, with caution, and only at their request.

- Avoid overhead obstructions and electric wires.
- Keep hands and feet clear of controls other than those in use.
- Move forklift or platform (or both) slowly, only for minor adjustments in horizontal positioning when personnel are on the platform, and only at their request.
- Mechanically secure the rotator (on forklifts equipped with rotators) to prevent movement.
- Engage the parking brake and block the wheels when the operator is not in the operating position.
- *Do not* exceed one-half of the capacity—as indicated on the nameplate of the forklift on which the platform is used—for the combined weight of the platform, load, and personnel.
- Make sure personnel remain on the platform floor.
- Prohibit use of railings, planks, or ladders on the platform for achieving additional reach or height.
- Make sure personnel and equipment on the platform *do not* exceed the available space.
- *Always* lower platform to ground level, or floor level, for personnel to enter and exit.
- *Do not* climb on any part of the forklift in attempting to enter and exit.

Do not allow anyone under 18 years of age to operate a forklift—it is against federal law; those over 18 *must* be trained and properly certified as required by federal regulations.

Inspection—always inspect a forklift before moving or using the equipment:

- Check for and report any leaks, including hydraulic fluid leaks.
- Check the fire extinguisher, horn, lights, backup alarm, and brakes to make sure they are all working properly.

Inspect load before attempting to move it.

- Check entire load, *ensuring* there are no broken or damaged pallets.
- *Make sure* pallets used are sturdy to avoid accidents and injuries from falling debris.
- *Make sure* to secure any loose items.
- *Make sure* to use a spotter, especially when a forklift is traveling in reverse. The spotter should have a clear view of everything in every direction, especially obstacles that obstruct the operator's view or cause distraction such as pallets or people in the path.

- Keep load low enough so as *not* to block the operator's view. At the same time, however, the forks *must not* be so low that they will scrape the ground, floor, or obstacles near ground level.
- *Do not* operate forklift while taking any medication that impairs response or perception, or causes drowsiness. Cold, sinus, and allergy medicines, even over-the-counter preparations, can cause severe drowsiness. *Never* operate heavy machinery or equipment when taking such medication. Pain relievers and other drugs, including prescription medicines, can impair reflexes and perception, which can be very dangerous. If in doubt about reaction to taking any medication, *do not* operate the forklift while taking it.
- *Do not* allow any person to walk underneath elevated forks, or to ride on any part of the forklift. Serious injuries can result from forklift accidents—and failing to follow these precautions can cause both, forklift operator and employer, to be held liable.
- *Follow carefully* forklift manufacturer's capacity guidelines.
- *Do not* overload the forklift.
- *Do not* tilt forward when a load is elevated; this can cause the forklift to tip over.
- *Do not* smoke while operating a forklift. Smoking can cause fire or an explosion; it can also be a distraction. Operators should also refrain from eating, drinking, using headphones, talking on cell phones, or anything else that could be a distraction while operating the forklift.
- Lower the forks; turn off the engine; engage the parking brake when the forklift is not in use.
- *Never* park the forklift where it will block entrances, exits, stairways, or fire lanes.
- Report any and all accidents, no matter how minor they may seem.
- Keep accident reports, even for those accidents that are not severe; some are required by insurance carriers. Keeping accurate accident reports helps to improve safety.

24.3 Forklift Safety Measures

Warning Devices

Every forklift truck must be equipped with an operator-controlled horn, whistle, gong, or other sound producing device(s).

The employer *must* determine if operating conditions require the forklift to be equipped with additional sound producing or visual (such as lights or blinkers) devices and is responsible for providing and maintaining such devices.

Safety Guards

The employer is responsible for ensuring that adequate guards are provided for forklifts and are used by operators. These include:

- *Overhead guards.* Provide forklifts with an overhead guard manufactured as specified by *ANSI B36.1*, para. 7.29, to protect the operator from falling objects.
  Do not consider the guards a substitute for good judgment and care in loading, handling, storage, etc.
  Provide a stronger guard, or one having smaller-sized openings, when specified by the operator—working with the forklift manufacturer.
  Reduce normal overhead guard height and vertical clearance under the guard, if necessary, where overhead obstructions limit the overall lowered height of the forklift to permit operation with a guard.

Note: *Users may operate forklift trucks without overhead guards,* only *if the vertical movement of the lifting mechanism is restricted to 72 in. (1.825 m) or less from the ground, or if forklift will operate in an area where:*

ONLY stable (preferably interlocked, unitized, or containerized) loads are handled.

There is protection against falling objects from adjacent high-stack areas.

Bottom of the top-tiered load is not *higher than 72 in. (1.825 m) and the top is not more than 120 in. (3.050 m) from the ground when tiered.*

- *Load backrest extensions.* If the type of load presents a hazard, the contractor must equip forklift trucks with a vertical load backrest extension manufactured in accordance with *ANSI B36.1*, para. 7.28.
- *Operator compartment guards.* For stand-up, end-controlled, narrow-aisle forklift trucks, more or less guarding may be required to enhance safe operation. Changes must be determined through cooperation between the contractor and manufacturer and only those changes approved by the manufacturer are allowed to be incorporated.

Aisles and Obstructions

Permanent aisles, roadways or passageways, floors, and ramps *must* be defined in some fashion or marked to conform with *ANSI Z535.2—Environmental and Facility Safety Signs.* Permanent or temporary protrusions of loads, equipment, material, and construction facilities into the usual operating area *must* be guarded, clearly, and distinctively marked, or clearly visible.

Lighting

- *Work areas.* Controlled lighting of adequate intensity *must* be provided in operating areas in conformance with *ANSI/IES RP7—Lighting Industrial Facilities.*
- *Forklifts.* Where operating conditions indicate, the forklift *must* be equipped with lights; auxiliary directional lights must be provided on the forklift where general lighting is less than 2 lumens per square foot.

Hazardous Locations

Forklift operators are responsible for determining the hazard classification of any particular atmosphere or location according to *ANSI/NFPA 505.*

Powered industrial trucks operated in hazardous areas and batteries used in hazardous areas *must be* approved for the type required by *ANSI/NFPA 505.*

Depending on the proposed type of forklift truck and area, approved forklifts *must be* built in compliance with one of the following standards:

UL 558—Safety for Internal Combustion Engine-Powered, Industrial Trucks (Forklifts)

UL 583—Safety for Electric Battery-Powered, Industrial Trucks (Forklifts)

Make sure all forklift trucks, and areas of use, are properly marked in accordance with *ANSI/NFPA 505.4.8.3.*

24.4 Safe Forklift Operation

It is against federal law for any person under the age of 18 to operate a forklift. Those over 18 must still be trained and properly certified per federal regulations.

To ensure safe operation of a forklift:

- Inspect the forklift before use. Check for and report any leaks, including hydraulic fluid leaks. Also check the fire extinguisher, horn, lights, backup alarm, and brakes to make sure they are all working properly.
- Inspect the load before attempting to move it. Look over the entire load, being careful to make sure there are no broken or damaged pallets. Sturdy pallets are required to avoid accidents

and injuries from falling debris. Also be sure to secure any loose items.

- Wear protective gear. Use the seatbelt that has been provided. You should also consider wearing a helmet and safety glasses for extra protection, even if they are not required at your place of employment.
- *Always* use a spotter when the forklift is traveling in reverse. The operator cannot possibly have a clear view of everything in every direction. A spotter can also be helpful when other obstacles obstruct view or cause distraction, such as when pallets or people are in the aisle.
- Keep load low enough to prevent blocking travel vision. However, the forks must not be so low that they will scrape the floor or obstacles near ground level.
- *Do not* operate a forklift while taking any medication that impairs response, perception, or causes drowsiness. Cold, sinus, and allergy medicines, even over-the-counter preparations, can cause severe drowsiness. *Never* operate heavy machinery or equipment when taking such medications.
- *Do not* allow any person to walk underneath elevated forks.
- *Do not* allow any person to ride on any part of the forklift.
- *Follow carefully* capacity guidelines.
- *Do not* overload the forklift.
- *Do not* tilt forward when load is elevated; this may cause the forklift to tip over.
- *Do not* smoke while operating a forklift—it can cause fire or explosion, and can be a distraction. Operators should also refrain from eating, drinking, using headphones, talking on cell phones, or anything else that could be a distraction while operating the forklift.

Make sure personnel know and understand the forklift's safety features and operating capabilities and limits—including stability principles, center of gravity, load centering, and the "stability triangle." (Table 24.1)

Design and construction of forklifts (powered industrial trucks) must be in compliance with the current revision of ANSI B56.1/ASME B58.1–1993—Safety Standards for Low lift and High lift Trucks (or the latest revision).

REMEMBER: *When the forklift is not in use, be sure to lower the forks, turn off the engine, and engage the parking brake. Never park the forklift where it will block entrances, exits, stairways, or fire lanes. Even though the lift is no longer in use, safety is still a concern.*

Report any and all accidents no matter how minor they may seem. Constantly look for ways to improve safety (Table 24.1).

Keep reports of accidents even if not severe. Having accurate incident reports helps in this effort.

When using front-end attachments (other than the manufacturers') the forklift must be marked identifying the attachment and listing the approximate combined weight of the forklift an attachment at maximum elevation with a centered load.

Safety Guidelines – See 29 CFR 1910.178 – Section N

- Only authorized employees *must* operate forklifts.
- *Do not* allow any riders on a forklift.
- *Do not* raise people on a forklift.
- *Do not* speed.
- *Always* drive up and back down ramps.
- *Do not* walk, stand, or work under the elevated portion of a forklift (even if it is not loaded).
- *Make sure* that the forklift has an overhead barrier to protect the operator from falling objects.
- *Always* work within the capacity limits of the forklift being used. Check with the manufacturer before modifying the operation or capacity limits of a forklift.
- *Do not* operate a forklift in areas within hazardous concentrations of acetylene, butadiene, hydrogen, ethylene, or diethyl ether, or other explosive environment.
- *Never* lift a load while moving the forklift; completely stop the forklift before raising its mast.
- *Make sure* that the top load sits squarely on the stack; an uneven load could topple.
- Travel with loads slightly tilted back to provide stability.
- Travel with loads at the proper height; a stable clearance height is usually 4 in. to 6 in. at the tips and 2 in. at the heels of fork blades.
- Lift stacked loads in the same manner as loads on the floor.
- *Always* drive the forklift with the load upgrade when ascending or descending a grade in excess of 10%.
- *Always* drive forklift in reverse if not able to see over a load.
- *Do not* try to look around a load and drive forward.
- *Always* lower the mast when preparing to leave the forklift unattended—neutralize the controls, shut off the power, and set the brakes. (The forklift is unattended when the operator is more than 25 ft away, or the forklift is out of the operator's view.)

Source: ANSI B56.6, *Safety Standard for Rough Terrain Forklift Trucks*.

Table 24.1 Forklift Safety Guidelines

24.5 Care and Maintenance of Forklifts

Perform preventative maintenance on forklifts regularly. Make sure that whoever performs the maintenance—whether in-house technicians or an outside service—knows the correct procedure to ensure proper machine maintenance (Table 24.2).

Visual checks	Operational checks
1. Tires, wheels, rims Not excessively worn No punctures	**1. Seat belt** If it's there, use it
2. Hydraulics No fluid leaks Piston free of debris	**2. Horn**
3. Forks and carriage No cracks or bends Any attachments securely fastened	**3. Brakes** Brings machine to a complete stop; parking brake holds machine in a fixed position
4. Overhead guard No cracks Nothing stacked on top or impeding vision	**4. Acceleration** Even acceleration does not stick
5. Mast chains Clear, links intact No excessive slack	**5. Steering** Responsive Minimal looseness
6. ID plate Visible and legible Lifting capacity enough to perform designated task	**6. Mast lift**
7. Battery Charge and in good condition Caps secure	**7. Mast tilt**
8. Propane tank If present, no leaks in connections Tank secured to forklift in designated position	**8. Back-up alarm, lights** If present, are they operational

Table 24.2 Forklift Maintenance Checklist

Forklift trucks *must not* be modified in such a way that will affect the capacity, stability, or safe operation of the forklift without the prior written approval of the original manufacturer (or its successor). If the original forklift manufacturer (or its successor) approves a modification, then the contractor *must* also make changes to the capacity plate, decals, tags, and operation and maintenance manuals. *Make sure* that all nameplates, caution decals, and tags are in place and that they are legible.

NOTE: *Make sure that any modifications involving rebuilding and repair of the basic unit are made in accordance with the manufacturer's established criteria and procedures. Be aware that any changes in load dimension may affect forklift capacity.*

Attachments

If the forklift is equipped with a front-end attachment(s), including fork extensions, the forklift *must* be marked to identify the attachment(s), show the weight of the forklift and attachment combination, and show the capacity of the forklift with attachment(s) at maximum elevation with the load (weight) laterally centered (weight value to be accurate within ±5 percent).

When using front-end attachments (other than the manufacturers') the forklift must be marked identifying the attachment and listing the approximate combined weight of the forklift and attachment at maximum elevation with a centered load.

NOTE: *Any items found to be defective will require immediate notification of equipment supervisor, and the forklift must be taken out of service until repaired.*

Batteries

In electric forklifts, batteries *must* comply with the minimum battery weight range shown on the forklift nameplate.

24.6 Inspection of Forklifts

Forklifts *must* be examined at least daily before being placed in service for that day. If a forklift is used daily on more than one shift, it *must* also be inspected after each shift (Table 24.3).

Tires

Are the tires cut or damaged?

Are the tires pumped up to the recommended pressure?

Mast

Is the mast still straight?

Are all the rollers on the mast still in place and turning?

Are the chains in good order and correctly adjusted?

Is the carriage damaged?

Is the backrest still in place?

Are the hydraulic cylinders, lift and tilt, leaking?

Are the forks (tires) worn, cracked, or bent?

Are the forks properly attached to the carriage?

Seating

Is the seat (or seats) broken or worn out?

Is the seat (or seats) firmly attached?

Are seat belts fitted, and in working condition?

Controls

Are the controls clearly marked?

Do the controls work properly?

Warning devices

Is the horn working?

Is the flashing light working?

Are the brake and turning lights (if fitted) working?

Hydraulic fluid

Are the hydraulic fluid levels adequate?

Capacity

Is a load plate fitted to the manufacturer's specifications?
(*Do not* use the forklift until this load plate is fitted.)

Brakes

Are both hand and foot brakes working properly?

Steering

Is the steering wheel moving smoothly?
(There should be no "slack" or "play" in the steering wheel—that is, no free movement in the steering wheel before the wheels start to turn.)

Source: Environmental Health and Safety Department, Texas A&M University.

TABLE 24.3 Forklift Inspection Checklist

Regulations and Standards

OSHA Regulations

1910.178 Subpart N—*Powered Industrial Trucks*

1926 Subpart O—*Motor Vehicles, Mechanized Equipment, and Marine Equipment*

1926.600 *Equipment*
1926.601 *Motor Vehicles*
1926.602 *Material Handling Equipment*

1926 Subpart W—*Rollover Protective Structures and Overhead Protection*

1926.1000 *Rollover Protective Structures (ROPS) for Material Handling*

1926.1001 *Minimum Performance Criteria for Rollover Protective Structures for Designated Loader, Crawler Tractors, etc.*

1926.1002 *Protective Frames (Roll-over Protective Structures – ROPS) for Wheel-Type Industrial Tractors Used in Construction.*

1926.1003 *Overhead Protection for Operators of Industrial Tractors.*

U.S. Consensus Standards

ANSI B56.1 *Safety Standard for Low Lift and High Lift Trucks*

ANSI B56.5 *Safety Standard for Rough Terrain Forklift Trucks*

NFPA 505 *Fire Safety Standard for Powered Industrial Trucks*

UL 558 *Safety for Internal Combustion Engine-Powered Industrial Trucks (including Forklift Trucks)*

UL 583 *Safety for Electric Battery-Powered Industrial Trucks (including Forklift Trucks)*

Industry Resources

Association of Equipment Manufacturers (*www.aem.org*)

Construction Equipment Association

Construction Equipment Council

Equipment Manufacturers Institute

CHAPTER 25

Portable Overhead Hoists

Portable overhead hoists, both hand-operated (chain) and motor-powered, are widely used in a variety of lifting and holding operations, especially where low overhead restricts the use of a crane or derrick. Their simplicity, dependability, and relatively low cost have made than standard material handling equipment for erection and maintenance work.

The most important consideration when selecting an overhead hoist is its capacity for the heaviest load to be lifted. Hand-operated hoists have a hand chain; motor-powered hoists may be electric or air (Fig. 25.1). The performance and physical characteristics that *must* be considered when selecting a hoist for a given use and installation include headroom height of lift, location, height of hand chain, type of suspension, and trolley clearances.

Basic consideration common to all types of hoists include initial cost, frequency of use, labor savings, safety portability, and maintenance requirements. Other considerations are environmental, such as moisture, heat, chemicals, and foreign material in the atmosphere that may require weatherproofing or special protection.

25.1 Chain Hoists

The simplest and least expensive type of chain hoist is the rigger's ratchet-type that has a mechanical advantage of approximately 15:1. A lightweight portable tool, it can be used for pulling at any angle

When loading, a driving pawl engages the ratchet and, by turning the lift wheel, causes tension on the chain. By shifting the direction lever to the unloaded position, the tension can be released one tooth at a time (Fig. 25.2).

The hand-chain hoist, suspended from a hook and operated by hand chain, is available in lift capacities to 50 tons, with almost no limit on its lift height. Although several types of hand-chain hoists may still be found on some job sites and in some maintenance shops,

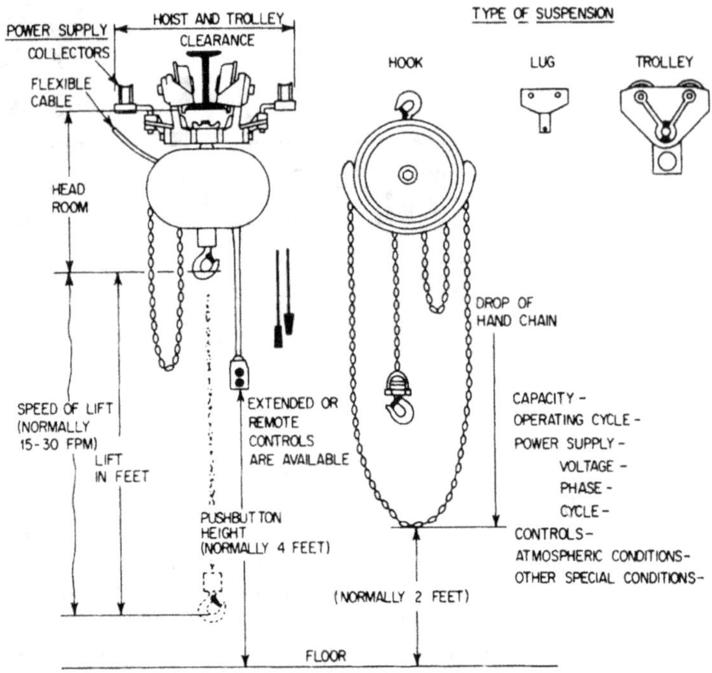

FIGURE 25.1 Chain hoist installation check diagram. (From Higgins and Morrow, *Maintenance Engineering Handbook*, 3d ed., McGraw-Hill 1979.)

but all the high-speed, spur-geared type are considered to be obsolete lifting devices.

Modern high-speed hand-chain hoists have relatively high mechanical efficiencies (65 to 80 percent) and mechanical advantages of approximately 22:1 (for low-capacity units). As hoist capacities increase, so does the mechanical efficiency, as the handwheel size is changed and reduction gearing is used (Fig. 25.3). To suspend a load at rest and during lowering, these hoists are usually fitted with self-energizing brakes, which prevent the descending load from running away on the lowering motion. The brake enables the load to be lowered at a very slow rate and permits the operator to position it precisely.

Separate hand and load chains operate over pocket wheels that are connected by a gear train. A one-way ratchet mechanism disengages the brake during hoisting. Pulling the hand chain continuously in the reverse direction overcomes the brake torque and permits lowering the load.

All chain hoists are designed with their lower hooks as the weakest parts of the device, the two hooks not being interchangeable. If the hoist is overloaded, the first indication is a spreading or opening of the lower hook. As designed, the inner contour of the hook is an arc of a simple circle, and any deviation from this circle is evidence of

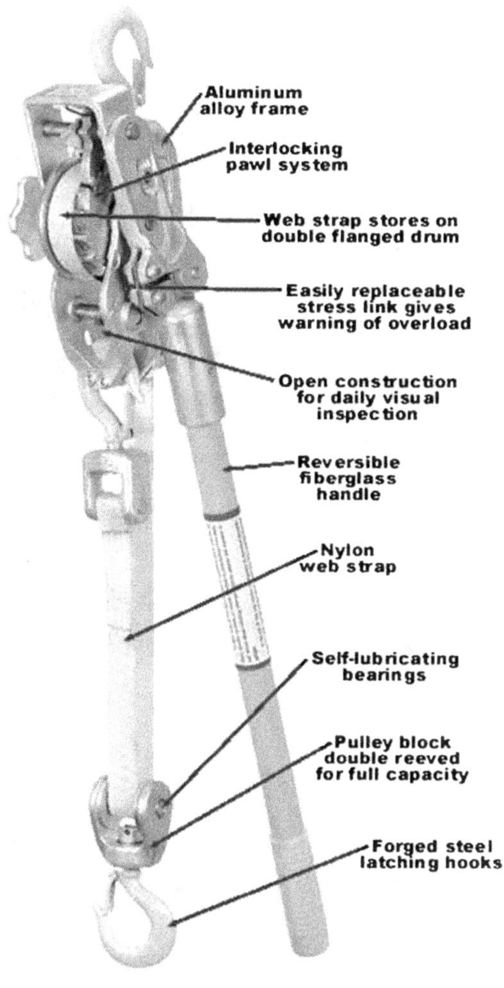

FIGURE 25.2 Ratchet lever hoist. (*Source:* Lug-All Corp., Morgantown, PA.)

overloading. If sufficiently overloaded, the hook will gradually straighten out until it finally releases the load; yet no damage should have been done to other than load-bearing parts of the hoist.

A distorted hook is positive evidence of overloading of great magnitude. All the hooks are rated at 1-ton capacity. Even an overload of 2000 lb does not cause noticeable distortion. An overload of 3000 lb on a 2000-lb hook causes very little spreading. Yet, it is not uncommon to see chain hoist in use with the hook opened up.

When a hook has been severely overloaded, it should be replaced by a new hook. *Never* attempt to force a spread hook back into shape. A new hook is too inexpensive to warrant taking any chances with an overloaded hook. If there is evidence of severe overloading, send the chain hoist to the maintenance shop for a complete internal examination and

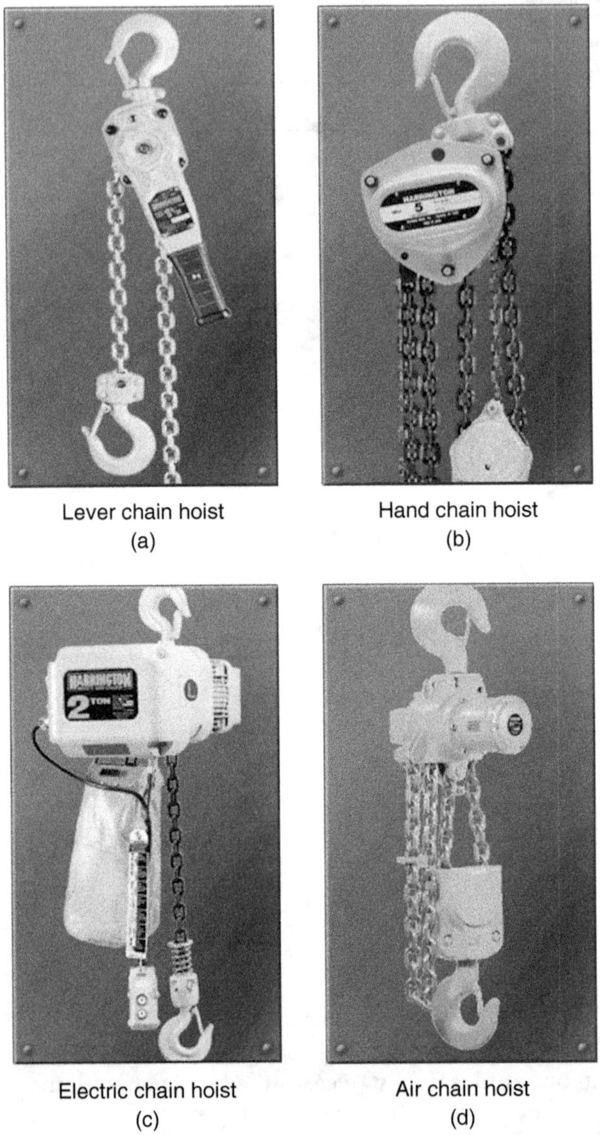

Figure 25.3 Typical chain hoists. (a) Lever chain hoist; (b) hand chain hoist; (c) electric chain hoist; (d) air chain hoist. (*Source:* Harrington Hoists, Manheim, PA.)

overhaul. Pay particular attention to the wear on the brake caused by excessive loading.

When inspecting a chain hoist, it is not only necessary to carefully examine the hooks and the general appearance of the chain; it is also extremely important to thoroughly examine the entire device. Among the many things to check is whether the load brake is receiving sufficient oil; if not, the brake's holding power will be reduced. Also ensure that all hoist parts subject to stress, such as the hooks, swivels, chain, sprocket, gears, and similar parts are made only of forged steel.

When using either a hand chain, or a lever pull, is almost impossible to estimate even roughly the load imposed on the hoist, and there is always a definite possibility of accidental overloading. If, however, only one person pulls on the hand chain, or only one person operates the level pull (without lengthening it by means of piece of pipe), then there is little danger of overloading the device.

Pullers, also known as come-alongs, are special lever-operated hoists that are much smaller and lighter than chain hoist of equal capacities. These devices are available in capacities from 3/4 to 6 tons, and have a mechanical advantage of approximately 25.1. This portable chain or wire hoist is used for short travel distance to lift or pull a load at any angle. A reversible ratchet mechanism in the lever permits short-stroke operation for both loading and unloading. An auto-load friction brake, or releasable ratchet, holds the load securely. A load-sensing device is available that warns the operator when the puller is being loaded beyond its rated capacity.

25.2 Electric-Powered, Wire Rope Hoists

These hoists have lifting capacities ranging from 1/8 to 20 tons, with lifting speeds from 2½ to 64 ft/min (Fig. 25.4). The electric hoist is essentially a drum or sprocket centered in a frame. Its drive motor is

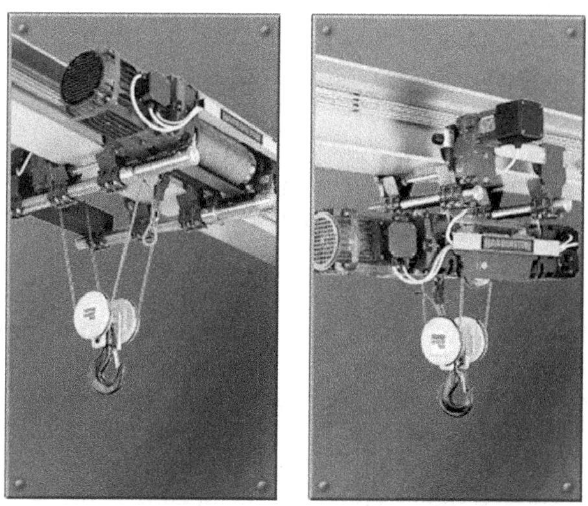

Ultra-low headroom hoist
(a)

Three-phase hoist
(b)

Figure 25.4 Electric wire rope hoists. (a) Ultra-low headroom hoist; (b) three-phase hoist. (*Source:* Harrington Hoists, Manheim, PA.)

located at one end of the frame, with the motor shaft passing through or alongside the drum or sprocket. Reduction gearing at the other end of the frame connects to the drive shaft.

Such hoists are either suspended by an integral hook or bolt-type lug, or attached to a trolley that rolls on an I-beam. The trolley may be the plain push type, geared hand-chain operated, or motor driven.

Two independent braking devices—electric and spring-loaded—are used in the electric hoist. When the current is shut off the electric brake is released, causing the spring-loaded disk brake to engage. When current is turned on, a solenoid trips the spring system to release the brake.

Electric hoists equipped with a brake of the type used in a hand-hoist rely on the motor to drive the load downward to release the brake. This type of brake generates considerable heat, which must be dissipated through an oil bath system. Because this brake does not act when the hoist is lifting a load, the hoist must also be provided with an auxiliary hand-released or electrically released friction brake. Today's electric hoists have push-button-type control of the dead-man type, stopping the hoist instantly upon release.

Safety limit switches for electric hoists control both up and down travel as well as prevent the load hook from jamming against the bottom of the hoist or the chain from running out of the hoist. Upper-limit switches are standard on all electric hoists, while lower-limit switches are standard on chain hoist and optional on wire rope hoists.

The hoist motor acts as a generator when lowering a load, putting current back into the line to control the lowering speed.

Electric motors for powering this type of hoist may be:

- Single-phase, 115-V units that can be plugged into conventional three-prong receptacles
- Three-phase, 210- or 460-V, single speed units
- Three-phase, 230- or 460-V, two speed units

All new hoists have integral load protectors to prevent overloading.

25.3 Air-Powered Hoists

This type of hoist is similar to the electric hoist, except that it operates on an air-powered motor. Although smaller and lighter than electric hoists of equal capacity, air hoists are more expensive because they require mufflers for reasonably quiet operation and normally are fitted with automatic lubricators in the air supply. As with electric hoists, air hoists are also available in two models: *roller or link chain*, having lift capacities to 1 and 3 tons, and *wire rope,* having lift capacities to 15 tons.

They may also be suspended by hook or lug, or attached to a trolley. An air hoist's horizontal movement usually is limited to about 25 ft because of its trailing air hose. For longer distances, a runway can be fitted with a series of air ports along the system, which may be opened and closed as the hoist moves along.

An air hoist is equipped with a brake system that interlocks with the controls and automatically holds the load in neutral. Moving the control releases the brake, either mechanically or by air pressure.

Motors used for air hoists may be either rotary vane or piston types. Piston motors are more costly; however, they provide better starting and low-speed performance and are preferable for larger-capacity models.

Air hoists have infinitely variable speeds, controlled simply by an adjustment of the air valve. When severely overloaded, the air motor automatically stalls, thus preventing hoist damage. Because its motor is remotely located from the hoist itself, an air hoist is particularly suitable for safe use in an explosive atmosphere.

25.4 Safe Operating Practices

There are many aspects to safe hoist operation, but the most important is knowledge about the hoist, the load, and about safe operating practices and the training and communication that support that knowledge.

Safe hoist operation begins with proper hoist selection. The hoist must be matched to the application, and hoist capacity is of primary importance. It is critical that the hoist selected has a capacity that exceeds the weight of the load. *Make sure* the hoist's load chain is long enough to reach the load. Consider a powered hoist if the load has to be lifted for a long distance or lifted repeatedly.

Operator qualification is also important. In addition to good hand/eye coordination, depth perception, and spatial orientation, a good operator should exhibit a willingness to perform maintenance operations and demonstrate safe work habits.

Operator Training

Critical to safety, operator training should be specific to the type of hoist the operator will be using, including information about lift capacity as well as inspections and maintenance, slip clutches, load limit devices, braking mechanisms, and wear limits. Training should also include a discussion of balanced lift points and safe rigging practices.

Hoist Attachments

Slings or other attachments should be seated in the saddle of the hook, and hook latches should be present and functioning properly. The hoist's load chain should *never* be used as a sling. Instead, *make sure* the chain is straight and properly seated in the load sheave and, unless the hook is specifically designed for point loads, avoid tip loading.

Loads

Always lift loads slowly at first, to make certain everything is seated and operating properly. Loads should always be lifted vertically, avoiding the temptation to side pull a load, which places additional stress on the hoist and risks uncontrolled load swings.

Avoid using the hoist's travel limits to stop operation. These limits usually are not designed for regular everyday usage—they are intended for emergency use.

Trolley

When the hoist is coupled to a trolley, the operator *must* exercise care to prevent crashing the trolley into the end stops on the beam. This increases stress on the hoist and will cause dangerous load swings. Jogging of the hoist's motor should be minimized as this will generate heat in the motor's windings, which could lead to motor failure.

Communication

Another aspect of hoist safety that *must* be emphasized is communication, particularly in noisy environments where lifting operations require a hoist operator and a signal person (normally the rigging or hookup person) to use hand signals or voice communication.

Hand Signals

The contractor *must* document and post hand signals, and the signals *must* be agreed on by the hoist operator and the signal person. Furthermore, the operator should only respond to hand signals from the designated signal person, regardless of who gives them, except to obey a stop signal. Effective voice communication requires that the operator and the signal person must be able to hear each other to ensure that the signals for hoist operation are clearly communicated, understood, and executed (Fig. 25.5).

Before giving the signal to lift a load, operators need to be aware of their surroundings, making sure they have a solid foundation for executing a manual lift and that they and others are clear of the load before lifting. They also need to communicate their intention to begin lifting to employees in the immediate vicinity of the lift, and to pay close attention to the task at hand, never leaving a load unattended or suspended.

Hoist: With forearm vertical, forefinger pointing up, move hand in small horizontal circles.

Lower: With arm extended downward, forefinger pointing down, move hand in small horizontal circles.

Bridge travel: Arm extended forward, hand open and slightly raised, make pushing motion in direction of travel.

Trolley travel: Palm up, fingers closed, thumb pointing in direction of motion, jerk hand horizontally.

Figure 25.5 Standard Hand Signals for Controlling Overhead Hoists and Cranes. *(Continued)*

Stop: Arm extended, palm down, move arm back and forth horizontally.

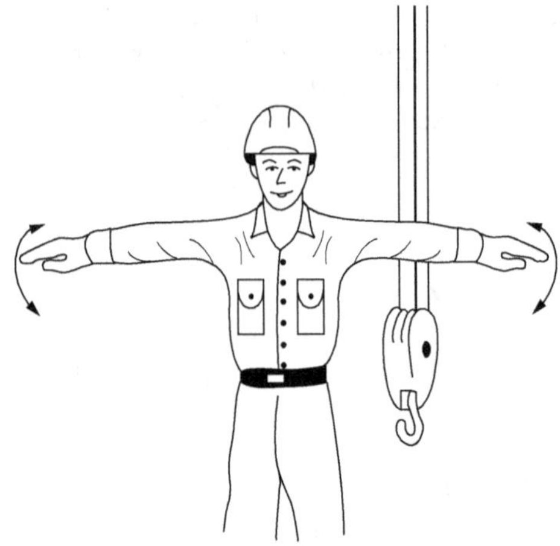

Emergency stop: Both arms extended, palms down, move arms back and forth horizontally.

Multiple trolleys: Hold up one finger for block marked "1" and two fingers for block marked "2." Regular signals follow.

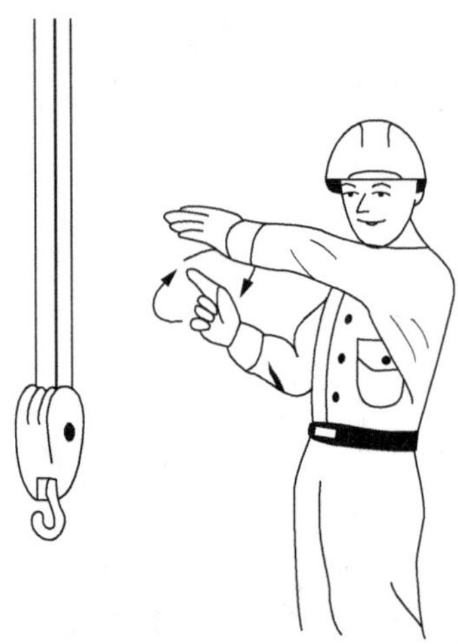

Move slowly: Use one hand to give any motion signal and place other hand motionless in front of hand giving the motion signal.

FIGURE 25.5 Standard Hand Signals for Controlling Overhead Hoists and Cranes. (*Source:* F&G Industries, Terrell, TX.) (*Continued*)

25.5 Operational Tests[1]

New, reinstalled, altered, repaired, and modified cranes *must* be tested by a qualified person prior to initial use to comply with current standards. A contractor-appointed, qualified worker should supervise subcontractors conducting the operational tests, which *must* include:

- Lifting and lowering
- Trolley travel
- Bridge travel
- Hoist-limit devices
- Travel-limit devices
- Locking and indicating devices (if provided)

Preoperational Inspection

At the start of each work shift, the hoist operator *must* test hoist operation before making lifts with any hoist.

- Test the upper limit switch—slowly raise the unloaded hook block until the limit switch trips.
- Visually inspect the hook, load lines, trolley, and bridge as closely as possible from the operator's station.
- Test the lower-limit switch (if provided).
- Test all direction and speed controls for both bridge and trolley travel.
- Test all bridge and trolley limit switches, where provided, if operation will bring the equipment in close proximity to the limit switches.
- Test the pendant emergency stop.
- Test the hoist brake to verify there is no drift without a load.
- If provided, test the bridge movement alarm.
- Lock out and tag for repair any hoist that fails any of the above tests. *Do not* return the hoist to service until necessary maintenance is completed.

25.6 Maintenance and Repair of Hoists

In addition to inspections, regular, ongoing maintenance is critical to hoist safety. This can help prevent any lost-time accidents resulting from hoist failure and prolong the life of the hoist. For more detailed

[1]*Source*: Texas A&M University, College Station, TX (*www.tamu.edu*).

information about hoist safety and maintenance refer to *ANSI/ASME B30.16* and *HMI* (*Material Handling Industry of America*).

Modern hand- and electric-chain hoists require a minimum of maintenance, provided they are used properly according to the manufacturer's recommendations. It is essential, however, to provide correct and adequate routine maintenance for all types of lifting equipment to ensure safe, efficient use of such devices. Moreover, preventive maintenance is extremely important for reducing maintenance costs and improving equipment reliability and production output.

Preventive maintenance *must* take into account both the type of service for which the equipment is used, and whether the equipment is used regularly or is frequently idle. Inspection and testing procedures will depend on these factors. The types of service are classified as:

Normal—hoist operation with randomly distributed loads within capacity, or uniform loads up to 65 percent of capacity (not more than 15 percent for hand-operated hoists and 25 percent for electric- or air-powered hoists) during a single work shift.

Heavy—hoist operation within rated capacity that exceeds normal service.

Severe—hoist operation involving normal or heavy service in adverse environmental conditions.

Initially, a preventive maintenance program may seem costly, but equipment downtime for periodic inspection only may be intolerable from a production standpoint. A key to holding down costs is a good record-keeping system for planning, scheduling, and training maintenance personnel. Not only does preventive maintenance reduce the corrective maintenance workload, but as preventive maintenance takes over, the time of the corrective workload shifts from "when you must do it" to "when you want to do it." Thus work can be done more efficiently and at a considerably lower cost.

Preventive maintenance consists of inspecting equipment and keeping records to show wear or other deterioration of parts. Such records may include the basis of routine inspection and alerting the hoist operator to the need for major overhaul before breakdown.

All hoist-operating mechanisms are subject to wear and *must* be inspected at regular intervals to assure that worn or damaged parts are removed from service before they become unsafe.

- After operation, clean the chain block and keep it in a dry place to prevent rust and corrosion.
- Clean the chain block annually by purging the parts in kerosene and applying grease to them. It is advisable that the cleaning work should be done by skilled hands.

- "O" marks on the two disk gears should be aligned.
- Stick the rollers of both left and right bearings to the inner race of the bearings on the chain sprocket shaft journal, and then put them into the outer race of the bearings on the side plates.
- After cleaning and repair, the chain block should be subjected to nonload and heavy load tests. If it works normally put it into operation.
- For convenience of maintenance and dismantling, open one of the links of the hand.

Lubrication

It is essential to set up a regular lubrication schedule for all moving parts of a hoist for which lubrication is specified. The method of lubrication should be checked for proper delivery of lubricant.

Be sure to follow the manufacturer's recommendations as to the quantity and type of lubrication to be used, as well as the lubrication frequency. *Do not* substitute lubricants that are not recommended by the hoist manufacturer.

Hoist Repairs

Hoist manufactures give assembly and disassembly instructions as well as the location and identification of all parts in their maintenance and parts manuals. When replacement parts are required, *make sure* to include nameplate information, especially the serial number of the hoist.

25.7 Inspection of Overhead Hoists

As vital as overhead hoists can be to rigging operations, they are often some of the more neglected equipment in a facility or on the worksite, many times getting attention only after they fail or are no longer functioning. Preventative maintenance and frequent inspection can prevent costly downtime and potentially dangerous situations (Tables 25.1 and 25.2).

Hand chain blocks, once used as permanently installed facilities for routine manufacturing and maintenance work, were generally very heavy, very robust, and relatively expensive—but would last a lifetime. Today, however, they are usually regarded as portable tools, and price and light weight are considered more important than durability.

Consequently, they are more likely to be dropped, badly stored, and generally abused than their permanently installed predecessors. As a result, in-service inspection must take this into consideration. The more portable these manually operated rigging tools become, the more attention they require while in service.

As with any equipment about to be inspected, the object must be reasonably clean. However, caution must be the watchword regards the methods used. Both hand chain blocks and lever hoists have a friction brake, and there is lubricant on adjacent bearings, gears, etc.

Using a pressure washer can remove essential lubricant and cause it to migrate into the brake. Also, as many components will corrode and seize if left wet, careless cleaning can create more problems than it solves.

If the equipment is already wet, allow it to dry without the application of direct heat. *Never* immerse the equipment in lubricant, diesel, or similar substances in an attempt to displace water. Inspect it after it has dried in case any corrosion occurs which might cause it to malfunction.

When inspecting the equipment, try to be methodical. It's always a good idea to work from the top down. Look at:

- *Shape of the top hook and the fit of the safety catch.* If the hook has been on too large a connection or is point loaded it can cause distortion, even within its safe working load (SWL).
- *Condition of the block or hoist body.* In particular, look for distortion of the frame and missing or damaged gear covers which might cause the mechanism to jam. Many modern blocks and hoists have frames comprising two plates and spacer studs. If the spacer studs become loose the frame will rack.

Hand Chain Block

Check the hand chain wheel for damage and also the guide or casing which prevents the hand chain from being displaced.

REMEMBER: *If the chain is displaced in service it can fall and cause serious injury.*

Check the condition of the chain and, in particular, the link that joins the ends. This is often not welded, and the join may be not be smooth. Bear in mind the possibility of injury to the operative's hands. Excessive wear is unlikely, but bent and stretched links are possible and generally arise if the chain has become snagged on the load.

Lever Hoists[1]

Check the condition of the lever, the reversing ratchet, and the free wheel mechanism. While it is certainly a bad practice, it is not unknown for hoist operators to extend the lever with a scaffolding tube, which almost inevitably results in damage.

[1]*Source: Hoist Magazine*, Sidkup, Kent, UK (*www.hoistmagazine.com*).

To secure the slack end anchor for the load chain:

- On a hand chain—block the chain by anchoring it back to the frame
- On a lever hoist—it has a stop on the chain

Whatever method is used, it is absolutely essential that it is effective.

Many people do not realize just how much force can be applied to the anchor or stop in the event that the chain is fully run out. It can easily be as much as twice the SWL, and here's why: Usually when the chain is fully run out it is because the operative didn't realize it and kept lowering off, thus opening the brake. However, as the load doesn't descend, the brake stays open. At this point the load on the anchor is the same as that on the hook. If the operative then tries to lower the load further, torque is added to the load wheel, resulting in more load on the anchor or stop.

Depending on the design and state of maintenance, some blocks and hoists can require almost as much effort to start lowering as to lift, so the operative will pull quite hard before eventually realizing what has happened. If the anchor or stop is not adequate, the chain will run through with potentially disastrous consequences. This is why the British Standard for hand chain blocks specifies an anchor strength of 2.5 times the SWL of the chain. Unfortunately, among those devices that do not adhere to this requirement, we often see slack end stops that are little more than a glorified key ring or a piece of bent mild steel bar tack welded at the ends, or a mild steel bolt substituted for the anchor on a block.

Condition of the Load Chain

Check for wear between the bearing surfaces of the links, bent links, stretched links, and for corrosion. Wear generally occurs gradually and is usually revealed by an operational test before the loss of strength becomes a problem. Bent or stretched links and corrosion, however, can occur at any time. Bent links are usually the result of abusing the chain by loading it across an edge or, worse still, using it as a sling by wrapping it around the load and back hooking.

The chain used in a block or hoist has a finer tolerance on the pitch than sling chain, ensuring it can accurately mate with the pockets in the load wheel. Bent or stretched chain will therefore not mate properly and can lack articulation. Also, the bottom hook always has a swivel so that if the load rotates, it will not twist the chain. Wrapping and back hooking renders the swivel ineffective.

When the block or hoist is hung up with the chain run out, the links should appear to line up without twist. A simple articulation test is to lower the chain to the floor and look for locking between links.

Condition of the Bottom Hook and Safety Catch

As for the top hook, check for distortion and that the safety catch is present and functional. Check also that the swivel is free. This is one component where the inspector can easily do some basic maintenance, removing any debris, such as cement, and adding a drop of oil.

Having completed the above steps it is always worth doing an operational test, preferably with a light load of about 5 percent of the SWL. Check that it operates smoothly in both directions. When lowering, blocks and hoists rely on the load to pull the brake in. Generally 5 percent of the SWL is the minimum they can be guaranteed to sustain but this can increase due to corrosion and lack of maintenance. If slippage occurs at this load, the equipment should be withdrawn from service for maintenance.

If the load chain jumps or makes a cracking noise during lifting or lowering, this is an indication that the chain is not mating accurately with the load wheel. There are several possible reasons, including the wrong load chain, worn load chain, worn load wheel, or a buildup of debris in the bottom of the load wheel pockets. The latter can be cured by removing the debris, but if this indication occurs the equipment should be withdrawn from service.

25.8 Inspection Scheduling

Initial Inspection

Prior to the initial use and on regular intervals, hoists *must* be inspected by a designated person to verify compliance with *ASME/ANSI B30.16*. The specific inspection requirements are found in Table 1 and Table 2 of the Standard. The inspections are classified as "frequent" inspections which do not require documentation and "periodic" inspections which require documentation. The interval between inspections depends on the service of the hoist.

The owner's manual, specific to the hoist, is a good source for inspection and maintenance requirements—and should be based on the requirements of this Standard (Tables 25.1 and 25.2).

Preoperational Inspection

The operator *must* perform a preuse inspection, if any deficiencies are found the equipment will be taken out of service and the supervisor notified.

- Visually inspect all load chains for gross damage that may be a hazard.
- Make sure hoist is properly secured.
- Verify that load does not exceed the load limit on the warning label.
- Check that motions are smooth and regular with no hesitations, vibration, binding, unusual noise, or other irregularity.
- Make sure chain is not kinked or damaged.

Hand chain hoists	Normal service	Heavy service	Severe service
Visual Frequent inspections	Monthly (1)	Weekly (1)	Daily (1)
• All functional operating mechanisms for maladjustment and unusual sounds • Hoist braking system for proper operation • Hooks in accordance with *ASME B30.10* • Hook latch operation, if used • Load chain in accordance with *ASME B30.10 (*para. 16-2.5.1 or 16-2.6.1) • Load chain reeving for compliance with hoist manufacturer's recommendations • Periodic Inspection *ASME B30.10* (para. 16-2.1.3)			
Record frequent inspections	Yearly	Semiannually	Quarterly
• Evidence of loose bolts, nuts, or rivets • Evidence of worn, corroded, cracked, or distorted parts such as load blocks, suspension housing, hand chain wheels, chain attachments, clevises, yokes, suspension bolts, shafts, gears, bearings, pins, rollers, and locking and clamping devices • Evidence of damage to hook retaining nuts or collars and pins, and welds or rivets used to secure the retaining numbers • Evidence of damage or excessive wear of load sprockets, idler sprockets, or hand chain wheel • Evidence of worn, glazed, or oil-contaminated friction discs; worn pawls, cams or ratchet; corroded, stretched, or broken pawl springs in brake mechanism • Evidence of damage of supporting structure or trolley, if used • Label or labels required by *ASME B30.10* (para. 16-1.1.4) for legibility • End connections of load chain			

Notes:

1 - By operator or other designated person with records not required.

2 - Visual inspection by a designated person making records of conditions to provide the basis for a continuing evaluation.

3 - As in Note 2 unless conditions indicate that disassembly should be done to permit detailed inspection.

Source: ASME B29.24, *Roller Load Chains for Overhead Hoists,* The American Society of Mechanical Engineers.

Table 25.1 Inspection Schedule for Hand Chain-Operated Hoists

Annual Inspection

The annual inspection may be performed with the hoist in its normal location and does not require the hoist to be dismantled. Covers and other items normally supplied to allow inspection of components *must* be opened or removed for these inspections. Check:

Electric- or air-powered hoists	Normal service	Heavy service	Severe service
Visual frequent inspections	**Monthly (1)**	**Weekly (1)**	**Daily (1)**
• All functional operating mechanisms for maladjustment and unusual sounds • Limit devices for operation • Hoist braking system for proper operation • Air lines, valves, and other parts for leakage • Hooks in accordance with *ASME B30.10* • Hook latch operation, if used • Hook rope in accordance with *ASME B30.10* (para. 16-2.4.a) • Load chain in accordance with *ASME B30.10* (para. 16-2.5.1 or 16-2.6.1) • Rope or load chain reeving for compliance with hoist manufacturer's recommendations			
Record frequent inspections	**Yearly**	**Semiannually**	**Quarterly**
• Requirements of frequent inspections • Hoist rope in accordance with *ASME B30.10* (para. 16-2.4.1a) • Evidence of loose bolts, nuts, or rivets • Evidence of worn, corroded, cracked, or distorted parts such as load blocks, suspension housing, chain attachments, clevises, yokes, suspension bolts, shafts, gears, bearings, pins, rollers, and locking and clamping devices • Evidence of damage to hook retaining nuts or collars and pins, and welds or rivets used to secure the retaining numbers • Evidence of damage or excessive wear of load sprockets, idler sprockets, and drums or sheaves • Evidence of excessive wear on motor or load brake • Electrical apparatus for signs of pitting or any deterioration of visible controller contacts • Evidence of damage of supporting structure or trolley, if used • Function labels on pendant control stations for legibility • Label or labels required by *ASME B30.10* (para. 16-1.1.4) for legibility • End connections of rope or load chain			

Notes:

1 - By operator or other designated person with records not required.
2 - Visual inspection by a designated person making records of conditions to provide the basis for a continuing evaluation.
3 - As in Note 2 unless conditions indicate that disassembly should be done to permit detailed inspection.

Source: The American Society of Mechanical Engineers

TABLE 25.2 Inspection Schedule for Electric- or Air-Powered Hoists

- Operating mechanisms for maladjustment and listen for unusual sounds that may indicate problems
- Tightness of bolts, nuts, and rivets
- For excessive wear, corrosion, cracks, or distorted parts in the following:
  Load blocks
  Suspension housings
  Hand chain wheels
  Chain attachments
  Clevises
  Yokes
  Suspension bolts
  Shafts
  Gears
  Bearings
  Pins
  Rollers
  Locking and clamping devices
- For damage or excessive wear on hook-retaining nuts or collars and pins and welds or rivets used to secure the retaining members
- Load sprockets, idler sprockets, hand chain wheel, and drums or sheaves for excessive wear or damage
- Hand chain–operated hoists for evidence of worn, glazed, or oil-contaminated friction disks; worn pawls, cams, or ratchets; and corroded, stretched, or broken pawl springs in braking mechanism
- Supporting structure or trolley for evidence of damage
- For presence of legible warning labels
- End connections load chains for evidence of wear, corrosion, cracks, damage, or distortion
- Welded link hoist chain
- Hooks

Visual and Record Inspections

Daily, frequent, and periodic inspection routines *must* be a regular part of hoist operation. Inspection should include an examination of the chain for wear, twists, excessive dirt, broken links, and proper lubrication. Hooks should be inspected for deformations, cracks, damage, and properly operating latches (Tables 25.3 to 25.5).

The hoist operator should conduct daily inspections, while qualified inspectors should conduct frequent and periodic inspections—at intervals determined by the severity and length of hoist usage. It is the hoist inspector's responsibility to alert maintenance workers of an inspection's findings. Hoists that do not pass inspection need to be tagged OUT OF SERVICE and *removed* from the hoisting area.

Inspection of hand- and electric-operated hoists is a *critical* part of preventive maintenance. The standard that most specifically addresses the requirements of overhead hoists is *ASME/ANSI B30.16 Overhead Hoists (Underhung)* (Tables of Inspection Schedules: Tables 25.1 and 25.2; Tables of Troubleshootong Guide: Tables 25.3, 25.4, and 25.5).

Problem	Probable cause	Check/remedy
Hoist is hard to operate in either direction.	Load chain worn long to gauge, thus binding between liftwheel and chain guide.	Check gauge of chain. Replace if worn excessively.
	Load chain rusty, corroded, or clogged up with foreign matter such as cement or mud.	Clean by tumble polishing or solvent. Lubricate with penetrating oil and graphite. (SC-46 and SC-146)
	Load chain damaged.	Check chain for gouges, nicks, bent or twisted links. Replace if damaged.
	Liftwheel clogged with foreign matter or worn excessively, causing binding between the liftwheel and chain guide.	Clean out pockets. Replace if worn excessively.
	Hand-chain worn long to gauge, thus binding between handwheel and cover.	Check gauge of chain.
	Handwheel clogged with foreign matter or worn excessively, causing binding of chain between the handwheel and cover.	Clean out pockets. Replace if worn excessively.
	Liftwheel or gear teeth deformed.	Excessive overload has been applied. Replace damaged parts.
Hoist is hard to operate in the lowering direction.	Brake parts corroded or coated with foreign matter.	Disassemble brake and clean thoroughly. (By wiping with a cloth—not by washing in a solvent.) Replace washers if gummy, visibly worn, or coated with foreign matter. Keep washers and brake surfaces clean and dry.
	Chain binding.	See first three check/remedy items in this table.
Hoist is hard to operate in the hoisting direction.	Chain binding.	See first three check/remedy items in this table.
	Chain twisted. (3-ton capacity and larger)	Rereeve chain, or on 3- and 4-ton unit, if both chains are twisted, capsize hook block through loop in chain until twists are removed. Caution—do not operate unit in hoisting direction with twisted chain or serious damage will result.
	Overload.	Reduce load or use correct capacity unit.

TABLE 25.3 Troubleshooting Guide for Spur-Geared Hoists

Problem	Probable cause	Check/remedy
Hoist will not operate in either direction.	Liftwheel gear key or friction hub key missing or sheared.	Install or replace key.
	Gears jammed.	Inspect for foreign material in gear teeth.
Hoist will not operate in the lowering direction.	Locked brake due to a suddenly applied load, shock load, or load removed by other means than by operating unit in the lowering direction.	With hoist under load to keep chain taut, pull sharply on hand chain in the lowering direction to loosen brake.
	Chain binding.	See first three check/remedy items in this table.
	Lower hook all the way out. Load chain fully extended.	Chain taut between the lift wheel and loose end screw. Operate unit in hoisting direction only.
Hoist will not operate in the hoisting direction.	Chain binding.	See first three check/remedy items in this table.
Hoist will not hold load in suspension.	Lower hook or load side of chain on wrong side of liftwheel.	Lower hook must be on same side of liftwheel as upper hook. Refer to assembly. Rereeve chain.
	Ratchet assembled in reverse.	Ratchet must be assembled as shown.
	Pawl not engaging with ratchet.	Pawl spring missing or broken pawl binding on pawl stud. Replace spring and clean so pawl operates freely and engages properly with ratchet. Do not oil.
	Ratchet teeth or pawl worn or broken.	Replace pawl and/or ratchet.
	Worn brake parts.	Replace brake parts which are worn.
	Oily, dirty, or corroded brake friction surfaces.	Disassemble brake. Clean thoroughly. (By wiping with a cloth—not by washing in a solvent.) Replace washers if gummy, visibly, worn, or coated with foreign matter. Keep washers and brake surfaces clean and dry.

Source: L.R. Higgins, *Handbook of Construction Equipment Maintenance*, McGraw-Hill, New York.

Table 25.3 Troubleshooting Guide for Spur-Geared Hoists (*Continued*)

Problem	Probable cause	Check/remedy
Hook does not respond to the control station.	No voltage at hoist—main line or branch circuit switch open; branch line fuse blown or circuit breaker tripped.	Close switch, replace fuse or reset breaker.
	Phase failure (single phasing, three-phase unit only)—open circuit, grounded or faulty connection in one line of supply system, hoist wiring, reversing contactor, motor leads or windings.	Check for electrical continuity and repair or replace defective part.
	Upper or lower limit switch has opened the motor circuit.	Press the "other" control and the hook should respond. Adjust limit switches.
	Open control circuit—open or shorted winding in transformer, reversing contactor coil or speed selecting relay coil, loose connection or broken wire in circuit; mechanical binding in contactor or relay; control station contacts not closing or opening.	Check electrical continuity and repair or replace defective part.
	Wrong voltage or frequency.	Use the voltage and frequency indicated on hoist identification plate. For three-phase dual-voltage unit, make sure the connections at the conversion terminal board are for the proper voltage.
	Low voltage	Correct low-voltage condition.
	Brake not releasing—open or shorted coil winding; armature binding.	Check electrical continuity and connections. Check that correct coil has been installed. The coil for three-phase dual-voltage unit operates at 230 volts when the hoist is connected for either 230-volt or 460-volt operation. Check brake adjustment.
	Excessive load.	Reduce loading to the capacity limit of hoist as indicated on the identification plate.
Hook moves in the wrong direction.	Wiring connections reversed at either the control station or terminal board (single-phase unit only).	Check connections with the wiring diagram.
	Failure of the motor reversing switch to effect dynamic braking at time of reversal (single-phase unit only).	Check connections to switch. Replace a damaged switch or a faulty capacitor.
	Phase reversal (three-phase unit only).	Refer to installation instructions.

TABLE 25.4 Troubleshooting Guide for All-Electric Hoists

Problem	Probable cause	Check/remedy
Hook lowers but will not raise.	Excessive load.	Reduce loading to capacity limit of hoist as indicated on the identification plate.
	Open hoisting circuit—open or shorted winding in reversing contactor coil or speed selecting relay coil; loose connection or broken wire in circuit; control station contacts not making; upper limit switch contacts open.	Check electrical continuity and repair or replace defective part. Check operation of limit switch.
	Motor reversing switch not operating (single-phase unit only).	Check switch connections and actuating finger and contacts for sticking or damage. Check centrifugal mechanism for loose or damaged components. Replace defective part.
	Phase failure (three-phase unit only).	Check for electrical continuity and repair or replace defective part.
Hook raises but will not lower.	Open lowering circuit—open or shorted winding in reversing contactor coil or speed selecting relay coil; loose connection or broken wire in circuit; control station contacts not making; lower limit switch contacts open.	Check electrical continuity and repair or replace defective part. Check operation of limit switch.
	Motor reversing switch not operating (single-phase unit only).	Check switch connections and actuating finger and contacts for sticking or damage. Check centrifugal mechanism for loose or damaged components. Replace defective part.
Hook lowers when hoisting control is operated.	Phase failure (three-phase unit only).	Check for electrical continuity and repair or replace defective part.
Hook does not stop promptly.	Brake slipping.	Check brake adjustment.
	Excessive load.	Reduce loading to the capacity limit of hoist as indicated on the identification plate.
Hoist operates sluggishly.	Excessive load.	Reduce loading to the capacity limit of hoist as indicated on the identification plate.
	Low voltage.	Correct low voltage condition.
	Phase failure or unbalanced current in the phases (three-phase unit only).	Check for electrical continuity and repair or replace defective part.
	Brake dragging.	Check brake adjustment.

Table 25.4 Troubleshooting Guide for All-Electric Hoists (*Continued*)

Problem	Probable cause	Check/remedy
Motor overheats.	Excessive load.	Reduce loading to the capacity limit of hoist as shown on the identification plate.
	Low voltage.	Correct low-voltage condition.
	Extreme external heating.	Above an ambient temperature of 104°F, the frequency of hoist operation must be limited to avoid overheating of motor. Special provisions should be made to ventilate the space or shield the hoist from radiation.
	Frequent starting or reversing.	Avoid excessive inching, jogging, or plugging. This type of operation drastically shortens the motor and contactor life and causes excessive brake wear.
	Phase failure or unbalanced current in the phases (three-phase unit only).	Check for electrical continuity and repair or replace defective part.
	Brake dragging.	Check brake adjustment.
	Motor reversing switch not opening start winding circuit. (Single-phase unit only.)	Check switch connections and actuating finger and contacts for sticking or damage. Check centrifugal mechanism for loose or damaged components. Replace defective part.
Hook fails to stop at either or both ends of travel.	Limit switches not opening circuits.	Check switch connections, electrical continuity, and mechanical operation. Check the switch adjustment. Check for a pinched wire.
	Shaft not rotating.	Check for damaged gears.
	Traveling nuts not moving along shaft—guide plate loose; shaft or nut threads damaged.	Tighten guide plate screws. Replace damaged part.
Hook stopping point varies.	Limit switch not holding adjustment.	Check switch connections, electrical continuity, and mechanical operation. Check switch adjustment. Check for pinched wire. Check for damaged gears. Tighten guide plate screws. Replace damaged part.
	Brake not holding.	Check the brake adjustment.

Source: From L.R. Higgins, *Handbook of Construction Equipment Maintenance,* McGraw-Hill, New York.

Table 25.4 Troubleshooting Guide for All-Electric Hoists (*Continued*)

Problem	Probable cause	Check/remedy
Hoist will not operate at slow speed in either direction.	Open circuit.	Open or shorted motor winding loose or broken wire in circuit, speed-selecting contactor stuck in opposite speed mode. Replace motor, repair wire, and/or repair speed-selecting contactor.
	Phase failure.	Check for electrical continuity and repair or replace defective part.
Hoist will not operate at fast speed in either direction.	Open circuit.	Open or shorted motor winding, loose or broken wire in circuit, speed-selecting contactor stuck in opposite speed mode. Replace motor, repair wire, and/or repair speed-selecting contactor.
Hook will not raise at slow speed.	Open speed-selecting circuit.	Open or shorted winding in speed-selecting contactor coil. Loose connection or broken wire in circuit. Mechanical binding in contactor. Control station contacts not making or opening. Replace coil; repair connection, contactor, or control station.
Hook will not lower at slow speed.	Phase failure.	Check for electrical continuity and repair or replace defective part.
Hook will not raise at fast speed.	Excessive load.	Reduce loading to capacity limit of hoist as indicated on the identification plate.
	Phase failure.	Check for electrical continuity and repair or replace defective part.
	Brake not releasing.	Check electrical continuity and connections. Check that correct coil has been installed. The coil for three-phase dual-voltage unit operates at 230 volts when the hoist is connected for either 230- or 460-volt operation. Check brake adjustment.
Hook will not lower at fast speed.	Phase failure.	Check for electrical continuity and repair or replace defective part.
	Brake not releasing.	Check electrical continuity and connections. Check that correct coil has been installed. The coil for three-phase dual-voltage unit operates at 230 volts when the hoist is connected for either 230- or 460-volt operation. Check brake adjustment.
Hook moves in proper direction at one speed—wrong direction at other speed.	Phase reversal.	Wiring reconnected improperly. Interchange two leads of motor winding that is out of phase at the speed-selecting relay.

Source: From L.R. Higgins, *Handbook of Construction Equipment Maintenance,* McGraw-Hill, New York.

Table 25.5 Troubleshooting Guide for Two-Speed Electric Hoists

Regulations and Standards

OSHA Regulations

1926 Subpart N—*Cranes, Derricks, Hoists, Elevators, and Conveyors*

1926.554 *Overhead Hoists*
1926.553 *Base-mounted Drum Hoists*

U.S. Consensus Standards

ASME/ANSI – 2003 *Performance Standards*

HST - 1M *Electric Chain Hoists*
HST - 3M *Manually Level Hoists*
HST - 4M *Overhead Electric Hoists*
HST - 5M *Air Chain Hoists*
HST - 6M *Air Wire Rope Hoist with Trolley*

CHAPTER 26

Personnel/Material Hoists

Towers for platform-type personnel/material hoists are constructed of tubular steel similar to tubular scaffolding, but of much heavier members. These towers can be erected to great heights, alongside tall buildings. But they *must* be properly guyed or anchored to the building.

Diagonal bracing *must* be used to form triangles on all four sides of a tower. This includes both loading and unloading sides, where bracing is commonly omitted to permit passage or handling of material at the landings.

Where cross-bracing is impractical, provide knee braces at the upper corners of the landing openings to afford at least some rigidity.

Two basic types of motor-drive hoist are in use today:

- *Rack-and-pinion.* Preassembled units consisting of base, frame, cage, and all necessary power and control components.
- *Hydrostatic (wire rope).* Tower with separate power pack containing prime power (electric motor or gas engine), hydraulic pump with control valve, and the necessary electric controls.

26.1 Erecting Hoist Towers

A contractor approved, competent hoist coordinator *must* supervise and approve the method of work for erecting the hoist. The hoist is at highest risk during installation because it is normally most unstable at this stage. Seemingly minor mistakes can compound and result in a failure of the hoist installation—usually with personal injury.

Before erecting a hoist on site, carefully consider where to spot the tower, keeping in mind its accessibility as well as its position relative to the building/structure. *Make sure* that there is sufficient clearance from the building/structure, as well as from other obstructions such as at all landing levels and columns. For proper layout, a concrete pad should be placed before the hoist is delivered to the site.

As the hoist is erected, attach the power to the building/structure by special telescoping tie-in assemblies that lock into wall anchors. These should be placed at the side of the building/structure at 25-ft intervals and 20 ft for tower guide rails (Fig. 26.1). Each tie-in requires only two mounting plates on the tower structure, located on either a vertical (wall) or a horizontal (floor) surface. The tie-in length is infinitely adjustable between minimum and maximum specifications.

If a crane is used to load or unload tower sections from their carrier, the pickup point is the hook eye, located inside and at the bottom of the tower base.

If a fork lift is used, pickup is at the same point, but a short cable should be looped over the tower section and the lift fork.

IMPORTANT: *Always check a tower hoist as soon as it is received at the site for any damage which may have occurred during shipment. Report immediately any damage observed to the shipper. Check also for any shortage of parts or components against the packing slip.*

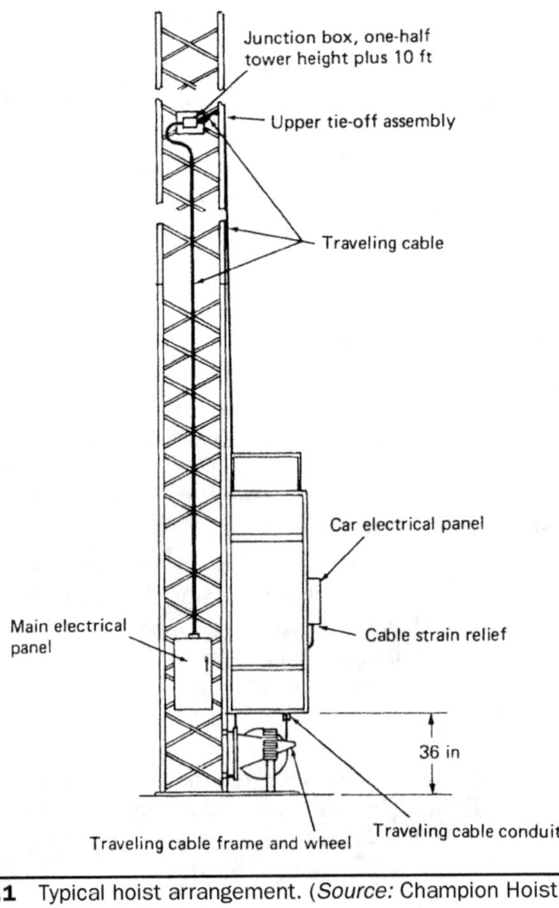

FIGURE 26.1 Typical hoist arrangement. (*Source:* Champion Hoist Company.)

26.2 Rack-and-Pinion Hoists

If the tower is to be raised without a crane, use a stiff leg and hoist gin pole.

- Place the tower base section on the foundation pad and bolt it in place using only the anchor bolts located within the tower. Secure the anchor bolts under the cage later.
- Level the tower, grouting it as required, and tighten the anchor bolt securely.
- Connect the traveling cable to the main power panel and to the cage control panel. Leave the wheel assembly with the cable on the ground so that the cable can be played out during erection.
- Place the aligning pin in the hole at the top of the gear rack mounted on the base tower.
- Bolt the lifting bale to the top of one tower section and, using a gin pole, hoist the first tower section atop the base section. Fasten the two sections with bolts.
- Make the first wall-tie at a point no farther than 25 ft from the ground. Position additional wall-ties no farther apart than 25 ft.
- Plumb the tower during erection, making necessary wall-tie adjustments to maintain a maximum ±½ in. from the true vertical.

Testing Rack-and-Pinion Hoist Installation

Following erection of the hoist, an independent, approved, competent person *must* be present at the testing of the hoist by the erection personnel to conduct a separate thorough examination of the hoist installation.

Once all the tests and examinations are completed, the hoist erection supervisor hands over the control of the hoist to the hoist site operator and check to ensure the operator is familiar with both the operation of the hoist and the daily/weekly maintenance checks required for the particular hoist.

Operation

After having erected the tower base section and connected the hoist's traveling cable, check to see that the oil level in the electric motor reducer is correct.

If a building gate interlock system or ground enclosure is used, it must be plugged into its proper receptacle. If such a system is not used, the jumper furnished must be plugged into its receptacle.

- Connect the hoist to the proper electric supply and turn on the main power panel.
- Check the disk of the phase-reversal relay and *make sure* that the electric hookup is in phase, allowing the motor to run in

the proper direction. Detach the UP/DOWN control switch from its mounting plate on top of the cage so that the cage can be operated from the ground.

- Pull the red mushroom switch to the ON position and close the gates.
- Check that all limit switches are electrically closed. Standing outside the cage, run up (a distance of) 5 or 6 ft. Turn off the main power switch.
- Place a block under the cage, of sufficient size and strength to support the cage, so that it cannot possibly be lowered. Then place the buffer support in position and the springs into their holders. Place and tighten the remaining base anchor bolts. Finally remove the blocks, turn power on, and lower the cage to a point several inches above the buffer springs.
- Continue erecting the hoist tower sections. Carry the UP/DOWN control switch to the top of the cage and run the unit up high enough so that another section can be added to the tower.
- *Make sure* to insert aligning pins in each section.
- Repeat this procedure until the tower is erected to the desired height, with wall-ties attached at proper intervals.
- Use a surveyor's transit to ensure true vertical alignment during erection.
- Check out the travel limit switches and controls according to the manufacturer's instructions.

26.3 Safe Operating Practices

For safe operation of the hoist, *only* an approved hoist operator holding a current state certificate for hoist operation *must* operate the hoist at all times.

A competent hoist supervisor *must* be informed of the requirements for the hoist on a daily basis, so that large deliveries of materials requiring a hoist to move them on the site should be given priority, as limited access to the hoist may result in unsafe use of the hoist—such as overloading—to get the job done.

Make sure that each morning, or at the start of a new shift, the operator makes a solo run before attempting to carry personnel, checking for safe and proper hoist operation:

- Motor safety
- Running of unit (power and smoothness)
- Oil level in gear reducer
- Lubrication as necessary
- Securely tightened anchor bolts, tower bolts, wall-tie bolts, traveling cable, guides, and sheaves

Check	Inspect for	Frequency
Electrical connectors	Loose connectors or clamps	Weekly
Electric cable	Severed insulation, broken wires	Daily
Traveling cable	Severed insulation, kinking and movement over sheaves and through guides	Daily
Limit-switch arms and rollers	Arms locked firmly to switch shafts, rollers move freely	Daily
Pinion gear-and rack	Proper pinion-and-rack clearance	Weekly

Table 26.1 Hoist Safety Schedule Checklist

Check all safety points before and after the first day of operating the hoist, and weekly thereafter (Table 26.1).

Overspeed or Drop Test

This test *must* be performed before placing the hoist in service and each time the tower height is increased. A drop test is a safe and simple procedure, but it should only be performed by an erection supervisor, or a qualified operator.

The overspeed device is designed to function when the cage reaches a downward speed in excess of 160 ft/min. To perform the test:

- Load the cage with weight as prescribed by local safety codes. If no code exists, use a load equal to the specified capacity of the hoist.
- Release the brakes slowly, allowing the cage to rest on the buffer springs. When the cage comes to rest on the springs, release the brakes fully. Turn power on.
- Run the hoist up as far as the control cable will allow while standing on the ground outside the cage.
- Release the control to permit the cage to fall free. After the cage has fallen a few feet, the overspeed governor should be activated, catching the cage and locking it to the tower. If the governor does not catch after the cage has fallen 5 to 6 ft, immediately return the control to the UP position. The motors will stop the cage. As soon as the cage stops falling, release the control and allow the cage to fall a few more feet. Repeat this sequence until the cage rests on the buffer springs.
- Check for, and locate, the cause of the governor malfunction before repeating the test.
- Reset the hoist brakes before putting the hoist into operation.

26.4 Maintenance of Rack-and-Pinion Hoists

The hoist supervisor *must* appoint a hoist operator to carry out the daily/weekly check on the hoist—thus ensuring that changes in the operation of the hoist can be monitored effectively—and a qualified maintenance service provider called in to correct problems as soon as they are apparent.

If a problem is found that affects the safe operation of the hoist, the hoist operator *must* isolate the hoist to prevent it from being used before informing the hoist supervisor of the problem so that repairs can be conducted by qualified maintenance service personnel.

The hoist supervisor *must* be kept informed of the frequency of the maintenance program, and make the hoist available for sufficient time for the required maintenance tasks to be completed correctly by the maintenance personnel.

Normal routing maintenance procedures apply to the hoist—as they would to any similar piece of equipment. Instruct operator and service and maintenance personnel in proper hoist operation and maintenance.

Lubrication Schedule

Follow the manufacturer's instructions for the hoist lubricating schedule. *Make sure* to check the motor pinion and cage rollers every week (Table 26.2).

26.5 Troubleshooting Rack-and-Pinion Hoists

If problems arise, first isolate each problem area. Mechanical problems are usually accompanied by noise. No evidence of action of movement usually indicates electrical problems.

After isolating the problem area, refer to the manufacture's troubleshooting chart or other available source data (Table 26.3).

Item	Schedule
Gear reducer	Keep to proper level, change yearly
Rack	On erection and monthly, or as required by use
Erection jib gear box	Change yearly
Erection jib pulley	As required
Overspeed governor	Weekly (fitting provided)
Gate rollers	As required

Table 26.2 Typical Lubricating Schedule for Rack-and-Pinion Hoists

Indication	Reason	Cause	Remedy
Green indicator lamp on cage does not light	Not receiving power	Check incoming power ON/OFF switch, and disconnect switch, fuses, and emergency stop switch; check gate limit switches; check contactor operation	
Motor will not stop	Not receiving power	Gate and overspeed switches open; final limit switch open	Check slide gate, split gate, overspeed switch
Car will not go up	Ac contactor not energized	Gate switch open; UP switch in hand control defective	Replace UP switch in hand control
Car will not go down	Ac contactor not energized	Gate switch open; DOWN switch in hand control defective	Replace DOWN switch in hand control
Safety mechanism activated	Governor switch tripped open	Reset safety mechanism after ascertaining that any danger that had activated the mechanism has been identified and remedied	

Table 26.3 Electrical Troubleshooting of Rack-and-Pinion Hoists Checklist

26.6 Hydrostatic Drive (Wire Rope) Hoists

Installation of this type of hoist tower follows the same procedures as those for the rack-and-pinion type, except that after the tower base has been securely attached to the foundation pad, the hoist platform must be rigged up.

- Place the wire rope spool on the platform and reeve the cable according to the manufacturer's instructions.
- Next, install the power pack in an area that will not interfere with hoist operation.
- Run the hydraulic lines from the power pack to the base unit, and *make sure* that quick disconnects are firmly seated.
- *Never* attempt to run the power pack unless all hydraulic hoses are properly connected. Failure to do so may cause a hose to burst at the extreme pressure that a totally confined pump is capable of producing.
- *Make sure* that, if a gasoline engine powers the system, the engine crank case is filled to its proper level with oil; and that the fuel tank is filled with a good grade of gasoline, as specified in the manufacturer's engine maintenance manual.
- Check the hydraulic fluid level in the upper tank. If it is not within 2 in. of the filler cup, fill it with the proper fluid as specified on the tank.
- Check that both shut-off valves beneath the upper tank are fully opened.

26.7 Safe Hydrostatic Drive Hoist Operating Practices

Start the engine and observe the vacuum gage. If the gage reading exceeds 10 in. Hg, allow the unit to idle until the reading drops below 10 in. If the reading is greater than 25 in. Hg, shut off power immediately and check for the cause.

Slowly wind the rope onto the drum, lifting the platform to take up slack. *Make sure* to straighten the rope on the drum keeping each wrap held tightly against the previous wrap. As the platform rises, increase the drum speed by rotating the speed control clockwise.

Never drive the platform downward into the base structure. This will cause the rope to unspool and tangle. If there is a second platform on the hoist tower, repeat startup steps for the other platform. Assemble the cage enclosure according to the manufacturer's instruction.

Broken-Rope Safety Test

To run a broken-rope safety test:

- Slowly run the platform down to the bottom of the rower and allow the lifting rope to slack.
- Visually check to see that the unit's safety shoes lift and engage tower legs. If the shoes do not engage, consult the manufacturer's troubleshooting chart and correct before proceeding.
- Slowly and carefully take the slack out of the rope and run the platform sufficiently high to clear the bottom structure. Shoes should now drop free of the tower legs.
- Place a test load—equal to the hoist-lifting capacity—on the platform.
- Raise the load 15 to 20 ft and tie the platform to the tower, using the locking chain.
- Ease the platform down until it is supported by the locking chain, then continue to run it down until able to pull a loop into the lifting rope.
- Splice a short length of ½-in. rope in the lifting rope loop and tie securely with *fist-grip clamps* (*do not* use regular cable clamps).
- Run the platform up until the locking chain can be removed and the load is supported by the lifting rope.
- Hold the spliced loop of the rope and, using an oxyacetylene torch, cut the short length of the spliced rope. The platform should lock to the tower within a few inches of fall.
- Remove ends of the cut rope from the lifting rope and, being careful to keep layers straight, spool the rope onto the drum. Then lift the platform to release the shoes from tower.
- Lower the platform and remove the test load.

26.8 Maintenance of Hydrostatic Drive Hoists

Before beginning operation each day, inspect the hoist by easing the platform onto the bottom structure and allowing the lifting rope to go slack. Check:

- Hoist's safety shoes to ensure that they engage the tower legs.
- Shoes and linkage of brake rope safeties to ensure that they are clean, free, and properly lubricated.
- Deflector and cathead sheaves, with rope still slack, turning them by hand to determine freedom of rotation, wear, and any condition that may damage the rope. Correct any of these conditions immediately.
- Fist-grip clips for tightness.
- Rollers on upright carriage to determine freedom of rotation or wear. Frozen bearings should be replaced and signs of wear should be checked and corrected immediately.
- Drum for proper rope winding, by running the unit up to remove slack from the lifting rope.
- Rope for obvious defects, by running the platform to the top.

Repeat the checks for the other platform.

Maintenance Checklist

- *Fluid*. In general, a fluid change performed annually or after 2000 hours of operation, whichever occurs first, is adequate with a sealed reservoir system. A more frequent fluid change is required if the fluid has become contaminated by water or other foreign matter, or has been subjected to abnormal operation conditions.
- *Reservoir*. The reservoir *must* be checked daily for the proper fluid level and the presence of water in the fluid. If fluid must be added to the reservoir, use only filtered or strained fluid. Drain any water as required.
- *Hydraulic lines and fittings*. Visually check daily for any fluid leakage. Tighten, repair, or replace as required.
- *Heat exchanger*. If used, the heat exchange core and cooling fins should be kept clean at all times for maximum cooling and system efficiency. Inspect daily for any external blockage and clear as required.

Troubleshooting Hydrostatic Hoists

Hydrostatic drive-type hoist troubleshooting is divided into three categories: electrical, hydraulic, and mechanical. If a problem arises, isolate the problem area and correct the problem according to the manufacturer's instructions (Table 26.4).

Indication	Cause	Remedy
	A. General	
1. Electric motor will not start		
a. Not receiving power		Check incoming power ON/OFF switch, disconnect-switch fuses, and emergency stop switch.
b. Phase reversal relay not on	Improper phase	Switch two of incoming power lines.
c. Motor contactor not closed	Coil burned out	Replace.
	Transformer	Replace.
	Transformer fuse	Replace.
2. Gasoline engine will not start (Unit must never be run when filter gage reads above 10, except in extreme cold weather. Under such circumstances, do not allow unit to run more than 2 min before reading drops below 10. If necessary, adjustments should be made for cold-weather operation.)	Check emergency button. Consult engine manual furnished with hoist.	
a. High reading on filter gage	Shutoff valve closed	Open fully.
	Blockage in oil takeup circuit	Remove blockage.
	Excessively dirty oil filter	Close shutoff valve and replace filter cartridge. *BE SURE TO REOPEN VALVE BEFORE STARTING ENGINE.*
	Extreme cold weather	If prime mover is: i. *Gasoline engine,* remove fan blade. This will direct engine heat to oil heat exchanger to maintain warmer hydraulic oil temperatures. ii. *Electric motor,* disconnect fan (when used) and blanket oil cooler radiator to maintain warmer hydraulic oil temperatures.

3. Platform/carriage will not go up		
a. Pump not being activated	Control valve not electrified	See electrical troubleshooting in equipment manual.
	Valve electrified but not stroking (listen for whine in pump)	If no whine, pump is not stroking.
b. Hydraulic motor improperly fed	Quick disconnect not fully seated	Reseat firmly.
c. No output from pump	Blockage in oil uptake	Open any restriction and check pump for damage.
	Damaged motor	Repair or replace.
d. Platform overloaded	Pressure-relief valve activated to relieve overload	Reduce platform load.
4. Platform goes up too slowly or will not lift rated capacity		
a. Brake not releasing	Brake adjusted too tight	Readjust so that brake just holds full load.
b. Lifting cable is binding	Lifting rope off sheaves	Replace rope on sheave and check for damage to rope and sheave.
	Sheave frozen	Repair or replace sheave; check for damage to rope.
c. Insufficient pump output	Engine speed insufficient	See performance chart for proper speed.
	Throttle valve defective	Replace.
	Pump defective	Locate and repair cause.
	Pump being starved (should show high reading on filter gage)	Immediately shut down system. Be sure shutoff valve is fully opened. Problem is in takeup portion of system; continued operation will damage pump.
d. Pump not fully stroked	Speed control set too low	Readjust.
	Pump control valve receiving low voltage	Check voltage.
e. Brake dragging	Improper adjustment	Readjust with full load.
5. Platform/carriage will not come down		
a. Broken-rope safety mechanism activating	Spring adjusted too tight	Readjust only after it is definitely determined that improper adjustment is cause.

TABLE 26.4 Comprehensive Troubleshooting Guide for Hydrostatic Hoists (Electrical Motor and Gasoline Engine) *(Continued)*

Indication	Cause	Remedy
	Obstruction causing carriage to catch on tower. (Once weight of carriage is removed from rope, safety mechanism will activate.)	Remove obstruction.
	Lifting rope binding in sheave	Free rope and carefully check for damage to rope and sheave.
b. Pump not being activated	See item A3a.	
6. Platform comes down, but slowly		
a. Insufficient oil flow through return circuit	Obstruction in return circuit	Remove any obstruction.
7. Platform/carriage slides down, will not hold load		
a. Brake not holding	Brake improperly adjusted	Readjust with full load.
b. Brake partially released	Brake valve leaking	Check for foreign material on seats or replace.
	Brake-release linkage is binding	Repair cause.
	Brake lining is wet	Dry with air. Brief period of operation should dry out lining. In severe case, tighten brakes so that they drag, then operate a few times so that heat of friction dries lining. Readjust to proper setting.
	Brake lining worn out	Since brake is not used to stop load, but only to hold it, lining should not wear out. If it does, replace lining, but be sure to check for brake drag.
8. Engine will not return to idle		
a. Throttle is being activated	Throttle-valve linkage out of adjustment	Readjust.
	Cylinder displacement due to throttle-valve leakage	Repair or replace valve.

B. System Will Not Operate in Either Direction		
1. Low or zero charge pressure	System low on fluid	Locate and fix leaks causing the loss of fluid. Replenish fluid in reservoir to proper level. *Note: Use only recommended fluids.*
	Faulty control linkage	Check entire linkage, from control lever to pump arm, to make sure it is connected and free to operate as it should. Adjust linkage to pump arm. *Do not move pump arm to meet linkage.*
	Disconnected coupling	Check that coupling from engine to pump shaft or from motor shaft to driven mechanism is not slipping or broken.
2. High vacuum and low charge pressure	Filter or suction line from reservoir to charge pump plugged or collapsed	Replace or clean filter suction line.
3. Low or zero charge pressure. If problem is in charge pump, pressure will be low or zero when pump is in neutral; if in manifold, pressure will be low when in stroke.	Pressure-relief valve in charge pump or motor manifold damaged or open	Replace pressure-relief valve.
4. Zero charge pressure when pump is in neutral or when trying to go into stroke	Charge-pump drive key or shaft broken	Replace charge-pump assembly.
5. Pieces or flakes of brass in reservoir or filter; noisy unit (pump or motor)	Internal damage to pump or motor, or both	In fixed- or variable-displacement motors: remove charge pressure-relief-valve cap in motor manifold, remove charge-relief spring, insert solid shim of sufficient length in place of spring, and reinstall the cap. This will block relief valve in a fully closed position.

TABLE 26.4 Comprehensive Troubleshooting Guide for Hydrostatic Hoists (Electrical Motor and Gasoline Engine) (*Continued*)

Indication	Cause	Remedy
6. Neutral charge pressure maintained, but pump will not go into stroke; handle moves freely	Disconnected control valve internal	Disconnect control linkage at directional control arm, and move arm back and forth by hand. If it moves freely with no resistance, control valve should be removed and checked for broken or missing parts.
7. Neutral charge pressure normal, but pump will not go into stroke	Plugged control orifice	Remove bolts that hold control housing to pump and check orifice. *Caution: Do not allow orifice or O rings to fall into pump case.*
	C. System Operates in One Direction Only	
	Faulty control linkage	Check entire linkage to make sure it is connected and free to operate as it should. Adjust linkage to pump arm.
1. Loss of or lower than normal system pressure in one direction only	Faulty high-pressure relief valve	Switch two high-pressure relief valves. If system operates in direction it would not operate before, one of the relief valves is inoperative. Repair or replace damaged valve and reset system.
	Shuttle-valve spool jammed (located in motor manifold)	Remove and replace entire manifold assembly.
2. Loss of system pressure in one direction only; charge pressure might be higher than normal	One check valve faulty	Remove two check valves located in pump and cap under charge pump and check to see if: i. Poppet or ball is missing. ii. Valve seat is eroded or deformed. *Note:* If condition (i) exists, replace pump; if condition (ii) exists, replace both check valves.
3. Pump will not return to neutral	Control-valve-spool jammed, sticking in one direction, or misadjusted	Replace control valve.

D. Neutral Difficult or Impossible to Find		
	Faulty linkage	Disconnect control linkage at directional control arm. If system now returns to neutral, linkage to control is out of adjustment or binding in some way.
	Control valve out of adjustment	Replace displacement control valve or readjust according to manufacturer's recommended instructions.
	Servo cylinder out of adjustment	Remove pump and return to factory for readjustment.
E. System Operating Hot (motor-case temperature above 120°F)		
1. Low charge pressure	Fluid level low	Replenish fluid supply.
	Fluid cooler clogged	Clean cooler air passage.
2. High vacuum and low charge pressure	Clogged filter or suction line	Replace filter. Clean or replace suction lines.
3. Lower than normal system pressure in one or both directions	Excessive internal leakage	Check high-pressure relief valves; one may be stuck partially open.
a. Lower than normal charge pressure may drop to or near zero when maximum obtainable system pressure is reached	Refer to B.	
b. Loss of acceleration and power		Replace pump, motor, or both.
	Case drain lines improperly plumbed	Check plumbing, reinstall to proper alignment.
	Continued operation at high-pressure relief-valve setting	Consult operator's manual for proper machine operation.
F. Acceleration and Deceleration Sluggish		
	Air in system	Check for low fluid level. Check inlet system, filter suction line, and so forth, for leaks allowing air to be drawn into system.

TABLE 26.4 Comprehensive Troubleshooting Guide for Hydrostatic Hoists (Electrical Motor and Gasoline Engine) (*Continued*)

Indication	Cause	Remedy
	Control orifice plug partially blocked	Remove bolts that hold control housing to pump and check the orifice. *Caution: Do not allow O rings to fall into the pump case.* If orifice is clean, remove the charge pump and plug at charge-pressure-gage port and blow clean air through the passage between charge pump and control orifice port.
1. Pieces or flakes of brass in reservoir or filter; noisy unit (pump or motor)	Internal wear or damage in pump or motor, or both	In fixed- or variable-displacement motors, remove charge pressure-relief-valve cap in motor manifold, remove the charge-relief spring, insert solid shim in place of spring, and reinstall cap. This will block relief valve in a fully closed position.
	Engine lugs down	Consult engine manual.
	G. System Noisy	
1. Considerable amount of foam in reservoir; low and fluctuating charge pressure; spongy control	Air in system	Check for low fluid level. Check inlet system, filter, suction line, and so forth, for leaks allowing air to be drawn into system.
	Plumbing not properly insulated	Make sure base or tubing is not touching any metal that can act as a sounding board. Insulate base and tubing clamps with rubber to absorb noise.

Source: Champion Hoist Company.

TABLE 26.4 Comprehensive Troubleshooting Guide for Hydrostatic Hoists (Electrical Motor and Gasoline Engine) (*Continued*)

26.9 Dismantling Hoist Towers

Before removing a hoist from the site the hoist supervisor *must* make sure that its removal does not increase any material handling risks, such as having to manually move components up stairs. If the removal of the hoist effects handling risks on the site, a risk assessment *must* be carried out, and a solution to reduce the risks arrived at before the hoist is removed.

If the risk cannot be reduced to a suitable level with the removal of the hoist, then the hoist *must not* be removed from the site. It may be that a smaller hoist is required to be installed in a more suitable location prior to the original hoist being moved so that the handling risk is not increased.

The dismantling of a hoist tower is essentially the reverse procedure of erecting the tower. Although dismantling has the same risks as its erection, normally they are considered to be lower. When dismantling tower sections, however, it is extremely important to follow recommended safety precautions, such as maintaining proper wall-ties throughout the operation. Always chain the platform to the tower while removing sections, and *never* ride the platform during tower dismantling.

The hoist supervisor *must* be especially careful to ensure that all safety precautions required for the safe dismantling of the hoist are effective during the work on site.

Regulations and Standards

OSHA Regulations

1926 Subpart N—*Cranes, Derricks, Hoists, Elevators, and Conveyors*

1926.550 *Cranes and Derricks*
1926.551 *Helicopters*
1926.552 *Material Hoists, Personnel Hoists, and Elevators*

U.S. Consensus Standards

ANSI/ASSE A10.4–2007 *Standard Safety Requirements for Personnel Hoists and Employee Elevators on Construction and Demolition Sites*

Industry Standards Resources

American Society of Safety Engineers (*www.asse.org*)
Association of Equipment Manufacturers (*www.aem.org*)
Hoist Manufacturers Institute (*www.mhia.org*)
Material Handling Industry of America (*www.mhia*)

National Association of Elevator Contractors (*www.nace.org*)

National Association of Elevator Safety Authorities International (*www.naesai.org*)

National Association of Tower Erectors (*www.natehome.com*)

National Council of Erectors, Fabricators, and Riggers (*www.wbdg.org*)

National Elevator Industry Inc. (*www.neii.org*)

National Erectors Association (*www.nea-online.org*)

Specialized Carriers & Rigging Association (*www.scranet.org*)

Steel Erectors Association of America (*www.seaa.net*)

CHAPTER 27

Helicopters

Where it may be impractical or uneconomical to use a crane of any sort as a lifting device, it may prove necessary to take advantage of the versatility of a helicopter. These flying cranes can reach points far higher than those attainable with the boom of a conventional crane or tower, and are also ideally suited for free egress into remote reaches. Otherwise, an extensive road network or access system may be required to bring in a crane, other equipment, or supplies.

Currently available commercial helicopters differ considerably from their military counterparts. Technology and advanced designs have combined to make the helicopter a tool that fulfills specific commercial needs at improved operating costs.

The high costs of these units, coupled with their restrictions imposed by weather, limited lift capacity, and area clearances, require that helicopter employment be closely examined before utilization is attempted. This high cost of use may be most economical, however, when considering short-duration lift requirements in unreachable situations. A decision to integrate the helicopter into construction operations will require careful analysis during the project estimating stage, encompassing the required support system, operation costs, and the most economical method of procurement.

27.1 Planning Lifting Operations

Using rotorcraft successfully on a construction job requires a sound project organization, a proper operational plan, and strict control, plus enforcement of operating procedures and on-site safety measures. Failure in any of these areas will produce a breakdown in coordinated efforts, increasing the amount of flight and ground time. These, in turn, will escalate costs and possibly delay the project's scheduled completion. Close coordination between construction supervisor, contractor, and helicopter operator is essential for the efficient use of rotorcraft on a construction site.

The first thing a contractor must do is educate construction crews on the sequence of tasks and specific safety requirements before introducing the helicopter into operations. *Make sure* that every member of the construction team understands the importance of executing each step on time, so that there will be no delays to subsequent tasks.

More so than with most equipment in the contractor's spread, project operations must be built around the helicopter—not the helicopter around the construction operations. Cost and availability make this essential.

The helicopter operator and the contractor should have a pre-job conference to coordinate all activities requiring use of the helicopter. This should include determining:

- Precautions to be in effect at the helicopter landing area, pickup areas, route to be flown, and deliver (setting) areas; arrangements for compliance with any other mutual requirements
- Design and strength of rigging and how it will attach to the load
- Accuracy of weights, structural strength of the loads, size and number of loads, and number of ground crews required for the operation
- Assignment of responsibility for clearing and securing pickup and setting sites
- Maximum time that helicopter can hover while ground crews are working beneath it
- Type and quantity of personal protection gear provided for the ground crews
- Type of scaffolding, if necessary, to be erected for ground crews to provide stable footing when attaching and unhooking the loads at elevated sites
- Working conditions that could be hazardous to ground crews, such as rotor wash, rain, static electricity, and gusty winds
- Operation of the rotorcraft in proximity to electric power lines, if applicable
- Clearance of nonessential personnel from the flight route
- Procedures for wetting down dusty and sandy areas
- Provision for a point of reference when the helicopter is hovering

A helicopter crew should consist of one or two pilots and two mechanics, depending on the helicopter model used. The mechanics act as signal persons, one stationed at the pickup point on the ground and the other at the drop-off point on the structure.

The helicopter operator and the signal person should determine the signal person's position, so that he or she can readily observe the load for pickup or delivery and still be seen or heard by the operator. They should also agree upon the type of communications to be used for the specific lift—hand signals, ground-to-air radio, intercom, or relay signals, or a combination of them—and alternative procedures to be followed if communications, or sight, are lost between operator and signal person.

NOTE: *Audible signals should be used as a warning to ground crews working with the helicopter during airlift operations.*

Finally, both operator and contractor should agree on any special identifying clothes or gloves that only the designated signal persons will wear. Signal persons should direct the pilot with hand signals to let the pilot know what control corrections are required.

Since union jurisdictional rules prevent helicopter crews from touching a load, the contractor should provide a minimum of two workers on the ground and four or more workers on the structure. For maximum accuracy in placing a load, the number of persons on the structure should be increased as the weight of the load increases.

27.2 Support Systems

Once it has been decided that helicopter support is desired for a project, it will be necessary to determine the type, size, and number of units required to effectively maintain the construction schedule. Helicopters are classified by gross weight (light, medium, or heavy) that has a direct correlation to their load-carrying capacity. Different tasks will require helicopters having different useful loads and different flying ranges. Even though a helicopter's useful load depends on its gross weight, the heavier the rotorcraft, the higher will be its operating costs.

Thus, it would not be economical to use a heavy helicopter on a job that could be done by a lighter rotorcraft at less cost; and on large complicated jobs requiring the support of many helicopters, a mix of different sized helicopters will often prove to be more economical than a single sized craft.

If helicopter support on a project is to be only temporary, the contractor should probably hire a helicopter operator, rather than add rotorcraft to his equipment spread.

Not all operators, however, do all types of work. Instead, they often specialize in contracts within a particular industry, such as offshore or pipeline construction. Others, of course, are limited in the kind and scope of work they do by the type (gross weight) of helicopters in their fleet. Then, too, many small helicopter operators usually work only locally. If a contractor's needs are great, a large operator might provide the best service.

Helicopter operators can be located through civil aviation authorities, helicopter manufacturers, or the Helicopter Association of America (HAA). Regardless of whether the rotorcraft is purchased, leased, or under contract, it must be certified by the Federal Aviation Administration (FAA) before it can be used. Make sure that each helicopter used displays its registration certificate, airworthiness certificate, and operating certificate in the craft, as required by the FAA.

Operation of helicopter cranes *must* comply with any applicable regulations of the FAA; and the weight of an external load *must not* exceed the manufacturer's rating.

27.3 Operations Briefing

Prior to each day's operation the contractor *must* conduct a briefing that sets forth the plan and operation of the pilot and ground personnel. But it is the helicopter operator who is responsible for size, weight, and manner in which loads are connected to the helicopter. If, for any reason, the helicopter operator believes the lift cannot be made safely, the lift *must not* be made.

- *Sling and tag lines*. Load *must* be properly slung; tag lines *must* be of a length that will not permit their being drawn up into rotors. Pressed sleeve, swedged eyes, or equivalent means *must* be used for all freely suspended loads to prevent hand splices from spinning open or cable clamps from loosening.
- *Cargo hooks*. All electrically operated cargo hooks *must* have the electrical activating device so designed and installed as to prevent inadvertent operation. In addition, these cargo hooks *must* be equipped with an emergency mechanical control for releasing the load. The hooks *must* be tested prior to each day's operation to determine that the release functions properly, both electrically and mechanically.
- *Hooking and unhooking loads*. When employees are required to perform work under hovering craft, a safe means of access *must* be provided for employees to reach the hoist line hook and engage or disengage cargo slings. Employees *must not* perform work under hovering craft except when necessary to hook or unhook loads.
- *Static charge*. Before ground personnel touch the suspended load, static charge *must* first be dissipated with a ground device; or all ground personnel *must* wear protective rubber gloves when touching the suspended load.
- *Ground lines*. Hoist wires or other gear, except for pulling lines or conductors that are allowed to "pay out" from a container or roll off a reel, *must not* be attached to any fixed ground structure, or allowed to foul on any fixed structure.
- *Signal systems*. Either radio or hand signal systems, between aircrew and ground personnel, *must* be understood and checked in advance of hoisting the load (Fig. 27.1).

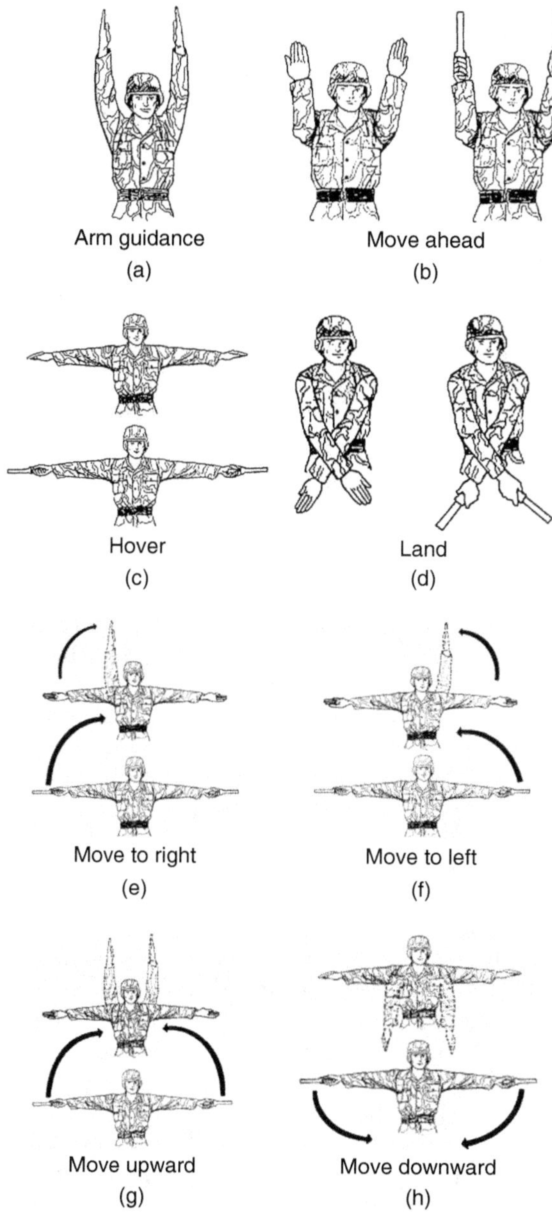

FIGURE 27.1 Helicopter hand signals. (a) Arm guidance; (b) move ahead; (c) hover; (d) land; (e) move to right; (f) move to left; (g) move upward; (h) move downward. (*Source:* Operation Manual, Tennessee Army National Guard, Nashville, TN.)

27.4 Operational Procedures

Before putting a helicopter to work, the operator must first advise the FAA of the job. Frequently an FAA inspector will be on site to make sure that the work is done in accordance with FAA regulations.

The contractor, in turn, is responsible for notifying local police of the planned operation and requesting police help when necessary to keep traffic moving or to keep spectators out of the area of operations.

If the job is in a downtown area, it is usually necessary for the contractor to obtain the approval of the city, and to have the help and cooperation of local police to block off streets around the area of operations.

Because of its unique vertical takeoff and landing capabilities, the helicopter requires no special landing facilities other than an open space with relatively level ground. With loads slung below the helicopter, it often is not necessary to land at all. The helicopter simply hovers over the desired location, releases the load, and flies off to complete the task. In some cases, it may be necessary during normal daily operations to land only to refuel.

A big concern of an operator who will not be operating from a home base, however, is where to base a helicopter during the support operation. A nearby airfield can often be used, or temporary hangar facilities can be set up on the construction site.

The length of contract and the amount of support needed will usually determine fleet maintenance requirements. Maintenance requirements, in turn, will determine facility needs.

Major considerations in determining flight operations are weather, altitude, and temperature, factors over which the pilot has no control, but which directly affect the aircraft's performance. The supervisor as well as the operator should be aware that engine efficiency is inversely proportional to temperature and altitude—as these two conditions increase, the engine power is reduced.

Weather conditions must always be considered prior to flight planning. The supervisor and the operator must demand that the capabilities of both the operator and the machine not be overextended in marginal weather.

Airlifting loads by helicopter calls for some preparations that are not normally required for conventional crane lifting, especially because of winds (comparable to a strong wind storm) that are created by the helicopter's rotors. Specific precautionary measures to be taken to prevent personnel injury and structural damage include:

- Clear the entire operational area of all debris, such as loose paper, plywood, sheet metal, tarpaulins, polyethylene paper, roofing-tar pots, insulation, and all other objects that could move in a strong wind.
- Properly secure all lumber and other large lightweight objects.

- Water down thoroughly both the helicopter landing area and the load pickup area if they are dusty.
- Make sure all wheel-type landing gear of helicopters is checked while on the ground or other supported area.
- All workers *must* wear eye goggles and hard hats (in accordance with OSHA regulations), using chin straps to secure the hard hat, when participating in helicopter operations. The contractor *must* supply this protective equipment.
- Make sure the landing area set aside for the helicopter is at least 200 ft in diameter, and is cleared of all obstacles such as poles, wires, trees, and sheds; position a wind indicator near the landing site.
- Make sure that, while the helicopter is on the landing site, the horizontal clearance between the main or auxiliary rotors and any obstacle is not less than 15 ft.
- Clear the helicopter's flight route from the pickup area to the setting area of all people, vehicles, and structures.
- Whenever approaching or leaving a helicopter with blades rotating, all employees *must* remain in full view of the pilot and keep in a crouched position. Employees *must* avoid the area from the cockpit or cabin rearward, unless authorized by the helicopter operator to work there.
- Unauthorized persons *must not* be allowed to approach within 50 ft of the helicopter when the rotor blades are turning.
- Sufficient ground personnel *must* be provided when required for safe helicopter loading and unloading operations.
- Constant, reliable communication *must* be maintained between the pilot and a designated employee of the ground crew who acts as a signal person during the period of loading and unloading. This person *must* be distinctly recognizable from other ground personnel.
- When visibility is reduced by dust or other conditions, ground personnel *must* exercise special caution to keep clear of main and stabilizing rotors.
- Employer *must* take precautions to eliminate, as far as practical, any reduced visibility.

27.5 Lifting Components

Slings and Tag Lines

A load to be lifted shall be properly slung; and tag lines must be of a length that will not permit their being drawn up into the rotors.

To prevent hand splices from spinning open or cable clamps from loosening, pressed sleeve, swedged eyes, or equivalent means shall be used for all freely suspended loads.

Cargo Hooks

All electrically operated cargo hooks shall have the electrical activating device so designed and installed as to prevent inadvertent operation. In addition, these cargo hooks *must* be equipped with an emergency mechanical control for releasing the load. The hooks shall be tested prior to each day's operation to determine that the release functions properly, both electrically and mechanically.

The primary cargo hook should be located as close to the helicopter as possible to prevent inadvertent entanglement of the hook and its suspension system with external fixed structures on the helicopter. The primary hook should include a quick-release device that enables the pilot to release the external load quickly during flight.

In addition to the primary control, an emergency release device (mechanical, electrical, hydraulic, pneumatic, explosive, or a combination of these) should be provided. The power source of this system must be independent of the normal primary hook release system power source.

The automatic hook release mechanism should be used only for specific operations where ground crews are not used, and should be activated only when actually placing a load.

Each lift should be made using a spreader bar that has four cables, with a safety hook on the end of each cable. The hooks should be attached to the suspended load by slings or by chokers attached to lifting points on the load.

Each load should have four pickup points on it. If the load does not have pickup points, and if it is designed so that eyebolts or clevices cannot be installed or J hooks used, the lift can be made using nylon straps. If there is any sizable number of loads without pickup points, the contractor should provide special rigging for such loads.

The length of sling legs should be such that no sling leg makes an angle from the vertical greater than 45°. The leg of multiple-leg slings should be connected at the top by an Apex fitting.

Rated load of the primary hook should be the maximum weight of load to be lifted, carried, pulled, or towed. This maximum weight should not exceed the weight approved by the FAA to be displayed on or adjacent to the hook. Rated loads for sling materials should be based on the catalog strength of the material used.

To determine the service break strength S of each leg of a sling, the calculations are:

Pendant (one-legged) sling or primary lifting cable	$S = 1.75nW$
Two-legged sling	$S = 1.32nW$
Three-legged sling	$S = 0.9nW$
Four-legged sling	$S = 0.7nW$

where n = design flight load factor for helicopter
W = rated capacity of entire sling
S = service break strength of each leg

The service break strength is actually the strength of the sling after 1 year of service. Thus, the new break strength may be higher than the service break strength, depending on the amount of deterioration caused by environmental exposure on each sling material.

To determine the new break strength NS, the formula is:

$$\text{New break strength} = \frac{\text{Service break strength}}{\text{Reduced strength factor}}$$

A.N.S.I.		**Number broken wires in running ropes**		**Number broken wires in standing ropes**	
No.	**Equipment**	**in one rope lay**	**in one strand**	**in one rope lay**	**at end connection**
B30.2	Overhead and Gantry Cranes	12	4	Not Specified	Not Specified
B30.4	Portal, Tower, and Pillar Cranes	6	3	3	2
B30.5	Crawler, Locomotive, and Truck Cranes	6	3	3	2
B30.6	Derricks	6	3	3	2
B30.7	Base Mounted Drum Hoists	6	3	3	2
B30.8	Floating Cranes, and Derricks	6	3	3	2
A10.4	Personnel Hoists	6*	3	2*	2
A10.5	Material Hoists	6*	Not Specified	Not Specified	Not Specified

Source: Hanes Supply Inc., Buffalo, NY (*www.hanessupply.com*).
*Also remove for 1 valley break.

TABLE 27.1 Reduced Strength Factors for Sling Materials

Reduced Strength Factor

The rope strength design factor is the ratio of the rated strength of the rope to its operating stress. If a particular rope has a rated strength of 100,000 lbs. and is working under an operating stress of 20,000 lbs., it has a rope strength design factor of 5. It is operating at one-fifth or 20 percent of its rated strength.

Reduced rope strength design factors frequently result in reduced service life of wire rope. Removal criteria:

Rope must be removed from service when diameter loss or wire breakage occurs as specified by OSHA (ANSI) Removal Criteria 5; *ANSI Safety Codes, Standards, and Requirements.*

OSHA requires monthly record keeping of wire rope condition.

NOTE: *Current industry recommendations and OSHA standards are based upon the use of steel sheaves. The manufacturer of plastic or synthetic sheaves or liners should be consulted for their recommendation on the safe application of their product, and possible revision in rope inspection criteria when used with their product.*

27.6 Personnel Safety

Safety, always important in construction, is even more so around helicopters. The operator and the contractor should work closely on developing and implementing a safety program, and devising a set of emergency procedures that must be adhered to. Where the helicopter is part of the productive process, that process may require alterations to include provisions for a contingency in case of an emergency.

Caution *must* be exercised to ensure protection of the rotorcraft from objects being blown or drawn into the rotor system or engine intakes. All items susceptible of creating such hazards must be secured or removed from the operating sites.

Personal protective equipment for employees receiving the helicopter load *must* consist of complete eye protection and hardhats secured by chinstraps.

The contractor *must* take every practical precaution to provide protection of the employees from flying objects in the rotor downwash. All loose gear within 100 ft of the place of lifting the load, depositing the load, and all other areas susceptible to rotor downwash *must* be secured or removed.

Personnel should be instructed well in advance about safe behavior when working around a helicopter. When entering or leaving a helicopter, each person should do so toward the front of the helicopter, never the rear. No one other than the pilot should enter the rotorcraft, except qualified personnel considered essential for a specific lift operation.

- Good housekeeping *must* be maintained in helicopter loading and unloading areas.
- Caution must be exercised to ensure protection of the rotorcraft from objects being blown or drawn into the rotor system or engine intakes. All items susceptible of creating such hazards must by secured or removed from the operating sites.
- Smoking, open flames, or other sources of ignition that could result in such fires being spread by the rotor downwash *must not* be permitted within 50 ft of fueling operations or storage areas. No one other than authorized personnel should be allowed within 50 ft of fueling operations.
- At least one 30-lb fire extinguisher, good for class A, B, and C fires, should be provided within 100 ft of refueling operations.
- While a rotorcraft is on a landing site, horizontal clearance between the main or auxiliary rotors and any obstacles should be not less than 15 ft.

Compliance with safety regulations, enforced by civil aviation authorities, is clearly the responsibility of the operator. However, the contractor should determine how, if at all, these regulations will affect construction activities

Regulations and Standards

OSHA Regulations

1910 Subpart N—*Materials Handling and Storage*

1910.183 *Helicopters*

1926 Subpart N—*Cranes, Derricks, Hoists, Elevators, and Conveyors*

1926.551 *Helicopters*

1926 Subpart V—*Power Transmission and Distribution*

1926.958 *External Load Helicopters (comply with provisions of 1925.551 Subpart N)*

U.S. Consensus Standards

ANSI/ASME – B30.12 *Handling Loads Suspended from Rotorcraft*
ASME/ANSI B30.16 – 1987 *Overhead Hoists (Underhung)*
ANSI/ASME B30.20 – 1985 *Below-the-Hook Lifting Devices*
ASME B30.21 – 1982 *Manual Lever Operated Hoists*

ANSI/ASME HST-1M – 1982 *Performance Standard for Electric Chain Hoist*

ANSI/ASME HST-2M – 1993 *Performance Standard for Hand Chain Manually Operated Chain Hoists*

ANSI/ASME HST-3M – 1985 *Performance Standard for Manually Lever Operated Chain Hoists*

ANSI/ASME HST-4M – 1985 *Performance Standard for Overhead Electric Wire Rope Hoists*

ANSI/ASME HST-5M – 1985 *Performance Standard for Air Chain Hoists*

ANSI/ASME HST-6M – 1986 Performance Standard for Air Wire Rope Hoists

ASME/ANSI B30.10 – 1987 *Hooks*

ANSI/ASME B30.9 – 1984 *Slings*

SECTION 8

Scaffolding and Ladders

CHAPTER 28

Scaffolding Systems

Practically all types of construction jobs require the use of scaffolds and ladders to enable workers to reach and work in locations that are otherwise inaccessible, and to do their jobs better and more efficiently. Scaffolds seldom are economical for work at heights of less than 30 ft, however, unless access to such work is impractical by ladder or any other means.

Scaffolding is a temporary framework used to support people and material in the construction or repair of buildings and other large structures, providing a safe place of work with safe access suitable for the work being done. Although scaffolding materials traditionally have been built-up timber structures and wood planking, most scaffolding today consists primarily of steel or aluminum tubing and couplers, and seamless steel or extruded aluminum tubing welded panels.

Tubes—either galvanized steel or aluminum alloy—come in a variety of lengths and standard diameter of 2 in. (48.3 mm). The chief differences between the two types of tubes are the lower weight of aluminum tubes—1.14 lb/ft (1.7 kg/m) as opposed to 3 lb/ft (4.4 kg/m)—and a greater flexibility, and so less resistance to force. Tubes generally are bought in 20-ft (6.3-m) lengths and can then be cut down to certain typical sizes.

Boards (planks) provide a working surface for users of the scaffold. They are seasoned wood and come in three thicknesses—1½, 2, and 2½ in. (38-, 50-, and 63-mm), standard width of 8 in. (225 mm), and maximum lengths of 9.25 to 11.75 ft (2.8 to 3.5 m). The board ends are protected by metal plates called hoop irons or sometimes nail plates. Steel or aluminum decking is also used, as well as laminated veneer lumber (LVL).

In addition, for the working platform there are sole boards that are placed beneath the scaffolding if the surface is soft or otherwise suspect. Although ordinary boards can be used, and may meet the minimum requirements, they can be too long and tougher boards may be necessary.

Scaffold couplers hold the tubes together. There are three basic types: right-angle couplers, putlog couplers, and swivel couplers. To join tubes end to end, joint pins (also called spigots) or sleeve couplers, or both, are used.

Other common scaffolding materials include base plates, ladders, ropes, anchor ties, reveal ties, gin wheels, and sheeting.

28.1 Types of Scaffolds

There are many different types of scaffolds in construction to match a variety of job applications. In general, these are divided into two broad categories: support or falsework, and access. (Only the access type of scaffold is covered in this handbook.)

- *Single-point.* A platform suspended by one rope from an overhead support and equipped with means to permit the movement of the platform to desired work levels. The most common among these is also known as a bosun's chair. The maximum intended load for these single-point adjustable suspension scaffolds is 250 lb.
- *Two-point.* Adjustable suspension scaffolds, also known as swing-stage scaffolds, are perhaps the most common type of suspended scaffold. They are hung by ropes or cables connected to stirrups at each end of the platform and hoisted by hand-operated rope blocks and falls, by steel wire rope manual hoisting mechanisms, or by electrically- or air-powered machines at platform levels.
- *Multipoint.* Adjustable scaffold consisting of a platform (or platforms) suspended by more than two ropes from overhead supports and equipped with means to raise and lower the platform(s) to desired work levels. Multipoint adjustable scaffolds *must* be suspended from metal outriggers, brackets, wire rope slings, or hooks.
- *Needle beam.* Consists of a platform suspended from needle beams, usually attached on one end to a permanent structural member. Scaffold support beams *must* be installed on edge. Ropes or hangers *must* be used for supports and *must* be securely attached to needle beams (*Exception: One end of the scaffold may be supported by a permanent structural member.*) Support connections *must* be arranged to prevent the needle beam from rolling or becoming displaced.

28.2 Scaffolding Safety Guidelines

Guidelines for safe erecting, dismantling, and using scaffolds, developed jointly by the Scaffolding Industry Association and the Scaffold, Shoring and Forming Institute, cover frame scaffolds, system scaffolds, tube- and clamp-scaffolds, and rolling scaffolds.

All scaffold users *must* read and comply with the following common sense guidelines, which are designed to promote safety in the erecting, dismantling, and use of scaffolds. These guidelines do not purport to be all-inclusive nor to supplant or replace other additional safety and precautionary measures to cover usual or unusual conditions.

If these guidelines in any way conflict with any state, local, or federal, or other government statute or regulation, said statute or regulation shall supersede these guidelines and it shall be the responsibility of each user to comply therewith.

Post these scaffolding safety guidelines in a conspicuous place and be sure that all persons who erect, dismantle, or use scaffolding are aware of them, and also use them in tool box safety meetings. Follow all state, local, and federal codes, ordinances, and regulations pertaining to scaffolding.

General Safety

- *Survey the job site.* A survey shall be made of the job site by a competent person for hazards, such as untamed earth fills, ditches, debris, high tension wires, unguarded openings, and other hazardous conditions created by other trades. These conditions should be corrected or avoided as noted in the following sections.
- *Inspect all equipment before using.* Never use any equipment that is damaged or defective in any way. Mark it or tag it as defective. Remove it from the job site.
- Erect scaffolds must be erected in accordance with design and/or manufacturers' recommendations.
- *Do not* erect, dismantle, or alter a scaffold unless under the supervision of a competent person.
- *Do not* abuse or misuse the scaffold equipment.
- Erected scaffolds should be continually inspected by users to be sure that they are maintained in safe condition. Report any unsafe condition to your supervisor.
- *Never* take chances! If in doubt regarding the safety or use of the scaffold, consult your scaffold supplier.
- *Never* use equipment for purposes or in ways for which it was not intended.
- *Do not* work on scaffolds if you feel dizzy or unsteady in any way.
- *Do not* work under the influence of alcohol or illegal drugs.

28.3 Erection and Use of Scaffolds

- Inspect scaffolds prior to use to ensure they have not been altered and are in safe working condition.
- Inspect erected scaffolds and platforms continuously (users responsibility).

- Exercise caution when entering or leaving a work platform.
- *Do not* overload scaffold. Follow manufacturer's safe working load recommendations.
- *Do not* jump onto planks or platforms.
- *Do not* use ladders or makeshift devices on top of working platforms to increase the height or provide access from above.
- Climb in access areas only and use both hands.
- Set scaffold base on an adequate sill or pad, to prevent slipping or sinking, and fixed to it where required. Any part of a building or structure used to support the scaffold should be capable of supporting the maximum intended load to be applied.
- Use adjusting screws or other approved methods instead of blocking to adjust to uneven grade conditions.

Frame Scaffolds

- Plumb and level all scaffolds as the erection proceeds. Do not force frames or braces to fit. Level the scaffold until proper fit can easily be made.
- Brace each frame or panel by horizontal bracing, cross bracing, diagonal bracing, or any combination thereof for securing vertical members together laterally. All brace connections shall be made secure, in accordance with the manufacturer's recommendations.

Tube- and Clamp-Scaffolds and System Scaffolds

- Erect posts should be erected plumb in all directions, with the first level of runners and bearers positioned as close to the base as feasible. The distance between bearers and runners shall not exceed manufacturer's recommendations.
- Plumb, level, and tie all scaffolds as erection proceeds.
- Fasten all couplers and connections securely before assembly of next level.
- Install vertical and horizontal diagonal bracing according to manufacturer's recommendations.

Continuous (Running) Scaffolds

- Tie continuous (running) scaffolds to the wall or structure at each end and at least every 30 ft of length when scaffold height exceeds the maximum allowable free standing dimension. Begin ties or stabilizers when the scaffold height exceeds that dimension, and repeat at vertical intervals not greater than 26 ft.

- Place the top anchor not lower than 4 times the base dimension from the top of the completed scaffold. Anchors must prevent scaffold from tipping into or away from wall or structure.
- Stabilize circular or irregular scaffolds in such a manner that completed scaffold is secure and restrained from tipping.
- Take specific precautions when scaffolds are partially or fully enclosed or subjected to overturning loads to ensure the frequency and accuracy of ties to the wall and structure.
- Check the scaffolding component to which ties are subjected for increased loads resulting from wind or overturning loads.

Free-Standing Scaffold Towers

- Restrain free-standing scaffold towers that exceed 4 times their minimum base dimensions vertically from tipping. (CAL/OSHA and some government agencies require a stricter ratio of 3 to 1.)
- *Do not* erect scaffolds near electrical power lines unless proper precautions are taken. Consult the power service company for advice.
- Provide a means of access to all platforms.
- *Do not* use ladders or makeshift devices on top of scaffolds to increase the height.
- Provide guardrails and mid-rails at each working platform level where open sides and ends exist, and toeboards where required by code.

Brackets and Cantilevered Platforms

- Install and use brackets for system scaffolds in accordance with manufacturer's recommendations.
- Seat correctly brackets for frame scaffolds—with side bracket parallel to the frames and end brackets at 90° to the frames. Brackets should not be bent or twisted from normal position. Brackets (except mobile brackets designed to carry materials) are to be used as work platforms only and should not be used for storage of material or equipment.
- Design, install, and use cantilevered platforms in accordance with manufacturer's recommendations.
- Install and use all scaffolding components in accordance with the manufacturer's recommended procedure.
- *Do not* alter components in the field.
- *Do not* intermix scaffold frames and components manufactured by different companies unless the component parts readily fit together and the resulting scaffold's structural integrity is maintained by the user.

Planking

- Provide working platforms with cover over scaffold bearer—as completely as possible. Only scaffold grade wood planking or fabricated planking and decking meeting scaffold use requirements shall be used.
- Check each plank prior to use to be sure plank is not warped, damaged, or otherwise unsafe.

NOTE: *Planking* must *have at least 12-in. overlap and extend 6 in. beyond center of support, or be cleated or restrained at both ends to prevent sliding off supports. Solid sawn lumber, LVL, or fabricated scaffold planks and platforms (unless cleated or restrained)* must *extend over their end supports not less than 6 in. nor more than 18 in. This overhang should not be used as a work platform.*

Putlogs and Trusses

- *Do not* cantilever or extend putlogs/trusses as side brackets without thorough consideration for loads to be applied.
- *EXTEND* putlogs/trusses at least 6 in. beyond point of support.
- Place recommended bracing between putlogs/trusses when the span of putlog/truss is more than 12 ft.

Rolling Scaffolds, Additional Guidelines

- Riding a rolling scaffold is very hazardous. The Scaffold Industry Association neither recommends nor encourages this practice. However, if you choose to do so, be sure to follow all state, federal, or other governmental guidelines.
- Casters with plain stems shall be attached to the panel or adjustment screw by pins or other suitable means.
- No more than 12 in. of the screw jack should extend between the bottom of the adjusting nut and the top of the caster.
- Wheels or casters should be provided with a locking means to prevent caster rotation and scaffold movement and kept locked.
- Joints should be restrained from separation.
- Use horizontal or diagonal bracing near the bottom and at 20-ft intervals measured from the rolling surface.
- Do not use brackets or other platform extensions without compensating for the overturning effect.
- The platform height of a rolling scaffold must not exceed 4 times the smallest base dimension. (CAL/OSHA and some government agencies require a stricter ratio of 3 to 1.)
- Cleat or secure all planks.

- Secure or remove all materials and equipment from platform before moving.
- Do not attempt to move a rolling scaffold without sufficient help. Watch out for holes in the floor and overhead obstructions; stabilize against tipping.

28.4 Dismantling Scaffolding[1]

- Check to assure scaffolding has not been structurally altered in a way that would make it unsafe and, if it has, reconstruct where necessary before commencing with dismantling procedures. This includes all scaffold ties.
- Inspect visually every plank prior to dismantling to be sure it is safe.
- Consider the effect removal of a component will have on the rest of the scaffold prior to that component's removal.
- *Do not* accumulate excess components or equipment on the level being dismantled.
- *Do not* remove ties until the scaffold above has been removed (dismantled).
- Lower dismantled components in an orderly manner; do not throw them form the scaffold.
- Stockpile dismantled equipment in an orderly manner.

These safety guidelines (*Codes of Safe Practice*) set forth common sense procedures for safely erecting, dismantling, and using scaffolding equipment. Equipment and scaffolding systems differ, however, and thus reference must always be made to the instructions and procedures of the supplier or manufacturer of the equipment.

Because field conditions vary and are beyond the control of the Scaffold Industry Association, safe and proper use of scaffolding is the sole responsibility of the user. Always follow safe erection procedures and use manufacturer manuals, and refer to *OSHA 1926 - Safety Standards for Scaffolds Used in the Construction Industry*.

28.5 Care and Maintenance of Scaffolds

Most system scaffold manufacturers provide daily and monthly inspection checklists that they recommend contractors follow. A competent person on staff who is trained and knows what to look for should pay attention to certain points of concern to keep a well-maintained scaffold:

---

[1]*Source:* Scaffolding Industry Association, Phoenix, AZ (*www.sia.org*); The Scaffold, Shoring and Forming Institute, Cleveland, OH (*www.ssfi.org*).

- Inspect scaffolding every time it is set up. Cross-bracing and the trueness of the scaffold are easy things to check while it is on the wall because problems are very easy to spot that way.
- Think about the materials used to manufacture the bulk of the adjustable scaffolding. Heavy-duty structural steel is the material of choice, resulting in durable scaffold systems that generally are not too cumbersome to store and transport when using the proper equipment.
- Wear on the scaffold's finish is unavoidable. Depending on the climate, and how much use one gets from a particular elevating scaffold system, sandblasting, and repainting may be necessary.
- If the steel is damaged in any way that compromises the strength or the ability to keep the scaffold straight and plumb, it needs to be taken out of service.
- Cross-braces and guardrails get damaged more than anything else, and it is the cross-bracing that really affects the trueness of scaffolding and the ability to keep it level and plumb.
- Consider the hand crank systems and other moving parts such as the winch and gears. For these components, reliability is more important than anything else. All of the winches should be relubricated. Keeping them well-lubricated makes it easier to continually raise and lower the platform.
- *Do not* neglect the braking systems on the adjustable scaffolding. Brake replacement will need to be done by someone certified to weld them on.
- Before and after every job, a contractor-appointed crew member should inspect the scaffolding using the manufacturer's inspection checklist.

WARNING: *Because some elevating scaffold systems have a reputation for being strong and durable, people may be tempted to push the limits.* Do not *do anything that is not in the safety installation manual and that you are not 100 percent sure of.*

28.6 Inspection of Scaffolding Systems

- Scaffolding must be inspected every day before work begins (Table 28.1).
- If any problems are found with the stability of the structure, repairs should be made before use.
- If cracks or warping is found on planks, the planks should be replaced before use.
- Any sharp edges, burrs, or other safety hazards should be eliminated.

• Does each scaffold and scaffold component support (without failure) its own weight and at least 4 times the maximum intended load?	Y	N	?
Note: The stall load of any scaffold hoist must not exceed 3 times its rated load.			
• Are all working levels on scaffolds fully planked or decked between the front uprights and the guardrail supports?	Y	N	?
• Are scaffold platform spaces 1 in. or less between adjacent units and the uprights?	Y	N	?
Note: Spaces up to 9 $^1/_2$ in. wide are permitted around uprights. If platforms are used only as walkways during erecting or dismantling, the employer may establish the space between planking as necessary to provide safe working conditions.			
• Are all scaffold platforms at least 18 in. wide?	Y	N	?
Note: Scaffold platforms less than 18 in. wide are permitted if wider platforms are not possible. Any person who uses these platforms, however, must be protected by guardrails or personal fall arrest systems.			
• Is the distance between the front edge of the scaffold platform and the face of the work 14 in. or less, unless guardrail systems are put along the front edge, or personal fall arrest systems are used?	Y	N	?
Note: Exceptions are permitted under special situations. Consult the OSHA regulations for details. The maximum distance from the face for plastering and lathing operations is 18 in.			
• Does the end of each scaffold platform extend over the centerline of its support at least 6 in. unless cleated or otherwise restrained by hooks or equivalent means?	Y	N	?
• If the scaffold platform is 10 ft or less in length, does the end of the scaffold platform extend 12 in. or less over its support?	Y	N	?
Note: The end of the scaffold platform may extend more than 12 in. over its support if the platform is designed and installed so that the cantilevered portion of the platform can support workers or materials without tipping, or a guardrail blocks worker access to the cantilevered end.			
• On scaffolds where planks are abutted to form a long platform, does each plank end rest on a separate support surface?	Y	N	?

TABLE 28.1 Scaffolding Inspection Checklist (*Continued*)

Note: Common support members, such as "T" sections, may be used to support abutting planks. Hook-on platforms designed to rest on common supports may also be used.			
• On scaffolds where platforms overlap to form a long platform, does the overlap occur over supports?	Y	N	?
• Is the overlap at least 12 in. unless the platform is nailed together or otherwise restrained to prevent movement?	Y	N	?
• At points of a scaffold where the platform changes direction, is this procedure followed?			
- Lay the platform that rests on a bearer at an angle other than a right angle.	Y	N	?
- Lay the platform that rests at right angles over the same bearer, on top of the first platform.	Y	N	?
• Is it prohibited to cover wood platforms on scaffolds with opaque finishes?	Y	N	?
Note: Platform edges may be covered or marked for identification. Wood platforms may be coated periodically with wood preservatives, fire-retardant finishes, and slip-resistant finishes; however, the coating may not obscure the top or bottom wood surfaces.			
• Do scaffold components from different manufacturers fit together without force and maintain the scaffold's structural integrity?	Y	N	?
• Are scaffold components from different manufacturers modified only if a competent person determines that the scaffold made of the 'mixed' parts is structurally sound?	Y	N	?
• Are scaffold components made of dissimilar metals used together only if a competent person has determined that galvanic action will not reduce the strength of any component to an unacceptable level?	Y	N	?
Criteria for supported scaffolds			
• Are supported scaffolds with a height to base width ratio of more than 4 to 1 restrained from tipping by guying, tying, bracing, or equivalents?	Y	N	?

Note: Install guys, ties, and braces at locations where horizontal members support both inner and outer legs. Install guys, ties, and braces according to the scaffold manufacturer's recommendations or at the closest horizontal member to the 4 to 1 height. Repeat vertically at locations of horizontal members every 20 ft or less thereafter for scaffolds 3 ft wide or less, and every 26 ft or less thereafter for scaffolds greater than 3 ft wide. Place the top guy, tie, or brace of completed scaffolds no further than the 4 to 1 height from the top. Install guys, ties, and braces at each end of the scaffold and at horizontal intervals 30 ft or less (measured from one end [not both] toward the other). Use ties, guys, braces, or outriggers to prevent tipping when there is an eccentric load, such as a cantilevered work platform.			
• Are supported scaffold poles, legs, posts, frames, and uprights placed on base plates and mud sills or other firm foundation?	Y	N	?
• Are footings level, sound, and rigid? Can they support the loaded scaffold without settling or displacement?	Y	N	?
• Is it prohibited to use unstable objects for supporting scaffolds and platform units?	Y	N	?
• Is it prohibited to use unstable objects as working platforms?	Y	N	?
• Is it prohibited to use front-end loaders and similar pieces of equipment to support scaffold platforms, unless they were designed by the manufacturer for such use?	Y	N	?
• Are supported scaffold poles, legs, posts, frames, and uprights plum and braced to prevent swaying and displacement?	Y	N	?
Access			
• Are ladders, stairs, ramps, or walkways provided to access scaffold platforms more than 2 ft above or below a point of access?	Y	N	?
Note: Cross-braces must not be used as a means of access.			
• Are portable, hook-on, and attachable ladders positioned to prevent the scaffold from tipping?	Y	N	?
• Are hook-on and attachable ladders positioned so the bottom rung is not more than 24 in. above the scaffold supporting level?	Y	N	?
• Are hook-on and attachable ladders designed for the scaffold in use?	Y	N	?

TABLE 28.1 Scaffolding Inspection Checklist (*Continued*)

• Do hook-on and attachable ladders have rung length of at least 11 $^1/_2$ in.?	Y	N	?
• Do hook-on and attachable ladders have uniformly spaced rungs with a maximum spacing between rungs of 16$^3/_4$ in.?	Y	N	?
• Is the bottom step of stairway-type ladders 24 in. or less above the scaffold supporting level?	Y	N	?
• Do stairway-type ladders have rest platforms at 12-ft maximum vertical intervals?	Y	N	?
• Do stairway-type ladders have a step width of at least 16 in.?		Y	N
Note: Mobile scaffold stairway-type ladders may have a minimum step width of 11 $^1/_2$ in.			
• Do stairway-type ladders have slip-resistant treads on all steps and landings?	Y	N	?
• Do ramps and walkways 6 ft or more above lower levels have guardrails?	Y	N	?
• Are ramps and walkways inclined with a slope less than 1 vertical to 3 horizontal 20° above the horizontal?		Y	N
• Do ramps and walkways that are steeper than 1 vertical in 8 horizontal have cleats 14 in. (or less) apart that are securely fastened to the planks to provide footing?	Y	N	?
• Are integral prefabricated scaffold access frames constructed for use as ladder rungs?	Y	N	?
• Do integral prefabricated scaffold access frames have rung lengths of at least 8 in.?	Y	N	?
• Do students and employees have fall protection if integral prefabricated scaffold access frames with rungs less than 11 $^1/_2$ in. are used as work platforms?	Y	N	?
• Are integral prefabricated scaffold access frames uniformly spaced within each frame section?	Y	N	?
• Do integral prefabricated scaffold access frames have a maximum spacing between rungs of 16 $^3/_4$ in.?	Y	N	?
Note: Nonuniform rung spacing caused by joining end frames together is allowed, provided the resulting spacing is 16 $^3/_4$ in. or less.			
• Do steps and rungs of ladder and stairway-type access line up vertically with each other between rest platforms?	Y	N	?

• Is the horizontal distance 14 in. (or less) and the vertical distance 24 in. (or less) between two surfaces used to provide direct access between them?	Y	N	?
• During erecting and dismantling of supported scaffolds, does a competent person provide and evaluate safe means of access?	Y	N	?
• During erecting and dismantling of supported scaffolds, are hook-on or attachable ladders installed as soon as they can be used safely?	Y	N	?
• During erecting and dismantling of supported scaffolds, are the ends of tubular welded frame scaffolds used as climbing devices for access only if the horizontal members are parallel, level, and 22 in. apart (or less) vertically?	Y	N	?
Use			
• Are scaffolds and scaffold components loaded below their maximum intended loads or rated capacities (whichever is less)?	Y	N	?
• Is the use of shore or lean-to scaffolds prohibited?	Y	N	?
• Does a competent person inspect scaffolds and scaffold components for visible defects before each work shift, and after any occurrence that could affect a scaffold's structural integrity?	Y	N	?
• Are parts of a scaffold that are damaged or weakened immediately repaired, replaced, braced, or removed from service until repaired?	Y	N	?
• Is the horizontal movement of a scaffold prohibited while workers are on the scaffold (unless the scaffold is designed for movement by a registered professional engineer, or is a mobile scaffold meeting OSHA standards)?			
• Are proper clearances between scaffolds and power lines always maintained?	Y	N	?
Note: Scaffolds and materials may be closer to power lines if such clearance is necessary, and only after the utility company or electrical system operator has been notified and has de-energized the lines, relocated the lines, or installed protective coverings to prevent contact with the lines.			

Source: Occupational Safety & Health Administration, Washington, DC (*www.osha.gov*).

Table 28.1 Scaffolding Inspection Checklist (*Continued*)

Regulations and Standards

OSHA Regulations

1910 Subpart D—*Walking working surfaces*

1910.28 *Safety Requirements for Scaffolding*

1910.29 *Manually Propelled Mobile Ladder Stands and Scaffolds (Towers)*

1915 Subpart E—*Scaffolds, Ladders, and Other Working Surfaces*

1915.71 *Scaffolds or Stagings*

1915.77 *Working Surfaces*

U.S. Consensus Standards

ANSI/ASSE A10.8 – 2001 *Scaffolding Safety Requirements*

CHAPTER 29

Swinging or Suspended Scaffolds

The swinging or suspended scaffold is available in a variety of shapes and sizes, ranging from the one-person bosun's chair to the articulated suspended platform-type scaffold.

29.1 Swinging Scaffolds

Most swinging scaffolds are two-person work platforms, made of wood or aluminum, suspended from lookouts that extend over the edge of flat-roofed buildings or from special U-shaped hangers on buildings that have wide cornices.

This type of scaffold may be hand operated using simple ropes and block and tackle, or it may be suspended from high-tensile steel wire ropes, relying on power-driven, remote-controlled winches to provide travel both vertically and horizontally (Figs. 29.1 and 29.2).

In its most familiar form, the swinging scaffold consists of a frame similar in appearance to a ladder with a decking of wood slats. The structure is supported near each end by a steel stirrup to which the lower block of a set of manila rope falls is attached. Its frame should be constructed of clear, straight-grained spruce rails and should have dimensions in accordance with *ANSI A10.8* (Table 29.1).

Scaffold Construction

The rungs of a swinging scaffold should be made of straight-grained oak, ash, or hickory—at least $1^1/_8$ in. (3.18 mm) with 7/8-in. (22.23-mm) tenons mortised into the side stringers at least 7/8 in. (22.23 mm) deep and spaced not more than 18 in. (457.2 mm) apart. The stringers should be tied together with tie-rods not less than ¼ in. (6.35 mm) diameter, passing through the stringers and riveted up tight against washers on both ends.

The platform flooring should be ½- × 3-in. (12.7- × 76.2-mm) strips spaced not more than 5/8 in. (15.88 mm) apart, except at the side

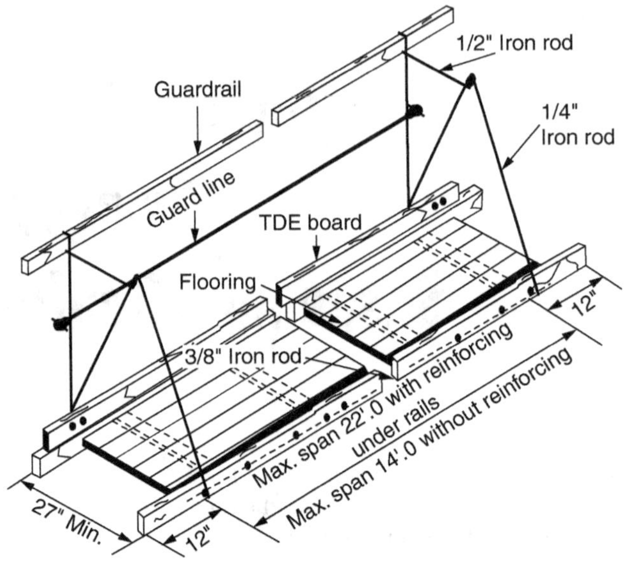

FIGURE 29.1 Swinging scaffold (should be suspended by wire). (*Source: U.S. Army Field Manual EM 385-1-1*, Department of the Army, Washington, DC.)

rails, where the space may be 1 in. (25.4 mm). The standard width of the scaffold is 28 in. (711.2 mm).

A 1- × 4-in. (25.4- × 101.6-mm) toeboard should be provided on the outboard side of such scaffolds. A guardrail not less than 2 × 4 in.

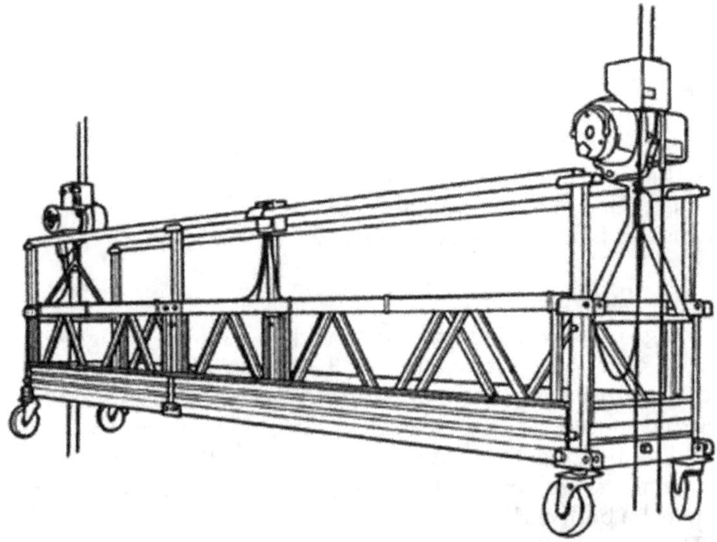

FIGURE 29.2 Modular-powered swing stage scaffolding system. (*Source:* Colgate Scaffolding and Equipment Corp., Bronx, NY.)

	Length of platform ft				
	12	14 and 16	18 and 20	22 and 24	28 and 30
Side stringers, minimum cross section (finished sizes)					
At ends	1¾ × 2¾	1¾ × 2¾	1¾ × 3	1¾ × 3	1¾ × 3½
At middle	1¾ × 3¾	1¾ × 3¾	1¾ × 4¼	1¾ × 5	
Reinforcing strip, minimum	A ⅛ × ⅞-in. steel reinforcing strip or its equivalent shall be attached to the side or underside, full length.				
Rungs	Minimum diameter 1⅛-in. with at least ⅞-in.-diameter tenons; maximum spacing 12 in. center to center.				
Tie rods					
Number, minimum	3	4	4	5	6
Diameter, minimum, in.	¼	¼	¼	¼	¼
Flooring, minimum finished size, in.	½ × 2¾	½ × 2¾	½ × 2¾	½ × 2¾	½ × 2¾

Source: From ANSI Std. A10.8.

TABLE 29.1 Maximum Length of Platforms

(50.8 × 101.6 mm) should be located not less than 36 in. (914.4 mm), nor more than 42 in. (1066.8 mm), above the platform floor—inserted into the sockets or loops in the stirrups provided for this purpose. When required, a midrail of 1 × 4 in. (25.4 × 101.6 mm) should be installed at all open sides on scaffolds more than 10 ft (3.05 m) above the ground or floor.

Additional stanchions should be provided to keep the span of the guardrail to 10 ft (3.05 m) or less. A screen or ½-in. (12.7-mm) mesh rabbit wire is recommended between the guardrail and the toeboard.

Plank-type Platforms

This type of platform should be constructed of not less than nominal 2 × 8 in. (50.8 × 203.2 mm) unspliced planks properly cleated together on the underside, starting 6 in. (152.4 mm) from each end, with intervals not to exceed 4 ft (1.22 m). It should never extend more than 18 in. (457.2 mm) beyond the hangers (see Table 29.1). A bar, securely fastened to the platform at each end, will prevent its slipping off the hanger. Limit spans between hangers to 10 ft (3.05 m). For beam-type platforms, side stringers should be made of lumber not less than 2- × 6-in. (50.8 × 152.4 m) set on edge (Fig. 29.3).

The span between hangers on this type of platform should not exceed 12 ft. Platform flooring of 1- × 6-in. lumber, properly nailed and spaced not more than ½ in. apart, should be supported on 2- × 6-in. stringer with a snug fit at intervals of not more than 4 ft, and securely nailed in place.

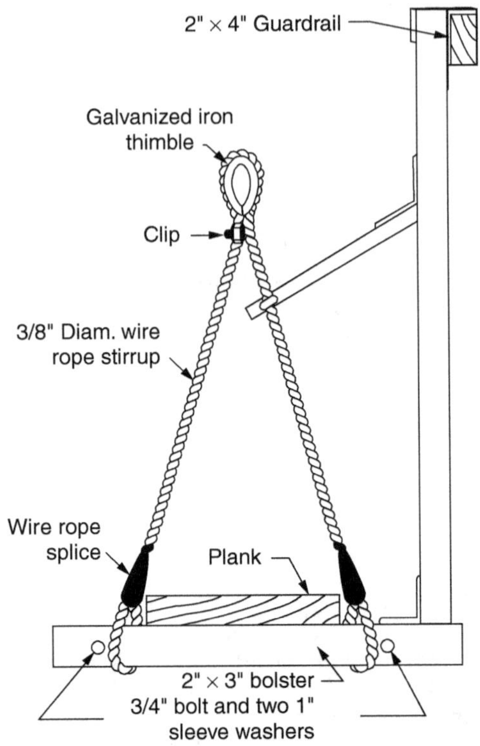

FIGURE 29.3 Typical hanger to use with plank scaffold. (*Source: U.S. Army Field Manual 55-125*, Department of the Army, Washington, DC.)

If light metal-type platforms are used, *make sure* that they are of tested design and approved for the particular scaffold duty intended. The metal stirrups that support the platform should be of wrought iron or mild steel design having a cross-sectional area capable of sustaining four times the maximum intended load. Stirrups should be placed between 6 and 18 in. from the ends of the scaffold and secured to it by U-bolts of adequate size.

Suspension Systems

A set of ¾-in. number 1 grade Manila rope falls—consisting of a double- and a single-pulley, 6-in. block—should be provided at each end of the scaffold. A safe means of supporting the upper locks is absolutely necessary. This may be in the form of U-shaped roof or cornice hooks, ½-in. wire rope slings, or other approved design.

Roof irons or hooks should be of wrought iron or mild steel of proper size and design, securely installed and anchored. A ¾-in. Manila rope tieback should be used to secure the cornice hook to a fixed anchorage on the roof.

Good practice includes wiring the hook on the lower pulley block to the eye of the stirrup, thus preventing accidental detachment of the hook. A mousing would ordinarily be used for such a condition; however, its use here would interfere with the hitch of the rope.

Instead, a hitch is made by holding a strain on the rope with one hand and pushing a bight of the slack part of the rope through the inverted *V* of the stirrup. The bight is then given a 180° twist and is placed over the bill of the hook. The strain on the live part of the rope forces the dead part into the *V*, into which it jams. Although a very simple hitch, it is dependable.

To ensure against the scaffold falling if a worker were to lose grip on the rope accidentally while raising or lowering the scaffold, it is necessary to install a special safety latch to the cheek of the lower block. The hauling part, or hand rope, is passed through the hole in the hinged plate, and the hitch is made in the manner just described. If the worker accidentally lets go of the rope, the safety latch will then raise and grab the rope, and the scaffold would be prevented from falling.

For special jobs, a box-type scaffold is often used that provides a two-member guardrail and toeboard on all sides.

When workers are using a swinging scaffold, *make sure* it is always secured to the building or structure to prevent it from moving and allowing the worker to fall between. On building walls, it is usually difficult to find something to lash a scaffold to, but it is often found practicable to provide a standard attachment for a window-cleaner's belt on a short length of Manila rope. To hold the scaffold to the wall, the device is simply attached to the special bolt at the side of the windows, and the rope is secured to the platform.

Safety Belts and Lifelines

Safety belts for each worker on the scaffold, on separately attached lifelines, are essential for proper worker safety. Only new rope, ¾- or 1-in. diameter, should be used, and it should be properly secured at the roof or upper part of the structure. Where the rope passes over sharp bends, such as copings or window sills, it should be padded to prevent abrasion.

Each worker should wear a 4-in. life belt or three-ply cotton (nylon) webbing with a ⅝-in. or ¾-in. rope tail line about 6 ft long. The tail line should be attached, as short as possible, to the hanging lifeline by a rolling hitch. This hitch can be readily slid up or down the hanging rope, yet if the worker falls, the hitch will jam and hold the worker. The rolling hitch is similar to the clove hitch, except that in tying the first part of the hitch, two wraps (instead on one) are made around the hanging line (see Chap. 11).

Nylon rope may be more suitable for lifelines because, in addition to being strong, it stretches and will stop a falling worker more gently.

The lifeline must reach to the ground, or other place of safety, and the worker *must* have the safety belt on and attached *at all times* particularly when the scaffold is being raised or lowered.

Trolley Cables

On special hanging scaffolds, trolley cables run along the upper hand-rail members. The worker attaches the snap hook of the safety belt tail line to the cable at the rear. This frees the worker to walk back and forth on half of the scaffold. If a rope should fail, and one end of the scaffold swings downward, the workers probably would remain within the scaffold railings. In the event they were thrown out, their life belts would keep them from falling to the ground.

Wire Rope Support

While Manila rope falls for supporting scaffolds are still used, it is more common to suspend such scaffolds from wire ropes, using winches to raise or lower the scaffold. Where the lift of the scaffold exceeds 100 ft, wire rope and winches *must* be used. Also, where acid solutions are used, fiber ropes *must never* be permitted, unless they are approved as acidproof.

Among the many advantages for using wire rope winch suspension systems are:

- Greater safety due to more positive inspections
- Ease of handling the scaffold because it can be winched up or down as desired
- Lower headroom because the scaffold can be raised closer to the overhead supports
- Less danger of failure when acid is used, such as in washing building walls

Load Testing

After a scaffold has been erected on a new job, it should be load-tested before workers risk their lives on it. Hoist the scaffold about 12 in. off the ground and apply a test weight equal to four workers for a period of 5 min. The test should be made every 10 days if the job continues for more than that time. A factor of safety of 4 should be used to calculate the strength of a swinging scaffold.

Where swinging scaffolds are suspended adjacent to each other, *never* place planks on the scaffolds to form a bridge between them, and *never* permit more than two persons to work on a scaffold at any one time. Occasionally it may be desirable to install wood bunters with rollers on a scaffold to hold it away from the building wall and to keep it from swinging or swaying.

Whenever a winch or a scaffold is to be removed to another location, *always* wind the cable on the winch properly. *Never* coil it on the ground to save time. A kink in the cable will weaken and may even ruin it.

The maximum load that can be suspended by one of these winches is 500 lb.

29.2 Suspended Scaffolds

Suspended scaffolds are most commonly used by bricklayers, working on new buildings, however, they are equally applicable for any heavy construction or repair work.

This type of scaffold consists of a number of outriggers, usually 7-in. steel or aluminum I beams located at roof level, from each of which two ½-in.-diameter wire ropes wind up on hand-operated winches on the scaffold platform.

The thrust-outs, or outriggers, of a suspended scaffold should not project more than 6 ft beyond the point where they bear on the support, unless used in pairs. The fulcrum point of the beam should rest on a secure bearing at least 6 in. in each horizontal dimension. *Make sure* that the beam is secured in place against movement and securely braced at the fulcrum point against tipping.

The inboard ends of the beams should be anchored to the roof steel by large U-bolts and anchor plates. These ends should be measured from the fulcrum point to the extreme point of support, should not be less than 1½ times the outboard end in length, and should be spaced not more than 10 ft apart. The suspension cables should be placed not more than 2 and 6 ft respectively, out from the bearing point of the beam.

The platform of a suspended scaffold consists of planks resting on putlogs, or bearers, which are supported by the winches. The width of the scaffold *must not* exceed 28 in. Planking should be 1 ¼-in. thick for spans up to 10 ft. These planks should be laid tightly and should extend to within 3 in. of the building.

Planking *must* be nailed or bolted securely to the putlogs, which should overlap by not more than 18 in. at each end. A standard guardrail, not less than 36 in. or more than 42-in. high, and 9-in. toeboard *must* be provided along the outer edge of the scaffold. A wire screen between them is recommended.

This type of scaffold equipment provides for a 2-in.-thick plant decking or roofing above the workers. Special hooks should be provided to hold the scaffold close to the building wall. The allowable loading on an outrigger is approximately 2000 lb.

To raise the scaffold, levers on the winches are operated up and down to rotate them. To lower the scaffold, depress the

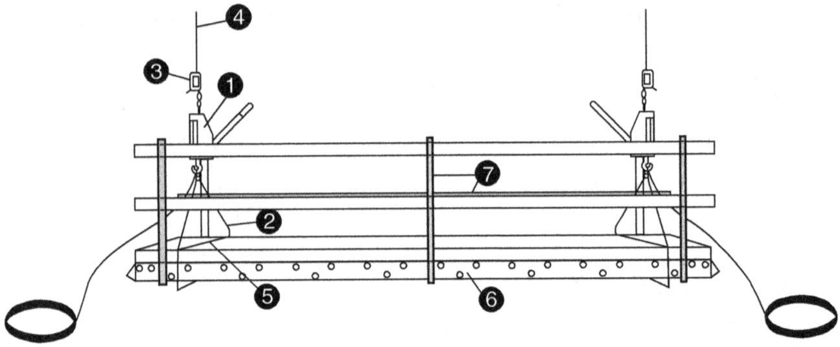

Illustration consists of:

1 1 Pair of griphoist mechanical hoists (TU-17)
2 1 Pair of griphoist stirrups (GHS24)
3 Pair of blocstop safety devices (B/S)
4 1 Pair of griphoist 60' wire rope assemblies (GHWR60)
5 1 Pair of steel wall bumpers with 'J' bolts (SSJB) (not shown on illustration)
6 1 Only – 24" wide × 24" long aluminum stage
7 2 Guard rail systems with toeboards (*Aluminum* 2" × 4"s *and center irons shown*)

FIGURE 29.4 Suspended scaffold. (*Source:* Colgate Scaffolding and Equipment Corp., Bronx, NY.)

ratchet handle with the driving pawl held out of engagement. Then replace the driving pawl and disengage the locking pawl so that the weight of the scaffold is sustained on the ratchet handle, which is allowed to rise, thus unwinding the cable from the drum (Fig. 29.4).

29.3 Safety Cage and Bosun's Chair

When a relatively minor scaffold job is required, a safe and convenient work platform from which to work may be provided by a one-person aluminum safety cage. This lightweight rig is more comfortable for the worker than any bosun's chair, yet is nearly as flexible in operation. These cages are often available with extensions attached to each side of the cage for use by two persons.

Safety cages are available in a variety of sizes and types, including modular work cages, fold-up work cages, and low-profile work cages, for one or two workers (Fig. 29.5).

In general, the cages are used with power-operated winches while the bosun's chair uses powered winches or blocks and falls (Fig. 29.6).

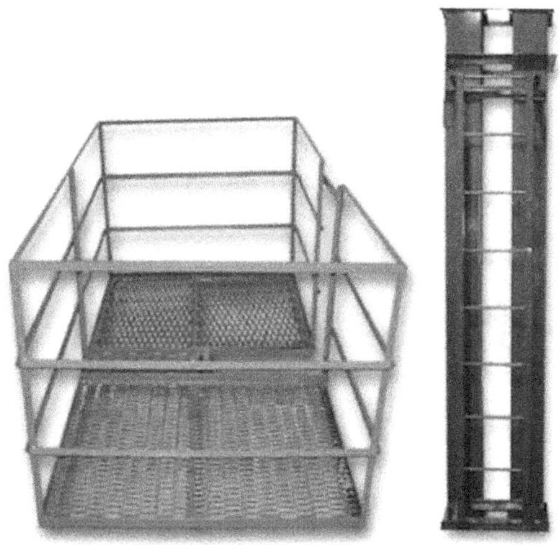

FIGURE 29.5 Safety cage and ladder. (*Source:* Airmaster, Winnepeg, Manitoba, Canada.)

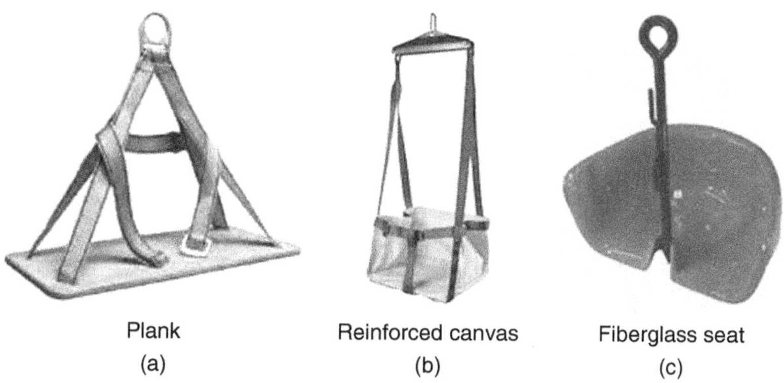

FIGURE 29.6 Bosun's chairs. (a) Plank; (b) reinforced canvas; (c) fiberglass seat. (*Source:* Airmaster, Winnepeg, Manitoba, Canada.)

Regulations and Standards

OSHA Regulations

1926 Subpart L – Appendix C—*Scaffolds*

1926.451 *General Requirements*

1926.452 *Additional Requirements Applicable to Specific Types of Scaffolds*

1926.453 *Aerial Lifts*

1926.454 *Training Requirements*

Appendix A *Scaffold Specifications*

Appendix B *Criteria for Determining the Feasibility of Providing Safe Access and Fall Protection for Scaffold Erectors and Dismantlers*

U.S. Consensus Standards

ANSI/SIA

A92.2–1990 *Vehicle-Mounted Elevating and Rotating Aerial Devices*

A92.3–1990 *Manually Propelled Elevating Aerial Platforms*

A92.5–1990 *Boom Supported Elevating Work Platforms*

A92.6–1990 *Self-Propelled Elevating Work Platforms*

A92.7–1990 *Airline Ground Support Vehicle-Mounted Vertical Lift Devices*

A92.8–1993 *Vehicle-Mounted Bridge Inspection and Maintenance Devices*

A92.9–1993 *Mast-Climbing Work Platforms*

CHAPTER 30

Stationary Scaffolds

Structural stability requires that a scaffold be capable of supporting itself and the applied load with an acceptable safety factor and without unacceptable deformation. There are many different types of scaffolds used in construction to match a variety of job applications. In general, however, these are divided into two broad categories: support or falsework, and access. (Only the access type of scaffold is covered in this handbook.)

The three basic types of stationary scaffolds are independent self-supporting, partially self-supporting, and cantilevered. They can be either job built-up or prefabricated systems.

Job built-up–type scaffolds are single- or double-pole structures made of wood or metal. The single-pole type is supported on one side by poles or uprights, on the other side by the wall or structure against which it is erected. The double-pole type is erected independently of the building or structure (Fig. 30.1).

These types of scaffolds must be built and inspected to conform with OSHA requirements, as well as state laws and local ordinances. If any portion of a wooden scaffold has been weakened or damaged by storm or accident, that scaffold should not be used until the necessary repairs have been made. Care must be exercised to prevent overloading any scaffold (Table 30.1).

30.1 Wood Pole and Planking Scaffolds

The woods most commonly used for job-constructed scaffolds are spruce, fir, Douglas fir, and southern yellow pine. Because it has the highest ratio of strength to weight, spruce is the preferred material for wood scaffolds. All lumber used for scaffold members and planking should be free from serious defects (see Chap. 4).

When erecting wood-pole scaffolds, *make sure* that all poles, or uprights, are accurately plumbed. Poles should bear on a foundation of sufficient size and strength to spread the load from the pole over a sufficient area to prevent settlement. Independent scaffolds should be set as near to the wall of the building as practicable.

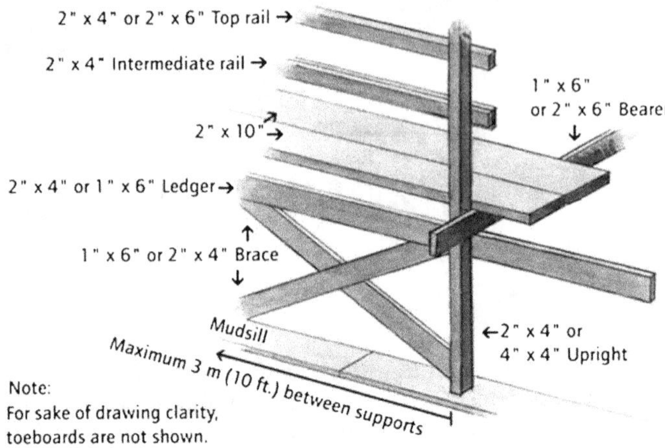

Single pole (wood, light duty) – Supported on its interior side by a structure or wall
(a)

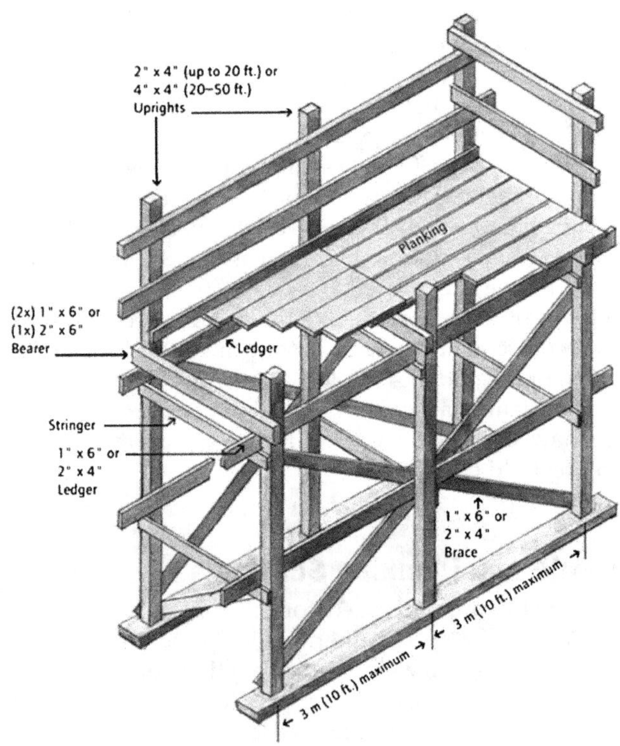

Double-pole (wood, light-duty)
Supported by double uprights independent of any structure.
(b)

FIGURE 30.1 Stationary scaffolds. Single pole (wood, light duty); supported on its interior side by a structure or wall. Double-pole (wood, light duty); supported by double uprights independent of any structure. (*Source:* Worker's Compensation Board, British Columbia, Canada.)

	YES	NO	COMMENTS
• Are scaffolds and scaffold components inspected before each work shift by a competent person?	❑	❑	________
• Have employees who erect, disassemble, move, operate, repair, maintain, or inspect the scaffold been trained by a competent person to recognize the hazards associated with this type of scaffold and the performance of their duties related to this scaffold?	❑	❑	________
• Have employees who use the scaffold been trained by a qualified person to recognize the hazards associated with this scaffold and know the performance of their duties relating to it?	❑	❑	________
• Is the maximum load capacity of this scaffold known and communicated to all employees?	❑	❑	________
• Is the load on the scaffold (including point loading) within the maximum load capacity of this particular scaffold?	❑	❑	________
• Is the scaffold plumb, square, and level?	❑	❑	________
• Is the scaffold on base plates and are mudsills level, sound, and rigid?	❑	❑	________
• Is there safe access to all scaffold platforms?	❑	❑	________
• Are all working platforms fully planked?	❑	❑	________
• Do planks extend at least 6 in. and no more than 12 in. over the supports?	❑	❑	________
• Are the planks in good condition and free of visible defects?	❑	❑	________
• Does the scaffold have all required guardrails and toeboards?	❑	❑	________
• Are 4:1 (height to width) scaffolds secured to a building or structure as required?	❑	❑	________

Note: Use this checklist for daily inspections of the scaffold. It is not all-inclusive and should be used as a starting point for the competent person to develop a checklist specific to the type of scaffold and jobsite conditions encountered.
Source: Electronic Library of Construction, OSHA, Washington, DC.

TABLE 30.1 Typical Stationary Scaffolding Inspection Checklist

All pole scaffolds should be securely guyed or tied to the building or structure. Where height or length is greater than 25 ft, the scaffold should be secured at intervals not greater than 25 ft vertically or horizontally.

When poles must be spliced, *make sure* that abutting ends are squared, and that the bottom end of an upper member rests squarely on the upper end of the lower member. Wood splice plates, not less than 4 ft long, should be used on at least two adjacent sides. Splices should overlap the abutted ends equally and should be of the same width, and of a cross-sectional area not less than that of the pole.

ANSI A10.8 provides the minimum sizes and maximum spacings for numbers of single-pole and independent-pole scaffolds (Tables 30.2 and 30.3).

	Light duty		Medium duty	Heavy duty
Uniformly distributed load	Not to exceed 25 lb/ft^2	Not to exceed 25 lb/ft^2	Not to exceed 50 lb/ft^2	Not to exceed 75 lb/ft^2
Maximum height of scaffold	20 ft	60 ft	60 ft	60 ft
Poles or uprights	2×4 in.	4×4 in.	4× 4 in.	4 × 4 in.
Pole spacing, longitudinal	6 ft	10 ft	8ft	6 ft
Maximum width of scaffold	5 ft	5 ft	5ft	5 ft
Bearers or putlogs			2× 9 in. or 3× 4 in.	2 × 9 in. or 3 × 5 in. (rough)
Bearers or putlogs to 30-ft width	2×4 in.	2 × 4 in.		
Bearers or putlogs to 5-ft width	2×6 in. or	2 × 6 in. or		
	3×4 in.	3 × 4 in. (rough)		
Spacing of bearers or putlogs			3ft 4 in.	6 ft
Ledgers	1×4 in.	1¼ × 9 in.	2× 9 in.	2 × 9 in.
Vertical spacing of horizontal members	7 ft	7 ft	9 ft	6 ft 6 in.
Bracing, horizontal	1×4 in.	1 × 4 in.	1× 6 in. or 1 ¼×4	2 × 4 in.
Bracing, diagonal	1×4 in.	1 × 4 in.	1×4 in.	2 × 4 in.
Tie-ins	1×4 in.	1 × 4 in.	1×4 in.	1 × 4 in.
Planking	1¼×9 in. (rough)	2 × 9 in.	2×9 in.	2 × 9 in.
Toeboards	4 in. high (minimum)	4 in. high (minimum)	4 in. high (minimum)	4 in. high (minimum)
Guardrail	2×4 in.	2×4 in.	2×4 in.	2×4 in.

*All members except planking are used on edge.

Source: After ANSI A10.8.

TABLE 30.2 Minimum Nominal Size of Single-Pole Scaffolds and Maximum Spacing of Members

	Light duty		Medium duty	Heavy duty
Uniformly distributed load	Not to exceed 25 lb/ft^2	Not to exceed 25 lb/ft^2	Not to exceed 50 lb/ft^2	Not to exceed 75 lb/ft^2
Maximum height of scaffold	20 ft	60 ft	60 ft	60 ft
Poles or uprights	2 × 4 in.	4 × 4 in.	4 × 4 in.	4 × 4 in.
Pole spacing, longitudinal	6 ft	10 ft	8 ft	6 ft
Pole spacing, transverse	6 ft	10 ft	8 ft	8 ft
Ledgers	1¼ × 4 in.	1¼ × 9 in.	2 × 9 in.	2 × 9 in.
Vertical spacing of horizontal members	7 ft	7 ft	6 ft	4 ft 6 in.
Spacing of bearers			8 ft	
Bearers			2 × 9 in. (rough) or 2 × 10 in.	2 × 9 in. (rough)
Bearers to 3-ft span	2 × 4 in.	2 × 4 in.		
Bearers to 10-ft span	2 × 6 in. or 3 × 4 in.	2 × 9 (rough) or 3 × 8 in.		
Bracing, horizontal	1 × 4 in.	1 × 4 in.	1 × 6 in. or 1¼ × 4 in.	2 × 4 in.
Bracing, diagonal	1 × 4 in.	1 × 4 in.	1 × 4 in.	2 × 4 in.
Tie-ins	1 × 4 in.	1 × 4 in.	1 × 4 in.	1 × 4 in.
Planking	1¼ × 9 in.	2 × 9 in.	2 × 9 in.	2 × 9 in.
Toeboards	4 in. high	4 in. high (minimum)	4 in. high (minimum)	4 in. high (minimum)
Guardrail	2 × 4 in.	2 × 4 in.	2 × 4 in.	2 × 4 in.

*All members except planking are used on edge.

Source: After ANSI A10.8.

TABLE 30.3 Minimum Nominal Size of Independent-Pole Scaffolds and Maximum Spacing of Members

Ledgers are horizontal members that extend along the length of the scaffold. They should be long enough to extend over two pole spaces. Ledgers should *never* be spliced between poles, but should always overlap the poles by at least 4 in. Wood blocks should be nailed to the pole below the overlapped ledgers to help support them.

Putlogs, or bearers, are horizontal members that extend across the width of the scaffold. They should be long enough to project over the ledgers beyond the inner and outer rows of poles by at least 3 in. for proper support.

Putlogs should be set on the ledgers, with their greater dimension vertical and nailed to the poles. Every wooden putlog on single-pole scaffolds should be reinforced with a 3/16- × 2-in. steel strap—or its equivalent—secured to its lower edge throughout the entire length.

Bracing

Pole scaffolds require full diagonal bracing across the entire face of the scaffold to prevent the poles from moving parallel to the wall of the building, or from buckling. Independent-pole scaffolds require cross bracing between inner and outer set of poles.

The free ends of all pole scaffolds should be cross-braced. *Make sure* that all braces are spliced at the poles. All diagonal braces should be installed so that they create a number of true triangles. These triangles may all be arranged parallel, or they may be placed in a zigzag manner; or a long diagonal brace extending two panels high may be used, provided it is secured to the intervening horizontal members. The diagonal should start at the ground level, and not at a point one panel above the ground, as is occasionally done. If the scaffold is extra long with respect to its height, two or more diagonals may be used, placed either parallel or in an inverted *V* (Fig. 30.2).

If a scaffold is higher than it is wide, bracing may be arranged so that the upper brace starts at the same elevation as the top of the lower brace. On all such scaffolds, diagonal bracing should be provided at every second pole and should continue from the ground to the top of the scaffold.

Scaffolds having a height greater than three times its width should be secured against overturning by means of ½-in. steel cables attached to the outer poles at two points. The cables should extend at about a 45° angle in a horizontal plane to the building column or other adequate supports. Such tie-ins should be provided at every second or third panel vertically.

Where the scaffold is placed against or adjacent to an irregularly shaped wall, the vertical load from the upper sections of poles should be transferred to the lower sections by diagonal bracing. The primary system of bracing of such a scaffold, however, should be kept continuous. If a balcony is bracketed out from a scaffold, then its bracing should be clamped to the projecting horizontal member.

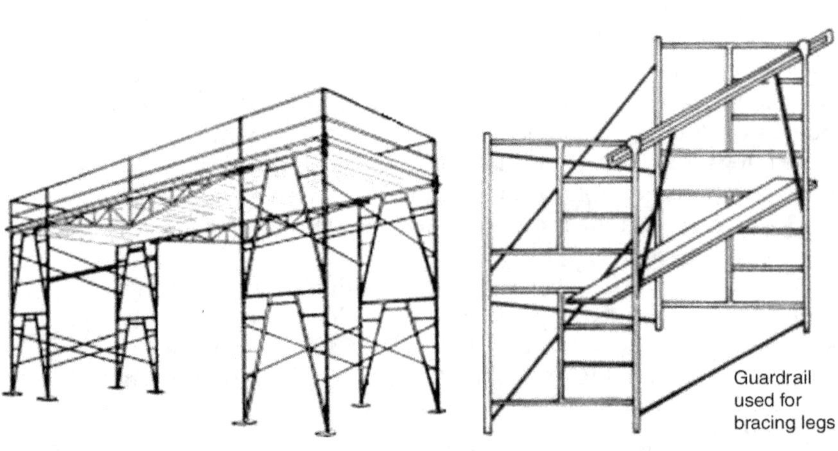

FIGURE 30.2 Diagonal bracing of scaffolds. (*Source:* Rolls Scaffolds, Ventura/Oxnard, CA.)

A gantry is sometimes built across the entrance to a building to prevent obstruction by scaffold poles. In such a case, the entire structure must be subdivided into triangles to form the necessary truss.

Joining

Scaffolding lumber is usually joined with nails, although other fastening devices may also be used, depending on the duration of time the structure will be in use at a particular site.

Never use nails in tension, that is, when stress on members tends to pull them away from each other. Nails should be used, however, where wood members are subject to shearing stress, such as when they tend to slide on each other. To develop full strength, at lest one-half the length of a nail must be driven into the main member to which the secondary member is being nailed.

The *minimum number* of proper size nails for securing a board or plank 4-in. wide is two; 6-in. wide, three; 8-in. wide, four; and 10- or 12-in. wide, five.

A common method of joining wood framework involves toenailing—driving a nail or a group of nails on a slant through the end or edge of an attached member and into a main member. Toenailing provides joints of greater strength and stability than ordinary end nailing.

Tests indicate that the maximum strength of toenailed joints under lateral and uplift loads is obtained by:

- Using the largest nail that will not cause excessive splitting
- Allowing an end distance (the distance from the end of the attached member to the point of initial nail entry) of approximately one-third the length of the nail
- Driving the nail at a slope of 30° with the attached member
- Burying the full shank of the nail, but avoiding excessive mutilation of the wood from hammer blows.

For built-up scaffolding inside a room, extending from wall to wall, diagonal braces can be omitted, provided the scaffold is wedged against the walls to prevent its collapse.

A scaffold member or members supporting concentrated loads should be properly reinforced or braced. When a load is applied, the horizontal member will bend under the weight and the feet of the poles will tend to spread. To overcome this, a tie member should be provided to hold the feet in place, or bracing should be added to prevent the bending of the horizontal member under load.

Independent-pole scaffolds require cross-bracing between inner and outer sets of poles. The free ends of all pole scaffolds should be cross-braced. Nails driven into lead holes with a diameter slightly smaller than the nail have somewhat higher withdrawal resistance than nails driven without lead holes. Lead holes also prevent or reduce splitting of the wood, particularly for dense species.

Planking

Scaffold decking planks should be laid with their edges close together so that tools and material cannot fall through. Each plank should overlap its end supports by at least 12 in., and planks should overlap each other at the bearers by not less than 24 in.

Where the ends of planks abut each other to form a flush floor, the butt joint should be located at the centerline of a pole. Abutted ends should rest on separate bearers. *Make sure* to provide intermediate beams, where necessary, to prevent planks from becoming dislodged because of deflection. Nail or cleat plank ends to prevent dislodging.

When moving platforms to the next level leave the old platform undisturbed until the new putlogs or bearers have been set in place, ready to receive the platform planks.

If a scaffold must turn a corner, *make sure* the platform planks are laid to prevent tipping. First lay planks that meet the corner putlog at an angle so that they extend over the diagonally placed putlog, far enough to have a good safe bearing without danger of tipping. Then place the planking that runs in the opposite direction at right angle so as to extend over and rest on the first layer of planking.

Guardrails and Toeboards

Every scaffold erected 10 ft or more above the ground *must* have a guardrail and a toeboard along the open sides and ends of the work level or platform.

Guardrails of 2- × 4-in. lumber or larger should be installed not less than 36 in. and not more than 42 in. high. Midrails and toeboards should be of 1- × 4-in. lumber with supports set at intervals not greater than 10 ft.

Alternative guardrails material includes:

- 1¼- × 1¼- × 1¼ -in. structural angle iron
- 1- × 0.07-in. steel wall tubing
- 1.990- × 0.068-in. aluminum wall tubing

When persons are required to work or pass under scaffolds, ½-in. wire mesh should be installed between toeboard and guardrail along the scaffold's entire opening. To protect works on a scaffold from being struck by material or tools dropped by workers from above, erect a tight roof of 2-in. planks above the work platform.

30.2 Metal Tubular Scaffolds

Galvanized steel or aluminum alloy tubular scaffolding has displaced wood scaffolding for most operations. Not only is the metal scaffolding stronger and safer to use than wood scaffolding, but it can usually

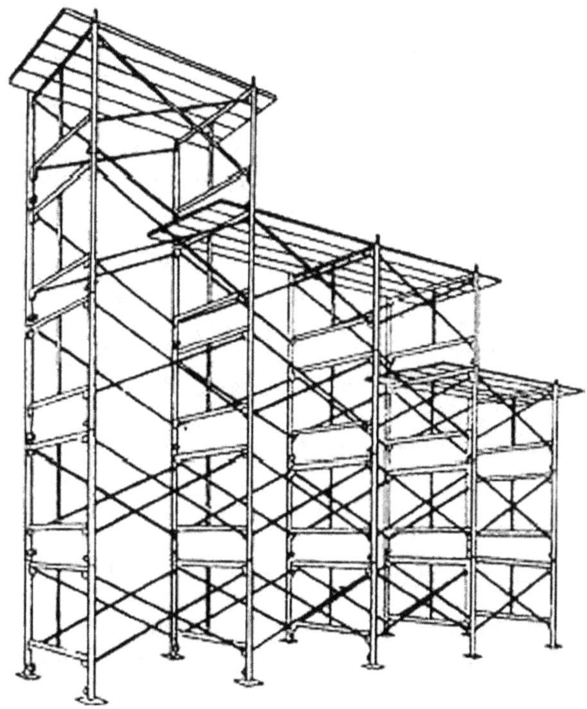

Figure 30.3 Tiered stationary scaffolding.

be erected and dismantled more easily, is not subject to deterioration under exposure to harsh weather or handling, and provides a greater measure of safety from major fire when used with fireproofed planks (Fig. 30.3).

The type of metal tube to be used for optimum efficiency will depend on a number of key factors, including:

- *Load-carrying ability*. Not only the strength when resisting steadily applied stresses, but also deformations due to those stresses have to be considered, as well as temperature change, fatigue, and stress fluctuation.
- *Corrosive resistance*. In some circumstances, this is of major importance because the useful working life of a tube has considerable economic significance. Corrosion may perhaps reduce safety factors sufficiently to cause collapse.
- *Ease of handling*. Most scaffolds are erected by manpower alone, and this factor governs the maximum acceptable weight of tubular components.

- *Availability*. A certain type of tube may not be greatly desired for a specific purpose, but it may due in short supply or when cost is a consideration, or both. The nominal 2-in. diameter, mild-steel tube is the most commonly used type.
- *Beam strength*. In general the safe UD load equals twice the safe central point load based on safe bending stresses. Shear strength is not generally a criterion for tubes.

Deflection, however, is definitely worthy of consideration. As bearers, 2-in. tubes are extremely flexible, and used where light loads could permit long spans without exceeding the safe bending stresses. It may be found that excessive deflection is the limiting design factor. From the handling aspect, the 2-in. nominal mild-steel tube at its maximum length of about 20 ft and mass of 62.5 lb represents about the maximum that one person can handle safely.

Ties and Lateral Stability

The problem of providing efficient lateral ties applies to may types of scaffolds, but the most important application serves for putlog and independent scaffolds because of their great height-to-width ratio.

The two main criteria when considering the efficiency with which a scaffold is laterally tied are the efficiency of each tie and the location and number of ties used. With regard to the first criterion, all ties should be positive and capable of resisting a push or pull. Consequently, a normal load-bearing coupler should be of ample strength for connecting the tie tube to the scaffold. If, however, a positive horizontal force is required, it must be calculated and sufficient strength ensured through the use of safety couplers if necessary.

Ties formed by connecting to an adjustable shuttering prop wedged between floors are not reliable because of possible creep and shrinkage of the timber packing, unless they are checked at frequent intervals. If used in conjunction with a hard rubber pad, such as neoprene, which is not affected by temperature and humidity changes, this type of tie could prove reliable.

It is imperative that the maximum specified spacing for ties not be exceeded. Although many scaffolds are erected successfully with ties that are insufficient and inadequate according to the prescribed requirements, this would appear to indicate that the specifications are somewhat conservative. Strict adherence to them, however, will ensure safety of the systems and the workers using the scaffolding. Nonadherence means gambling on uncalculated favors, which could pose serious dangers and problems. These factors include:

- Limited, but completely unknown tying effects obtained from putlog blades in the case of putlog scaffolds
- Short life of most scaffold structures, reducing the risk of failure

- Reduced strength of standard because of increased flexibility between excessively spaced tie positions or the positions not adjacent to node points
- Ledger bracing increasing stiffness in the vertical direction

Protection Fans and Nets

These structures are intended to catch falling debris and workers. They are usually constructed as cantilevers of suitable width, slope, and vertical positioning all around high scaffolds or buildings that are either under construction or being demolished. In addition, they protect persons below from falling objects.

The actual catching surface *must* be securely attached to the fan structure, be capable of stopping the falling objects, and may consist of the following alternatives:

- Scaffold boards, plywood, or heavier board, if necessary
- Corrugated iron sheeting
- Industrial safety nets in accordance with ASTM specifications

The nets may consist of 4-in. mesh netting, over a 20-ft span, designed to catch a person falling up to 20 ft, or of ¾-in. mesh designed to catch debris.

Apart from the wind loading, fans should be designed for light access loads. At heights the uplift due to wind may be considerable and should be checked and allowed for where necessary.

It is customary for fans to slope down toward the building, and in the case of large corrugated sheeting fans, the possibility of large rainfall runoff may have to be dealt with.

Loading Platforms

Where high independent scaffolds are in use, it is often convenient to raise building materials by crane and deposit them onto a platform attached to the scaffold at some convenient height. On large tower block construction, there may be dozens of such loading platforms at various positions. Also, on some buildings load where no normal scaffolding is in use it may be required to provide a loading platform projecting from the building itself.

In either case, the loading is usually determined in the form of an equivalent uniformly distributed load allowing for dynamic effects, and then designed by tracing the forces through a triangulated cantilever truss structure.

Corrosion

The problem is especially noted in steel tubes. The average useful life of a 2-in. nominal steel tube varies between 10 and 20 years. Several factors influence the life of a steel tube.

- *Degree of exposure to which the tube is subjected.* This depends on atmospheric conditions and the length of time in those conditions. Although scaffolding structures are generally considered temporary in nature, it is possible that a particular scaffold may be required to stand for a year or more in an environment exposed to bad corrosive conditions such as high rainfall or sulfurous or salty atmosphere. During this time, the corrosion rate could be so great as to reduce the structural safety of the scaffolding appreciably.
- *Variation of corrosion rate over the tube surface.* Assuming an exposed steel surface, the normal corrosion process involves the formation of pits, which join up and gradually reduce the tube-wall thickness. But this reduction does not occur uniformly over the surface. A tube tends to become more corroded at its ends, in extreme cases producing an almost knife-like edge. Not only the outside, but also the inside surface of the tube corrodes, due to trapped moisture through rainfall or condensation. In addition, the attachment of couplers on a tube may result in increased local corrosion due to the local trapping of small pockets of moisture. There is little that can be done to counteract these variable corrosion rates over the tube surface, except that the inside surface may be protected by using tight rubber or plastic bungs in the tube ends.
- *Contact with other materials.* This can also result in local corrosion. Galvanic corrosion occurs between steel and aluminum alloy where used in the same scaffold. Gypsum plaster, wood, brick, and concrete, combined with the presence of moisture, have been shown to cause local corrosion.

As a tube deteriorates, it is possible that it could still be used safely in working conditions—where the working stresses are adjusted to allow for the loss in strength. Both vary with the moment of inertia. For practical purposes, however, it can be said that the reduction in beam and strut strength of a tube varies directly with the wall thickness. Hence, if a corroded tube has only three-quarters of its original wall thickness, it is only three-quarters as strong as when in new condition.

Alloy tubes are not subject to corrosion because the alloy forms a protective oxide skin.

Protective Surface Finishes

Unless a tube is to be sealed by bungs, it is essential that any protective surface be applied inside and outside, otherwise, only half of the required protection is obtained.

- *Hot-dip galvanizing.* This is probably the most effective form of protection. Descaled tubes are dipped into a bath of molten zinc and give a uniform coating.

- *Zinc spraying*. Fine zinc particles are sprayed onto the outside surface of a tube.
- *Painting*. Because of the large amount of abrasive wear and tear, and the handling to which tubes are subjected, the application of paint *cannot* be considered in the same light as for structural steelwork. If paint is used, however, probably one coat of paint applied frequently will provide better protection than several coats applied less frequently.

Tubular Metal Scaffolding Types

The three basic types of tubular metal scaffolding generally used today are:

- *Tube and coupler*. Consists of straight tube members in steel or aluminum, in varying sizes and lengths, joined together by steel or aluminum couplers.
- *Welded sectional steel*. Consists of prefabricated modular welded steel frames and accessories.
- *Welded aluminum-alloy*. Used in folding sectional stairway- and ladder-type scaffolds. Because of its lightness, mobility, and ease of erection, this type of scaffolding is most suitable for light-duty work, especially where the equipment requires frequent erection and dismantling.

30.3 Tube-and-Coupler Scaffolding

Four basic components comprise the tube-and-coupler system: baseplates, interlocking tubing or pipes, bolt-activated couplers for making right-angle connections, and adjustable couplers for making connections other than at right angles. Special casters can be added to the baseplates to provide rolling scaffold systems.

With the tube-and-coupler system, horizontal runners can be placed at any point on the vertical posts and bearers can be placed at any point on the runners. Such scaffolds can be erected to any required dimension or height, on any terrain (Fig. 30.4). This type of scaffold provides maximum versatility for scaffolding around odd shapes, such as processing works and refineries, or other structures having uneven exteriors and projections. The same member can be used to build storage racks of any size or capacity.

To provide additional strength, posts on tube-and-coupler scaffolds can be spaced more closely, and more adequate bracing can be provided.

All members can be used interchangeably for posts, putlogs, ledgers, bracing, and handrails. Posts are joined end to end by placing down the female end of a member and locking it to the male upper end of the post below, giving a 90° twist. Rigid 90° couplers secure the various members together. Swivel couplers are available if necessary.

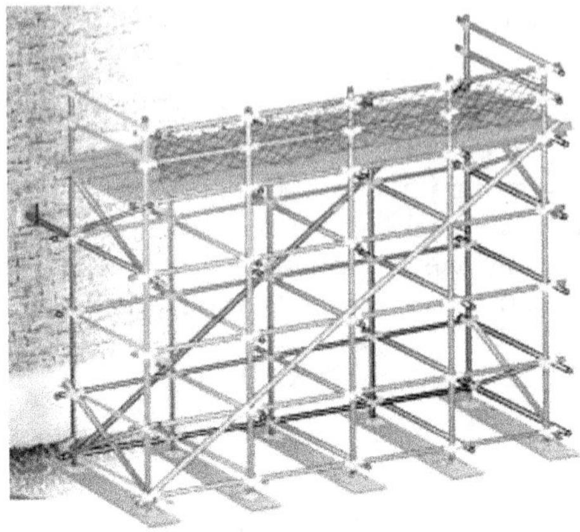

FIGURE 30.4 Tube-and-coupler scaffolding.

In erecting this type of scaffolding, the posts are often placed on steel bases that distribute the load to footing planks. The ledgers are then clamped to the posts at the desired heights, with putlogs or bearers clamped to the ledgers close to the posts.

The braces on the outboard face are clamped to the projecting ends of the putlogs, close to the posts. Cross-braces are installed at every fourth or fifth pair of posts. Usually the putlogs extend so as to bear against the wall of the building or other substantial support. Ties of $^3/_8$-in. wire rope should be used to lash the scaffolding to the structure.

As the scaffold extends upward with construction progress, planking can be removed from the lower levels and placed higher up. Handrail members should remain in place at all times to provide added rigidity to the scaffold.

Post splices should be located a short distance above the ledger couplings and staggered so as to occur alternately at different levels. Where longitudinal diagonal bracing is made continuous from the ground to the top of the scaffold and braces do not meet the putlogs, *make sure* to clamp the braces to the post by means of swivel couplers.

To check the strength of tubular steel scaffolding of this type, first determine the live or movable load that may be applied to the plant decking. *Do not* count less than 25 psi of deck (preferably no less than 50 psi). In a scaffold or staging more than one panel in length, each putlog carries a load on the decking equal to that on one-half the area of the panel on either side of it—or usually the load on the area of one panel.

For example, if the posts are spaced 5 ft $\times$ 5 ft, the area supported is 25 ft^2. This area, multiplied by 25 psf, gives a load of 625 lb uniformly distributed over the length of the span of the putlog. To this must be added the weight of the plank decking and toeboards, say 200 lb, which comes to a total load of 865 lb on the putlog.

Scaffold Erection

Posts for tube-and-coupler scaffolds should be accurately erected on suitable bases and maintained plumb. They should be spaced for the depth of the scaffold:

- No more than 6 ft apart, along its length.
- No further apart than 10 ft (light duty), 8 ft (intermediate duty), and 6 ft 6 in. (heavy duty).

And posts spaced:

- Not more than 6 ft × 8 ft apart, along the length of the scaffold; should have bearers of nominal 2 ½-in.-outside-diameter steel tube or pipe.
- Not more than 5 ft × 8 ft apart; should have bearers of nominal 2-in.-outside-diameter tube or pipe.

When other structural metals are used, they must be designed to carry an equivalent load.

All tube-and-coupler scaffolds should be limited in height and number of working levels, and should be constructed and erected to support four times the maximum intended loads. *ANSI A10.8* specifies the limits for light, intermediate, and heavy-duty tube-and-coupler scaffolds (Table 30.4). All tube-and-coupler scaffolds exceeding these limitations *must* be designed by a licensed professional engineer. Drawings and specifications *must* be prepared and copies made available to the contractor for inspection purpose.

Runners along the length of the scaffolds should be set *not more* than 6 ft 6 in. center to center, located on both the inside and the outside posts at even height. *Make sure* that the runners are interlocked to form continuous lengths and are coupled to each post. Locate the bottom runners as close to the base as possible.

All bearers should be installed transversely between posts, coupling them securely to the posts bearing on the runner coupler. When coupled directly to runners, keep bearers as close to the posts as possible. Bearers *must* be at least 4 in., but *not more* than 12 in. longer than the post spacing or runner spacing. They may be cantilevered for use as brackets to carry *not more* than two planks.

Cross-bracing *must* be installed across the width of the scaffold at least every third set of posts horizontally and every fourth runner vertically. Extend all such bracing diagonally from the inner and outer runners upward to the next outer and inner runners.

Longitudinal diagonal bracing *must* be installed at approximately a 45° angle from near the base of the first outer post, upward to the

Light Duty		
Uniformly distributed load	Not to exceed 25 lb/ft^2	
Post spacing, longitudinal	10 ft	
Post spacing, transverse	6 ft	
Working levels	Additional planked levels	Maximum height
1	8	125 ft
2	4	125 ft
3	0	91 ft
Medium Duty		
Uniformly distributed load	Not to exceed 50 lb/ft^2	
Post spacing, longitudinal	8 ft	
Post spacing, transverse	6 ft	
Working levels	Additional planked levels	Maximum height
1	6	125 ft
2	0	78 ft
Heavy Duty		
Uniformly distributed load	Not to exceed 75 lb/ft^2	
Post spacing, longitudinal	6 ft 6 in.	
Post spacing, transverse	6 ft	
Working levels	Additional planked levels	Maximum height
1	6	125 ft

Source: ANSI A10.8.

TABLE 30.4 Guidelines for Tube-and-Coupler Scaffolds

extreme top of the scaffold. Where a long length of scaffold permits, duplicate such bracing—at every fifth post. Install bracing also from the last post, extending back and upward to the first post, or to runners if conditions preclude attaching it to the posts.

Installing Planking and Toeboards

Make sure to tie and securely brace the entire scaffold against the structure at intervals *not* exceeding 30 ft horizontally, and 26 ft vertically. Guardrails, midrails, toeboards, and wire mesh *must* be installed in accordance with OSHA regulations (Fig. 30.5).

- Tie the two sidewalk brackets together in the first and last bays of a scaffold run by nailing the bracket planks to the bracket conversion plates.
- Platform planking is to be minimum 2- × 10-in. wide, full size, graded as scaffold plant, rough cut 7 ft.
- Platform planking *must* be 6 ft 11½-in. long (6 ft 11¼-in. minimum). Ends *must* be square.

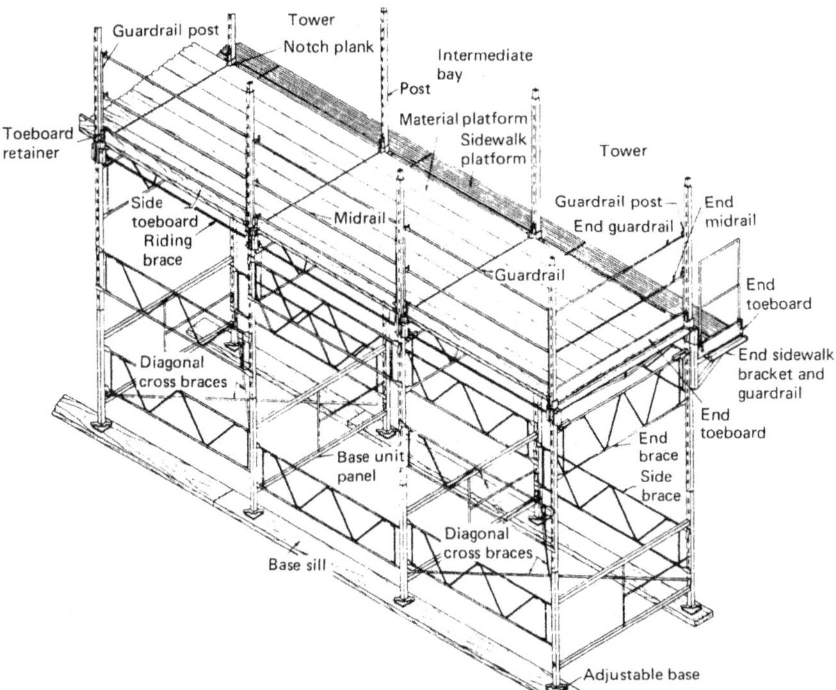

INSTRUCTIONS FOR PLANKING AND TOEBOARDS

1. Tie the two sidewalk brackets together in the first and last bays of a scaffold run by nailing the bracket planks to the bracket conversion plates.
2. Platform planking to be minimum 2 by 10 in. wide, full size, graded as scaffold plank, rough cut 7 ft.
3. Platform planking must be 6 ft 11½ in. long (6 ft 11¼ in. minimum). Ends must be square.
4. Notch the plank on material platform nearest to building at both ends, 1½ in. deep for 2 in. of its width.
5. Toeboard lumber must be minimum 1 in. nominal thickness and extend at least 4 in. above the platform.
6. Cut toeboard sections for the outside edge of the material platform, 6 ft 10 in. long.
7. Securely fasten end toeboards to the platform planking.

Figure 30.5 Typical scaffold guardrail, midrail, and toeboard installation. (*Source:* Patent Construction Systems, Paramus, NJ, *www.pcshd.com*.)

- Notch the plank material platform nearest to the building at both ends—1½-in. deep for 2 in. of its width.
- Toeboard lumber must be a minimum 1 in. nominal thickness and extend at least 4 in. above the platform.
- Cut toeboard sections for the outside edge of the material platform, 6 ft 10 in. long.
- Securely fasten end toeboards to the platform planking.

Access to all built-up scaffolds *must* be by one of the following accessories:

- Portable wood or metal ladders
- Access ladders positioned so that their use will *not* tend to tip the scaffold
- Hook-on or attachable metal ladder specifically designed for use in conjunction with proprietary types of scaffolds
- Step- or stair-type accessories designed specifically for use with proprietary types of scaffolds

Do not use the scaffold's cross-braces as a means of access or descent.

30.4 Welded Sectional Steel Scaffolding

This type of scaffolding consists of tubular steel frames—essentially two short poles, a putlog, and bracing members welded together as an integral unit.

To assemble, erect two frames opposite each other and place the removable cross-braces between them on the stud bolts, which are welded to, and form an integral pat of, the frame. The posts of the bottom frames rest on steel bases and are nailed to planks that distribute the load over the ground or floor surface.

After the first tier of panels has been erected, the second tier of frames is socketed onto the upper ends of the first tier.

The prefabricated frames are standard 5 ft wide and range in height from 3 to 10 ft (5-ft-high frames are the most commonly used size). Because of the numerous designs of frames by various manufacturers, the allowable load distributed on the putlog (the top member of the frame) varies considerably.

Likewise, because of the variation in the length of posts between the removable cross-braces, there is a wide range in the load that these frames can support. (Check the manufacturer's catalog before erecting such a system.)

Because frames are all interchangeable, unless the rigger knows exactly the load that each model will safety carry, the live loading should be limited to that permitted on the model having the lowest strength.

The safety factor on this type of scaffolding should be 4 to provide for possible overloading of the system by mistake. Because the dead load on the lower sections of built-up scaffold varies almost directly with the height to which the scaffold is erected, the live load is generally a guess.

For example, when a scaffold is about to be erected, the rigger may assume that one pallet of bricks will be placed on the decking in each panel. But if, through ignorance or carelessness, two pallets are placed in the panels, this will almost double the live load, and when the dolly used to move the pallets sets down the load, it may produce an additional impact load of as much as 100 percent.

The factor that usually limits the loading on a scaffold frame is the strength of the putlog or top horizontal member of the frame on which the planks rest. The total load on any frame is one-half the combined live and dead loads on the two adjacent spans of decking.

The total loading on any frame *must not* exceed 2000 lb on frames having diagonal bracing member, or 1600 lb on those having no diagonal braces. These figures are based on a safety factor of 4.

The 2-in. scaffold planks come in 13-ft lengths. Hence, regardless of the spacing of the frames, the same weight of planking will usually be used, with the amount of overlap varying as required. Thus the weight of the planking on each frame putlog will be about 250 lb. The weight of the frames and their removable bracing may be estimated at about 10 lb for steel and 5 lb for aluminum per foot of scaffold.

30.5 Welded Aluminum Alloy Scaffolding

Because of its lightness, mobility, and ease of erection, this type of scaffolding is most suitable for light-duty work, especially where the equipment requires frequent erection and dismantling.

These scaffolds are prefabricated from high-strength aluminum-alloy tubing and have internal stairways. They are often equipment with casters to permit easy movement for maintenance work.

The folding ladder types are constructed in one-piece base sections that speed erection and dismantling. Ladder-type base sections are available in widths of 29 in., or 4 ft 6 in., with spans of 6, 8, or 10 ft between frames. This type of scaffold has two diagonal braces and one horizontal brace that form integral parts of the folding unit. Intermediate extension and guardrail sections can be placed atop the folding unit, using individual end frames and braces.

A large folding-type scaffold features its upper sections as well as the base section being one-piece folding units. When the scaffold must be erected higher than recommended for a base of this size, outriggers that clamp to the legs of the base section can be used. Leg equipment includes a leveling mechanism to level the scaffold on uneven ground. Casters on the legs are locked at both wheel and swivel.

Folding scaffold sections, however, are heavier than individual components of demountable sectional scaffolding. Sectional aluminum stairway scaffolds are designed with end frames of various heights to provide different working levels, and adjustable bottom sections with casters, but without the folding feature intermediate sections, half-sections, and guardrail sections. All components are demountable to reduce weight, and to make it easier to erect and dismantle. Internal stairways are used. Outriggers may be used to increase the base area.

Folding sectional stairway and ladder scaffolds are used for outdoor cleaning and maintenance work—the former for high or heavy work, the latter for low to medium height and one-person jobs. Both types are especially suitable when the work is horizontal. When used indoors, they simplify work on walls and ceilings.

In general, the principles of sound design and construction practice referred to for mild-steel putlog and independent scaffolds are also applicable to scaffolding systems of aluminum alloys. The size and load-carrying capacity of alloy scaffolds, however, is considerably less than that for steel scaffolding, Therefore, when planning their use, the following factors *must always* be considered:

- *Flexibility*. Compared with steel, aluminum-alloy tubes are three times as flexible, which reduces their efficiency as struts and beams.
- *Inconvenience*. To avoid electrolytic corrosion, *only* alloy couplers can be used with alloy tubes, because it is essential to avoid mixing alloy and steel members on the same scaffolding system.
- *Convenience*. Alloy tubes weigh one-third as much as steel ones, and consequently are easier to transport and handle.
- *Corrosion*. Alloy tubes and fittings are less susceptible to corrosion than similar mild-steel tubes and fittings.

Regulations and Standards

OSHA Regulations

1926 Subpart L–Appendix C—*Scaffolds*

1926.451(g); 1926.451(h); 1926.452(w) *Requirements for Scaffold Toprails, and Midrails*

1926.452 *Requirements Applicable to Specific Types of Scaffolds*

Appendix A *Scaffold Specifications*

Appendix B *Criteria for Determining the Feasibility of Providing Safe access and Fall Protection for Scaffold Erectors and Dismantlers*

U.S. Consensus Standards

ANSI/SIA - A92.5; A 92.6 *Stability Test Requirements*

A92.2–1990 *Vehicle-Mounted Elevating and Rotating Aerial Devices*

A92.3–1990 *Manually Propelled Elevating Aerial Platforms*

A92.5–1990 *Boom Supported Elevating Work Platforms*

A92.6–1990 *Self-Propelled Elevating Work Platforms*

A92.7–1990 *Airline Ground Support Vehicle-Mounted Vertical Lift Devices*

A92.8–1993 *Vehicle-Mounted Bridge Inspection and Maintenance Devices*

A92.9–1993 *Mast-Climbing Work Platforms*

CHAPTER 31

Specialized Scaffolds

In addition to the swinging and the built-up scaffolds, a number of specialized scaffolds are commonly used for particular types of operations.

31.1 Outrigger Scaffold

This type of scaffold is frequently projected from a building window to enable workers to reach the upper portions of a wall or cornice.

Outrigger scaffolding consists of a pair of heavy timber beams, placed on edge, with their fulcrum points resting on a bearing plank on the windowsill. Rough planking, which forms a working platform, is laid tight to within 3 in. of the building wall, and nailed or bolted to the outrigger beams (Fig. 31.1). The outrigger beams should extend not more than 6 ft beyond the face of the wall. Their inboard ends, measured from the fulcrum point to the extreme point of support, should be not less than 1½ times longer than their outboard ends.

To resist the upward thrust of their inboard ends, outrigger beams are usually nailed to vertical struts that are wedged against bearing blocks at the floor and ceiling of the building. Cross-bridging between the outriggers at the fulcrum point resists any tendency of the planks to tip or to roll over. The entire supporting structure *must* be securely braced in both directions to prevent any horizontal movement.

The bearing plank on the windowsill should provide a surface of at least 6 in. in each horizontal direction to support each beam's fulcrum point. Guardrails, midrails, and toeboards *must* be installed on all open sides and ends of platforms, more than 10 ft above ground or floor.

Guardrails of 2- × 4-in. timber, or structural angle iron, steel, or aluminum tubing of equivalent strength, *must* be installed not less than 36 in., nor more than 42 in. high with a midrail. Both guardrails and midrail supports *must* be installed not more than 10 ft apart along the platform.

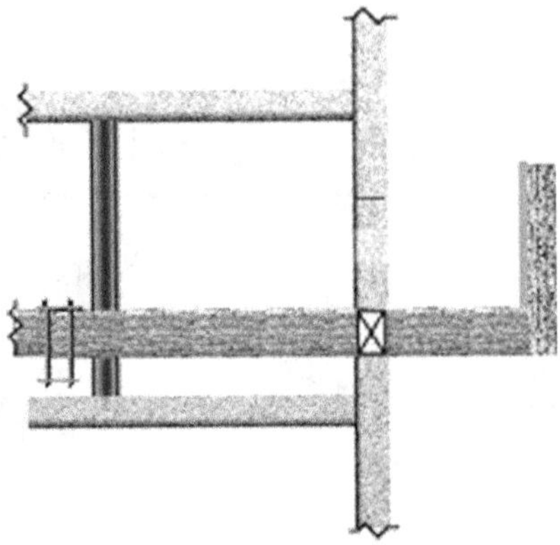

FIGURE 31.1 Outrigger scaffold.

	Light duty	Medium duty
Maximum scaffold load (lb/ft²)	26	60
Outrigger size (in.)	2 × 10	3 × 10
Maximum outrigger spacing (ft)	10	6
Planking, rough (in.)	2 × 9	2 × 9

TABLE 31.1 Specifications for Outrigger Scaffolding

Toeboards and midrail should be 1- × 4-in. lumber or equivalent, with the toeboard extending at least 3 ½-in. above the working surface. Where there is danger of material falling from the scaffold, a wire mesh or other enclosure *must* be provided between the guardrail and the toeboard (Table 31.1).

The ends of planks should not extend more than 18 in. beyond the outrigger beams. Where additional working levels are required to be supported by the outrigger method, the plans and specification of the outrigger and scaffolding structure *must* be designed by a licensed professional engineer.

31.2 Mason's Adjustable Multipoint Suspension Scaffold

This type of scaffold consists of a continuous wood work platform supported by bearers suspended by wire rope from overhead outrigger beams that are securely fastened or anchored to the structure's frame or floor system.

Each outrigger beam should be equivalent in strength to at least a standard 7-in., 15.3-lb steel I beam and should be at least 15-ft long.

Beams should project not more than 6 ft 6 in. beyond their bearing point. If channel iron outrigger beams are used in place of I-beams, they should be securely fastened together with the flanges turned out.

Where the overhead is greater than 6 ft 6 in., outrigger beams should be of stronger material, or multiple beams should be used, installed in accordance with approved design and instructions.

All outrigger beams should be set and maintained with their webs in a vertical position. They should rest on suitable wood bearing blocks.

The free end of the suspension wire ropes should be equipped with proper size thimbles and secured by splicing or other equivalent means, while the running ends should be securely attached to the drum of an approved hoisting machine.

Most systems are raised and lowered by a hand- or foot-operated cable winch, although power winches are available for the taller scaffolds. Adjustable scaffolds using a winch are not as fast or easy to adjust as mast climbing work platforms, however, they are less expensive and offer many of the same advantages.

Make sure that at least four turns of rope remain on the drum at all times.

Where a single outrigger beam is used, place the steel shackles of the clevis, with which the wire ropes are attached, to the outrigger beams directly over the hoisting drum. The suspended rope should be capable of supporting at least six times the intended load.

The scaffold work platform should be equivalent in strength to at least 2-in. planking capable of supporting a working load of at least 50 lb/ft^2. *Never* load the platform in excess of this figure. The maximum permissible span for 2- × 10-in. (nominal) [2- × 9-in. (rough)] planks supporting this load is 8 ft.

All parts of the scaffold, such as bolts, nuts, fittings, clamps, wire ropes, and outrigger beams and their fastenings, should be maintained in sound and good working condition. They should be inspected before each installation and periodically thereafter.

Guardrails and toeboards *must* be installed and protective wire mesh provided between guardrail and toeboard in accordance with specifications.

Make sure that each scaffold is installed or relocated in accordance with approved designs and instructions and supervised by a designated and competent individual.

31.3 Needle-Beam Scaffold

This type of light-duty scaffolding, consisting of needle beams supporting a platform, is used only for very temporary jobs, in particular for riveting or bolting operations on steel structures. No material should be stored on this type of scaffold (Fig. 31.2).

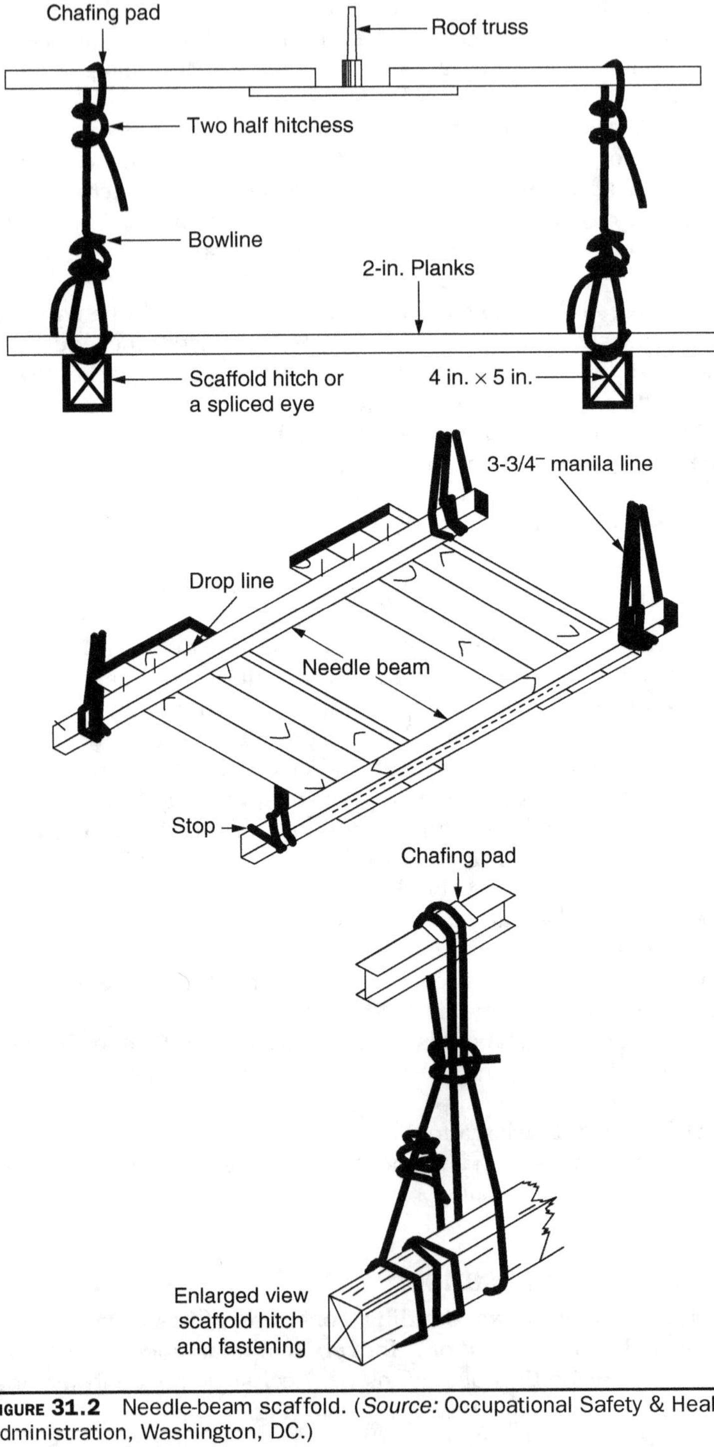

FIGURE 31.2 Needle-beam scaffold. (*Source:* Occupational Safety & Health Administration, Washington, DC.)

Two needle beams, 4 × 6 in. on edge, are placed parallel to each other in a horizontal or nearly horizontal plane, depending on circumstances, and no more than 10 ft apart. They are usually suspended by rope or hangers, near each end. A center support is always required.

The rope supports should be equivalent in strength to 1-in.-diameter No. 1 grade Manila rope. Attach the ropes in the needle beam by a scaffold hitch or a properly made eye splice to prevent the beams from rolling over on their sides under load.

Tie the loose ends of the rope with a bowline knot or a round turn and one-half hitch. Then extend the rope up over a structural member, down under the needle beam, and back over the structural member and secure it.

The river scaffold hitch is frequently used, although a clove hitch or rolling hitch is preferable. *Make sure* to take proper precautions to prevent the ropes from slipping off the ends of the needle beams—particularly if the beams are inclined.

Decking should be of 2-in. planks, having a length of at least 2 ft, and no more than 3 ft longer than the span between the needle beams. Planking spans should be not less than 3 ft, and not more than 8 ft, when using 2-in. planks. For spans greater than 8 ft, platforms should be designed based on requirements for the special span.

If the scaffold is not level, nail the planks to the beams or nail cleats on their undersides to engage the beams. One end of a needle-beam scaffold may be supported by a permanent structural member. *Make sure* that all unattached tools, bolts, and nuts used on needle-beam scaffolds are kept in suitable containers.

When working on a needled-beam scaffold 10 ft or more above the ground or floor, each worker *must* wear a safety belt with lanyard and fall-protection device, which will limit a fall not more than 6 ft. *Make sure* that the lanyard and the fall-prevention device are secured to a structural member other than the scaffold, or attached to a lifeline suspended from a structural member other than the scaffold.

31.4 Interior-Hung Scaffold

This type of scaffold, suspended from a ceiling or roof structure, consists of either a wood or a steel tube-and-coupler–type system (Fig 31.3).

For hanging wood scaffolds, the following minimum nominal size material is recommended (Table 31.2).

Both wood and steel tube-and-coupler–type scaffolds *must* be designed to sustain a uniformly distributed working load up to heavy-duty scaffold loads, with a safety factor of 4. Install guardrails and midrails (when required) and toeboards at all open sides on all scaffolds more than 10 ft above ground or floor. Provide wire mesh between guardrail and toeboard.

When wire rope is used to support such scaffolds, *make sure* that the rope is wrapped at least twice around the supporting members

FIGURE 31.3 Interior-hung scaffold. (*Source:* Occupational Safety & Health Administration, Washington, DC.)

Members	Dimensions
Supporting beams	2 × 9 in. on edge
Planking	2 × 9 in. or 2 × 10 in.
Maximum planking span	7 ft for heavy-duty applications
	10 ft for light- or medium-duty applications

TABLE 31.2 Dimensions for Interior-Hung Scaffolds

and twice around bearers of the scaffold. Secure each end of the wire rope with at least three standard wire rope clips. The suspension rope should be capable of supporting at least six times the intended load.

Inspect and check all overhead supporting members for proper strength.

31.5 Carpenter's Bracket Scaffold

These scaffolds consist of triangular frames of wood not less than 2 × 3 in. in cross section, or of equivalent-strength metal brackets (Fig. 31.4). Each bracket should be attached to the structure by means of one of the following:

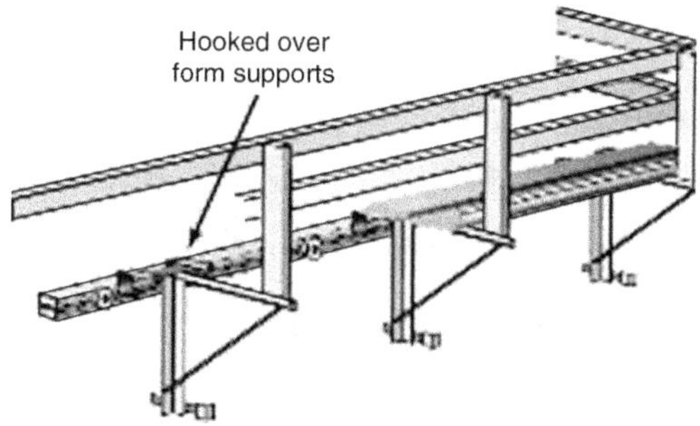

FIGURE 31.4 Carpenter's bracket scaffold. (*Source:* Occupational Safety & Health Administration, Washington, DC.)

- Bolt, not less than ½-in. diameter that extends through the inside of the building or structure wall
- Metal stud attachment device
- Welding to the structure
- Hooking over a well-secured and adequately strong supporting member

Make sure that brackets are spaced not more than 10 ft apart. Use at least two 2- × 9-in. (nominal size) wood planks for the work platform. Planks should extend not more than 18 in. and not less than 6 in. beyond each end support. Install guardrails and midrails (when required) and toeboards on all open side when a scaffold is being used more than 10 ft above ground or floor. Provide wire mesh between guardail and toeboard.

Never allow more than two persons to occupy any given 10 ft of a bracket scaffold at any one time. *Make sure* that tools and materials do not exceed 75 lb in addition to workers.

31.6 Bricklayer's Square Scaffold

This type of scaffold consists of square shapes of framed structural wood that support a wooden platform. The wood frames should be stacked no higher than three tiers, one square resting directly above the other (Fig. 31.5).

Each upper tier stands on a continuous row of planks laid across the next lower tier and is nailed down or otherwise secured to prevent displacement. The squares should be not wider than 5 ft and no higher than 5 ft; and they should be set not more than 5 ft apart for medium-duty scaffolds and not more than 8 ft apart for light-duty scaffolds (Table 31.3).

Make sure each square is reinforced on both sides of each corner with 1- × 6-in. gusset braces. Place bracing of 1- × 8-in. lumber on both

FIGURE 31.5 Bricklayer's square scaffold. (*Source:* Occupational Safety & Health Administration, Washington, DC.)

Members	Dimensions (in.)
Bearers or horizontal members	2 × 6
Legs	2 × 6
Braces at corners	1 × 6
Braces, diagonally from center frame	1 × 8

TABLE 31.3 Dimensions for Bricklayer's Scaffolds

sides, running from center-to-center of each member, to ensure scaffold strength and rigidity.

Extend 1- × 8-in. bracing from the bottom of each square to the top of the next square on both the front and the rear sides of the scaffold to provide sufficient strength and rigidity. Before use, check that the scaffold is level and resting on a firm foundation. Platform planks should be at least 2 × 9 in. (nominal size) with their ends overlapping the bearers of the squares. Support each plank on at least three squares.

31.7 Miscellaneous Scaffolds

Horse Scaffold

Wood horses may be used to construct light- or medium-duty scaffolds, provided they are spaced not more than 6 ft apart for medium-duty and not more than 8 ft apart for light-duty applications. The horses support the work platform and may be arranged in tiers—not more than two tiers high and not more than 10 ft above ground (Fig. 31.6).

When arranged in tiers, each horse should be placed directly over the horse in the tier below, with the legs of the upper horses nailed

Figure 31.6 Horse scaffold. (*Source:* Occupational Safety & Health Administration, Washington, DC.)

Members	Dimensions (in.)
Horizontal members of bearers	3 × 4
Legs	1¼ × 4 1/8
Longitudinal braces between legs	1 × 6
Gusset brace at top of leg	1 × 8
Half-diagonal braces	1¼ × 4 1/2

Table 31.4 Dimensions for Horse Scaffolds

down to the platform planks of the tier below to prevent displacement or thrust. *Make sure* that each tier is cross-braced substantially (Table 31.4).

Never use horses or parts that have become weak or defective. Immediately remove such items from service and dismantle the horse or replace damaged parts.

Ladder-Jack Scaffold

This is a light-duty type scaffold consisting of a work platform supported by brackets attached to ladders. Ladder jack scaffold should be limited to light-duty applications and a height of 20 ft above floor or ground.

Make sure that the ladder jack is designed and constructed to bear on the side rails in addition to the ladder rungs. If it bears on rungs only, *make sure* to provide at least 10 in. of bearing area on each rung.

Ladders used in conjunction with ladder jacks should be paced, fastened, held, or equipped with devices to prevent slipping. The ladders *must* be of the heavy-duty type designed and constructed in accordance with *ANSI A14.2*.

Wood platform planks should not be less than 2-in. nominal thickness with support spans not greater than 8 ft. *Make sure* that platform planks, whether metal or wood, overlap the bearing surface by not less than 12 in. Platforms should be not less than 18 in. wide. *Do not* permit more than two persons to occupy a ladder jack scaffold at one time.

When working on a ladder jack scaffold 10 ft or more above the ground or floor, each worker *must* wear a safety belt tied to a lanyard and a fall prevention device that will limit the fall to 6 ft. *Make sure* the lanyard and fall-protection device are secured to a structural member, or attached to a lifeline suspended from a structural member other than the scaffold.

Window-Jack Scaffold

A window jack scaffold consists of a work platform supported by a bracket or jack that projects through a window opening. The window jacks should be designed and constructed so as to provide a secure grip on the window opening and be capable of supporting the design load (Fig. 31.7).

This type of scaffold should be used only for the purpose of working at the window opening through which the jack is placed. *Never* support planks between one window jack and another, nor use window jacks to support other elements of scaffolding. *Make sure* that window jacks have suitable guardrails, or provide each worker with approved safety belts and attached lifelines. *No more* than one person should work on a window jack scaffold at a time.

Bosun's Chair

The bosun's chair is a suspended seat designed to accommodate one worker in a sitting position. The chair consists of wood planking not less than 12 × 24 in. and 1-in. thick, reinforced on the underside by

Figure 31.7 Window jack scaffold. (*Source:* Occupational Safety & Health Administration, Washington, DC.)

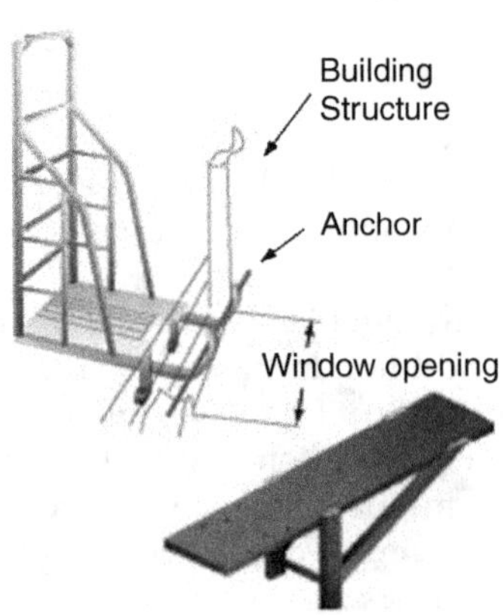

cleats securely fastened to prevent the board from splitting. Seat slings of 5/8-in.-diameter fiber ropes are reeved through four seat holes so as to cross each other on the underside of the seat.

Tackle should consist of the correct size ball bearing or bushed blocks and properly splices No. I-grade Manila rope. When a worker in a bosun's chair is welding or working with an open flame, seat slings should be at least 3/8-in. wire rope.

Make sure that roof irons, hooks, or the objects to which the tackle is anchored are securely installed. When tiebacks are used, they should be installed at right angles to the face of the building and securely fastened to a chimney, standpipe, skylight, or other part of the structure.

31.8 Manually Propelled Mobile Scaffold

These portable rolling scaffolds constructed of tubular welded metal members are supported by casters that permit ease of travel (Fig. 31.8). The casters of the mobile tower are designed to hold the scaffold in position. They should have rubber or similar resilient tires with a minimum of 5-in. diameter wheels, unless specific design requirements dictate the use of other materials. If the casters have plain stems, they should be secured in the scaffold to prevent them from falling out.

Free-standing mobile scaffold towers should not be erected any higher than four times the minimum base dimension, including any outrigger frames that may be used to increase stability. *Make sure* that the towers are properly braced with cross-bracing as well as horizontal bracing, in accordance with standard practice. Platforms for mobile scaffolding should be tightly planked for the full

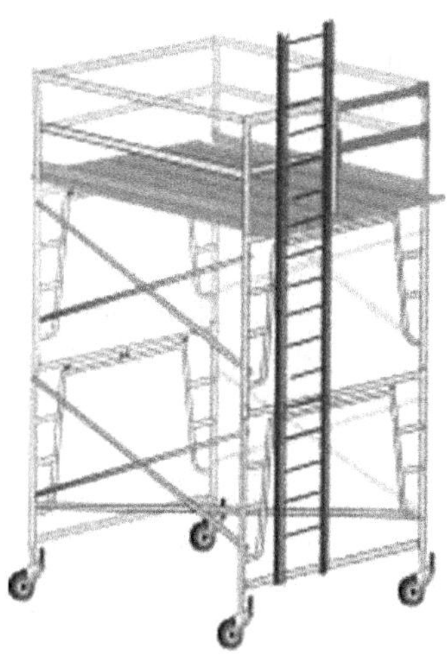

FIGURE 31.8 Manually propelled mobile scaffold. (*Source:* Occupational Safety & Health Administration, Washington, DC.)

width of the scaffold, except for necessary entrance openings. *Make sure* that platforms are securely fixed in place to prevent any horizontal movement. For scaffolds more than 10 ft above the ground or floor, provide guardrails, midrails, and toeboards, with wire mesh between guardrail and toeboard.

For proper access and egress, a ladder or stairway *must* be affixed or built into the scaffold and so located that when in use it will not have a tendency to tip the scaffold. These can be:

- Portable wood or metal ladders.
- Separate attachable or built-in ladders with regularly space steps or rungs, having maximum variations between adjacent rungs of 2 in. Spacing between rungs may be up to 16½ in. If such spacing is necessitated by practical limitations of the equipment in conjunction with which the ladder is being use.
- Step- or stair-type accessories specifically designed for use with proprietary types of scaffolds.

Landing platform intervals should not exceed 35 ft.

When mobile scaffolds are used, *make sure* that they rest on suitable footing and that they stand plumb. Always lock casters or wheels before using a mobile scaffold to prevent any movement.

When necessary to level the mobile scaffold platform, use screw jacks or other means in each leg section for adjusting the height. *Make sure* that at least l6 in. of the screw jacks are in each scaffold leg, with no more than 12 in. extended.

When moving a mobile scaffold, always apply the force as close to the base as practical. *Make sure* that the tower is stabilized during movement from one location to another. Move scaffolds only on floors free of obstructions and openings.

Workers are permitted to ride on manually propelled mobile scaffolds *only* if guardrails, midrails, and toeboards are installed on all open sides of the platform, and *only* when:

- The floor or surface is within 1½ degrees of level concrete, free from pits, holes, or obstructions.
- The minimum dimensions of the scaffold base, ready for rolling, are at least one-half the scaffold's height. Outriggers, if used, should be installed on both sides of the staging and then can be calculated as part of the base dimension.
- All tools and materials are secured, or removed from the platform, before the mobile platform is moved.
- Workers on the mobile scaffold are advised and aware of each movement in advance.

31.9 Safe Use of Mobile Scaffolds

General guidelines for safe use of mobile scaffolds:

- Post a sign indicating that only authorized staff are allowed to use mobile scaffold. Mobile scaffold must be used only on a hard level surface and must not be located closer than 1 m to any slab edge, penetration, or other stepdown unless a fixed fence, rail, or suitably high upturn is in place.
- Check for physical defects before every use, including the access ladder. (Refer to manufacturer's manual if available.)
- Ensure the scaffold is level.
- Ensure all sections are pinned or appropriately secured.
- Keep scaffold loads to a minimum (including tools and other equipment) and remove when the scaffold is not in use. *Do not* exceed the scaffold's working load (refer to manufacturer's manual).
- Heavy tools, equipment, and supplies must be hoisted up (rather than carried up by hand).
- Leave all casters locked or chocked against any possible movement. Casters shall comply with *ANSI 1576.2* and shall be capable of being locked when work is being performed from the scaffold.
- Where the casters incorporate adjustable legs, the gradient of the surface shall not exceed 5°, unless provision is made to take the load off the casters during use of the scaffold.
- Where practicable, portable barriers and lights should be placed around the scaffold while in use.
- Ensure guardrails and toeboards are in place on all open sides.
- Never move the scaffold while someone is on it.
- Secure materials before moving the mobile scaffold.
- Remove the loads including tools after using the scaffold.

Regulations and Standards

OSHA Regulations

1926 Subpart L–Appendix C—*Scaffolds*

1926.451 *General Requirements*

1926.452 *Additional Requirements Applicable to Specific Types of Scaffolds*

Appendix A *Scaffold Specifications*
Appendix B *Criteria for Determining the Feasibility of Providing Safe Access and Fall Protection for Scaffold Erectors and Dismantlers*

U.S. Consensus Standards

ANSI/SIA

A92.3-1990 *Manually Propelled Elevating Aerial Platforms*
A92.6-1990 *Self-Propelled Elevating Work Platforms*

CHAPTER 32

Safe Scaffolding Practices and Glossary

32.1 Scaffolding Safe Practices

All scaffolding used on a worksite *must* be erected, maintained, and disassembled by an experienced crew under the direction of a competent worker—in accordance with the manufacturer's specifications and OHSA regulations.

Each employee *must* be protected by an approved safety life belt attached to a lifeline. The lifeline should be securely attached to substantial members of the structure (not the scaffold), or to securely rigged lines that will safely suspend the employee in case of a fall. In order to keep the lifeline continuously attached, with a minimum of slack, to a fixed structure, the attachment point of the lifeline should be appropriately changed as the work progresses.

- All personnel working on scaffolding are required to wear fall protection.
- Scaffolding over 125 ft high should be designed by a professional engineer.
- Scaffolding under125 ft high should be designed by competent and experienced personnel.
- Scaffolds *must* be erected on sound, rigid footing, capable of supporting four (4) times the intended load—without settling or displacement.
- An access ladder or equivalent form of safe access should be provided.
- Guardrails and toe boards *must* be installed on all open sides and ends of platforms more than 10 ft above the ground or

floor, except needle beam scaffolds and floats. Scaffolds 4 to 10 ft in height, having a minimum dimension in either direction of less than 45 in., should have standard guardrails installed on all open sides and ends of the platform.

- There *must* be a screen with maximum 1/2-in. openings between the toe board and the guardrail, where the persons are required to work or pass under the scaffold.
- Scaffold planking *must* overlap at least 12 in. and be secured from movement.
- Scaffold planking *must* extend over their end supports not less than 6 in. nor more than 12 in.
- All planking should be of Scaffold Grade or equivalent as recognized by approved grading rules for the species of wood used.
- The maximum permissible spans for 2- × 1-in., or wider, planks are listed in Table 32.1.
- Employees *must not* work on scaffolding during bad weather.
- When hoisting materials onto scaffolding, it shall have a tag line.
- Overhead protection shall be provided to all employees working on or near erected scaffolding.
- Tools, materials, and debris *must not* be allowed to accumulate on scaffolding.
- Scaffolding *must* be secured to a permanent structure.
- Scaffolds cannot be used for storage of stone or other heavy materials.
- Runners have to be erected along the length of the scaffolding, both inside and outside.
- Bearers should be installed between posts and securely coupled to the post bearing on the runner coupler.
- All scaffolding should be cross braced.
- All scaffolding should be longitudinal and braced.
- Scaffolding should be tied and securely braced against the building at intervals no greater than 30 ft horizontally and 26 ft vertically.

Full thickness undressed lumber	Nominal thickness lumber				
Working load (psf)	25	50	75	25	50
Permissible span (ft)	10	8	6	8	6

TABLE 32.1 The Maximum Permissible Span for 1¼- × 9-in. or Wider Plank of Full Thickness is 4 ft, with Medium Loading of 50 p/f

- All access ladder or equivalent safe access shall be provided.
- All scaffolding and accessories having any defective parts *must* be immediately replace or repaired.

32.2 Scaffolding Glossary

Bearer Horizontal member of a scaffold upon which the platform rests and which may be supported by ledgers.

Bosun's chair Seat supported by slings attached to a suspended rope, designed to accommodate one worker in a sitting position.

Brace Tie that holds one scaffold member in a fixed position with respect to another member.

Bricklayer's square scaffold Scaffold composed of framed wood squares that support a platform; limited to light and medium duty.

Carpenter's bracket scaffold Scaffold consisting of wood or metal brackets supporting a platform.

Cleat Weighted block used to secure scaffold on even or uneven grounds.

Coupler Device for locking and aligning together the components of a tubular metal scaffold.

Crawling board or Chicken ladder Plank with cleats spaced and secured at equal intervals for use by a worker on roofs, not designed to carry any material.

Dado Type of cut used to make depressions in wood, used to interlock members.

Design working load (maximum intended load) Total of all loads, including the weight of the people, materials, equipment, and platform.

Double-pole (Independent-pole) scaffold Scaffold supported from the base by a double row of uprights, independent of support from the walls, and constructed of uprights, ledgers, horizontal platform bearers, and diagonal bracing.

Float (Ship) scaffold Scaffold hung from overhead supports by means of ropes and consisting of a substantial platform having diagonal bracing underneath, resting on and securely fastened to two parallel plank bearers at right angles to the span.

Guardrail Barrier along the exposed (open) sides and ends of platform to prevent persons from falling.

Handrail Rail connected to a ladder stand running parallel to the slope or top step, or both, and serving the purpose of assisting the user in maintaining balance, while climbing and descending.

Heavy-duty scaffold Scaffold designed and constructed to carry a working load not to extend 75 $lb/in.^2$

Horse scaffold Scaffold for light or medium duty, composed of horses supporting a work platform.

Interior-hung scaffold Scaffold suspended from the ceiling or roof of a structures.

Ladder jack scaffold Light-duty scaffold supported by brackets attached to ladders.

Ladder stand Mobile fixed-size, self-supporting ladder consists of a wide flat tread ladder in the form of stairs; and may include handrails.

Lean-to-scaffold Type of scaffold kept in place by leaning it against a wall or other solid structure.

Ledger (Stringer) Horizontal scaffold member that extends from post to post supporting the putlogs or bearers, and forming a tie between posts.

Light-duty scaffold Scaffold designed and constructed to carry a working load not to exceed 25 lb/in.^{2}

Manually propelled Mobile Scaffold: Portable rolling scaffold supported by casters.

Mason's adjustable multipoint suspension scaffold Scaffold having a continuous platform supported by bearers suspended by wire rope from overhead supports, so arranged and operated as to permit the raising or lowering of the platform to desired working positions.

Medium-duty scaffold Scaffold designed and constructed to carry a working load not to exceed 50 lb/in.^{2}

Mobile scaffold (tower) Light-, medium-, or heavy-duty-scaffold mounted on casters or wheels.

Mobile work platform Fixed work level on casters or wheels, with bracing from platform to vertical frame.

Needle-beam scaffold Light-duty scaffold consisting of needle beams supporting a platform.

Outrigger scaffold Scaffold supported by outriggers or thrust-outs projecting beyond the wall or face of a building or structure, the inboard ends of which are secured inside the building or structures.

Putlog Scaffold member upon which the platform rests.

Rate load Official guideline supplied by manufacturers for maximum loads for hoists and scaffolds.

Runner Lengthwise horizontal bracing or bearing members or both.

Scaffold Temporary elevated platform and its necessary vertical, diagonal, and horizontal members used for supporting workers and materials, also known as scaffold tower.

Single-point adjustable-suspension scaffold Manually or power-operated unit designed for light-duty use, supported by a single wire rope from an overhead support, so arranged and operated as to permit the raising or lowering of the platform to desired positions.

Single-pole scaffold Scaffold with platforms resting on putlogs or cross beams, the outside ends of which are supported on ledgers secured to a single row of posts or uprights and the inner ends of which are supported on or in a wall.

Stone setter's adjustable multiple-point suspension scaffold Swinging-type scaffold having a platform supported by hangers suspended at four points so as to permit the raising or lowering of the platform to the desired working position by the use of hoisting machines.

Tenon Protruding part of a piece of wood used to connect joints.

Toe board Barrier at platform level erected along the exposed sides and ends of a scaffold platform to prevent materials and equipment from falling.

Tube-and-coupler scaffold Assembly consisting of tubing that serves as posts, bearers, braces, ties, and runners, a base supporting the posts, and uprights and couplers that serve to join the various members, usually used in fixed locations.

Tubular fabricated sectional folding scaffold Sectional metal scaffold that folds. Either of ladder frame or of inside stairway design, substantially built of prefabricated sections consisting of end frames, platform frames, inside inclined stairway frame and braces, or hinge-connected diagonal and horizontal bearers with intermediate members. Panels or frames are braced with diagonal or cross braces.

Two-point suspension scaffold (swinging) Scaffold whose platform is supported by hangers (stirrups) at two points, suspended from overhead supports so as to permit the raising or lowering the platform to the desired working position by tackle or hoisting machines.

Walkway Part of a scaffold that is kept clear and used solely form moving around the scaffolding.

Window jack scaffold Scaffold whose platform is supported by a bracket or jack and projects through a window opening.

(*Source:* Occupational Safety & Health Administration, Washington, DC.)

Regulations and Standards

OSHA Regulations

1926 Subpart L—*Scaffolds*

OSHA *Standards for Scaffolds Used in the Construction Industry*

1910.25; 1910.28; 1910.30

1926.451; 1926.452

Appendix A *Scaffold Specifications*

1962.1052

1910.28 *OSHA Safety Requirements for Scaffolding*

Subpart D—*Walking-Working Surfaces*

U.S. Consensus Standards

ASME/ANSI B76 *Standards of Pipes and Fittings*
ANSI/ASSEA10.8–2006 *Safety Requirements for Scaffolding*
ANSI/SIA A92.3–2006 *American National Standards for Manually Propelled Elevating Platforms*

ASTM *International Standards for Steel Pipes, Tubes, and Fittings*

Industry Consensus Standards Resources

American Ladder Institute (*www.americanladderinstitute.org*)
Calculating Reliability of Wood Scaffold Planks
Forest Products Laboratory (*www.fpl.fs.fed.us*)
Scaffolding Industry Association (*www.scaffold.org*)
Scaffolding, Shoring & Forming Institute (*www.ssfi.org*)
Steel Tube Institute of North America (*www.steeltubeinstitute.org*)
Truss Plate Institute (*www.tpinst.org*)

CHAPTER 33

Portable Ladders

Portable ladders are one of the handiest, simplest tools used. Because of their effectiveness, ladders are used by many different people to perform many different tasks. Ladders basically are uncomplicated; however, planning and care are still required to use them safely.

Although most portable ladders used today are still made of wood, both metal and reinforced-plastic ladders are in wide use for a variety of applications.

33.1 Types of Portable Ladders

Portable ladders are designed as "one-person" equipment with the proper strength to support the worker as well as his tools and materials. Ladders are constructed under three general classes:

- *Type IA Industrial.* Heavy-duty; load capacity not more than 300 lb.
- *Type I Industrial.* Heavy-duty; load capacity not more than 300 lb.
- *Type II Commercial.* Medium-duty; load capacity not more than 225 lb (suited for painting and similar tasks).
- *Type III Household.* Light-duty; load capacity of 200 lb.

Stepladders

Stepladders are portable ladders made from steel, wood, aluminum, or fiberglass. They come in an A-shape model, with either one side having steps or both sides having wide steps that are generally 3 to 10-in. deep, and lined with corrugated rubber or have slip-resistant texture.

The foot of the ladder should have slip-resistant molded footpads or rubber cups for better grip. Stepladders range from 1½ to 20 ft and can carry up to 375 lb of weight depending on the model.

Some stepladders have a high railing for support at the top and wheels for rolling at the bottom. A few have an adjustable height of about 1 ft that can be incremented or reduced per requirements.

A platform on the stepladder provides an ideal working place to keep tools and materials. Most stepladders can accommodate accessories for use while working, including paint can/bucket hanger, conduit holder, tool holder, and ladder caddy.

A special type of stepladder—telescopic stepladder—folds up to the length of a step using a proprietary lock-and-release mechanism for opening and closing the ladder.

Even though stepladders provide a better standing space, they must be used carefully. There should not be any extra loading on the side, because they can easily overturn.

Portable Rung Ladders

There are four principal types of rung ladders: single-rung, extension, sectional, and trestle.

Single-Rung Ladders

As with the steps on stepladders, the rungs on these ladders *must* be parallel, level, and uniformly spaced. Wood ladder rungs should be bored through the side rails and should extend at least flush with the outside rail surface. The shoulder of each rung should be forced firmly against the side rails and the tenon secured in place with a nail or its equivalent to prevent the rung from turning and to maintain its position on the side rail (Table 33.1). With metal and plastic ladders, rungs should be rigidly connected to side rails to ensure specified strength.

Round rungs of group 1 wood should be *not less* than $1^1/_8$-in. diameter for lengths up to 36 in. between side rails, and $1^1/_4$-in. diameter for lengths over 36 in. between side rails—up to and including 72 in.

Tenons should be not less than $^7/_8$-in. diameter. When wood rungs between side rails are 28 in. or longer, they should also be provided with center bearing.

Oval rungs or rungs of any other cross section may be used in wood ladders, provided they are secured by nails at each end, and have at least the same strength and bearing as round rugs of the same length.

When measured along the side rails, no wood rung ladder or section should be more than 4 in. shorter than the specified length.

All metal parts and fittings, including hooks and nonslip bases, should be securely belted, riveted, or attached by equivalent fasteners.

A single, portable rung ladder is a non-self-supporting type, nonadjustable in length, consisting of but one section. Its size is designated by the overall length of the side rail. Single, rung ladders longer than 30 ft *do not* meet standard safety code requirements (Table 33.2).

Extension

An extension ladder is a non-self-supporting ladder consisting of two or more sections that travel in guides or brackets so arranged as to

	Type I						Type II					
	12-ft length or less		14- and 16-ft length		18- and 20-ft length		3–8-ft length		10-ft length		12-ft length	
	Thickness, in.	Depth, in.	Thickness, in.	Depth, in.	Thickness, in.	Depth, in.	Thickness, in.	Depth, in.	Thickness, in.	Depth, in.	Thickness, in.	Depth, in.
Side rails	25/32	3¼	25/32	3½	1 1/16	3½	¾	2⅝	¾	2⅝	¾	3
Back legs	25/32	2¼	25/32	2⅝	1 1/16	2¼	¾	1⅝	¾	1¾	¾	2
Steps	25/32	3⅝	25/32	4¼	25/32	4¼	¾	3½	¾	3½	¾	3⅝
Tops	25/32	5½	25/32	5½	25/32	5½	¾	5	¾	5	¾	5

Notes 1. Flat steps should be secured with at least two 6-d nails at each end. Each step should be reinforced by a steel rod not less than 0.160 in. in diameter, which is fastened through metal washers on each end. A truss block should be fitted between the rod and the center of each step. As an alternative, a metal angle firmly secured to the steps and side rails may be used to provide the required bracing.

2. All steps 3⅝ in. wide and 27 in. or more long, and all steps 4¼ in. wide and 32 in. or more long, should be braced at each end with a metal angle securely attached to the step and side rails.

3. The back legs should be reinforced by a rivet that passes through the depth of the leg above the hinge point to prevent splitting, or by metal plates or collars at the hinge point.

4. Minimum width of step or tread for metal and plastic ladders:

Type I ladders 3 in.

Type II ladders 3 in.

Type III ladders 2½ in.

Note Metal and plastic ladders should be fitted with slip-resistant material at the bottom of the four rails.

Table 33.1 Dimensions for Wood Step Ladders

Ladder length, ft	Thickness, in.	Depth, in.
Up to and including 16	1⅛	2½
Over 16 up to and including 22	1¼	2¾
Over 22 up to and including 30	1¼	3

Notes 1. Smaller side rails are acceptable in all ladders of this type if they are reinforced by a steel wire, rod, or strap that runs the length of the side rails and is adequately secured to them.

2. The width between side rails at the base, inside to inside, should be at least 11½ in. for single wood ladders up to and including 10 ft. This minimum width should increase by at least ¼ in. for each additional 2 ft of ladder length.

3. Metal or plastic single ladders should not exceed the following lengths:

Type I	30 ft
Type II	24 ft
Type III	16 ft

Table 33.2 Dimensions for Single, Rung Ladders

Overall length of ladder, ft	Side rails for two-section ladders, in.	
	Thickness × depth, group 2 wood	Thickness depth, group 3 wood
16	$1\frac{1}{16} \times 2$	$1\frac{1}{8} \times 2$
20	$1\frac{1}{16} \times 2$	$1\frac{1}{8} \times 2\frac{1}{4}$
24	$1\frac{1}{16} \times 2\frac{1}{2}$	$1\frac{1}{8} \times 2\frac{1}{2}$
28	$1\frac{1}{16} \times 2\frac{3}{4}$	$1\frac{1}{8} \times 2\frac{3}{4}$
32	$1\frac{1}{8} \times 2\frac{3}{4}$	$1\frac{5}{16} \times 3$
40	$1\frac{5}{16} \times 2\frac{3}{4}$	$1\frac{3}{8} \times 3$
44	$1\frac{5}{16} \times 3$	$1\frac{3}{8} \times 3\frac{1}{4}$
48–52		$1\frac{3}{8} \times 3\frac{3}{4}$
56–60		$1\frac{5}{8} \times 3\frac{3}{4}$

Notes 1. Smaller side rails are permitted when reinforced by a steel wire, rod, or strap running the length of the side rails, adequately secured to the rails. Minimum dimensions of side rails are based on the extension ladder's working length, which is the size of the ladder less the minimum overlap.

2. Minimum distance between side rails of the bottom section (inside to inside):

Ladder length, ft	Minimum distance, in.
Up to and including 28	$14\frac{1}{2}$
Over 28 up to and including 40	16
Over 40 up to and including 60	18

3. Minimum overlap of wood extension ladder side rails:

Ladder length, ft	Minimum overlap
Up to and including 36	3
Over 36 up to and including 48	4
Over 48 up to and including 60	5

TABLE 33.3 Dimensions for Wood Extension Ladders

permit the user to adjust the overall length. The size of an extension ladder is designated by the sum of the lengths of the sections, measured along the side rails.

Two-section wood and metal extension ladders longer than 60 ft *do not* meet standard safety requirements; three-section metal ladders longer than 72 ft *do not* meet standards safety requirements (Tables 33.3 and 33.4).

Extension ladders should be equipped in such a manner that the ladder cannot be used with an overlap less than that specified in the tables. The fly section of metal or plastic extension ladders should contain the following warning in letters *not less* than $^{1}/_{8}$-in. high:

CAUTION: *This ladder section is not designed for separate use.*

If the fly section of a metal extension ladder incorporates a stationary lock or locks, which by their location eliminate a rung in section, the section should be permanently marked:

Ladder duty rating	Two-section ladder, ft	Three-section ladder, ft
Type I, heavy duty	Up to and including 60	Up to and including 72
Type II, medium duty	Up to and including 48	Up to and including 60
Type III, light duty	Up to and including 32	

Notes 1. All metal extension ladders should be clearly marked to indicate both the total nominal length of the sections and the maximum nominal extended length or maximum nominal working length.

2. The minimum clear width between side rails of the uppermost (fly) section of metal extension ladders shall be not less than 12 in. The minimum clear width between side rails at the bottom of the base section shall be not less than the following dimensions:

Ladder length, ft	Minimum clear width, in.
Up to and including 16	12½
Over 16 up to and including 28	14
Over 28 up to and including 40	15
Over 40 up to and including 72	18

3. Minimum overlap of metal or plastic extension ladder side rail when fully extended:

Ladder length, ft	Two-section ladder total overlap, ft	Three section ladder total overlap, ft
Up to and including 36	3	6
Over 36 up to and including 48	4	8
Over 48 up to and including 72	5	10

TABLE 33.4 Dimensions for Metal Extension Ladders

CAUTION: *This ladder section is not designed for separate use.*

Guide irons should be securely attached and so placed on the base section of the ladder as to prevent the upper section from tipping or falling out while it is raised or lowered , or while the extension is in use.

An extension ladder's rope and pulley should be securely attached to the ladder so as not to weaken the rungs or the side rails (Table 33.5). Rope should be of sufficient length for proper operation. Where cable is used, on three-section metal ladders the cable should be at least $^3/_{16}$-in. diameter.

Ladder material	Pulley diameter, in.	Rope diameter, in.	Breaking strength, lb
Wood	1¼	5⁄16	560
Metal or plastic	1	¼	560

TABLE 33.5 Minimum Pulley and Rope Specifications

Length, ft	Thickness, in.	Depth, in.
Up to and including 21	1⅛	2¾
Over 21 up to and including 31	1⅛	3⅛

TABLE 33.6 Minimum Dimensions of Side Rails, Group 2 or Group 3 Woods

Sectional

These ladders are not as common as the other types; they are used in particular by window cleaners. Sectional ladders are defined as non-self-supporting portable ladders, nonadjustable in length, and consist of two or more sections of ladder, so constructed that the sections may be nested to function as a single ladder. The size of such ladders is designated by the overall length of the assembled sections (Table 33.6).

Ladders of this type should have either straight sides that converge slightly toward the top of each section, or flaring sides at the bottom of the bottom section, with the top section having side rails that converge to a width not less than 4 in. The minimum width between side rails of the bottom section should be 11 in.

Individual sections are joined by means of a groove, in the bottom end of the upper section's rails, which sets firmly over extensions outside the side rails of the next lower section's top most rung. A groove in the top end of the lower section's rails sets firmly under the bottom rung, inside the side rails of the section immediately above.

The distance between the topmost rung of one section and the bottom rung of the next section above should be *not less* than 12 in. The grooved ends of the sections should be reinforced with a metal plate (*not less* than 16 gage) secured to the rail by a rivet through the depth of the rail. Rungs and grooves should fit securely, without binding or unnecessary play.

Trestle and Extension-Trestle Ladder Scaffold

A trestle ladder is a self-supporting portable ladder, nonadjustable in length, consisting of two sections hinged at the top to form two equal angles with the base. Its size is designated by the length of the side rails, measured along the front edge of the ladder.

The combination extension-trestle ladder is an adjustable self-supporting portable ladder that consists of a trestle ladder base and a vertically adjustable single ladder, with suitable means for locking the ladders together. The size of this type of ladder is designated by the length of the trestle ladder base (Fig. 33.1).

Trestle ladders, or extension sections, or base sections of extension trestle ladders longer than 20 ft *do not* meet standards safety requirements (Tables 33.7 and 33.8). Trestle ladders and base sections of extension trestles should open so that the spread of the trestle at the

Figure 33.1 Trestle ladder. (*Source*: Werner Co., Greendale, PA.)

Length, ft	Thickness, in.	Depth, in.
Up to and including 16	1 5/16	2 3/4
Over 16 up to and including 20	1 5/16	3

Table 33.7 Minimum Dimensions for Trestle Ladder Side Rails, or Extension-Trestle Ladder Base Sections, Group 2 or Group 3 Woods

Length, ft	Thickness, in.	Depth, in.
Up to and including 12	1 5/16	2 1/4
Over 12 up to and including 16	1 5/16	2 1/2
Over 16 up to and including 20	1 5/16	2 3/4

Table 33.8 Minimum Dimensions for Trestle Ladder Side Rails, or Extension-Trestle Ladders with Parallel Sides, Group 2 or Group 3 Woods

Length, ft	Side rails Thickness, in.	Depth, in.	Diameter Rung, in.	Tenon, in.
Up to and including 12	1⅝	3⅝	1⅜	1
Over 22 up to and including 40	1⅝	4½	1⅜	1

TABLE 33.9 Minimum Dimensions for Mason's Ladder Side Rail

bottom, inside to inside, is at least 5½-in. per foot of ladder length. Side rail width at the base of these ladders should be at least 21 in. for all ladders and sections up to and including 6 ft. For longer lengths, side rails should be *not less* than 12 in.

All rungs should be parallel and level, spaced *not less* than 8 in. nor more than 18 in. apart on trestle ladders or base sections of extension-trestle ladders; and *not less* than 6 in. nor more than 12 in. apart on extension sections.

Special Ladders

Two special ladders used on construction sites are the mason's ladder and the cleat-type ladder.

- *Mason's ladder.* A single ladder that should not be supplied in sizes longer than 40 ft. The width between side rails at the bottom rung, inside to inside, should be *not less* than 12 in. for all mason's ladders up to and including 10 ft in length. This should be increased by at least ¼-in. for each additional 2 ft of ladder length. Rungs of mason's ladders should be spaced *not less* than 8 in. nor more than 12 in. apart (Table 33.9).
- *Cleat ladders.* Also of the single type, should not be supplied in sizes longer than 30 ft. Wood side ails of these ladders should be *not less* than 1⅝-in. deep (nominal 2- × 4-in.). Wood cleats should be inset into the side rails *not less* than ½-in. If the cleats are attached directly to the side rails, they should be fastened by three 10-d wire nails to each rail, with filler blocks of the same thickness as the cleats should be securely attached to the edge of the rail for the full length between cleats.

33.2 Job-Made Wood Ladders

All ladder component surfaces *must* be finished to avoid injury to employees and to prevent snagging of clothing. *Make sure* working length is not greater than 24 ft; and that fasteners are driven full length and countersunk not more than ⅛ in.

NOTE: *Fasteners include plain-shank and helically threaded steel nails. Staples and wood screws of equivalent shank withdrawal, head pull-through, and bending/shear resistance (as determined by test data or published formulas and tabulated values) may also be used.*

Materials

Lumber used in job-made ladders is stress grade—lumber that has been assigned allowable stress (allowable stress design) or reference strengths (load resistance factor design) values. It is identified by the grade-mark or certificate of inspection issued by a lumber inspection bureau or agency accredited by the Board of Review of the American Lumber Standard Committee. The grade-mark specifies the grade, species, and dryness of the lumber.

Wood parts for job-made ladders are seasoned to moisture content of not more than 19 percent. Cleats are nominal 2- × 4-in. stress-grade dimension lumber.

- *Double-cleat ladder.* A job-made ladder with two side rails and a center rail connected with continuous cleats. It allows personnel to climb and descend at the same time.
- *Single-cleat ladder.* A ladder consisting of a pair of side rails connected by cleats, rungs, or steps.
- *Stress-grade lumber.* Assigned allowable stress (allowable stress design) or reference strengths (load resistance factor design) values; identified by the grade-mark or certificate of inspection issued by a lumber inspection bureau or agency accredited by the Board of Review of the American Lumber Standard Committee. The grade-mark specifies the grade, species, and dryness of the lumber (Table 33.10).

Side Rails

Side rails *must* be uniform throughout the length of climb, and *must* have a minimum clear distance between rails at least:

16 in. but not greater than 20 in. for single-cleat ladders
18 in. but not greater than 22 in. for double-cleat ladders

If splicing is required to obtain the necessary ladder length, *make sure* that the resulting side rail doesn't have more than one splice—located as close to the top point of bearing as possible—and is equivalent in strength to a one-piece side rail made of the same material.

The required side rails are spliced using bolts with a nut and lock washer below the nut. Use either common steel bolts with a 1-in. diameter, $^3/_{32}$-in. thick steel washer under the bolt head; or ½-in. diameter carriage bolts.

Species	Minimum grade
Aspen	Select Structural
Beech-Birch-Hickory	Select Structural No. 2
Cottonwood	Select Structural
Douglas Fir-Larch	No. 2
Douglas Fir-Larch (north)	No. 1/No. 2
Douglas-Fir-Larch (south)	No. 2
Eastern Hemlock-Tamarack	Select Structural
Eastern Softwoods	Select Structural
Eastern White Pine	Select Structural
Hem-Fir	No. 2
Hem-Fir (north)	No. 1/No. 2
Mixed Maple	Select Structural
Mixed Oak	No. 2
Northern Red Oak	No. 2
Northern Species	Select Structural
Red Maple	No. 2
Red Oak	No. 2
Redwood	No. 1
Spruce-Pine-Fir	No. 1/No. 2
Spruce-Pine-Fir (south)	No. 1
Southern Pine	No. 2 (nondense)
Western Cedars	Select Structural
Western Woods	Select Structural
White Oak	No. 2
Yellow Popular	Select Structural
MSR	1200f-1.2E
MEL	M-7

Note: The allowable stress in bending after adjustment for size (i) shall not be less than 1200 psi (pound-force per square inch)—the corresponding reference strength for load and resistance factor design; (ii) shall not be less than 3.05 ksi (kips-force per square inch).

Source: Washington State Department of Labor and Industries, Tacoma, WA.

Table 33.10 Acceptable Stress-Grade Lumber for Job-Made Ladders

Nominal-dimension lumber		
Working length (feet)	**Spliced side rail**	**Continuous side rail**
12 or less	2 × 4	2 × 4
14	2 × 4	2 × 4
16	2 × 4	2 × 6
18	2 × 4	2 × 6
20	2 × 6	2 × 6
22	2 × 6	2 × 6
24	2 × 6	2 × 6

Source: Washington State Department of Labor and Industries, Tacoma, WA.

TABLE 33.11 Minimum Rail Size for Single-Cleat Ladders

Cleats

Make sure cleats are parallel and level when the ladder is in position to be used, and that they are evenly spaced throughout the length of the ladder from the base to the top point of bearing (Table 33.11). The distance from the top of a cleat to the top of an adjacent cleat *must* be at least 8 in. but not greater than 12 in. Cleats on double-cleat ladders are continuous and extend the full width of the ladder (Table 33.12)

Nominal-dimension lumber		
Working length (ft)	**Spliced side rail**	**Continuous side rail**
12 or less	2 × 4	2 × 4
14	2 × 4	2 × 6
16	2 × 6	2 × 6
18	2 × 6	2 × 6
20	2 × 6	Stresses exceed capacity of 2 × 6 rails
22	2 × 6	Stresses exceed capacity of 2 × 6 rails
24	2 × 6	Stresses exceed capacity of 2 × 6 rails

Source: Washington State Department of Labor and Industries, Tacoma, WA.

TABLE 33.12 Minimum Rail Size for Double-Cleat Ladders

Length, ft	Thickness, in.	Width, in.
Up to and including 20	25/32	3
Over 20 up to and including 30	25/32	3¾

Table 33.13 Minimum Dimensions for Wood Cleats, Group 1 Woods

Attaching Cleats

Cleats *must* be attached to the narrow face of each side rail by three 3¼-in. long 12-d common nails, or an equivalent set of fasteners (Table 33.13).

- Use filler blocks between cleats.
- *Do not* cut side rails to insert cleats.
- *Make sure* filler blocks is the same thickness as the cleats; butt tightly against the underside of each cleat,; and attached to the side rails by three 3¼-in. long 12-d common nails, or an equivalent set of fasteners.

33.3 Ladder Specifications

Most ladder specifications are based primarily on data contained in the safety requirements for portable ladders published by the American National Standards Institute (ANSI):

- A14.1 *Safety Requirements for Portable Wood Ladders*
- A14.2 *Safety Requirements for Portable Metal Ladders*
- A14.5 *Safety Requirements for Portable Reinforced Plastic Ladders*
- A14.3 *Safety Requirements for Fixed Ladders*

Not only should these standards serve as a basis for purchase requirements, but they should also be used for instructing workers in ladder safety practices.

Before putting any new ladders into use, a contractor-appointed competent worker should first check all the dimensions of the ladder, including the sizes of the side rails and rungs to make sure they are not less than the minimum allowed by the standards.

Next, examine the ladder's hardware to determine whether it is of ample strength.

- Avoid malleable iron and cast iron for parts subject to bending or tensile stress.
- *Make sure* all hardware is securely attached.
- Check if rungs and steps are tightly fitted into the side rails and secured against turning.
- *Make sure* that special *not used* for industrial or commercial purposes.

- Determine if spreader bars on stepladders are designed to prevent serious finger-pinching hazards, as well as to prevent the ladder from folding up accidentally if pushed along the floor. The moving parts should operate freely without binding or undue play.
- *Make sure* bucket shelves, an integral part of a stepladder, are fastened so that they can be folded up when the ladder is closed.
- *Make sure* that extension ladders have their guide brackets long enough to engage the full width of the side rails on the other section of the ladder.
- *Make sure* the rope and pulley are of ample strength.

NOTE: *Near the lower end of the upper section of extension ladders, a rung is often omitted at the location of the locks. A special offset rung may be necessary, however, because the sections of an extension ladder are frequently separated and used independently.*

- Examine thoroughly the ladder material, especially wooden ladders.

  All parts of wooden ladders *must* be free from splinters and defects.

  All parts of metal and plastic ladders must be fee from burs, cracks, and defects.

ANSI standard *A 14.1* classifies the various species of wood into four groups on the basis of the mechanical properties acceptable for use in ladder construction (Table 33.14).

33.4 Care and Maintenance of Portable Ladders

Proper care and regular maintenance of ladders will ensure their continued safety and serviceability. Protect wood ladders from deterioration due to exposure to the elements, by regular application of a specially selected, clear coating that will provide suitable protection for exterior conditions.

Portable ladders must be maintained in good condition at all times—free of oil, grease, and other slipping hazards—and *must* be inspected at regular, frequent intervals. Visually inspect all ladders in regular use, at *not less* than 3-month intervals or when first used on a new job. Check:

- Side rails—free from splits, bruises, and the like
- Rungs—sound condition

Group 1		
White ash	Hickory	Red oak
Beech	Locust*	White oak
Birch	Hard maple	Pecan
Rock elm	Red maple	Persimmon

The allowable fiber stress in bending for these species, when used for side rails, must not exceed 2150 lb/in.² Group 1 woods may be substituted for those in group 3 provided the dimensions are not more than 10% smaller for each cross section; or, if the thickness remains unchanged, the width is not more than 15% smaller when used edgewise (as in a rail), nor 25% smaller when used flatwise (as in a tread).

Group 2		
Douglas fir (coastal)	Western larch	Southern yellow pine

The allowable fiber stress in bending for these species, when used for side rails, must not exceed 2000 lb/in.² Group 2 woods may be substituted for those in group 3 provided the dimensions are not more than 7½% smaller for each cross section; or, if the thickness remains unchanged, the width is not more than 11% smaller when used edgewise (as in a rail), nor 20% smaller when used flatwise (as in a tread).

Group 3		
Red alder	Douglas fir (Rocky Mountain)	Poplar
Oregon ash	Noble fir	Redwood*
Alaskan cedar*	Gum	Eastern spruce
Port Orford cedar*	West coast hemlock	Sitka spruce
Cypress*	Oregon maple	Sycamore
Soft elm	Norway pine	Tamarack

The allowable fiber stress in bending for these species, when used for side rails, must not exceed 1600 lb/in.²

Group 4		
Aspen	Idaho white pine	Eastern hemlock
Basswood	Ponderosa pine	Soft maple
Butternut	Cottonwood	Lodgepole pine
Incense cedar*	White fir	Northern white pine
Western red cedar*	Sugar pine	

The allowable fiber stress in bending for the species, when used for side rails, must not exceed 1375 lb/in.² Group 4 woods may be substituted for those in group 3 provided the dimensions are at least 5% greater for each cross section; or, if the thickness remains unchanged, the width is at least 7½% greater when used edgewise (as in a rail), or 15% greater when used flatwise (as in a tread).

Notes 1. Where resistance to decay is required, the species having the most durable heartwood, as indicated by an asterisk(*), should be used; or the wood should be given a treatment with a preservative before using a ladder under conditions favorable to decay.

2. All minimum dimensions and specifications for side rails and flat steps are based on group 3 species. Substitution of species in other groups is permitted only when used in sizes that provide at least equivalent strength, as specified.

3. All minimum dimensions and specifications for rungs and cleats are based on group 1 species. However, cleats may be made of wood from any other group, provided that the cross-sectional dimensions specified for group 1 species are increased by the following factors:

Species	Each dimension	Width only (thickness unchanged)
Group 1	1.00	1.00
Group 2	1.03	1.05
Group 3	1.11	1.19
Group 4	1.17	1.26

TABLE 33.14 Classification of Various Species of Wood Acceptable for Use in Ladders

- Rung joints—tightness
- Reinforcing wire—taughtness
- Hardware—good condition
- Extension ladder—two halves move freely; rope in good condition
- Joints—tightness between side rails and rungs
- Metal pulley bearings—lubricate frequently

To test the tightness of joints, stand the ladder upright on a flat surface and at right angles to a wall. Hold the inner side rail firmly against the wall and attempt to lift the outer side rail clear of the floor. If it lifts more than 0.01 in. (3 mm) examine the fit of the rungs in the side rails and make good any defects.

Alternatively, the ladder may be laid on the edge of any suitable level surface with the top of the lower side rail against a stop; push horizontally on the upper side rail and measure the relative movement between the side rails.

Portable ladders having structural defects such as:

- Broken, split or missing rungs, cleats or steps
- Broken or split side rails
- Missing or loose bolts, rivets or fastening
- Defective ropes, or
- Any other structural defect

Shall not be used and either shall be tagged as unusable if kept on the premises or shall be removed from the worksite.

Step ladders may be tested the same way but, owing to the inward slope of the side rails, the outer side rail will not reach the floor. Use a rule upright against a mark on the side rail to measure any movement.

Wood Ladder Maintenance Checklist

Wood ladders should be coated for protection *only* with a clear sealer varnish, shellac, linseed oil, or wood preservative.

- *Do not* paint wood ladders, because the paint could hide defects.
- Check carefully for cracks, rot, splinters, or broken uprights, braces, or rungs, or loose joints and bolts.
- *Never* subject ladders to unnecessary dropping, jarring, or misuse. Ladders, as with any tool, are designed for a specific purpose or use. Therefore, any variation from his use is a mishandling of equipment.

- Inspect ladders periodically, initially upon receipt, as well as each time before use.
  Step ladders for:
  Wobbly (from side strain)
  Loose or bent hinge spreaders
  Stop on hinge spreaders broken
  Loose hinges
  Broken, split, or worn steps
  Extension ladders for:
  Loose, broken, or missing extension locks
  Defective locks that do not seat properly while extended
  Worn or rotted rope
  Loose steps or rungs (considered loose if they can be moved at all with the hand)
  Cracked, spilt, or broken uprights, braces, or rungs
  Slivers on uprights, rungs, or steps
  Damaged or worn nonslip bases

- Remove from service immediately any ladders that have fallen or sustained a heavy blow and show any defects. Only a competent person should be allowed to make permanent repairs—replacing any worn or broken members.
- *Do not* pin or bind broken or damaged side rails or rungs; this is a dangerous practice that *must not* be tolerated.
- Tag all ladders that are to be repaired with the message DANGEROUS—DO NOT USE.
- *Make sure* that, on wood ladders, the joint between steps or rungs and side rails is tight; on metal ladders, check for side rail dents or bends, or excessively dented rungs.
- *Never* attempt to straighten or use a bent ladder.
- Keep steps and rungs free of grease, oils, and other slippery substances.
- Make sure all bolts and rivets are in place and tight before using a ladder.
- *Never* use a ladder if any bolts or rivets are missing.
- Check hardware, fittings, and accessories frequently; *make sure* they are in proper working condition and that all moving parts operate freely without binding or undue play. Lubricate metal bearings or locks, wheels, and pulleys regularly.
- Inspect ropes or cables frequently; replace any that are worn or frayed.
- Check ladder safety shoes or padded feet; if they are excessively worn, *do not* use the ladder until shoes or feet are repaired.

Metal Ladder Maintenance Checklist

Steel or aluminum ladders should be inspected for rough burrs and sharp edges before use. Aluminum ladders are very corrosion resistant, but exposing them to fertilizer can cause damage.

- Inspect closely for loose joints and bolts, faulty welds, and cracks.
- *Make sure* all hooks and locks on extension ladders are in proper working condition.
- Replace worn or frayed ropes on extension ladders at once.
- Destroy the ladder if rails, braces, or other parts of any metal ladder are bent.
- Visually inspect for damage all metal ladders that are exposed to excessive heat, then test them for reduced deflection and possible loss of strength. (In doubtful cases, refer to the manufacturer.)
- *Never* use a metal ladder near exposed electric equipment or overhead distribution wires, crane ways, trolley conductors, or switchboards.

Plastic Ladder Maintenance Checklist

Fiberglass ladders should have a surface coat of lacquer. Ladders that are scratched beyond normal wear should be lightly sanded before applying a coat of lacquer.

- Protect plastic reinforced ladders from excessive exposure to ultraviolet rays (sunlight), which may cause damage to glass fiber. Where this condition is evident, wash down the ladder with a suitable commercial solvent (or liquid detergent solution) compatible with the ladder's composition—acetone, isopropyl alcohol, methyl ethyl ketone, or the like.
- Allow the ladder to air dry and handle the ladder as little as possible. Brush or spray the ladder with a good commercial grade of acrylic lacquer, epoxy coating, or other suitable material. One or two coats may be required depending on the degree of glass fiber prominence.
- Provide adequate ventilation, prohibit smoking, and *make sure* workers wear the proper protective equipment when using such solvents and coatings.
- Reduce the possibility of glass fiber prominence on plastic ladders that must be stored outdoors for any extended period by applying a coat of good commercially available nonslip paste wax, or other suitable substance on the fiberglass side rails of the ladders.
- Carry out a semiannual coating program to ensure the ladder's longevity.

- Maintain plastic ladders in a clean condition at all times to ensure the retention of the original electric insulating characteristics.
- Remove all surface buildup of dirt, dust, grease, grime, and carbonaceous and other conductive materials that provide a ready path for electric currents and thus create a potential danger for the user.

Storage and Transportation of Ladders

Ladders should *always* be stored in a well-ventilated location off the ground and out of the weather.

- *Never* store wood ladders near radiators, stoves, steam pipes, or other places subjected to excessive heat or dampness.
- *Always* support ladders at sufficient points to prevent sagging. Supports at one-fifth of the ladder length from each end will be satisfactory for moderate length ladders. For longer ladders, additional supports are necessary.
- *Do not* store ladders as a platform on which to stack other gear.
- Tie down ladders securely to a truck rack for transporting; *make sure* such ladders are supported adequately to prevent sagging.
- Secure the feet from pivoting about the ladder while the vehicle is in motion. The ladder truck rack should be designed to clamp the ladder into a fixed position and prevent any chafing that might be caused by horizontal and vertical movement of the ladder with respect to the truck rack, the truck, or other ladders or individual sections.
- Keep to a minimum any ladder overhang beyond supporting points, which should be of hardwood or rubber-covered iron pipe to reduce chaffing and the effects of road shock.
- Tie down the ladder to each support point to greatly reduce any damage that may be caused by road shock.
- *Never* transport ladders suspended by one rail; such a method causes rungs to become loose in the side rails and will repeatedly ruin a ladder. Securely stow or fasten ladders carried on vehicles to prevent damage from vibration or due to falling from the vehicle in the event of a sudden stop.

33.5 Safe Use of Ladders and Stairways

Ladder safety is a lifesaving program. Ladder accidents usually are caused by improper selection, care, or use, not by manufacturing defects. Accident prevention, however, requires proper planning, correct ladder selection, good work procedures, and adequate ladder maintenance (Table 33.15).

Ladder hazards • Using a ladder having -Missing or broken parts -Too low a weight rating -Too short for the job • Using a ladder as a working platform • Using a metal ladder near electrical wires • Objects falling from ladders
Ladder accidents • Wrong ladder used to do a job • Ladder failure due to physical condition of ladder or condition of its supporting surface • Ladder improperly used • Ladder not the right tool for the job • Ladder not safely used—without personal protective equipment
Preventing accidents • *Never* hand-carry loads on ladder. • *Always* carry tools up or down on a belt or hoist them in a bag or bucket. • *Always* wear shoes with clean, nonskid soles. • *Always* face the ladder while climbing up or down. • *Always* hold onto the side rails with both hands. • Always keep centered when on the ladder so that any belt buckle lies in between side rails. • *Never* stand on the ladder's top three rungs. • *Never* use the top step of a stepladder as a step. • Replace top rung of a portable ladder with chain or rope to reduce rocking when working on cylindrical poles and columns.
Safe ladder set-up • Get help to move ladders too heavy to move alone. • Position ladders (moving) to safely reach without losing balance. • Space ladder base 1 ft out for every 4 ft it reaches. • Use nonskid feet on the ladder or spurs to prevent it from slipping on hard, smooth surfaces. • *Make sure* that side rails extend at least 36 in. above the bearing point, when used to gain access to a roof, floor, or other level walking surface.

TABLE 33.15 Hazards of Ladders and Stairways, and Accident Prevention (*Continued*)

Check ladders before use
• Damaged side rail may cause one side of a ladder to give way. • *Make sure* step ladders are securely spread open. • *Never* use a folding step ladder in an unfolded position. • *Make sure* both extension ladder locks hold adequately to prevent overloading a rail. • Use nonskid pads to prevent a ladder from slipping on hard, smooth surfaces. • *Never* work near power lines. Electrical shock can occur with metal or wet wooden ladders. Not only is shock itself dangerous, but it can cause falls resulting in injury.
Avoid dangerous work practices
• *Do not* step on the top two stepladder steps; or top three straight ladder rungs. • *Never* climb the cross-bracing on the rear section of a step ladder (unless it is designed for such use and provides steps on both the front and rear section). • *Never* place ladders on unstable bases, such as barrels, truck beds, boxes, or uneven ground; or in doorways or passageways. • *Do not* walk a ladder while standing on it. • *Do not* use a metal ladder near power lines. • Keep stairway treads and walkways free of dangerous objects, debris, tools, and materials. • Correct slippery conditions on stairways and walkways immediately. • *Make sure* that treads cover the entire step and landing; stairways having four or more risers or rising more than 30 in. *must* have at least one handrail.

Table 33.15 Hazards of Ladders and Stairways, and Accident Prevention (*Continued*)

Portable ladders having structural defects such as broken or missing rungs, cleats or steps, broken or split rails, corroded components, or other faulty or defective components *must* be marked DEFECTIVE and immediately tagged with DO NOT USE, then withdrawn from service until repaired or destroyed.

Some of the more common hazards involving ladders, such as instability, electrical shock, and falls, can be predicted and prevented.

- *Always* use the correct ladder for the task. Have a competent person visually inspect a ladder before use.
- *Make sure* ladders are long enough to safely reach the work area.

- Tag or mark immediately DO NOT USE any damaged or defective ladders for repair or replacement; or destroy them.
- *Never* load ladders beyond the maximum intended load or beyond the manufacturer's rated capacity.
- *Make sure* the load rating can support the weight of the user, including materials and tools.
- Avoid using ladders with metallic components near electrical work and overhead power lines.

33.6 Inspection of Portable Ladders

A department supervisor or designated responsible person *must* inspect all ladders for visible defects on a semiannual basis—and after any incident that could affect their safe use. The person performing the inspection *must* complete a Portable Ladder Inspection Checklist.

The following checklist (Table 33.16) covers regulations issued by the Occupational Safety & Health Administration (OSHA) under the construction standards *29 CFR 1926.1050* to *1926.1060*. It applies to portable ladders used at temporary worksites associated with construction, alteration, demolition, or repair work including painting and decorating.

• Are ladders or stairways provided at all points of access that are elevated 19 in. or more, and no ramp, runway, sloped embankment, or personnel hoist is provided?	Y	N	N/A
• Does a competent person provide ladder training that teaches users how to recognize hazards and procedures for minimizing these hazards?	Y	N	N/A
• Is ladder retraining provided when necessary?	Y	N	N/A
• Can ladders support the load they are expected to carry?	Y	N	N/A
• Are ladder rungs, cleats, and steps parallel, level, and uniformly spaced when the ladder is in position for use?	Y	N	N/A
• Are rungs, cleats, and steps of portable ladders (other than step stools and extension trestle ladders) spaced at least 10 but not more than 14 in. apart (as measured between center lines of the rungs, cleats, and steps)?	Y	N	N/A

TABLE 33.16 Portable Ladder Inspection Checklist (*Continued*)

• Are rungs, cleats, and steps of step stools at least 8 but not more than 12 in. apart (as measured between center lines of the rungs, cleats, and steps)?	Y	N	N/A
• Are rungs, cleats, and steps of the base section of extension trestle ladders at least 8 but not more than 18 in. apart (as measured between center lines of the rungs, cleats, and steps)?	Y	N	N/A
• Are rungs, cleats, and steps of the extension section of extension trestle ladders at least 6 but not more than 12 in. apart (as measured between center lines of the rungs, cleats, and steps)?	Y	N	N/A
• Is the clear distance between side rails for all portable ladders at least 11½-in.?	Y	N	N/A
• Are the rungs and steps of portable metal ladders corrugated, knurled, dimpled, coated with skid-resistant material, or otherwise treated to minimize slipping?	Y	N	N/A
• Are ladders prohibited from being tied or fastened together to provide longer sections (unless they are designed for such use)?	Y	N	N/A
• Is a metal spreader or locking device provided on each stepladder to hold the front and back sections in an open position when the ladder is being used?	Y	N	N/A
• Are ladders surfaced to prevent injury from punctures or lacerations, and to prevent snagging of clothing?	Y	N	N/A
• Is it prohibited to coat wood ladders with any opaque covering, except for identification or warning labels that are placed on only one face of a side rail?	Y	N	N/A
• Do portable ladders extend at least 3 ft above the upper landing surface for which the ladder is used to gain access?	Y	N	N/A
Note: As an alternative, secure the ladder at its top to a rigid support that will not deflect. Use a grasping device (such as a grabrail) to mount and dismount the ladder. The extension should never be such that the ladder deflection under load would, by itself, cause the ladder to slip off its support.			
• Are ladders maintained free of oil, grease, and other slipping hazards?	Y	N	N/A

Table 33.16 Portable Ladder Inspection Checklist (*Continued*)

• Are ladders loaded at or below the maximum intended load for which they were built, or at or below the manufacturer's rated capacity?	Y	N	N/A
• Are ladders only used for the purpose for which they were designed?	Y	N	N/A
• Are non-self-supporting ladders used at an angle such that the horizontal distance from the top support to the foot of the ladder is approximately $^1/_4$ of the working length of the ladder (the distance along the ladder between the foot and the top support)?	Y	N	N/A
• Are ladders used only on stable and level surfaces, unless secured to prevent displacement?	Y	N	N/A
• Are ladders used on slippery surfaces *only* when they are secured or provided with slip-resistant feet to prevent displacement?	Y	N	N/A
Note: Do not use slip-resistant feet as a substitute for care in placing, lashing, or holding a ladder on surfaces such as flat metal or concrete that cannot be prevented from becoming slippery.			
• Are ladders secured to prevent displacement, especially in busy, high-traffic areas?	Y	N	N/A
Note: As an alternative, a barricade may be used to keep the activities or traffic away from the ladder.			
• Is the area around the top and bottom of ladders kept clear?	Y	N	N/A
• Is the top of a non-self-supporting ladder placed with the two rails supported equally, unless it has a single support attachment?	Y	N	N/A
• Is it prohibited to move, shift, or extend ladders while they are occupied?	Y	N	N/A
• Do ladders have nonconductive side rails if they are used where they could contact exposed energized electrical equipment?	Y	N	N/A
• Is standing on the top or top step of a stepladder prohibited?	Y	N	N/A
• Is climbing on the cross-bracing on the rear section of stepladders prohibited?	Y	N	N/A
Note: This is allowed if the ladder is designed and provided with steps for climbing on both front and rear sections.			

Table 33.16 Portable Ladder Inspection Checklist (*Continued*)

• Are ladders inspected periodically by a competent person and after any incident that could affect their safe use?	Y	N	N/A
• Are portable ladders with structural defects (a) immediately marked in a manner that readily identifies them as defective, (b) tagged with do not use or similar language, or (c) withdrawn from service until repaired?	Y	N	N/A
Note: Structural defects include broken or missing rails, corroded components, or other faulty or defective components.			
• Does a ladder that is repaired meet its original design criteria before it is returned to use?	Y	N	N/A
• Do all students and employees face the ladder when moving up or down the ladder?	Y	N	N/A
• Do all students and employees use at least one hand to grasp the ladder when moving up or down the ladder?	Y	N	N/A
• Is it prohibited to carry any object or load that could cause a person to lose balance and fall?	Y	N	N/A

Source: Occupational Safety & Health Administration, Washington, DC.

TABLE 33.16 Portable Ladder Inspection Checklist (*Continued*)

33.7 Temporary Ladders and Portable Ladders Glossary

Cages Guards that are fastened to the side rails of the fixed ladder or to the structure to encircle the climbing space of the ladder for the safety of the person who must climb the ladder.

Double-cleat ladder Ladder with a center rail to allow simultaneous two-way traffic for employees ascending or descending.

Fixed ladder Ladder that cannot be readily moved or carried because it is an integral part of a building or structure.

Ladders Equipment usually consisting of two side rails joined at regular intervals by cross-pieces called steps, rungs or cleats, on which a person may step in ascending or descending.

Midrails Rail approximately midway between the guardrail and platform, used when required, and secured to the uprights erected along the exposed sides and ends of platforms.

Permanent structure Permanent structure can be either a building or permanent fixture, such as a smokestack.

Platform Working surface for persons, elevated above the surrounding floor or ground; such as an extended step or landing breaking a continuous run of stairs.

Point of access All areas used by employees for work-related passage from one area or level to another.

Rungs Ladder cross-pieces of circular or oval cross section on which a person may step in ascending or descending.

Safety devices Devices other than a cage or well, designed to eliminate or reduce the possibility of accidental falls and which may incorporate such features as life belts, friction brakes, and sliding attachments.

Stairrail Vertical barrier erected along the unprotected sides and edges of a stairway to prevent employees from falling to lower levels.

Wells Permanent complete enclosure around a fixed ladder, which is attached to the walls of the well. Proper clearances for a well will give the person who must climb the ladder the same protection as a cage.

Regulations and Standards

OSHA Regulations

1910.29 *Manually Propelled Mobile Ladder Stands and Scaffolds (tower)*

1915 Subpart E—*Scaffolds, Ladders, and Other Working Surfaces*

1915.71 *Scaffolds Staging*

1915.77 *Working Surface*

1926 Subpart X—*Stairways and Ladders*

1926.1050 *Scope, Application, and Definitions Applicable to this Subpart*

1926.1051 *General Requirements*

1926.1052 *Stairways*

1926.1053 *Ladders*

Appendix A *Ladders*

U.S. Consensus Standards

A14.1–2000 *Safety Requirements for Wood Ladders*

A14.2–2000 *Safety Requirements for Portable Metal Ladders*

A14.3–2000 *Safety Requirements for Fixed Ladders*

A14.4–2000 *Safety Requirements for Job-Made Ladders*

A14.7–2000 *Safety Requirements for Mobile Ladders*

Industry Standards Resources

National Safety Council,

Job-Made Ladders, Data Sheet Number 1-568-76, 1976

Accident Prevention Manual for Industrial Operations, 9th Ed., 1988

Scaffold Industry Association *(www.scaffold.org)*

Scaffolding, Shoring and Forming Institute (*www.ssfi.org*)

SECTION 9

Protective Equipment and Devices

CHAPTER 34

Personal Protective Equipment

Personal protective equipment (PPE) for eyes, face, head, and extremities, protective clothing, respiratory devices, and protective shields and barriers *must* be provided, used, and maintained in a sanitary and reliable condition wherever it is necessary by reason of hazards of processes or environment, chemical hazards, radiological hazards, or mechanical irritants encountered in a manner capable of causing injury or impairment in the function of any part of the body through absorption, inhalation, or physical contact.

Where employees provide their own protective equipment, the employer shall be responsible to assure its adequacy, including proper maintenance and sanitation of such equipment. All personal protective equipment *must* be of safe design and construction for the work to be performed.

Even though it is not required by OSHA, a hazard assessment of all work areas should be conducted so that proper protective equipment can be selected. When choosing PPE, you should consider such hazards as heat, impact, chemicals, compression, electrical, light/radiation, punctures, and dust.

34.1 Bodily Protection

Implementing a PPE program is essential to ensure the proper use of such equipment on the construction site. The program should address the hazards present; the selection, maintenance, and use of protective equipment and clothing; the training of employees; and monitoring of the program to ensure its ongoing effectiveness.

PPE provided by the company *must* be appropriate for the work, give adequate protection, and be worn by workers. The types of protective equipment and clothing used *must* be approved in accordance with the manufacturer's instructions; *must* be reasonably comfortable and *not* unnecessarily awkward, and *must* be maintained in safe and sanitary condition.

Any employee-owned protective equipment or clothing *must* comply with the same requirements as company-owned PPE.

Head Protection

Employers *must* post hardhat areas where there is a potential for objects falling from above, for bumps to worker's heads from fixed objects, or of accidental head contact with electrical hazards; workers *must* wear hardhats in those areas. (Workers *must* confine their hair where there's a risk of injury from entanglement in moving parts, or a risk of contamination by combustible or toxic substances.)

All hardhats *must* comply with *ANSI Z 89.1 1981, Requirements for Protective Headwear for Industrial Workers.* Old hardhats (particularly those purchased prior to 1985) *must* be replaced. Hardhats and their suspensions and liners *must* be maintained in good condition, and hard hats *must* be:

- Routinely inspected for dents, cracks, or deterioration
- Replaced after a heavy blow or electrical shock

Workers exposed to electrical shock and burns *must* wear hardhats that meet the applicable *ANSI* standard requirements:

- Low voltage (600 V or less)—Class A or D
- High voltage (above 600 V)—Class B

Eye and Face Protection

Employers *must* provide and employees *must* use suitable eye or face protection (or both) when they are at risk of eye injury due to flying particles, hazardous substances, projections, or injurious light rays such as during welding, cutting, grinding, nailing, or when working with concrete or harmful chemicals, or both. Workers *must* wear safety glasses or face shields when exposed to any electrical hazards including work on energized electrical systems or when exposed to any electrical hazards.

All eye and face protection used on the construction site *must* meet the requirements of *ANSI Z 87.1 1979 Standard Practice for Occupational and Educational Eye and Face Protection.*

- *Impact-resistant safety glasses* with side shields, or impact-resistant goggles, *must* be used by employees for chipping, grinding, sawing, drilling, and other operations where there is a danger of flying fragments, chips, or other particles.
- *Splash resistant goggles* (either unventilated, or having indirect ventilation) *must* be used by employees working with acids and other hazardous liquid chemicals.

- *Welding goggles with filter lenses or plates* appropriate for the specific type of welding *must* be used by employees when doing welding to screen out harmful light and ultraviolet rays.
- *Laser safety goggles* that protect against the specific wavelength and power of the laser *must* be used by employees exposed to laser beam use.
- *Safety glasses or goggles* having suitable corrective lenses, or goggles that fit over the employee's own glasses, *must* be provided to employees who use vision-correcting glasses and need job site eye protection. Contact lenses *must not* be used on the site unless medically approved precautionary measures are taken.

Hand Protection

Employers *must* provide, and workers *must* use, appropriate hand protection when there is a potential risk of cuts, burns, harmful physical or chemical agents, or radioactive material. Hand protection is not required if gloves might become caught in moving parts or machinery. Gloves should fit snugly with correct types, including:

- *Impermeable gloves* to prevent skin contact with hazardous substances, and replace used gloves as required. (To determine the appropriate glove for the substance, contact the glove supplier or manufacturer.)
- *Gloves having vibration-damping material in palms and fingers* when working with vibrating power tools (jackhammers, chain saws, etc.) that *must* fit properly and don't increase the grip force required to control the tool.
- *Nonflammable gloves with gauntlets* when welding or assisting with welding.
- *Insulated rubber gloves* for live high voltage electrical work; rubber gloves *must* be protected by outer canvas or leather gloves, and *must not* be used to replace other required safety measures.

Foot Protection

Workers should wear work shoes or boots with slip-resistant and puncture-resistant soles. Appropriate protective footwear used on a construction site *must* comply with the requirements of *ANSI Z 41.1 1967, Safety-Toe Footwear* and include:

- *Steel-toed safety shoes or boots (or both)* when exposed to potential foot injuries from crushing or penetrating actions, hot surfaces, falling objects, or hazardous substances.

- *Steel covering over the whole foot, not just the toes* when using jackhammers.
- *Rubber boots* when working with concrete or in water.

Body Protection

Workers *must* wear clothing appropriate for the work being done, *never* any loose clothing around machinery in which it might become entangled.

Employers *must* provide, and workers *must* use, appropriate body protection when necessary, depending on the hazard, that can protect against:

- *Toxic substances, steam, oil, water, and extreme heat or cold.* An apron, coveralls, or a full body suit.
- *Asbestos, lead, and other regulated carcinogens.* Protective clothing as required by the specific applicable OSHA standards.
- *Welding.* Leather aprons, and shirts with long sleeves and collars, as well as required head, face, eye, hand, foot, and respiratory protection.
- *Near vehicular traffic hazards (when working on foot).* Bright orange warning garments (shirts, vests, jackets).
- *Rainy weather.* Orange or yellow rain gear.
- *During hours of darkness.* Reflectorized clothing.
- *Life preservers.* When working over or near water, if not continuously protected by railings, nets, or safety belts (U.S. Coast Guard–approved personal flotation devices, including ring buoys, lifesaving boats, and other safety devices for emergency rescues).

Cleanup

Workers *must* wash promptly and thoroughly after exposure to injurious substances, regardless of the type of protective clothing or equipment that has been used.

Clothing *must* be promptly removed and not worn until cleaned when it becomes saturated or impregnated with flammable liquids, corrosive substances, irritants, oxidizing agents, or other hazardous chemicals.

34.2 Hearing Protection

OSHA 1926.101 requires protection against the effects of noise exposure when 8-hour time-weighted average sound levels exceed a permissible exposure limit (PEL) of 90 decibels (dBA) measured on the A scale of a sound level meter set at slow response. The exposure level is raised 5 dB for every halving of exposure duration.

Duration per day (hours)	Sound level dBA slow response
8	90
6	92
4	95
3	97
2	100
1½	102
1	105
½	110
¼ or less	115

Source: OSHA 29 CFR 1910/95; 1926.52; 1926.101.

TABLE 34.1 Permissible Exposure Standards to Occupational Noise in Construction

Employers *must* provide, and workers *must* use, hearing protection devices wherever it is not feasible to reduce the noise exposure (level times duration) to within the *PEL* specified in Table 34.1.

- Hearing protection devices inserted in the ear must be fitted by competent persons.
- Plain cotton is not an acceptable protective device.

NOTE: *When the daily noise exposure includes two or more periods of noise exposure of different levels, their combined effect should be considered, rather than the individual effect of each.*

If the sum of the fractions $C(1)/T(1) + C(2)/T(2) = C(n)/T(n)$ exceeds unity, the mixed exposure should be considered to exceed the limit value, where

$C(n)$ = total time of exposure at a specified noise level
$T(n)$ = total time of exposure permitted at that level

Exposure to impulsive or impact noise should *not* exceed 140 dB peak sound pressure level.

34.3 Respiratory Protection

Respirators are used to reduce or eliminate occupational disease caused by exposure to air contaminated with harmful dusts, fog, fumes, mists, gases, smokes, spray, or vapors. The primary objective is to reduce or eliminate the atmospheric contamination. The devices are common items of personal protective equipment found on construction work sites, ranging from simple particulate dust masks to supplied-air abrasive blasting hoods. Any respirator used, however, *must* be certified by the National Institute for Occupational Safety and Health (NIOSH).

As far as feasible, prevention of atmospheric contamination should be accomplished, by engineering control measures. Dependent on the type of operation and contaminant to be controlled, engineering controls could include:

- Mechanical local or general exhaust ventilation
- Dust collection systems
- Powered tools equipped with shrouds and high efficiency particulate (HEPA) vacuums
- Enclosing cabs on machinery to limit exposure to contaminants
- Utilizing wet methods for cutting or in dust suppression

When engineering controls are not feasible, or while they are being implemented, appropriate respiratory protection must be used. Before providing respirators for worker use, the contractor *must* establish and implement a written respiratory protection program with worksite-specific procedures. The written program should include the following:

- Procedures for selecting respirators for use in the workplace
- Medical evaluation of employees required to use respirators
- Procedures for fit-testing tight-fitting respirators
- Procedures for proper use in routine and emergency situations
- Procedures and schedules for cleaning, disinfecting, storing, inspecting, and otherwise maintaining respirators
- Procedures to ensure adequate air quality, quantity, and flow of breathing air for atmosphere-supplying respirators
- Training of employees in the respiratory hazards to which they are potentially exposed
- Training of employees in the proper use and limitations of respirators
- Procedures for regularly evaluating the effectiveness of the program.

The implementation of an effective respiratory protection program is vital. A respirator only functions properly if it is the correct respirator for the contaminant it is designed for and if it fits properly. A respirator also must be cleaned, stored, and maintained properly.

Regulations and Standards

OSHA Regulations

1910.132 *General Requirements*

1910.133 *Eye and Face Protection*

1910.135 *Head Protection*

1910.136 *Occupational Foot Protection*

1910.138 *Hand Protection*

Subpart E—*Personal Protective and Life Saving Equipment*

1926.62 Appendix C *Medical Surveillance Guidelines*

1926.65 Appendix A *Personal Protective Equipment Test Methods*

1925.65 Appendix B *General Description and Discussion of the Levels of Protection and Protective Gear*

1926.65 Appendix C *Compliance Guidelines*

1926.95 *Criteria for Personal Protective Equipment*

1926.96 *Occupational Foot Protection*

1926.100 *Head Protection*

1926.101 *Hearing Protection*

1926.102 *Eye and Face Protection*

1926.103 *Respiratory Protection*

U.S. Consensus Standards

ANSI Z 87.1–2003 *Practice for Occupational and Educational Eye and Face Protection*

ANSI Z 89.1–2003 *Personal Protection—Protective Headwear for Industrial Workers*

ASTM F2413–2005 *Standard Specification for Performance Requirements for Foot Protection*

CHAPTER 35
Protective Devices

35.1 Safety Belts, Harnesses, and Lanyards

A variety of personal restraining devices meant to be worn and used by employees working on the construction, repair, or demolition of structures are available to protect the worker from falls and to minimize the risk of severe physical injury from a fall. These include safety belts, harnesses, lanyards, lifelines, and drop lines (Fig. 35.1).

Most details of fall protection can be found under *Subpart M 1926.500 Floor and Wall Openings. 1926.104* covers the specifics of anchorage, test requirements, required material for manufacturing of belts, lifelines and lanyards, and load testing. When looking for equipment that meets or exceeds requirements in this section, look for *ANSI*-rated equipment:

- ANSI A10.14 Fall Protective Devices
- ANSI A14.3 Ladder Safety Systems

Refer to *ANSI Z359.1-1992 Harnesses and Lanyards*, which is more strict than the *A10.14*. However, the *A10.14* is the general guideline used for fall protective devices. This standard states that if any of fall protective equipment is subjected to a fall, it *must* be immediately removed from service and not used again. It also states that fall arrest systems should allow for a fall of no greater than 6 *ft*.

35.2 Life Jackets, Ring Buoys, and Lifesaving Skiffs

Even though a worker may be a swimmer, a fall into the water may cause injury—by the fall itself, by striking another object, due to a strong current, or due to water that is too cold, making survival difficult. Protection, prior to occurrence, is the simplest solution to this problem.

- Employer *must* provide fall protection for personnel working above water if the distance from the walking/working surface to the water's surface is 6 ft (1.8 m) or more by the use of

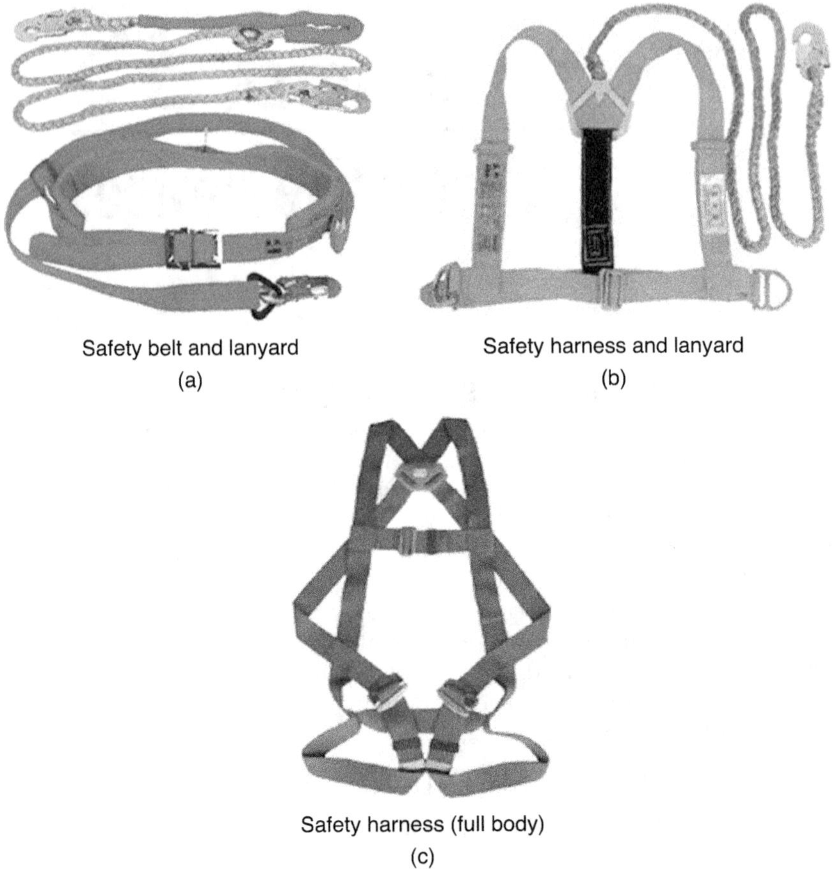

FIGURE 35.1 Typical fall systems. (a) Safety belt and lanyard; (b) safety harness and lanyard; (c) safety harness, full body. (*Source:* Tessco Technologies Inc., Hunt Valley, MD.)

guardrail systems, safety net systems, or personal fall arrest systems (*OSHA 1926.501.b.1*).

- Personnel working over or around water, where the danger of drowning exists;
- *Must Always* wear U.S. Coast Guard–approved life jackets or work vests. These jackets and vests must be inspected for defects after each use (Table 35.1).
- When fall protection is provided on walking/working surfaces located above water—effectively removing the drowning hazard—employees *do not* need to wear U.S. Coast Guard–approved life jackets or buoyant work vests.

Approval category	Equipment
160.002	Kapock life preserver
160.003	Cork life preserver (*obsolete*)
160.004	Balsa wood PFD (*obsolete*)
160.005	Fibrous glass PFD
160.053	Unicellular plastic foam work vest
160.055	Unicellular plastic foam PFD
160.077	Hybrid inflatable PFD
160.155	Life jacket (Solas)
160.176	Inflatable life jacket (Solas)

* PFD, personal flotation device.
Source: U.S. Coast Guard, Washington, DC.

TABLE 35.1 Ring Life Buoys and Vests Approved by the U.S. Coast Guard

- When using safety nets as fall protection, however, U.S. Coast Guard–approved life jacket or buoyant work vests are usually required.
- Employer *must* provide ring buoys and a lifesaving skiff, even if using fall protection eliminates drowning hazard (*OSHA 1926.106c* and *106d*)
- Ring buoys, with at least 90 ft of line, *must* be readily available for emergency rescue operations. Distance between ring buoys must not exceed 200 ft. At least one lifesaving skiff must be immediately available at locations where employees are working over or adjacent to water.

35.3 Safety Nets

When working more than 25 ft above the ground or water surface, and using a ladder, safety line, scaffold, platform, or temporary floor is impractical, the use of safety nets is allowed. These nets must extend 8 ft beyond the edge of the work surface and be installed as close under the work surface as practical, but can not exceed 25 ft. The nets must meet or exceed *ANSI A10.11-1989, Personal and Debris Nets*. This guideline stipulates mesh size, impact resistance and breaking strength (Table 35.2; Fig. 35.2).

Minimum net distance requirements	
Vertical distance from working level to horizontal plane of net	Minimum required horizontal distance of outer edge of net from edge of working surface
Up to 5 tf	8 ft
5 ft to 10 ft	10 ft
More than 10 ft	13 ft
Safety nets must be installed with sufficient clearance to prevent contact with the surface or structures under them when subjected to an impact force equal to the drop test	

Source: 1926.502 (c)(1) OSHA, Washington , DC.

TABLE 35.2 Safety Net Minimum Vertical and Horizontal Required Distances from Work Surfaces

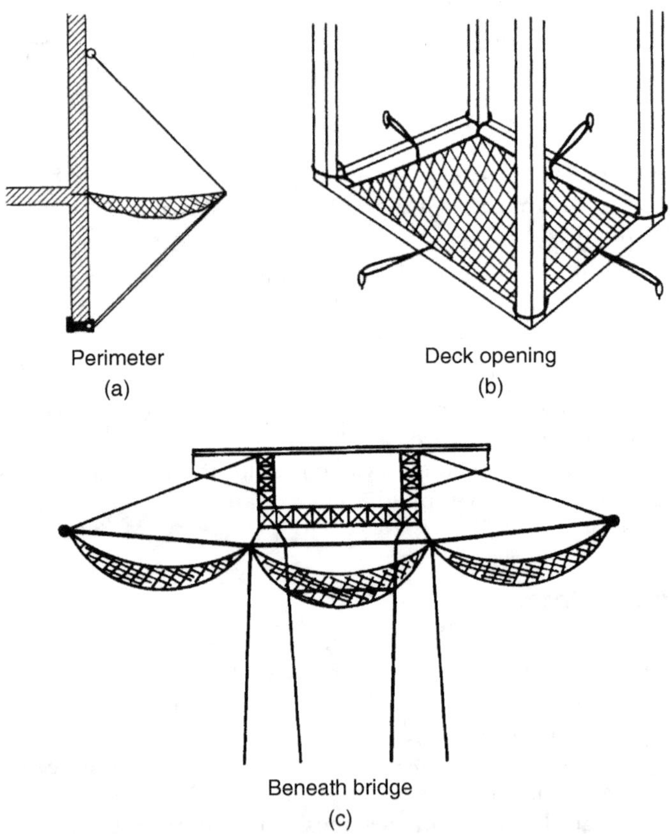

FIGURE 35.2 Safety nets. (a) Perimeter; (b) deck opening; (c) beneath bridge. (*Source:* Sinco Netting Solutions, Middletown, CT.)

Regulations and Standards

OSHA/USCG Regulations

1926.104 *Safety Belts, Lifelines, and Lanyards*

1926.105 *Safety Nets*

1926.106 *Working Over or Near Water*

1926.500 Subpart M—*Floor and Wall Openings*

USCG Approval Category 160.150 *Ring Life Buoy (Solas)*

U.S. Consensus Standards

ANSI A10.14 *Fall Protective Devices*

ANSI Z359.1 *Harnesses and Lanyards*

USCG *Personal Flotation Devices*

SECTION 10

Safety, Health, and Security

CHAPTER 36

Personal Safety and Health

Studies of general construction worksites have shown that the four hazards most commonly reported are falls, electrical, struck by, and caught in. These types of hazards can be prevented by developing and implementing an effective safety and health management system that focuses on the four elements.

36.1 Safety and Health Management Program

Key to a successful system is the development of a policy that leads employees in making decisions favoring safety. Begin by ensuring that employees understand that safety is at least equally if not more important than production, and that they are not to place themselves or workers around them in dangerous situations.

Demonstrate management leadership by requiring that all managers and supervisors (superintendents or foremen) have an active role in worksite safety, such as performing safety meetings/training sessions and inspecting the site.

- Provide adequate resources for employees to prevent hazards (falls, electrical, caught in, and struck by). Some examples of resources needed to prevent exposure to hazards are summarized in Table 36.1.
- Implement an accountability system that holds all personnel responsible for working safely on the construction site (identify unsafe acts and immediately counsel employees on the safe methods of performing the tasks). If an employee continues to work unsafely, you must consider discipline.
- Involve all the trades in discussing hazards and implementing methods to prevent falls, electrical, caught in, and struck by hazards.

Management of worker safety and health protection is a decisive factor in reducing the extent and severity of work-related injuries and illnesses and related costs. An effective occupational safety and health program includes four main elements: management, worksite analysis, hazard prevention and control, and safety and health training.

Management

- State clearly a worksite policy on safe and healthful work and working conditions, so that all personnel with responsibility at the site fully understand the priority and importance of safety and health protection in the organization.
- Establish and communicate a clear goal for the safety and health program and define objectives for meeting that goal so that all members of the organization understand the results desired and measures planned for achieving them.
- Provide visible top management involvement in implementing the program so that all employees understand that management's commitment is serious.
- Arrange for and encourage employee involvement in the structure and operation of the program and in decisions that affect their safety and health so that they will commit their insight and energy to achieving the safety and health program's goals and objectives.
- Assign and communicate responsibility for all aspects of the program so that managers, supervisors, and employees in all parts of the organization know what performance is expected of them.
- Provide adequate authority and resources to responsible parties so that assigned responsibilities can be met.
- Hold managers, supervisors, and employees accountable for meeting their responsibilities—so that essential tasks will be performed.
- Review program operations at least annually to evaluate their success in meeting the goals and objectives so that deficiencies can be identified and the program and/or the objectives can be revised when they do not meet the goal of effective safety and health protection.

Worksite Analysis

A practical analysis of the work environment involves a variety of worksite examinations to identify existing hazards and conditions and operations in which changes might occur to create new hazards. Unawareness of a hazard stemming from failure to examine the worksite is a sign that safety and health policies and/or practices are ineffective. Effective management actively analyzes the work and worksite to anticipate and prevent harmful occurrences. To identify all existing and potential hazards:

- Conduct comprehensive baseline worksite surveys for safety and health and periodic comprehensive update surveys and involve employees in this effort.
- Analyze planned and new facilities, processes, materials, and equipment.
- Perform routine job hazards analyses.
- Assess risk factors of ergonomics applications to workers' tasks.
- Conduct regular site safety and health inspections so that new or previously missed hazards and failures in hazard controls are identified.
- Provide a reliable system for employees to notify management personnel about conditions that appear hazardous and to receive timely and appropriate responses and encourage employees to use the system without fear of reprisal. This system utilizes employee insight and experience in safety and health protection and allows employee concerns to be addressed.
- Investigate accidents and "near miss" incidents so that their causes and means of prevention can be identified.
- Analyze injury and illness trends over time so that patterns with common causes can be identified and prevented.

36.2 Hazard Prevention and Control

Where feasible, workplace hazards are prevented by effective design of the job site or job. Where it is not feasible to eliminate such hazards, they must be controlled to prevent unsafe and unhealthful exposure. Elimination or control must be accomplished in a timely manner once a hazard or potential hazard is recognized (see Table 36.1 for Hazards Prevention Checklist).

Specifically, as part of the program, employers should establish procedures to correct or control present or potential hazards in a timely manner; and they should include the following:

- Use engineering techniques where feasible and appropriate.
- Establish, at the earliest time, safe work practices and procedures that are understood and followed by all affected parties. Understanding and compliance are a result of training, positive reinforcement, correction of unsafe performance, and, if necessary, enforcement through a clearly communicated disciplinary system.
- Provide personal protective equipment (PPE) when engineering controls are infeasible.

Falls	Electrical	Caught in	Struck by
Scaffolding equipped with railing and toeboards	Ground fault circuit interruptors	Adequate guarding of all exposed moving parts	Training of all heavy equipment operators
Proper housekeeping of all work surfaces	Electrical test equipment to ensure proper wiring	Inspection of all machines and power of all circuits tools	Daily inspections of the vehicles
Fall arrest/ prevention equipment	Employee training clearance distances and PPE	Operator training in the use of machines and power tools	Provision of safety equipment, back-up alarms, flashing lights
Employee training in fall protection, scaffolding, aerial lifts, and use of ladders	Provision and use of appropriate PPE (gloves, hard-hats, etc.)	Preventive maintenance program for equipment	Safe working practices (work rules that limit use of equipment to qualified operators)
Daily inspections focus on conditions and behavior	Daily inspections of electrical equipment used on-site	Safe work practices—lockout/tag-out	Provision and use of PPE when working around flying particles
Safe work practices—fall prevention plans	Procedures ensured grounding procedure for equipment and lockout/tag-out	Standard operating procedures	Training for personnel who work around equipment

Source: Maryland Occupational Safety & Health, Div. of Labor & Industry, Baltimore, MD.

TABLE 36.1 Hazards Prevention Checklist—Falls, Electrical, Caught in, Struck by

- Use administrative controls, such as reducing the duration of exposure.
- Maintain the facility and equipment to prevent equipment breakdowns.
- Plan and prepare for emergencies, and conduct training and emergency drills, as needed, to ensure that proper responses

to emergencies will be "second nature" for all persons involved.

- Establish a medical program that includes first aid onsite as well as nearby physician and emergency medical care to reduce the risk of any injury or illness that occurs.

Safety and Health Training[1]

Training is an essential component of an effective safety and health program, helping identify the safety and health responsibilities of both management and employees at the site. Training is often most effective when incorporated into other education on performance requirements and job practices. The complexity of training depends on the size and complexity of the worksite as well as the characteristics of the hazards and potential hazards at the site.

- *Employee training:* to ensure that all employees understand and are aware of the hazards to which they may be exposed and the proper methods for avoiding such hazards
- *Supervisory training:* to understand the key role they play in job site safety and to enable them to carry out their safety and health responsibilities effectively, including:
  Analyzing the work under their supervision to anticipate and identify potential hazards
  Maintaining physical protection in their work areas
  Reinforcing employee training on the nature of potential hazards in their work and on needed protective measures through continual performance feedback and, if necessary, through enforcement of safe work practices
  Understanding their safety and health responsibilities

Keep records of all safety and health training. Records can provide evidence of the employer's good faith and compliance with OSHA standards. Documentation also can supply an answer to one of the first questions an accident investigator will ask: "Was the injured employee properly trained to do the job?"

36.3 Hazard Analysis and Safety Training

Analyzing hazards is a key function of integrating safety and environmental management into construction worksite operations. Maintaining safety requires identifying work tasks and their

[1]*Source:* Occupational Health and Safety Administration, Washington, DC.

potential hazards, and then establishing controls to mitigate risks before starting work.

Hazards Analysis

Implement a system to identify safety hazards on the construction worksite by:

- Involving experts (e.g., someone with knowledge of safety engineering) in the evaluation of the worksite. In some cases, engineers may be required to design the scaffolds.
- Planning for the project by reviewing the safety needs in relation to preventing *falls, electrical, caught in,* and *struck by* hazards prior to the commencement of construction activities (planning needs to be proactive, not reactive).
- Performing frequently documented self-inspections of the construction site to identify *falls, electrical, caught in,* and *struck by* hazards; and track hazards that need to be corrected.
- Implementing an effective preventive maintenance program that requires frequent inspection of all tools and equipment used on the site (i.e., crane inspections, construction vehicles, power tools, scaffolds, lifts).

Safety Training

All managers, supervisors (superintendents, foremen), and tradesmen need safety training that is appropriate for the tasks being performed and their areas of responsibility.

Management training *must* be based on a manager's responsibilities; managers also need to have an understanding of the importance of providing visible management support and the goals and how they are measured.

- *Tradesman or laborers* on construction sites may need safety and health training to prevent falls, electrical, struck by, and caught in accidents:

  Fall protection, electrical safety, use of ladders, hazard communication, confined space entry, aerial lifts, forklift operation, and respiratory protection

  Use of equipment including powder-actuated tools, construction vehicles, cranes/derricks/hoists, scaffolding assembly, lifting, rigging operations, accident prevention, and safe work procedures

  Emergency preparedness, policy, goals, responsibilities, accountability system, and resources

- *Superintendent or foreman* should have an understanding of:
  - Accident investigations
  - Worksite hazard analysis (safety inspections)
  - Training methods
  - Accountability and discipline reinforcement
  - Behavioral based safety
- *Superintendent or foreman* may also need "competent person" training. OSHA requires that a "competent person" be involved in the following activities, among others:
  - Scaffold erection
  - Crane lifting/hoisting
  - Fall prevention
  - Accident prevention

36.4 Worksite Inspection

Designate a *competent person* to perform construction worksite safety inspections—one who's capable of identifying existing and predictable hazards in the surroundings, or working conditions, which are unsanitary, hazardous, or dangerous to employees.

The designated competent person *must* be authorized to take prompt corrective measures to eliminate them, at his or her discretion, based on that person's training and knowledge of the process.

The inspector's decision *must* be deemed appropriate and final until such time as the issues of concern are addressed and resolved to the inspector's satisfaction, as well as that of the Director of Safety and Superintendent or Manager of Construction (or both).

Only the competent person in charge of a particular site, however, may authorize the start-up of a pending worksite that has been halted as a result of a local agency violation or an OSHA violation, which was so noted by that competent individual.

Inspection Frequency

- Perform daily worksite inspections of at least the following items:
  - Scaffolding (fall hazards)
  - Excavation/trenching (falls, electrical, caught in, struck by hazards)
  - Accident prevention
  - Cranes (struck by and caught in hazards)
  - Fall protection (fall hazard)
- Conduct daily project hazard analyses and communicate it to all personnel involved with the tasks, including:

Identifying work to be performed
Analyzing the hazards associated with the phase of construction
Determining resources needed
Taking steps to eliminate possible hazards

Inspection Program

- *Identification*. Implement a system to identify safety hazards on the construction site:
  Involve experts (e.g., someone with the knowledge of safety engineering) in the evaluation of the site. In some cases, engineers may be required to design your scaffolds.
  Plan for the project by reviewing the safety needs in relation to preventing falls, electrical, caught in, and struck by hazards prior to the commencement of construction activities (planning needs to be proactive, not reactive).
  Perform frequently documented self-inspections of the construction worksite to identify falls, electrical, caught in, and struck by hazards.
- *Analysis*. Perform a project hazard analysis daily and communicate it to all personnel involved with the tasks, consisting of:
  Identifying work to be performed
  Analyzing hazards associated with the phase of construction
  Determining resources needed
  Taking steps to eliminate possible hazards
- *Tracking*. Track all potential hazards that need to be corrected, among them:
  Scaffolding (fall hazards)
  Excavation/trenching (falls, electrical, caught in, struck by hazards)
  Cranes (struck by and caught in hazards)
  Fall protection (fall hazard)
- *Prevention and control*. Prevent or control all safety hazards on construction sites immediately through:
  Elimination of the hazard
  Engineering controls (e.g., guards)
  Work practices (i.e., developing procedures for lockout/tag-out, ensured equipment grounding, hazard communication and fall prevention plans), and PPE
  Implementing an effective preventative maintenance program that requires frequent inspection of all tools and equipment used on the site (i.e., crane inspections, construction vehicles, power tools, scaffolds, lifts)

Regulations and Standards

OSHA Regulations

1926 Subpart M—*Fall Protection*

1926.501 *Duty to Have Fall Protection*

1926.502 *Fall Protection Systems Criteria and Practices*

Appendix A *Determining Roof Widths – Non-Mandatory Guidelines for 1926.501(b)(10)*

Appendix B *Guardrail Systems – Non-Mandatory Guidelines for 1926.502(b)*

Appendix C *Personal Fall Arrest Systems – Non-Mandatory Guidelines for 1926.502(d)*

Appendix D *Positioning Device Systems – Non-Mandatory Guide for 1926.502(e)*

Appendix E *Sample Fall Protection Plan – Non-Mandatory Guidelines for 1926.502(k)*

Electronic Library of Construction Occupational Safety and Health (*www.cdc.gov/elcosh*)

U.S. Consensus Standards

ANSI A10.33 2004 *Safety and Health Program Requirements for Multi-Employer Projects*

Industry Consensus Standards and Guidelines

Construction Safety Council (*www.buildsafe.org*)

Associated General Contractors (*www.agc.org*)

American Society of Safety Engineers (*www.asse.org*)

National Institute for Occupational Safety and Health (*www.cdc.gov/niosh/elcosh*)

Occupational Safety & Health Administration (*www.osha.gov*)

CHAPTER 37

Project Safety and Security

Construction sites contain a variety of hazardous conditions such as work areas, tools, equipment, machines, and stockpiles of materials that are potential obstacles and dangers to pedestrians when not correctly cordoned off from public use.

Accidents, however, don't just happen. They are caused by someone or something working unsafely or protected improperly. Controlling accidents on a worksite is the result of management and employees working together for a common objective—with leadership originating with management.

37.1 Safe Access onto Construction Site

To reduce potential safety and access problems at or near construction sites, maintain a continuous route for all pedestrians at all times. It is not acceptable simply to close a sidewalk without identifying an alternate circulation route—to enable pedestrians to bypass the construction site without retracing their steps or going significantly out of their way.

Temporary Route

Consider the needs of pedestrians with disabilities or use unofficial alternatives (such as using an adjacent grass surface); when a temporary route is established, it must be accessible to people with disabilities.

Use information sources (e.g., signs, audible information, and electronic information sources) to provide advance warning to pedestrians of the presence of the sidewalk construction site and to clearly mark alternate circulation routes. It is particularly important to ensure that all information sources are available to people with vision and cognitive impairments who may not be able to access signs or written information sources traditionally used.

NOTE: *Information sources must conform to ADAAG for items such as finish and contrast, raised and Braille characters, character proportion and height, pictorial symbol signs (pictograms), and mounting height; many people with cognitive impairments will be unable to read or understand written text and graphics.*

- *Make sure* pedestrians are aware of construction site location and the impact on the circulation route at each intersection or cross street location prior to the construction site so they may alter their routes.
- Discourage contractors from blocking sidewalks and parking lanes to provide contractor parking.
- Indicate duration of the construction work and any changes to the regular circulation route at the proposed site at a reasonable amount of time before the construction takes place so that pedestrians who use the route on a regular basis have sufficient time to plan and learn alternate routes to their destination.
- Delineate the alternate circulation route location and any instructions required for its use (e.g., altered crosswalk signal locations).
- Consider using traffic lanes to continue pedestrian access because most pedestrians will walk in the street; Jersey barriers can provide protection for pedestrians from traffic, while parking lanes can provide a temporary pathway.
- Consider closing off one lane of the street to traffic if pedestrians need to be diverted off of the sidewalk at a site location, allowing the outside (curb) lane for motorists to be used as the alternate pedestrian circulation route.

Barriers[1]

Construct ground level, solid, continuous barriers to prevent pedestrians from entering the construction, site either intentionally or unintentionally, thereby eliminating obstacles and potential hazards that will be present if access to the construction site is permitted.

Barriers defining the alternate route should be a minimum of 36 in. (915 mm) in height and continuous with the ground surface around the entire perimeter of the construction site or the entire length of the alternate circulation route having:

- No breaks or gaps along the full length of the barrier
- Bottom rail, solid and continuous between 10 mm (4 in.) and 305 mm (12 in.) in height

[1]*Source:* Federal Highway Administration, Washington, DC.

- High contrast color and material
- Temporary ramps and boardwalks, as required, to ensure a smooth and continuous surface that complies with ADAAG
- Level landing, at least 60 in. × 60 in. (1.525 m × 1.525 m) in size, at the top and bottom of any slopes greater than 5 percent
- Smooth transition area from the permanent to the temporary route
- Flagging tape, ribbon, or signs to identify a site without a solid barrier

Make sure that all supervisors, contractors, and workers at the site are sensitized to the potential pedestrian conflicts that may occur, so that they can be alerted to changing hazards and conditions that might impact pedestrian safety (e.g., the delivery of a new load of materials) and provide any assistance that pedestrians may require.

37.2 Hazard Awareness

Construction sites pose a unique hazard for workers in or around the project. Workers *must* be made aware of activities around them and be aware of safety precautions at all times to prevent an accident from occurring.

Safety procedures designed specifically for the particular construction site are a *must* to ensure the safety of employees and to protect the assets of the project owner.

Elements that *must* be contained in each construction site and security plan include, among others:

- Site-specific safety rules and procedures for the types of risks expected to be encountered on the site
- Routine inspections of construction sites; at a minimum:
  Visual inspection of all tools or equipment by the user before use; daily inspections of the worksite to identify potential safety hazards
  Monthly inspections of the worksite and equipment including such items as fall protection, lanyards, tools, personal protective equipment, and respiratory protection
- Training for each employee in safe work procedures and practices
- Availability of first aid, emergency equipment, and services at the construction site for emergency transportation
- Procedures to prevent theft, vandalism, and other losses at construction sites
- Periodic testing and evaluation of the plan to identify potential problems and correct them in an expedient manner

37.3 Construction Site Safety

Construction site hazards that OSHA most frequently cites are listed below.

Scaffolding

When scaffolds are not erected or used properly, fall hazards can occur.

- *Make sure* that scaffolds are set on sound footing.
- *Take out* of service damaged parts that affect the strength of the scaffold.
- *Do not* alter scaffolds.
- *Make sure* all scaffolds are fully planked.
- *Never* move scaffolds horizontally while workers are on them unless scaffolds are designed to be mobile and workers have been trained in the proper procedures.
- *Do not* permit persons to work on scaffolds when covered with snow, ice, or other slippery materials.
- *Do not* erect scaffolds or move them within 10 ft of power lines.
- *Prohibit* persons to work on scaffolds in bad weather or high winds unless a competent person has determined that it is safe to do so.
- *Never* use ladders, boxes, barrels, buckets, or other makeshift platforms to raise work height.
- *Do not* allow extra material to build up on scaffold platforms.
- *Do not* load scaffolds with more weight than they were designed to support.

Fall Protection

A number of factors are often involved in falls, including unstable working surfaces, misuse of or failure to use fall protection equipment, and human error. Studies have shown that using guardrails, fall arrest systems, safety nets, covers, and restraint systems can prevent many deaths and injuries from falls.

- Consider using aerial lifts or elevated platforms to provide safer elevated working surfaces.
- Erect guardrail systems with toeboards and warning lines or install control line systems to protect workers near the edges of floors and roofs.
- Cover over floor holes or use safety net systems (or both) or personal fall arrest systems (body harnesses).

Electrical Safety

Work on new and existing energized (hot) electrical circuits is prohibited until all power is shut off and grounds are attached.

- *Make sure* an effective lockout/tagout system is in place.
- *Make sure* frayed, damaged or worn electrical cords or cables are promptly replaced.
- *Make sure* all extension cords have grounding prongs.
- Protect flexible cords and cables from damage; *avoid* sharp corners and projections.
- Use only extension cord sets with portable electric tools and appliances that are the three-wire type and designed for hard or extra-hard service. (Look for the letters S, ST, SO, or STO imprinted on the casing.)
- *Make sure* all electrical tools and equipment are maintained in safe condition, checked regularly for defects, and taken out of service if a defect is found.
- *Do not* bypass any protective system or device designed to protect employees from contact with electrical energy.
- Locate and identify all overhead electrical power lines.
- *Make sure* that ladders, scaffolds, equipment, or materials never come within 10 ft of electrical power lines.
- All electrical tools *must* be properly grounded unless they are of the double insulated type.
- Prohibit use of multiple plug adapters.

Floor and Wall Openings

Floor/wall openings accidents mainly involve open-sided floors with the most common causes being the lack of guards, inadequate covering, and failure to use personal lifelines.

- *Make sure* all floor openings (12 in. or more) are guarded by a secured cover, a guardrail or equivalent on all sides (except at entrances to stairways).
- Install toeboards around the edges of permanent floor openings (where persons may pass below the opening).

Elevated Surfaces

Fall protection is required wherever employees are working on unguarded surfaces more than 10 ft above a lower level, or at *any* height above dangerous equipment; and includes use of lifelines, safety belts or harnesses, and lanyards.

- Post signs when appropriate, showing the elevated surface load capacity.
- Install standard guardrails around surfaces elevated more than 48 in. above the floor or ground.
- Install standard 4-in. toeboards around all elevated surfaces (beneath which people or machinery could be exposed to falling objects).
- Provide a permanent means of entry and exit with handrails to elevated storage and work surfaces.
- Pile, stack, or rack all material in a way that prevents it from tipping, falling, collapsing, rolling, or spreading.

37.4 Construction Site Security

Construction project managers/contractors are responsible for taking all necessary precautions to protect the safety of all persons who might be in the proximity of the construction worksite. Therefore, it is the construction manager/contractor's responsibility to maintain security on the construction worksite at all times.

Perimeter Fencing, Gates, and Locks

- Minimize the number of access points to the construction site and, where feasible, situate access points in locations that are highly visible from an adjacent street.
- Install fencing around the site perimeter adjacent to the streets.
  Must be a minimum of 6 ft high with blackout heavy-duty plastic screening.
  Do not cover vehicle and pedestrian gates.
  Do not cover fencing for 20 ft on each site of the gates.
- Secure vehicle and pedestrian gates and openings after hours.
  Use padlocks for securing gates that are designed to prohibit cutting of shackle.
  Secure locking gates with minimum $^3/_8$ in. thick, 30 grade coil chain, or minimum $^5/_{16}$-in. cable.

Signage

- Post address signs at all vehicle entrances listing street name and number and using letters and numbers at least 6 in. high.
- Install NO TRESPASSING signs listing state penal code section number.
- Post signs with a 24-hour emergency phone number, to notify person who can respond to the site in questions—at the main entrance gate and on the exterior of the onsite office trailer or building.

Site Lighting

- Illuminate the following locations during hours of darkness:
  - Vehicle gate locations
  - All open centralized storage areas for building materials or building fixtures
  - Trailers, temporary buildings, or containers (use for offices or for storage of building materials or fixtures)
- Provide a minimum light level of 1-ft-candle at the ground for a 15-ft radius at required locations.
- Install light fixtures at a minimum of 18 ft. high, with tempered or polycarbonate lens approved for wet locations (*UL 2572*).
- Where required security lighting would affect occupied residential properties:
  - Use motion sensors to control light fixtures.
  - Modify elements when first approved by the Building Department.

Office Trailers and Temporary Buildings

- Provide a deadbolt lock and an auxiliary locking device on doors, using a slide bolt or hasp, with a protective device to prohibit cutting off the padlock.
  - Attach bolt or hasp with nonremovable bolts from the exterior and padlock with a minimum $^1/_2$-in. thick shackle with heel and toe locking.
- Secure windows with either:
  - Steel bars, minimum $^1/_2$-in. round or 1-in. × $^1/_4$ in. flat, spaced not more than 5 in. apart, secured with bolts that are nonremovable from the exterior; or
  - Steel grate mesh, minimum $^1/_8$ in. thick, securely attached using a means resistant to remove from the exterior

Storage Containers

- Locate storage containers away from the perimeter fence in a fairly conspicuous area on the construction site.
- Secure all doors using a hasp or slide bolt with a protective device to prohibit cutting off the padlock and attached with nonremovable bolts from the exterior.
  - Padlocks should have a minimum $^1/_2$-in.-thick shackle with heel-and-toe locking.
- Render exterior hinge pins nonremovable by design or welding.

Motorized Equipment

- Render forklifts inoperable after operating hours, making them unable to start, operate, or move; remove and secure forks.

- Maintain an onsite record for all motorized equipment with a minimum 15-in. wheel diameter, listing the equipment manufacturer, model, license plate number, vehicle identifi-cation number (VIN), and product identification number (PIN).

Regulations and Standards

OSHA Regulations

Subpart G—*Signs, Signals, and Barricades*

1910.151 *Construction First Aid and Safety*
1926.200 *Accident Prevention Signs and Tags*
1926.203 *Definitions Applicable to Subpart G*

U.S. Consensus Standards

ANSI 10.33-98 *(R04) Safety and Health Program Requirements for Multi-Employer Projects*

ANSI 308.1-98 *Minimum Requirements for Workplace First Aid Kits*

ISEA/ANSI 308.1-03 *Minimum Requirements for Workplace First Aid Kits*

ANSI 2535.1 *Safety Color Code*
ANSI 2535.2 *Environmental and Facility Safety Signs*
ANSI 2535.3 *Criteria for Safety Symbols*
ANSI 2535.5 *Accident Prevention Tags*

CHAPTER 38
Fire Safety

Construction worksite fires not only delay project completion dates, but also result in substantial monetary losses and, in the worst cases, may result in injuries and deaths.

38.1 Major Causes of Construction Site Fires

The major causes for fire on construction sites include:

- Accumulation of loosely packed or scattered building materials (or both) that are usually highly combustible
- Inadequate insulation/protection and overloading (or both) of electric wiring
- Carelessness in welding work
- Lack of safety measures and supervision to control burning of rubbish
- Smoking and carelessness in the disposal of smoking materials
- Improper use and storage of dangerous goods
- Lack of fire protection system due to the noncompletion of the service installations or improper maintenance (or both)

38.2 Fire Prevention Guidelines

To mitigate the fire risk and minimize the possibility of fires at construction sites, contractors and site workers should observe the following guidelines:

- Stack combustible material on site in an orderly manner and to the minimum, providing ample spaces separating each stack both from the top and the sides.
- Store dangerous goods in approved structures or containers.
- Observe safety precautions at all times when handling dangerous goods.

- Practice good housekeeping to reduce the chances of a fire, the rate of fire spread, and potential injuries and fatalities.
  - Store materials neatly.
  - Remove discarded, combustible packing materials, wood chips, saw dust, and the like.
  - Maintain an unobstructed exit route at all times.
- Exercise extreme care when carrying out any welding work, and prevent sparks flying from open flame work.
- Remove all combustible material from the welding area; maintain a portable fire extinguisher at immediate readiness.
- Display prominently NO SMOKING signs in the construction area, especially where highly inflammable adhesives or solvents are used.
- Properly insulate and protect all electric wiring; install leakage circuit breakers to trip the supply when overloading of the circuit occurs.

38.3 Emergency Procedures

- Display written emergency procedures at prominent locations on the site and provide all employees with a copy of the procedures.
- Provide a means of giving warning of fire—including bells, whistles, horns—that can be readily identified as being a fire alarm.
- Maintain clear access to the site and buildings at all times.
- Maintain clear signs posted in prominent positions throughout the site, indicating locations of access routes, escape routes, and locations of fire extinguishers.
- Install fire detection systems that automatically detect a fire, initiating an alarm and terminating other action as appropriate.
- Post telephone numbers for local fire, police, and emergency medical services in prominent site locations containing location of site, project office telephone number, and name of designated fire safety office.

38.4 Fire Prevention/Protection Checklist

This checklist covers fire protection and prevention for construction sites including selected Occupational Safety & Health Administration (OSHA) construction regulations under *29 CFR 1926* that apply to temporary worksites associated with construction, alteration, demolition, and repair including painting and decorating.

Inspection procedures *must* comply with the OSHA or EPA standard, or with a nonregulatory recommendation.

General Requirements

- Is access to firefighting equipment maintained at all times at construction sites?
- Is firefighting equipment conspicuously located at construction sites?
- Is firefighting equipment periodically inspected and maintained in operating condition at construction sites?
- Is a fire extinguisher, rated not less than 2A, or a 55-gal open drum of water with two fire pails provided for each 3000 ft^2 of protected building area?
- Is the travel distance to the fire extinguisher or drum of water 100 ft or less?

NOTE: *A $^1/_2$-in. diameter garden-type hose line (100 ft in length or less and equipped with a nozzle) may be substituted for a 2A-rated fire extinguisher, provided it can discharge at least 5 gal/min with a minimum hose stream range of 30 ft horizontally. The garden-type hose lines must be mounted on conventional racks or reels and must be able to reach all points in the area.*

- Are one or more fire extinguishers, rated not less than 2A, provided on each floor?
- At construction sites involving multistory buildings, at least one fire extinguisher is located adjacent to the stairway?
- Are fire extinguishers and water drums protected from freezing?
- If more than 5 gal of flammable or combustible liquids or 5 lb of flammable gas is present, a fire extinguisher, rated not less than 10B, is provided within 50 ft?

NOTE: *This requirement does not apply to the integral fuel tanks of motor vehicles.*

- Are portable fire extinguishers selected according to the classes of anticipated fires and size and degree of hazards?

Fire Prevention

- Is internal-combustion-engine-powered equipment located so that the exhausts are well away from combustible materials?

- When internal combustion engine exhausts are piped outside the building under construction, is a clearance of at least 6 in. maintained between such piping and combustible material?
- Is stability maintained when combustible materials are piled? Are piles lower than 20 ft?
- Are weeds and grass kept down and a regular procedure provided for the periodic cleanup of the entire area?
- Are outdoor combustible materials stored more than 10 ft from a building or structure?
- Are indoor materials stored so that they do not obstruct or adversely affect the means of exit?
- Are indoor materials stored, handled, and piled to minimize the spread of fire and permit convenient access for firefighting?
- Are indoor materials stored so that a clearance of at least 36 in. is maintained between the top level of stored materials and the sprinkler deflectors?
- Is proper clearance maintained around lights and heating units to prevent ignition of combustible materials?
- Is a clearance of at least 24 in. maintained around the path of travel of fire doors, unless a barricade is provided?
- Are materials stored more than 36 in. away from a fire door opening?

Temporary Heating Devices

- Are temporary heating devices stored away from flammable and combustible materials?

38.5 Selecting Portable Fire Extinguishers

An adequate number of portable fire extinguishers *must* be provided and located in conspicuous positions throughout the worksite in accordance with OSHA regulations and NFPA standards.

General Requirements

- Provide portable fire extinguishers; mount, locate, and identify them so that they are readily accessible to employees without subjecting employees to possible injury.
- Use only approved portable fire extinguishers to meet OSHA requirements.

- *Do not* provide or make available in the workplace portable fire extinguishers using carbon tetrachloride or chlorobromomethane extinguishing agents.
- *Make sure* that portable fire extinguishers are maintained in a fully charged and operable condition and kept in their designated places at all times except during use.
- Remove from service all soldered or riveted shell self-generating soda acid or self-generating foam or gas cartridge water type portable fire extinguishers operated by inverting the extinguisher to rupture the cartridge or to initiate an uncontrollable pressure generating chemical reaction to expel the agent.

Selection and Distribution

- Provide portable fire extinguishers for employee use, selected and distributed based on the classes of anticipated workplace fires and on the size and degree of hazard which would affect their use.
- Distribute portable fire extinguishers for use by employees on Class A fires so that the travel distance for employees to any extinguisher is 75 ft (22.9 m) or less.
- Distribute portable fire extinguishers for use by employees on Class B fires so that the travel distance from the Class B hazard area to any extinguisher is 50 ft (15.2 m) or less.
- Distribute portable fire extinguishers used for Class C hazards on the basis of the appropriate pattern for the existing Class A or Class B hazards.
- Distribute portable fire extinguishers or other containers of Class D extinguishing agent for use by employees so that the travel distance from the combustible metal working area to any extinguishing agent is 75 ft (22.9 m) or less. Portable fire extinguishers for Class D hazards are required in those combustible metal working areas where combustible metal powders, flakes, shavings, or similarly sized products are generated at least once every 2 weeks.

Fire Extinguisher Size and Spacing

The National Fire Protection Association (NFPA) recommends that:

- *Class A.* Locations (such as offices, classrooms, and assembly halls) that contain mainly Class A combustible materials *must* have one 2A extinguisher for every 3000 ft^2. (Standard for Portable Fire Extinguishers, NFPA 10/2007) (Table 38.1).

Class A fire hazard—Extinguisher size and spacing
The National Fire Protection Association (NFPA) recommends that locations such as offices, classrooms, and assembly halls that contain mainly Class A combustible materials have one 2A extinguisher for every 3000 square feet. [Standard for Portable Fire Extinguishers (NFPA 10/2007), Table 6.2.1.1, Fire Extinguisher Size and Placement for Class A Hazards].
OSHA requires that all employees have access to an extinguisher within 75 ft travel-distance [*29 CFR 1910.157(d)(2)*].
Note: Uniformly spaced standpipe systems or hose stations connected to a sprinkler system for emergency use can be used instead of Class A portable fire extinguishers, if they meet the respective requirements of (*29 CFR 1910.158*) or (*29 CFR 1910.159*), provide total coverage of the area to be protected, and employees are trained at least annually in their use [*29 CFR 1910.157(d)(3)*].

TABLE 38.1 Class A Fire Hazards—Extinguisher Size and Spacing

- *Class B.* Locations that contain Class B flammables, such as workshops, storage areas, research operations, garages, warehouses, or service and manufacturing areas *must* have an extinguisher accessible to all employees—within 50 ft travel-distance (Table 38.2).
- *Class C.* Locations where energized electrical equipment is used. The extinguisher size and spacing is based on its Class A or B hazard [*29 CFR 1910.157(d)(5)*].

Hazard	Extinguisher	Spacing
Light (Low)—Small amounts of flammable liquids	5-B 10-B	30′ 50′
Ordinary (Moderate)—The total amount of flammable liquids are present in greater amounts than expected under low-hazard locations, including garages, workshops, or support service areas.	10-B 20-B	30′ 50
Extra (High)—Locations where flammable liquids are present and used in large quantities. including areas used for storage, production, woodworking (finishing), and vehicle repair, or where painting, dipping, and coating, operations are performed with flammable liquids.	40-B 80-B	30′ 50

Source: Adapted from Standard for Portable Fire Extinguishers, *NFPA 10* (2007).

TABLE 38.2 Class B Fire Hazards—Extinguisher Size and Placement

- *Class D.* Locations where combustible metal powders, flakes, shavings, or similarly sized materials are generated at least once every 2 weeks must install Class D portable fire extinguishers not more then 75 ft from the hazard [*29 CFR 1910.157(d)(6)*].

38.6 Inspection, Maintenance, and Testing of Fire Extinguishers

The employer *must* inspect, maintain, and test of all portable fire extinguishers in the workplace.

- Inspect visually, once a month. Portable extinguishers or hoses used
- *Make sure* that portable fire extinguishers are subjected to an annual maintenance check. (Stored pressure extinguishers do not require an internal examination.)
- Record the annual maintenance date and retain this record for 1 year after the last entry or the life of the shell, whichever is less.
- *Make sure* the records are available upon request.
- *Make sure* that alternate equivalent protection is provided when portable fire extinguishers are removed from service for maintenance and recharging.

Hydrostatic Testing

- *Make sure* that portable extinguishers are hydrostatically tested by trained persons, with suitable testing equipment and facilities; at the intervals listed in Table 38.3; in addition to an external visual examination, the employer shall ensure that an internal examination of cylinders and shells to be tested is made prior to hydrostatic tests.
- *Make sure* that hydrostatic tests are performed on extinguisher hose assemblies equipped with a shut-off nozzle at the discharge end of the hose. The test interval must be the same as specified for the extinguisher on which the hose is installed.

Extinguisher shells, cylinders, or cartridges that fail a hydrostatic pressure test or are not fit for testing *must* be removed from service and from the workplace.

Maintenance and Testing Frequency[1]

Extinguishers *must* be subjected to maintenance not more than 1 year apart or when specifically indicated by an inspection (*NFPA 10*).

[1]*Source*: National Fire Protection Association, Quincy, MA.

Container type	Flammable liquids		Combustible liquids		
	IA	IB	IC	II	IIIA
Glass	500 mL (1 pt)	1 L (1 qt)	4 L (1.1 gal)	4 L (1.1 gal)	20 L (5 gal)
Metal (other than DOT drums) or approved plastic	4 L (1.1 gal)	20 L (5 gal)	20 L (5 gal)	20 L (5 gal)	20 L (5 gal)
Safety cans	10 L (2.6 gal)	20 L (5 gal)	20 L (5 gal)	20 L (5 gal)	20 L (5 gal)
Metal drum (DOT spec.)	N/A	20 L (5 gal)	20 L (5 gal)	227 L (60 gal)	227 L (60 gal)
Polyethylene (DOT Spec. 34, UN 1H1, or as authorized by DOT exemption)	4 L (1.1 gal)	20 L (5 gal)	20 L (5 gal)	227 L (60 gal)	227 L (60 gal)

TABLE 38.3 Maximum Allowable Container Capacity (Flammable and Combustible Liquids)

- Disassemble, annually, stored pressure type fire extinguishers containing a loaded stream agent.
  Subject them to complete maintenance procedures.
  Prior to disassembly, fully discharge the extinguisher to check the operation of the discharge valve and pressure gauge.
  Conduct testing on all carbon dioxide hose assemblies. Hose assemblies found to be nonconductive *must* be replaced.
- Empty, every 6 years, stored pressure extinguishers that require a 12-year hydrostatic test and subject them to the applicable maintenance procedures. When the applicable maintenance procedures are performed during periodic recharging or hydrostatic testing, the 6-year requirement shall begin from that date.
- Use only a halon closed recovery system to remove agent from halogenated agent extinguishers.
- Remove and service test, at intervals, hoses stored on racks and reels; and in hose houses not exceeding 5 years after purchase date and every 3 years thereafter.
- Physically inspect in-service hose—at least annually—unracked, unreeled, or unrolled.

Training and Education[2]

Where portable fire extinguishers are provided for employee use in the workplace, an educational program *must* also be provided—upon initial employment and at least annually thereafter—to familiarize employees with the general principles of fire extinguisher use and the hazards involved with incipient stage firefighting.

Employees, who have been designated to use firefighting equipment, as part of an emergency action plan, *must* be provided training in the use of the appropriate equipment upon initial assignment to the designated group of employees and at least annually thereafter.

38.7 Flammables and Combustibles

Flammable Liquid

Any liquid having a closed-cup flash point below 100°F (37.8°C) as determined by NFPA specified test procedures and apparatus. Flammable liquids classified as Class I include:

- *Class I liquid.* Any liquid that has a closed-cup flash point below 100°F (37.8°C); Reid vapor pressure not exceeding 40 psia (2068.6 mm Hg) at 100°F (37.8°C)—as determined by *ASTM D 323, Standard Method of Test for Vapor Pressure of Petroleum Products (Reid Method).* Class I liquids shall be further classified as follows:

  Class IA liquids. Having flash points below 73°F (22.8°C) and boiling points below 100°F (37.8°C).

  Class IB liquids. Having flash points below 73°F (22.8°C) and boiling points at or above 100°F (37.8°C).

  Class IC liquids. Having flash points at or above 73°F (22.8°C), but below 100°F (37.8°C).

Combustible Liquid

Any liquid that has a closed-cup flash point at or above 100°F (37.8°C) as determined by NFPA specified test procedures and apparatus (Table 38.2).

- *Combustible liquids,* classified as Class II or Class III, include:

  Class II liquid. Having a flash point at or above 100°F (37.8°C) and below 140°F (60°C).

  Class IIIA liquid. Having a flash point at or above 140°F (60°C), but below 200°F (93°C).

  Class IIIB liquid. Having a flash point at or above 200°F (93°C).

[2]*Source:* Occupational Safety & Health Administration, Washington, DC.

Regulations and Standards

OSHA Regulations

Subpart F—*Fire Protection and Prevention*

1910.157 *Portable Fire Extinguishers*
1926.150 *Fire Protection*
1926.151 *Fire Prevention*
1926.152 *Flammable and Combustible Liquids*
1926.153 *Liquefied Petroleum Gas (LP-Gas)*
1926.154 *Temporary Heating Devices*
1926.155 *Definitions Applicable to Subpart F*
1926.156 *Fixed Extinguishing Systems, General.*
1926.157 *Fixed Extinguishing Systems, Gaseous Agent.*
1926.158 *Fire Detection Systems*
1926.159 *Employer Alarm Systems*

U.S. Consensus Standards

ANSI/NFPA 10 *Standard for Portable Fire Extinguishers*
ANSI/NFPA 20 *Standard for the Installation of Centrifugal Fire Pumps*
ANSI/NFPA 22 *Standard for Water Tanks for Private Fire Protection*
ANSI/UL *Standard for Rating and Fire Testing Fire Extinguishers*

Industry Consensus Standards and Guidelines

NAFED *Portable Fire Extinguisher Selection Guide*
NFPA 10-07 *Standard for Portable Fire Extinguishers*
NFPA 196 *Standard for Fire Hose*
NFPA 198 *Standard for the Care of Fire Hose*

APPENDIX A
Bibliography

Rigging

Codes and Standards—Cranes and Hoists—*ASME*

Introduction for Fall Protection—*ASSE*

Rigging Manual for Iron Workers—*IABSCIW*

Safety Requirements for Rigging Qualifications and Responsibilities—*ASSE*

Strength of Materials, Block and Tackle, Tackle Maintenance—*The Crosby Group*

The Art of Knotting and Splicing—*United States Naval Institute*

The Splicing Handbook: Techniques for Modern and Traditional Ropes—New England Ropes—*New England Ropes*

Wire Rope Sling User's Manual—*AISI*

Wire Rope User's Manual—*WRTB*

Building Codes

ASTM Standards—Building Codes

Building Construction and Safety Code—*NFPA*

Code of Standard Practice and Manual of Steel Construction-Allowable Stress Design—*AISC*

Code of Standard Practice for Steel Buildings and Bridges (2005)—*AISC*

Standards—Construction and Demolition

Construction-OSHA 1926—Subpart E

Demolition-OSHA 1926—Subpart T

National Standard for Construction and Demolition Operations: Standards in Building Codes—*ASTM*

Safety Requirements for Demolition Operations—ANSI A10.4 0 2007

Design Loads

Design Load on Structures during Construction—ASCE 57-02

Minimum Design Loads for Building and Other Structures—*ASCE*

Lighting

Outdoor Lighting Code Handbook—*IDSA*

Lifting and Hoisting Equipment

Internal-Combustion-Engine-Power Industrial Trucks (Forklifts)

Electric-Battery-Powered Industrial Trucks—*UL*

Personnel Hoists, Safety Requirements for—*ANSI*

Metal Structural Members

Aluminum Design Manual (2005 Edition)—*AA*

Steel Erection—Safety Requirements—*ANSI*

Steel Construction Manual, Thirteenth Edition—*AISC*

Code of Standard Practice for Steel Buildings and Bridges (2005)—*AISC*

Specification for Structural Joints Using ASTM A325 or A490 Bolts (2004)—*RCISC*

OSHA

Construction Safety and Health Compliance—*OSHA 1926*

Compliance with OSHA Worker Right Regulations—*OSHA*

Hazard Communications Standard—*OSHA*

Plant Safety Regulations and Guide—*OSHA*

Occupational Safety and Health Standards—*OSHA* 1910

Safety and Health

Construction Safety—*OSHA*

Conveyors and Related Equipment, Safety Standards for

Engineering and Management System, Safety

Environmental Control, Safety and Health

Industrial Safety—*OSHA*

Manual of Accident Prevention for Construction—*AGC*

Scaffolding

Falsework for Construction Purposes—*CSA*

National Standard for Construction and Demolition Operations:

Scaffolding Safety Requirements; Steel Erection—Safety Requirements—*ANSI*

Scaffolding Safety Requirements—*ANSI*

Recommended Procedures for Compression Testing of Welded Frame

Scaffolds and Shoring Equipment—*SSFI*

Security

Security and Fire Alarm Systems

Security Planning and Design—Joseph A. Durkin, *AIA*

Signaling Systems

Signaling Systems—*NFPA*

Welding

Structural Welding Code—AWS D1.1/D1.1 M—2004

Welding and Cutting, Safety in—*ANSI*

Welding Code for Aluminum, Structural—*AWS*

Welding Code for Steel, Structural—*AWS*

Welds, Guide for visual Inspection of—*AWS*

Wood

Annual Book of Standards Vol. 04.10 Wood—*ASTM*

Engineered Wood Handbook—*APA*

Guide to Good Practice for Handling, Installing and Bracing of Metal Plate Connected Wood

Lumber Storage (TG-5)—*WWPA*

Manual for Engineered Wood Construction—*AWC*

National Design Specifications for Wood Construction—*AFPS*

Recommendations of Handling and Erecting Wood Trusses

Bracing Wood Trusses: Commentary and Recommendations—*TPI*

Recommended Design Specification for Temporary Bracing of Metal Plate

Connected Wood Trusses, DSB-89—*TPI*

Southern Pine Use Guide—*SPC*

Storage and Erection, AITC 111—*AITC*

Timber Construction Manual (4th Edition)—*AITC*

Trusses (BCSI 1-03)—*WTCA*

Wood Engineering Handbook—*FPL*

Wood Handbook: Wood as an Engineering Material-*FPL-113—FPL/USDA*

Wood Trusses, Design Specifications for Metal Plate Connected

APPENDIX B
Acronyms

AA	Aluminum Association
ACA	American Chain Association
ACMA	American Chain Manufacturers Association
AGC	Associated General Contractors of America
AHS	American Helicopter Society
AIA	American Institute of Architects
AISC	American Institute of Steel Construction
AISI	American Iron and Steel Institute
AIST	Association for Iron and Steel Technology
AITC	American Institute of Timber Construction
ALA	American Lighting Association
ALI	American Ladder Institute
APA	The Engineered Wood Association
ASCE	American Society of Civil Engineers
ASIS	American Society for Industrial Security
ASME	American Society of Mechanical Engineers
ASSE	American Society of Safety Engineers
AWS	American Welding Society
AWCI	American Wire Cloth Institute
AWPA	American Wire Producers Association
AWRF	Associated Wire Rope Fabricators
AWWA	American Water Works Association
CEA	Construction Equipment Association
CEMA	Conveyor Equipment Manufacturers Association
CI	Cordage Institute
CMAA	Crane Manufacturers Association of America
COA	Crane Operators Association
HAI	Helicopter Association International
HMA	Hoist Manufacturers Institute
IDSA	International Dark Sky Association

ISEA	Industrial Safety Equipment Association
MBCEA	Metal Building Contractors and Erectors Association
MHED	Material Handling Equipment Distributors
MHI	Material Handling Industry
NAAA	National Alarm Association of America
NAEC	National Association of Elevator Contractors
NAESA	National Association of Elevator Safety Authorities
NAFED	National Association of Fire Equipment Distributors
NATE	National Association of Tower Erectors
NBFAA	National Burglar and Fire Alarm Association
NCCCO	National Commission for the Certification of Crane Operators
NEA	National Erectors Association
NECA	National Electrical Contractors Association
NEI	National Elevator Industry
NFPA	National Fire Protection Association
NFPA	National Forest Products Association
NSC	National Safety Council
SCRA	Specialized Carriers and Rigging Association
SIA	Scaffolding Industry Association
SJI	Steel Joists Institute
SSFI	Scaffolding, Shoring and Forming Institute
SSPC	The Society for Protective Coatings
STI	Steel Tube Institute of North America
TPI	Truss Plate Institute
UL	Underwriter Laboratories Inc.
WEF	Water Environment Federation
WSTA	Web Sling and Tiedown Association
WRTB	Wire Rope Technical Board
WWPA	Woven Wire Products Association

APPENDIX C

OSHA Regulations

Construction Safety and Health—29 CFR 1926

Subpart A—*General*

1926.2 *Variances from Safety and Health Standards*
1926.4 *Rules of Practice for Administrative Adjudications for Enforcement of Safety and Health Standards*

Subpart B—*General Interpretations*

1926.16 *Rules of Construction*

Subpart C—*General Safety and Health Provisions*

1926.20 *General Safety and Health Provisions*
1926.21 *Safety Training and Education*
1926.22 *Recording and Reporting of Injuries. (Reserved)*
1926.23 *First Aid and Medical Attention*
1926.24 *Fire Protection and Prevention*
1926.25 *Housekeeping*
1926.26 *Illumination*
1926.27 *Sanitation*
1926.28 *Personal Protective Equipment*
1926.29 *Acceptable Certifications*
1926.31 *Incorporation by Reference*
1926.34 *Means of Egress*
1926.35 *Employee Emergency Action Plans*

Subpart D—*Occupational Health and Environmental Controls*

1926.50 *Medical Services and First Aid*
1926.51 *Sanitation*
1926.52 *Occupational Noise Exposure*

1926.53 *Ionizing Radiation*
1926.54 *Nonionizing Radiation*
1926.55 *Gases, Vapors, Fumes, Dusts, and Mists*
1926.56 *Illumination*
1926.57 *Ventilation*
1926.59 *Hazard Communication*
1926.61 *Retention of DOT Markings, Placards and Labels*
1926.62 *Lead*
1926.64 *Process Safety Management of Highly Hazardous Chemicals*
1926.65 *Hazardous Waste Operations and Emergency Response*

Subpart E—*Personal Protective and Life Saving Equipment*

1926.95 *Criteria for Personal Protective Equipment*
1926.96 *Occupational Foot Protection*
1926.97 *(Reserved)*
1926.98 *(Reserved)*
1926.99 *(Reserved)*
1926.100 *Head Protection*
1926.101 *Hearing Protection*
1926.102 *Eye and Face Protection*
1926.103 *Respiratory Protection*
1926.104 *Safety Belts, Lifelines, and Lanyards*
1926.105 *Safety Nets*
1926.106 *Working Over or Near Water*
1926.107 *Definitions Applicable to this Subpart*

Subpart F—*Fire Protection and Prevention*

1926.150 *Fire Protection*
1926.151 *Fire Prevention*
1926.152 *Flammable and Combustible Liquids*
1926.153 *Liquefied Petroleum Gas (LP-Gas)*
1926.154 *Temporary Heating Devices*
1926.155 *Definitions Applicable to this Subpart*
1926.156 *Fixed Extinguishing Systems, General*
1926.157 *Fixed Extinguishing Systems, Gaseous Agent*
1926.158 *Fire Detection Systems*
1926.159 *Employee Alarm Systems*

Subpart G—*Signs, Signals, and Barricades*

1926.200 *Accident Prevention Signs and Tags*
1926.201 *Signaling*
1926.202 *Barricades*
1926.203 *Definitions Applicable to this Subpart*

Subpart H—*Materials Handling, Storage, Use, and Disposal*

1926.250 *General Requirements for Storage*
1926.251 *Rigging Equipment for Material Handling*
1926.252 *Disposal of Waste Materials*

Subpart I—*Tools–Hand and Power*

1926.300 *General Requirements*
1926.301 *Hand Tools*
1926.302 *Powers-Operated Hand Tools*
1926.303 *Abrasive Wheels and Tools*
1926.304 *Woodworking Tools*
1926.305 *Jacks–Lever and Ratchet, Screw, and Hydraulic*
1926.306 *Air Receivers*
1926.307 *Mechanical Power-Transmission Apparatus*

Subpart J—*Welding and Cutting*

1926.350 *Gas Welding and Cutting*
1926.351 *Arc Welding and Cutting*
1926.352 *Fire Prevention*
1926.353 *Ventilation and Protection in Welding, Cutting, and Heating*
1926.354 *Welding, Cutting, and Heating in Way of Preservative Coatings*

Subpart K—*Electrical*
GENERAL

1926.400 *Introduction*

INSTALLATION SAFETY REQUIREMENTS

1926.402 *Applicability*
1926.403 *General Requirements*
1926.404 *Wiring Design and Protection*
1926.405 *Wiring Methods, Components, and Equipment for General Use*
1926.406 *Specific Purpose Equipment and Installations*
1926.407 *Hazardous (classified) Locations*
1926.408 *Special Systems*

SAFETY-RELATED WORK PRACTICES

1926.416 *General Requirements*
1926.417 *Lockout and Tagging of Circuits*

SAFETY-RELATED MAINTENANCE AND ENVIRONMENTAL CONSIDERATIONS

1926.431 *Maintenance of Equipment*
1926.432 *Environmental Deterioration of Equipment*
1926.441 *Battery Locations and Battery Charging*

Subpart L—*Scaffolds*

1926.450 *Scope, Application, and Definitions Applicable to this Subpart*
1926.451 *General Requirements*
1926.452 *Additional Requirements Applicable to Specific Types of Scaffolds*
1926.453 *Aerial Lifts*
1926.454 *Training Requirements*

APPENDICES A THROUGH E TO SUBPART L

Subpart M—*Fall Protection*

1926.500 *Scope, Application, and Definitions Applicable to this Subpart*
1926.501 *Duty to Have Fall Protection*
1926.502 *Fall Protection Systems Criteria and Practices*
1926.503 *Training Requirements*
Appendix A *Determining Roof Widths*
Appendix B *Guardrail Systems*
Appendix C *Personal Fall Arrest Systems*
Appendix D *Positioning Device Systems*
Appendix E *Sample Fall Protection Plan*

Subpart N—*Cranes, Derricks, Hoists, Elevators, and Conveyors*

1926.550 *Cranes and Derricks*
1926.551 *Helicopters*
1926.552 *Material Hoists, Personnel Hoists, and Elevators*
1926.553 *Base-mounted Drum Hoists*
1926.554 *Overhead Hoists*
1926.555 *Conveyors*

Subpart O—*Motor Vehicles, Mechanized Equipment, and Marine Operations*

1926.600 *Equipment*
1926.601 *Motor Vehicles*
1926.602 *Material Handling Equipment*
1926.604 *Site Clearing*

Subpart P—*Excavations*

1926.650 *Scope, Application, and Definitions Applicable to this Subpart*
1926.651 *Specific Excavation Requirements*
1926.652 *Requirements for Protective Systems*
Appendix A *Soil Classification*
Appendix B *Sloping and Benching*
Appendix C *Timber Shoring for Trenches*
Appendix D *Aluminum Hydraulic Shoring for Trenches*
Appendix E *Alternatives to Timber Shoring*
Appendix F *Section of Protective Systems*

Subpart Q—*Concrete and Masonry Construction*

1926.705 *Requirements for Lift-slab Construction Operations*
1926.706 *Requirements of Masonry Construction*
Appendix A *References to Subpart Q*

Subpart R—*Steel Erection*

1926.750 *Scope*
1926.751 *Definitions*
1926.752 *Site Layout, Site-specific Erection Plan, and Construction Sequence*
1926.753 *Hoisting and Rigging*
1926.754 *Structural Steel Assembly*
1926.755 *Column Anchorage*
1926.756 *Beams and Columns*
1926.757 *Open Web Steel Joists*
1926.758 *Systems-engineered Metal Buildings*
1926.759 *Falling Object Protection*
1926.760 *Fall Protection*
1926.761 *Training*
Appendix G *Fall Protection Systems Criteria and Practices from 1926.502*
Non-Mandatory Guidelines for Complying with 1926.760(d)

Subpart S—*Tunnels and Shafts, Caissons, Cofferdams, and Compressed Air*

1926.800 *Underground Construction*
1926.803 *Compressed Air*

Subpart T—Demolition

1926.850 *Preparatory Operations*
1926.851 *Stairs, Passageways, and Ladders*
1926.852 *Chutes*
1926.853 *Removal of Materials Through Floor Openings*
1926.854 *Removal of Walls, Masonry Sections, and Chimneys*
1926.855 *Manual Removal of Floors*
1926.856 *Removal of Walls, Floors, and Material with Equipment*
1926.857 *Storage*
1926.858 *Removal of Steel Construction*
1926.859 *Mechanical Demolition*
1926.860 *Selective Demolition by Explosives*

Subpart V—*Power Transmission and Distribution*

1926.950 *General Requirements*
1926.951 *Tools and Protective Equipment*
1926.952 *Mechanical Equipment*
1926.953 *Material Handling*
1926.954 *Grounding for Protection of Employees*
1926.955 *Overhead Lines*
1926.956 *Underground Lines*
1926.958 *External Load Helicopters*
1926.959 *Lineman's Body Belts, Safety Straps, and Lanyards*

Subpart W—*Rollover Protective Structures and Overhead Protection*

1926.1000 *Rollover Protective Structures (ROPS) for Material Handling Equipment*
1926.1001 *Minimum Performance Criteria for Rollover Protective Structures for Designated Loaders*
1926.1002 *Protective Frames (roll-over protective structures, known as ROPS) for Wheel-type Tractors*
1926.1003 *Overhead Protection for Operators of Industrial Tractors*

Subpart X—*Stairways and Ladders*

- 1926.1050 *Scope, Application, and Definitions Applicable to this Subpart*
- 1926.1051 *General Requirements*
- 1926.1052 *Stairways*
- 1926.1053 *Ladders*
- 1926.1060 *Training Requirements*
- Appendix A *Ladders*

Subpart Z—*Toxic and Hazardous Substances*

- 1926.1101 *Asbestos*
- 1926.1103 *13 Carcinogens (4-Nitrobiphenyl, etc.)*
- 1926.1117 *Vinyl Chloride*
- Appendix A *Designation for General Industry Standards Incorporated into Body of Construction Standards*

APPENDIX D

Government Agencies

Construction Industry Compliance Assistance (CICA)
CICA is the primary source for plain language explanations of environmental rules for the construction industry. Information is provided free of charge.
Web: *www.cicacenter.org*

National Center for Manufacturing Sciences (NCMS)
3025 Boardwalk
Ann Arbor, MI
Tel: (734) 995-7962
Web: *www.ncms.org*

Environmental Protection Agency (EPA)
Ariel Rios Building
1200 Pennsylvania Avenue, N.W.
Washington, DC 20460
Tel: (202) 272-0167
Web: *www.epa.gov*

Forest Products Laboratory (FPL)
U.S. Forest Service (USDA)
One Gifford Pinchot Drive
Madison, WI 53726-2398
Tel: (608) 231-9200
Web: *www.fpl.fs.fed.us*

National Institute for Occupational Safety and Health (NIOSH)
(*Department of Health and Human Services*)
Centers for Disease Control and Prevention
1600 Clifton Road
Atlanta, GA 30333, USA
Tel: (404) 639-3311
Web: *www.niosh.gov.eg*

National Institute for Standards and Technology (NIST)
100 Bureau Drive, Stop 1070
Gaithersburg, MD 20899-1070
Tel: (301) 975-6478
Web: *www.nist.gov*

National Technical Information Service (NTIS)
5285 Port Royal Road
Springfield, VA 22161
Tel: (703) 605-6000
Web: *www.ntis.gov*

Occupational Safety and Health Administration (OSHA)
200 Constitution Avenue, NW
Washington, DC 20210
Tel: (202) 693-2000
Web: *www.osha.gov*

U.S. Government Printing Office (USGPO)
732 North Capitol Street, NW
Washington, DC 20401
Tel: (202) 512-0000
Web: *www.gpo.gov*

APPENDIX E

Construction Safety and Health Standards (ANSI, ASME, AWS, NFPA, SAE, UL)

Accident Prevention (ANSI)

Z35.1—Accident Prevention Signs, Specifications for

Z35.2—Accident Prevention Tags, Specifications for

Personal Protection (ANSI)

Head

Z89.1—Industrial Head Protection

Z90.1(a) and Z90.1(b)—Protective Headgear for Vehicular Users, Specifications for

Eye and Face

Z87.1—Eye and Face Protection, Practice for Occupational and Educational

Foot

Z41.1—Safety-Toe Footwear

Hand

Z105—Hand Protection Selection Criteria

Respiratory

Z88—Respiratory Protection

K13.1—Air-Purifying Respirator Canister and Cartridges, Identification of

Z110—Air Purifying Respiratory Protective Escape Devices

Fall Protection

Z365—Fall Protection Equipment

Safety Apparel

Z107—High-Visibility Safety Apparel and Headwear

Z207—High-Visibility Public Safety Vests

Worksite Safety (ANSI)

First Aid

Z308.1—Minimum Requirements for Workplace

Z87.1—Emergency Eyewash and Shower Equipment

Color Coding

Z53.1—Marking Physical Hazards, Safety Color Code for

Z53.1 Set A—Swatches for Highway Colors (mentioned in Z53.1)

Z53.1 Set B—Swatches for Highway Colors (mentioned in Z53.1)

Z535—Safety Color Code, Signs and Colors

Acoustics

S1.13—Sound Pressure Levels, Methods for Measurement of

S1.23—Designation of Sound Power Emitted by Machinery and Equipment

S1.26—Sound Power Levels of Noise Sources, Methods for Determination of

S1.4—Sound Level Meters

S5.1—Pneumatic Equipment, Measurement of Sound from

SR22—Noise Abatement and Control, Assessment and Recommendations (Report)

Lighting

ANSI/UL 94—Emergency Lighting Equipment and Power Equipment, Safety Standard for

Ventilation

Z117.1—Working in Tanks and Other confined Spaces, Safety Requirements

Floor and Wall Openings

A12.1—Floor and Wall Openings, Railings and Toeboard, Safety Requirements for

Gases, Compressed

CGA C-4—Marking Portable Compressed Gas Cylinders to Identify Material Contained

Compressors

ANSI/AME B19.3—Compressors for Process Industries, Safety Standard for

Fire Protection (ANSI)

Prevention/Protection

NFPA 101—Life Safety Code

Firefighting Equipment

UL 711—Fire Extinguishers, Rating and Fire Testing of, Safety Standards for

Construction Equipment (ASME)

Lifting Devices

B30.1—Jacks

B30.9—Slings

B30.10—Hooks

Cranes

B30.2—Cranes, Overhead and Gantry (Top Running Bridge, Single or Multiple Girder, Top-Running Trolley Hoist)

B30.4—Cranes, Portal, Tower, and Pillar

B30.5—Cranes, Mobile and Locomotive Truck Cranes

B30.11—Cranes, Monorails and Underhung Cranes

B30.3—Cranes, Hammerhead Tower

Derricks

B30.6—Derricks

B30.12—Rotorcraft, Handling Suspended Loads from

Hoists

B30.16—Hoists, Overhead (Underhung)

Drum Hoists

A90.1—Belt Manlifts

B30.7—Base-Mounted Drum Hoists

HST 1M—Hoists, Performance Standard for Electric Chain Hoists

Scaffolds and Platforms

A92.1—Mobile Ladder Stands and Scaffolds (Tower, Manually Propelled)

A92.2—Vehicle-Mounted Elevating and Rotating Aerial Devices

A92.3—Elevating Work Platforms, Manually Propelled

A92.5—Boom-Supported Elevated Work Platforms

A92.6—Work Platforms, Self-Propelled Elevating

A120.1—Powered Platforms for Exterior Building Maintenance, Safety Code for

Ladders

A-14.1—Wood Ladders, Safety Requirements for Portable

A14.2—Metal Ladders, Safety Requirements for Portable

A14.3—Fixed Ladders, Safety Requirements for

A14.4—Job-Made Ladders, Safety Requirements for

A14.5—Plastic Ladders, Safety Requirements for

Tools, Electric

UL 978—Stationary and Fixed Electric Tools, Safety Standards for

UL 1063—Machine-Tool Wires and Cables, Safety Standards for

Tools, Hand

B173.2—Hammers, Ball Peing, Safety Requirements for

B173.3—Striking Tools, Safety Requirements for Heavy

B173.7—Hatchers, Safety Requirements for

B173.8—Hammers, Riveting, Scaling and Tinners, Safety Requirements for

B209.7—Chisels, Safety Requirements for Rippling and Flooring/ Electricians'

Tools, Power

B7.5—Cutting-Off Machines, Safety Code for Construction, Use and Care of Gasoline-Powered, Hand-Held, Portable Abrasive

B165.1—Power-Driven Brushing Tools, Safety Requirements for Design, Care and Use of

B165.2—Brushing Tools Constructed with Wood, Plastic or Composition Hubs or Cores, Safety Requirements for Design, Care and Use of Power-Driven

Vehicles, Surface—Sound Levels (SAE)

SAE J1029—Machinery, Lighting and Marking of Construction and Industrial

SAE J1074—Engine Sound Level Measurement Procedure

SAE J1083—Starting or Moving Machines, Unauthorized

SAE J115—Safety Signs

SAE J1174—Operator, Safety Considerations for

SAE J1175—Bystander Sound Level Measurement Procedure for Small Engine-Powered Equipment

SAE J366—Trucks and Buses, Exterior Sound Level for Heavy

Welding and Cutting

ANSI Z49.1—Welding and Cutting, Safety in

NFPA 51B—Cutting and Welding Processes

AWS F1.1—Sampling Airborne Particles Generated by Welding and Allied Processes

UL 551—Transformer-Type Arc-Welding Machines, Safety Standard for

APPENDIX F
Standards Writing Resources

American National Standards Institute (ANSI)
25 West 43rd Street
New York, NY 10036
Tel: (212) 642-4900
Web: *www.ansi.org*

ANSI Headquarters
1819 L Street NW, Suite 600
Washington, DC 20036
Tel: (202) 293-8020
Web: *www.ansi.org*

ASTM International (ASTM)
100 Barr Harbor Drive
West Conshohocken, PA 19428-2959
Tel: (610) 832-9500
Web: *www.astm.org*

Canadian Standards Association (CSA)
5060 Spectrum Way, Suite 100
Mississauga, ON L4W 5N6, Canada
Tel: (800) 463-6727
Web: *www.csa.ca*

Construction Specifications Institute (CSI)
99 Canal Center Plaza, Suite 300
Alexandria, VA 22314-1588
Tel: (800) 689-2800
Web: *www.csinet.com*

International Code Council (ICC)
500 New Jersey Avenue NW, 20th Floor
Washington, DC 20001-2000
Tel: (202) 733-2348
Web: *www.iccsafe.org*

International Standardization Organization (ISO)
1 Rue de Varembe
CH-1211 Geneva 20, Switzerland
Tel: (41 22) 749-0111
Web: *www.iso.ch*

National Council of States on Building Codes and Standards (NCSBCS)
505 Huntmar Park Drive, Suite 210
Herndon, VA 20170
Tel: (703) 437-0100
Web: *www.ncsbcs.org*

Standards Council of Canada (SCC)
270 Albert Street, Suite 200
Ottawa, ON K1P 6N7, Canada
Tel: (613) 238-322
Web: *www.scc.ca*

APPENDIX G

Document Distributors

American National Standards Institute (ANSI)
1819 L Street NW, Suite 600
Washington, DC 20036
Tel: (202) 293-8020
Web: *www.ansi.org*

New York Offices
25 West 43rd Street–4th Floor
New York, NY 10036
Tel: (212) 642-4900
Web: *www.ansi.org*

ASTM International (ASTM)
(American Society for Testing and Materials)
100 Barr Harbor Drive
West Conshohocken, PA C 700
Tel: (610) 832-9500
Web: *www.astm.org*

Construction Specifications Institute (CSI)
99 Canal Center Plaza
Alexandria, VA 22314
Tel: (703) 684-0300
Web: *www.csi.org*

Document Automation and Production Services (DAPS)
Building 4, Section D
700 Robbins Avenue
Philadelphia, PA 19111
Tel: (215) 697-2971
Web: *www.assis.daps.dla.mil*

Document Center Inc. (DC)
111 Industrial Road, Suite 9
Belmont, CA 94002
Tel: (650) 591-7600
Web: *www.document.center.com*

Document Engineering Company Inc. (DEC)
15210 Stagg Street
Van Nuys, California 91405
Tel: (800) 645-7732
Web: *www.doceng.com*

First Source Building Code Library
30 Technology Parkway South
Norcross, GA 30092
Tel: (800) 395-1988
Web: *www.reeedfirstsource.somc/codes*

Global Engineering Documents (GED-IHS)
15 Inverness Way East
Englewood, CO 80112
Tel: (303) 397-7956
Web: *www.ihs.com*

International Code Council (ICC)
500 New Jersey Avenue, Floor 20
Washington, DC 20001-2000
Tel: (202) 733-2348
Web: *www.iccsafe.org*

Library of Congress
101 Independence Avenue SE
Washington, DC 20540
Tel: (202) 707-5000
Web: *www.loc.gov*

National Technical Information Service (NTIS)
5285 Port Royal Road
Springfield, Virginia 22161
Tel: (703) 605-6000
Web: *www.ntis.gov*

Standardization Document Order Desk
Building 4, Section D
700 Robbins Avenue
Philadelphia, PA 19111-5094
Tel: (215) 697-2667
Web: *www.farsite.hill.af.mil*

TechStreet
Thomas Scientific
36588 Treasury Center
Chicago, IL 60694-6500
Tel: (734) 913-3930
Web: *www.techstreet.com*

U.S. Government Printing Office
732 North Capitol St. NW
Washington, DC 20401
Tel: (202) 512-0000
Web: www.gpo.gov

APPENDIX H

Industry Resources

Aluminum Association (AA)
1525 Wilson Boulevard, Suite 600
Arlington, VA 22209
Tel: (703) 358-2960
Web: *www.aluminum.org*

American Chain Association (ACA)
6724 Lone Oak Boulevard
Naples, FL 34109-6834
Tel: (239) 514-3441
Web: *www.americanchainassn.org*

American Chain Manufacturer's Association (ACMA)
6724 Lone Oak Boulevard
Naples, FL 34109
Tel: (239) 514-3441
Web: *www.americanchainassn.org*

American Forest and Paper Association (AFPA)
(Formerly: National Forest Products Association)
1111 19th Street, NW, Suite 800
Washington, DC 20036
Tel: (202) 463-2700
Web: *www.afandpa.org*

American Helicopter Society (AHS)
217 N Washington Street
Alexandria, VA 22314-3538
Tel: (703) 683-6777
Web: *www.vtol.org*

American Institute of Architects (AIA)
1735 New York Avenue NW
Washington, DC 20006-5292
Tel: (202) 626-7300
Web: *www.aia.org*

American Institute of Steel Construction (AISC)
One E Wacker Drive, Suite 700
Chicago, IL 60601-1802
Tel: (312) 670-2400
Web: *www.aisc.org*

American Iron and Steel Institute (AISI)
1140 Connecticut Avenue NW, Suite 705
Washington, DC 20036
Tel: (202) 452-7100
Web: *www.steel.org*

American Institute of Timber Construction (AITC)
7012 S Revere Parkway, Suite 140
Centennial, CO 80112-3932
Tel: (303) 792-9559
Web: *www.aitc-glulam.org*

American Ladder Institute (ALI)
401 N Michigan Avenue
Chicago, IL 60611-4267
Tel: (312) 644-6610
Web: *www.americanladderinstitute.org*

American Lighting Association (ALA)
PO Box 420288
2050 Stemmons Freeway, Suite 10046
Dallas, TX 75342-0288
Tel: (214) 698-9898
Web: *www.americanlightingassoc.com*

American Society for Industrial Security (ASIS)
1625 Prince Street
Alexandria, VA 22314-2818
Tel: (703) 519-6200
Web: *www.asisonline.org*

American Society for Nondestructive Testing (ASNT)
P.O. Box 28518
1711 Arlingate Lane
Columbus, OH 43228-0518
Tel: (614) 274-6003
Web: *www.asnt.org*

American Society of Civil Engineers (ASCE)
1801 Alexander Bell Drive
Reston, VA 20191-4400
Tel: (703) 298-6000
Web: *www.asce.org*

American Society of Mechanical Engineers (ASME)
Three Park Avenue
New York, NY 10016-5990
Tel: (847) 699-2929
Web: *www.asme.org*

American Society of Safety Engineers (ASSE)
1800 Oakton Street
Des Plaines, IL 60018-2187
Tel: (708) 692-4121
Web: *www.asse.org*

American Water Works Association (AWWA)
6666 W Quincy Avenue
Denver, CO 80235-3098
Tel: (303) 794-7711
Web: *www.awwa.org*

American Welding Society (AWS)
530 NW LeJeuen Road
Miami, FL 33126
TeL: (305) 443-9353
Web: *www.aws.org*

American Wire Producers Association (AWPA)
801 Fairfax Street, Suite 211
Alexandria, VA 22314-1757
Tel: (703) 299-4434
Web: *www.awpa.org*

APA – The Engineered Wood Association (APA)
(Formerly: American Plywood Association)
7011 South 19th Street
Tacoma, WA 98466
Tel: (253) 565-6600
Web: *www.apawood.org*

Associated General Contractors of America (AGC)
2300 Wilson Boulevard #400
Arlington, VA 22201
Tel: (703) 548-3118
Web: *www.agc.org*

Associated Wire Rope Fabricators (AWRF)
PO Box 748
Walled Lake, MI 48390-0748
Tel: (248) 994-7753
Web: *www.awrf.org*

Association for Iron and Steel Technology (AIST)
186 Thorn Hill Road
Warrendale, PA 15086-7528
Tel: (724) 776-6040
Web: *www.aist.org*

Association of Crane and Rigging Professionals (ACRP)
PO Box 87907
Vancouver, WA 98687
Tel: (800) 690-3921
Web: *www.wrrc.com*

Association of Equipment Manufacturers (AEM)
6737 W Washington Street #2400
Milwaukee, WI 53214-5647
Tel: (414) 272-0943
Web: *www.aem.org*

Conveyor Equipment Manufacturers Association (CEMA)
6724 Lone Oak Boulevard
Naples, FL 34109-6834
Tel: (239) 514-3441
Web: *www.cemanet.org*

Cordage Institute (CI)
994 Old Eagle School Road, Suite 1019
Wayne, PA 19087-1866
Tel: (610) 971-4854
Web: *www.ropecord.com*

Crane Certification Association of America
PO Box 89707
Vancouver, WA 98687-7907
Tel: (360) 834-3805
Web: *www.ccaaweb.net*

Crane Manufacturers Association of America (CMAA)
(Section of Materials Handling Industry of America)
8720 Red Oak Boulevard, Suite 201
Charlotte, NC 28217-3992
Tel: (704) 676-1190
Web: *www.mhia.org*

Forest Products Laboratory (FSL)
(Forest Service – US Dept. of Agriculture)
One Gifford Pinchot Drive
Madison, WI 53726-2398
Tel: (608) 231-9200
Web: *www.fpl.fs.fed.us*

Forest Products Society
2801 Marshall Court
Madison, WI 53705-2295
Tel: (608) 231-1361
Web: *www.forestprod.com*

Helicopter Association International (HAI)
1635 Prince Street
Alexandria, VA 22314-2818
Tel: (703) 683-4646
Web: *www.rotor.com*

Hoist Manufacturers Association (HMA)
(Affiliate of Materials Handling Industry of America)
8720 Red Oak Boulevard, Suite 201
Charlotte, NC 28217-3992
Tel: (704) 676-1190
Web: *www.mhis.org*

International Dark Sky Association (IDSA)
3225 N First Avenue
Tucson, AZ 85719
Tel: (520) 293-3198
Web: *www.darksky.org*

Material Handling Equipment Distributors Association (MHEDA)
201 U.S. Highway 45
Vernon Hills, IL 60061-2398
Tel: (847) 680-3500
Web: *www.mheda.org*

Material Handling Industry of America (MHIA)
8720 Red Oak Boulevard, Suite 201
Charlotte, NC 28217-3992
Tel: (704) 676-1190
Web: *www.mhia.org*

Material Handling and Management Society (MHMS)
8720 Red Oak Boulevarrd, Suite 201
Charlotte, NC 28217-3992
Tel: (704) 676-1183
Web: *www.mhia.org*

National Alarm Association of America (NAAA)
PO Box 3409
Dayton, OH 45401-3409
Tel: ((937) 461-2208
Web: *www.naaa.org*

NACE International (NACE)
(Formerly National Association of Corrosion Engineers)
1440 South Creek Drive
Houston, TX 77084-4906
Tel: (800) 797-6223
Web: *www.nace.org*

National Association of Elevator Contractors (NAEC)
1298 Wellbrook Circle, Suite A
Conyers, GA 30012
Tel: (770) 760-9660
Web: *www.naec.org*

National Association of Elevator Safety Authorities (NAESA)
8805 N. 23rd Avenue #350
Phoenix, AZ 85021-4146
Tel: (602) 266-9701
Web: *www.naesai.org*

National Association of Fire Equipment Distributors (NAFED)
104 S Michigan Avenue, Suite 300
Chicago, IL 60603
Tel: (312) 263-8100
Web: *www.nafed.org*

National Burglar and Fire Alarm Association (NBFAA)
2300 Valley View Lane, Suite 230
Irving, TX 75062
Tel: (214) 260-5970
Web: *www.alarm.org*

National Council of Erectors, Fabricators, Street and Riggers (NCEFR)
10382 Main Street, Suite 200
Fairfax, VA 22030
Tel: (703) 591-1870
Web: *www.wbdg.org*

National Elevator Industry (NEI)
PO Box 838
1677 County Road 64
Salem, NY 12865-0838
Tel: (518) 854-3100
Web: *www.neii.org*

National Fire Protection Association (NFPA)
One Batterymarch Park
Quincy, MA 02169-7471
Tel: (617) 770-3000
Web: *www.nfpa.org*

National Safety Council (NSC)
1121 Spring Lake Drive
Itasca, IL 60143-3201
Tel: (630) 285-1121
Web: *www.nsc.org*

Naval Safety Center (USNSC)
375 A Street
Norfolk, VA 23511
Tel: (757) 444-3520
Web: *www.safetycenter.navy.mil*

Safety Equipment Distributors Association (SEDA)
2105 Laurel Bush Road, Suite 200
Bel air, MD 21015
Tel: (443) 640-1065
Web: *www.safetycentral.org*

Scaffolding Industry Association (SIA)
2001 E Campbell Avenue, Suite 101
Phoenix, AZ 85016
Tel: (602) 257-1144
PO Box 20574
Web: *www.scaffold.org*

Scaffolding, Shoring and Forming Institute (SSFI)
1300 Sumner Avenue
Cleveland, OH 44115-2251
Tel: (216) 241-7333
Web: *www.ssfi.org*

Specialized Carriers and Rigging Association (SCRA)
2750 Prosperity Avenue, Suite 620
Fairfax, VA 22031-4312
Tel: (703) 698-0291
Web: *www.sceanet.org*

Steel Erectors Association of America (SEAA)
2216 W Meadowbrook Road
Greensboro, NC 20407
Tel: (336) 294-8880
Web: *www.seaa.net*

Steel Joist Institute (SJI)
1173B London Links Drive
Forest, VA 24551
Tel: (434) 525-7377
Web: *www.steeljoist.org*

Steel Tube Institute of North America (STI/NA)
2000 Ponce de Leon, Suite 600
Coral Gables, FL 33134
Tel: (305) 421-6236
Web: *www.steeltubeinstitute.org*

Texas A&M University (TAMU)
Zachry Department of Civil Engineering
College Station, Texas 77843
Tel: (979) 845-2438
Web: *www.civil.tamu.edu*

The Association of Union Constructors (TAUC)
(Formerly: National Erectors Association)
1501 Lee Highway, Suite 202
Arlington, VA 22209-1109
Tel: (703) 524-3336
Web: *www.tauc.org*

Truss Plate Institute (TPI)
218 N Lee Street, Suite 300
Alexandria, VA 22314
Tel: (703) 683-1010
Web: *www.tpinst.org*

Underwriter Laboratories Inc. (UL)
333 Pfingsten Road
Northbrook, IL 60062-2096
Tel: (847) 272-8800
Web: *www.ul.co*

Water Environment Federation (WEF)
601 Wythe Street
Alexandria, VA 22314-1994
Tel: (703) 684-2400
Web: *www.wef.org*

Web Sling and Tiedown Association (WSTA)
2105 Laurel Bush Road, Suite 200
Bel Air, MD 21015
Tel: (443) 640-1070
Web: *www.wstda.com*

Welding Research Council (WRC)
3 Park Avenue 24th Floor
New York, NY 10156-5902
Tel: (212) 591-7956
Web: *www.wrc.org*

Wire Association International (WAI)
1570 Boston Post Road
PO Box 178
Guilford, CT 06437-0578
Web: *www.wirenet.org*

Wire Reinforcement Institute (WRI)
942 Main Street, Suite 300
Hartford, CT 06103
Tel: (800) 552-4974
Web: *www.wirereinforcementinstitute.org*

Wire Rope Technical Board (WRTB)
801 N Fairfax Street, Suite 211
Alexandria, VA 22314-1757
Tel: (703) 299-8550
Web: *www.wrtb@usa.com*

Index

Note: Page numbers referencing figures are italicized and followed by an *"f."* Page numbers referencing tables are italicized and followed by a *"t."*

C

G

J

N

O

Q

R

Y

Z

www.ingramcontent.com/pod-product-compliance
Lightning Source LLC
LaVergne TN
LVHW052203200926
848481LV00001B/19
9781265823252